The Essence of Dielectric Waveguides

C. Yeh • F. I. Shimabukuro

The Essence of Dielectric Waveguides

Springer

C. Yeh
California Advanced Studies
2432 Nalin Drive
Los Angeles
CA 90077
USA

F. I. Shimabukuro
California Advanced Studies
2432 Nalin Drive
Los Angeles
CA 90077
USA

ISBN 978-0-387-30929-3 e-ISBN 978-0-387-49799-0

Library of Congress Control Number: 2008923746

Printed on acid-free paper

9 8 7 6 5 4 3 2 1

springer.com

To
Our Families
Vivian, John, and Evelyn; Siblings-Dorothy, Richard, and Vicky
Karen, Susan and Lee

PREFACE

"It is our responsibility as scientists, knowing the great progress which comes from a satisfactory philosophy of ignorance, the great progress which is the fruit of freedom of thought, to proclaim the value of this freedom, to teach how doubt is not to be feared but welcomed and discussed; and to demand this freedom as our duty to all coming generations" —— *Richard Feynman, 1955* ——

First, as students from Cal Tech and MIT and then as researchers and teachers from other universities and industry, we are benefited greatly from the philosophy of learning practiced by these and other distinguished universities in the US, namely, learn and teach the fundamentals and not the fashions. Under this guiding light, this comprehensive book was formed, covering the most important modern topics on guided waves. As such, it may be used as a research reference book or as a textbook for senior undergraduate students or first-year graduate students. The lectures for an one-semester or one-quarter course on guided waves along surface wave structures can begin with a review of EM fundamentals (Chap. 2), and then move on to a discussion on the general important and relevant characteristics of these guided surface waves (Chap. 3). Then follows the rigorous analytic treatment for canonical structures (planar, circular, and elliptical) (Chaps. 4–8). By the end of these lectures, the students would have gained a very solid theoretical foundation on this subject. Then the fun part starts. The students can now learn how they may make use of their fundamental knowledge to treat the many modern up-to-date applications: linear and nonlinear wave propagation in fibers, solitons in fibers and WDM beams propagation in fibers (Chaps. 9 and 10), plasmon (sub-wavelength) waves (Chap. 12), waves in periodic structures (photonic structures) (Chap. 13), surface waves on metamaterial (artificial material) and other exotic (moving medium) structures (Chap. 14). Finally, the students can now be introduced to the many numerical approaches (Chap. 15) that can be used on the various guided wave structures, with the comforting knowledge that they possess the necessary theoretical foundation to correctly interpret the numerical data.

Substantial amount of the material of the text appears in book form for the first time. References are given to the original sources. However, unintentional oversight by us is unavoidable. For this the authors offer their apologies. It is curious to note that many popular references (with many citations in the literature) may not represent the papers published by the originators of the concepts. Special care has been taken by us not to follow this erroneous path. References are listed at the end of each chapter for clarity and ease of usage.

As far as nomenclatures and symbols are concerned, we have not been able to have a given symbol to represent a single unique entity throughout the whole book.

Instead, we only make sure that a given symbol clearly and uniquely represents a single entity in that chapter. Whenever possible, universally accepted nomenclatures are used to represent vector and scalar quantities.

It is with deep gratitude and great pleasure for us to acknowledge the significant guidance and encouragement given to us by Professors C. H. Papas, J. R. Whinnery, and R. W. Gould. We also wish to acknowledge with special thanks to Dr. Peter Siegel who introduced us to the field of terahertz research and who planted the seed for us to pursue the writing of this book. Throughout our professional careers, we benefited greatly from the many positive advice and encouragement from our colleagues. We express our deepest thanks and gratitude to them. Finally, we express our sincerest thanks to Marshall Kwong for his dedicated professional graphic arts work for this book, without which this book would be incomplete.

We greatly appreciate the careful reading and constructive comments by the reviewers.

C. Yeh
F. I. Shimabukuro
Los Angeles

CONTENTS

1

INTRODUCTION

The increasing capabilities of digital computation have altered the way electromagnetic problems are being solved. It is no longer necessary that analytical solutions be obtained. Many practical problems with complicated geometries for which there are no closed form analytic solutions can now be solved numerically. Nevertheless, understanding the fundamental behavior (the essence) of the solutions must still be gained from analytic solutions of canonical problems. In other words, correct interpretation of the numerical results must depend on knowing the essence of guided waves on certain related canonical structures. Therefore, the primary goal of this book is to provide an insight into this essence.

Review of the wave guiding structures over the whole electromagnetic spectrum shows that, for frequencies below 30 GHz, mostly metal-based structures are used, and for frequencies above 30 GHz, increasing skin-depth losses in metal requires that low-loss structures be made without the use of any metallic material. Hence, the importance of pure dielectric waveguides for carrying large bandwidth signals is established. See Fig. 1.1 for a display of spectral regions in which certain guiding structures are useful. It is seen that the useful spectrum for dielectric waveguides can span more than seven decades, from 10^9 to 10^{16} Hz.

1.1 Brief Historical Background

The concept of guiding electromagnetic waves either along a single conducting wire with finite surface impedance or along a dielectric rod/slab has been known for a long time. As early as 1899, Sommerfeld [1] conceived the idea of guiding a circularly symmetric TM[1] wave along a conducting wire with small surface resistivity. However, because of the large field extent outside the wire, this "open-wire" line remained a novelty with limited practical applications. In 1909, Sommerfeld treated the problem of an oscillating dipole above a finitely conducting plane [2]. He found theoretically that there existed not only a radiated wave due

[1]This notation and classification will be discussed in detail later.

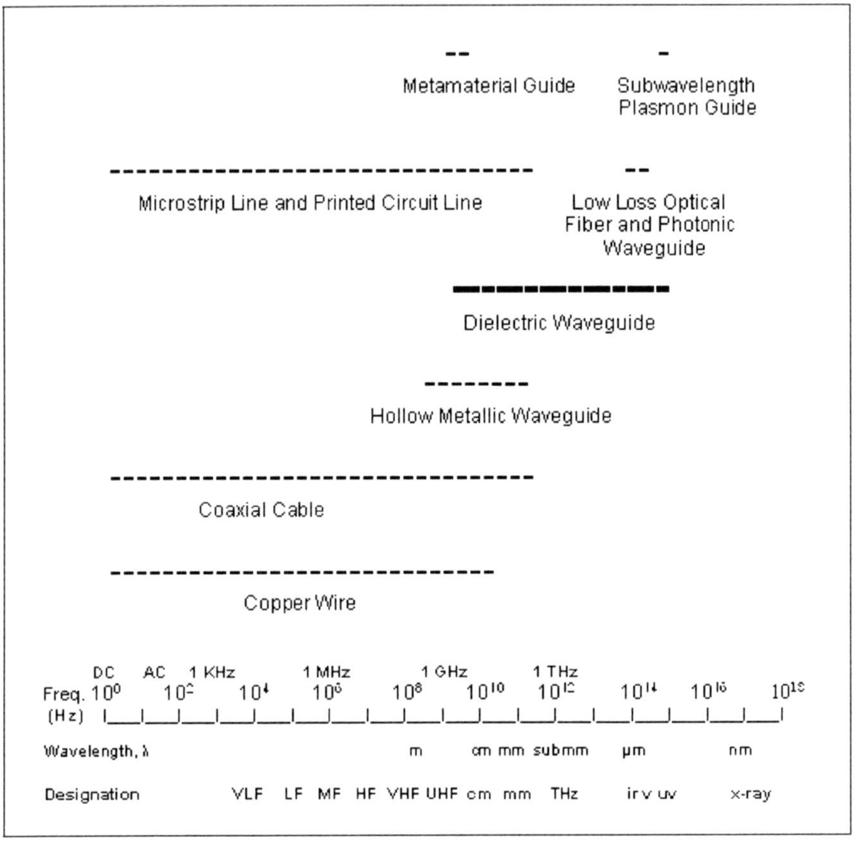

Figure 1.1. Spectral regions for various waveguides

to the oscillating dipole, but also a surface wave that traveled along the lossy surface [3]. This is the well-known "Sommerfeld Problem". In 1910, Hondros and Debye [4] demonstrated analytically that it was possible to propagate a TM wave along a lossless dielectric cylinder. Zahn [5] in 1915 and his two students, Ruter and Schriever [6], confirmed the existence of such a TM wave experimentally. Not until 1936 were the propagation properties of asymmetric waves on a round dielectric rod obtained by Carson et al. [7], who provided a complete mathematical analysis of this problem. It was noted in their paper that in order to satisfy the boundary conditions for the general case, a hybrid wave (i.e., the coexistence of longitudinal electric and magnetic fields) must be assumed. In other words, asymmetric TE and TM modes were inextricably coupled to each other along a circular dielectric rod. They also showed that (1) pure TE and TM waves could only exist in the circularly symmetric case and (2) there existed one and only one mode, namely the lowest order hybrid wave called the HE_{11} mode, which possessed no cutoff frequency[2] and could propagate at all frequencies. All other circularly symmetric or nonsymmetric modes had cutoff frequencies. The dispersion relations of these modes were also obtained in their paper, but no numerical results were given. In their paper they also mentioned that Southworth in 1920 accidentally observed a guided TM wave in a trough of water. Later, in 1936, Southworth [8] described more detailed experimental measurements on the phase velocity and attenuation of the circularly symmetric TM wave on a circular dielectric guide. Soon afterwards, in 1938, Schelkunoff [9] wrote a paper on the coupled transmission line representation of the waves and the impedance concept, which became the foundation for the development of microwave circuits.

In 1943, Mallach [10] published his results on the use of the dielectric rod as a directive radiator. He showed experimentally that the radiation pattern obtained by the use of the asymmetric HE_{11} mode produced only one lobe in the principal direction of radiation. Soon after Mallach's paper, Wegener [11] presented a dissertation in which the asymmetric HE_{11} mode, together with the lowest order circularly symmetric TE and TM modes, were analyzed in detail. Not only were the numerical results of the propagation constants for these waves obtained, but also their attenuation characteristics. He also obtained a few experimental points substantiating his theoretical results. Apparently, he was not aware of the earlier Carson, Mead, and Schelkunoff's work. Elsasser in 1949 [12], independent of Wegener's work, published his computation on the attenuation properties of these three lowest-order modes. In a companion paper, Chandler [13] verified experimentally Elsasser's results on the dominant HE_{11} mode. He found that the guiding effect was retained even when the rod was only a fraction of a wavelength in diameter. For this case, since the greater part of the guided energy was outside the dielectric rod, very little loss was observed. This was also the first time the cavity resonator technique for open dielectric structure was introduced to measure the attenuation constant of the HE_{11} mode. It should be noted that the formula used by Chandler to obtain the attenuation constant, α, from the measured Q value was approximately correct, since it was derived assuming that the propagating mode was a TEM mode. The correct formula relating α and Q for the hybrid HE_{11} mode was given by Yeh in 1962 [14].

In the mid-1940s, Brillouin summarized his research on wave propagation in periodic structures in a book (1946) [15]. In 1951, Sensiper [16] wrote his thesis on wave propagation on a helical wire waveguide, a periodic structure waveguide. In 1954, Pierce [17] also provided the results on the interaction of electron beam

[2]The cutoff frequency does not have the conventional definition as that for the metal waveguides. Here it is defined that at the cutoff frequency the open dielectric waveguide structure ceases to act as a binding medium for the guided surface wave, and the wave can no longer be guided by the structure.

with slow waves guided by a periodic structure. The fundamental theory on wave propagation in periodic media is well established by these publications. At about the same time, the increasing demand for higher bandwidth low-loss transmission lines for transcontinental television and long-distance phone transmission provided the incentive to find new ways to transmit microwaves efficiently. King and Schlesinger [18] studied the dielectric image line (1954), while Goubau experimented with a conducting wire coated with a thin sheath of dielectric material, a modified version of the Sommerfeld line (1950) [19]. High loss or instability of the guided field due to the large field extent hampered further development of these approaches. During the 1950s, significant amount of research on the excitation of surface wave problem was carried out (See Collin's book [20]). These investigations together with Sommerfeld's research provided the basic understanding of the problem of wave excitation on a dielectric structure. Another notable effort was the concentrated research undertaken by the Bell Laboratory investigators on the transmission of millimeter wave in a oversized circular conducting tube supporting the low-loss circularly symmetric TM wave. (This approach turned out to be not very fruitful due to high loss caused by the modal instability of the low-loss circularly symmetric TM mode in an oversized waveguide.) At that time, the Bell Laboratory group chose not to investigate dielectric fiber as a viable optical waveguide due to its high dielectric losses. History tells us that this was an unfortunate decision.

Observation of waveguide modes in optical fibers was first reported by Snitzer and Hicks in 1959, then later in 1961 by Snitzer and Osterberg, and by Kapany and Burke in 1961 [21]. In 1961, Snitzer restudied the problem of wave guidance along an optical fiber (a circular dielectric cylinder). He provided detailed numerical computations on several lower-order modes and obtained field configurations for these modes [22]. In 1962, Yeh [23] solved the unique-canonical problem of surface wave propagation on an elliptical dielectric waveguide. Unlike the circular cylinder case where each mode can be described by a single order of Bessel function, each surface wave mode for an elliptical dielectric cylinder would require infinite sums of all orders of Mathieu functions. In other words, the dispersion relation of each mode on an elliptical dielectric cylinder must be represented by an infinite determinant of all orders of Mathieu functions. In this case no pure TE or TM mode can exist on an elliptical dielectric cylinder; all modes must be of a hybrid type, that is, the HE type. Yeh not only provided the complete analytical solution to this problem, he also obtained numerical solutions on the propagation constants as well as the attenuation constants for the dominant modes. Experimental verifications were also obtained by him. Independently, at about the same time, Lynkimov et al. [24] gave an analytic solution to this problem, but no experimental or detailed numerical results were given. One notes that the use of elliptical fiber is one way of making a polarization-preserving fiber [25]. In 1965, Bloembergen [26] wrote a book summarizing his research on wave propagation in nonlinear dielectrics. His work on nonlinear dielectric became the backbone of the later discovery of solitons in optical fibers.

Two events changed the tempo and direction of research on the optical fiber as a viable information transmission link: (1) Kao and Hockham [27] in 1964 recognized that if the impurities in optical fiber can be eliminated, the fiber may become a very low-loss transmission waveguide for optical signals; and (2) Kapron et al. [28] in 1970 minimized these impurities in fused silica, resulting in the successful making of optical fiber with optical transmission losses of approximately 20 dB km^{-1}. These events awakened the researchers in the communication communities throughout the world. Major efforts were started in the U.S. (Bell Telephone Laboratories, Corning Glass Works, the Naval Electronics Laboratory Center in San Diego, and the Naval Research Laboratory), in the United Kingdom (Standard Telecommunications Laboratories and the British Post Office), in Japan (Nippon Electric Company and the Nippon Sheet Glass Company), and in Germany (AEG-Telefunken, Schott Glass Company, and the Siemens Company). Three major types of optical glass fibers were in contention: the solid core single-mode fiber, the liquid core fiber, and the solid core parabolic-index-variation multimode fiber. Because of the superior dispersion property (i.e., high bandwidth behavior) of the solid core single-mode fiber, it is now universally accepted as the long-distance fiber. At that time, the hope for an all-optical communication system also ignited a significant amount of research in integrated optical circuits, that is, planar imbedded optical dielectric waveguides [29]. Because the index of refraction of the core region and that of the cladding region of an optical fiber are quite close, Snyder in 1969 [30] and Gloge in 1971 [31] provided a new look on the modes that can exist in this so-called "weakly guiding" fiber. Since the 1980s, the emphasis of the research communities has been towards finding ways to increase the bandwidth capacity and to decrease the loss behavior of a single mode fiber. The use of WDM (Wavelength Division Multiplexed) scheme [32] and solitons [33,34] has provided the much sought-after improvement. From the 1990s until now, we find an explosion of novel dielectric waveguides due to the discovery of new materials.

The revolution in digital processing started in the 1950s finally took flight in the 1960s due to the rapid advances in the use of integrated circuits in digital computers. The impact has been incredible and far-reaching. Many heretofore unsolvable engineering or scientific problems could now be solved using a relatively straightforward numerical computational approach. Much advances in the application of numerical techniques to guided wave problems were therefore developed in the period from 1965 to 1980. For example, Yee in 1966 [35] developed the FDTD (Finite Difference Time Domain) algorithm to solve the Maxwell equations numerically; Mur in 1981 [36] developed an effective absorbing boundary condition for FDTD; Yeh and Wang in 1972 [37] made use of the two-point boundary value numerical approach to solve the graded-index fiber problem; Yeh and Lindgren [38] also found an efficient numerical way to solve the many layered guided wave structure problem; Yeh et al. in 1978 [39] and, few months afterwards, Feit and Fleck [40] applied the beam propagation method to treat the problem of wave propagation in single-mode or multimode fibers; Yeh et al. in 1975 [41] became the first group who successfully adopted the finite element technique to solve a large variety of single-mode optical waveguides; and Mariki and Yeh in 1985 [42] perfected the 3D TLM (Transmission-Line Matrix) technique based on the Schelkunoff's impedance concept to solve the arbitrarily shaped dielectric waveguide problem. Several numerical approaches (e.g., FDTD, Finite Element Method, Beam Propagation Method) have already been developed into commercial software packages where a given problem is viewed as a "blackbox" having input data (that specify the problem parameters) and output data (that provide numerical results). There is no need to understand the physics or engineering aspects of the problem. The increasing importance of these numerical approaches to treat guided waves in complex dielectric structures in such a mechanical manner is the reason why there is a need to write this book on the essence of dielectric waveguides. A more thorough discussion of these numerical techniques will be given in the chapter on numerical methods.

Although by the mid-1960s, most of the fundamental concepts of guided wave propagation on linear dielectric structures have been uncovered and understood, it is the explosive revolutionary applications of these concepts in the modern world that establish the importance of understanding the essence of dielectric (surface) waveguides. Optical fibers, which are basically dielectric waveguides, are now routinely used as high-bandwidth communication links. Integrated optical circuits, also basically dielectric waveguides, are in the process of being used exclusively for super-speed computers. Recently, the pursuit of high data rate optical integrated circuits that are compatible with electronic integrated circuits has succeeded in the development of silicon based optical integrated circuits, sources, modulators, and detectors. The only remaining unexploited spectral region is the terahertz band. This is now being actively explored. It appears that, because of the high loss of metallic material in this spectrum, dielectric waveguides may be the only viable option for terahertz links. It should be pointed out that low-loss material in the terahertz region has yet to be found. Lack of suitable low-loss material in the terahertz spectrum means that the traditional optical fiber approach cannot be used to design a low-loss terahertz waveguide. Yeh and Shimabukuro in 2000 [43] found that the configuration of a high dielectric constant waveguide structure could affect greatly the loss behavior of the dominant TM-like mode. Hence, very low loss terahertz waveguide may be designed using this discovery. Other modern application areas for dielectric waveguides include the photonic crystal waveguide [44–47], basically an air or dielectric core surrounded by periodic dielectric structures; surface plasmon polaritons guides [48–50], basically a type of Sommerfeld guide; left-handed material (metamaterial) waveguide, that is a dielectric waveguide whose core region is made with artificial dielectrics with negative permittivity and negative permeability [51, 52]. The surface plasmon waveguide is of special interest in nanostructure research because of the subwavelength property of its guided wave. The peculiar behavior of waves guided by artificial metamaterial structure provides unique opportunity to invent new applications.

1.2 Scope of this Book

The plethora of dielectric waveguides and its vast modern applications mean that it is not possible to write an all-encompassing book on dielectric waveguides. Therefore, our goal is to write a "back to the basics" book that provides the foundation of dielectric waveguides that is useful, clear, and easy to understand.

Chapter 2 presents the fundamental electromagnetic equations with new insight in boundary conditions, classification of fields, the impedance concept, and the scalar-wave approach. Then, an over-all view of dielectric waveguides without delving directly into the specific solution of a given dielectric guided wave structure is presented in Chap. 3. The concepts given there are universally applicable to any dielectric waveguide. New and unique treatment on attenuation has been included.

Specific canonical dielectric guided wave structures will be treated in Chaps. 4–6. They are the planar, circular cylindrical, and elliptical cylindrical structures. Classical analytic modal solutions will be given and explained. The emphasis is to show how one may understand the wave guiding characteristics of a complex, perhaps more practical, dielectric structure from the knowledge of the fundamental solutions from these canonical structures. Approximate approaches for the rectangular dielectric waveguide structure and other structures with no known analytic solutions and inhomogeneous dielectric waveguides are considered in Chaps. 7 and 8.

Subsequent chapters (Chaps. 9–14) will deal with modern applications. Chapters 9 and 10 deal with linear or nonlinear optical fiber structures, where WDM propagation and WDM solitons will be emphasized. Chapter 11 deals with low-loss structures in the terahertz/millimeter wave region. Plasmon (subwavelength) waveguides are treated in Chap. 12. Chapter 13 deals with photonic crystal waveguides. Other uncommon structures, such as metamaterial structure, moving medium waveguide, and anisotropic material structures, are treated in Chap. 14.

Finally, a brief description of several important numerical techniques with examples will be given in Chap. 15.

References

1. A. Sommerfeld, "Über die fortplanzung elektrodynamisches wellen langes eines drahtes," Ann. der Phys. Chem. **67**, 233 (1899)
2. A. Sommerfeld, "An oscillating dipole above a finitely conducting plane," Ann. der Physik, **28**, 665 (1909); Ann. der Physik **81**, 1135 (1926)
3. J. Zenneck, "Propagation of plane EM waves along a plane conducting surface," Ann. der Physik **23**, 846 (1907)
4. D. Hondros and P. Debye, "Elektromagnetische wellen in dielektrischen drahtes," Ann. der Physik **32**, 465 (1910); D. Hondros, "Elektromagneticsche wellen in drahtes," Ann. der Physik **30**, 905 (1909)
5. H. Zahn, "Detection of electromagnetic waves along dielectric wires," Ann. der Physik **49**, 907 (1916)
6. H. Ruter and O. Schriever, "Elektromagnetische wellen an dielektrischen drahten," Schriften des Naturalwissenschaftlichen vereines fur Schleswig-Holstein **16**, 2 (1916)
7. J. R. Carson, S. P. Mead, and S. A. Schelkunoff. "Hyperfrequency waveguides-mathematical theory," Bell Syst. Tech. J. **15**, 310 (1936)
8. G. C. Southworth, "Hyperfrequency waveguides – general considerations and experimental results," Bell Syst. Tech. J. **15**, 284 (1936)
9. S. A. Schelkunoff, "The impedance concept and its application to problems of reflection, refraction, shielding and power absorption," Bell Syst. Tech. J. **17**, 17 (1938)
10. Mallach, "Dielektrische richtstrahler," Bericht des V. I. F. S (1943)
11. G. F. Wegener. "Ausbreitungsgeschwindigkeit wellenwiderstand und dampfung elektromagnetischer wellen an dielektrischen Zylindern," Dissertation, Air Material Command Microfilm ZWB/FB/RE/2018, R8117F831 (1946)
12. W. M. Elsasser, "Attenuation in a dielectric circular rod," J. Appl. Phys. **20**, 1193 (1949)
13. C. H. Chandler. "An investigation of dielectric rod as waveguides," J. Appl. Phys. **20**, 1188 (1949)
14. C. Yeh, "A relation between α and Q," Proc. IRE, vol. **50**, 2143 (1962)
15. L. Brillouin, "Wave Propagation in Periodic Structures," Dover, New York (1953)
16. S. Sensiper, "Electromagnetic wave propagation on helical conductors," Research Laboratory for Electronics, Mass. Inst. of Tech., Tech. Rept No. 194, May 16 (1951); Proc. IRE **43**, 149 (1955)

17. J. R. Pierce, "Theory and Design of Electron Beams," D. Van Nostrand, Princeton (1950)

18. D. D. King, "Dielectric image line," J. Appl. Phys. **23**, 699 (1952); D. D. King and S. P. Schlesinger, "Losses in dielectric image lines," IRE Trans. Microw. Theory Tech. **MTT-5**, 31 (1957)

19. G. Goubau, "Surface waves and their application to transmission lines," J. Appl. Phys. **21**, 1119 (1950); G. Goubau, "Single conductor surface wave transmission lines," Proc. IRE **39**, 619 (1951)

20. R. E. Collin, "Field Theory of Guided Waves," McGraw-Hill, New York (1960)

21. E. Snitzer and J. W. Hicks, "Optical wave-guide modes in small glass fibers. I. Theoretical," J. Opt. Soc. Am. **49**, 1128 (1959); E. Snitzer and H. Osterberg, "Observed dielectric waveguide modes in the visible spectrum," J. Opt. Soc. Am. **51**, 499 (1961); N. S. Kapany and J. J. Burke, "Fiber optics IX. Waveguide effects," J. Opt. Soc. Am. **51**, 1067–1078 (1961)

22. E. Snitzer, "Cylindrical dielectric waveguide modes," J. Opt. Soc. Am. **51**, 491 (1961)

23. C. Yeh, "Elliptical dielectric waveguides," J. Appl. Phys. **33**, 3235 (1962); C. Yeh, "Attenuation in a dielectric elliptical cylinder," IEEE Trans. Antenn. Propag. **AP-11**, 177 (1963)

24. L. A. Lynbimov, G. I. Veselov, and N. A. Bei, "Dielectric waveguide with elliptical cross-section," Radio Eng. Electron. (USSR) **6**, 1668 (1961)

25. R. B. Dyott, "Elliptical Fiber Waveguides," Artech House, Boston (1995)

26. N. Bloembergen, "Non-linear Optics," W. A. Benjamin, New York (1965)

27. K. C. Kao and G. A. Hockham, "Dielectric fiber surface waveguides for optical frequencies," Proc. IEEE **113**, 1151 (1966)

28. F. P. Kapron, D. B. Keck, and R. D. Maurer, "Radiation losses in glass optical waveguides," Appl. Phys. Lett. **17**, 423 (1970)

29. D. Marcuse, Ed., "Integrated Optics," IEEE Press, New York (1973)

30. A. W. Snyder, "Asymptotic expressions for eigenfunctions and eigenvalues of a dielectric or optical waveguide," IEEE Trans. Microw. Theory Tech. **MTT-17**, 1130 (1969)

31. D. Gloge, "Weakly guiding fibers," Appl. Opt. **10**, 2252 (1971)

32. G. P. Agrawal, "Fibre-Optic Communication Systems," Wiley, New York (2002)

33. A. Hasegawa and T. Tappert, "Transmission of stationary nonlinear optical pulses in dispersive dielectric fibers. I. Anomalous dispersion," Appl. Phys. Lett. **23**, 142 (1973)

34. C. Yeh and L. A. Bergman, "Existence of optical solitons on wavelength division multiplexed beams," Phys. Rev. **E 60**, 2306 (1999)

35. K. S. Yee, "Numerical solution of initial boundary value problems involving Maxwell's equations in isotropic media," IEEE Trans. Antenn. Propag. **14**, 302 (1966)

36. G. Mur, "Absorbing boundary conditions for finite-difference approximation of the time-domain electromagnetic field equations," IEEE Trans. Electromag. Compat. **23**, 1073 (1981)

37. C. Yeh and P. Wang, "Scattering of obliquely incident waves by inhomogeneous fibers," J. Appl. Phys. **43**, 3999 (1972)

38. C. Yeh and G. Lindgren, "Computing the propagation characteristics of radially stratified fibers – An efficient method," Appl. Opt. **16**, 483 (1977)

39. C. Yeh, L. Casperson, and B. Szejn, "Propagation of truncated gaussian beams in multimode or single-mode fiber guides," J. Opt. Soc. Am. **68**, 989 (1978)

40. M. D. Feit and J. D. Fleck, "Light propagation in graded-index optical fibers," Appl. Opt. **17**, 3990 (1980)

41. C. Yeh, S. B. Dong, and W. Oliver, "Arbitrarily shaped inhomogeneous optical fiber or integrated optical waveguides," J. Appl. Phys. **46**, 2125 (1975); C. Yeh, K. Ha, S. B. Dong, and W. P. Brown, "Single-mode optical waveguides," Appl. Opt. **18**, 1490 (1979)

42. G. E. Mariki and C. Yeh, "Dynamic 3D TLM analysis of microstrip-lines on anisotropic substrates," IEEE Trans. Microw. Theory Tech. **MTT-33**, 789 (1985)

43. C. Yeh, F. Shimabukuro, and P. H. Siegel, "Low-loss terahertz ribbon waveguides," Appl. Opt. **44**, 5937 (2005); C. Yeh, F. Shimabukuro, P. Stanton, V. Jamnejad, W. Imbriale, and F. Manshadi, "Communication at millimeter–submillimeter wavelengths using a ceramic ribbon," Nature **404**, 584 (2000)

44. E. Yablonovitch, "Inhibited spontaneous emission in solid-state physics and electronics," Phys. Rev. Lett. **58**, 2059 (1987)

45. S. John, "Strong localization of photon in certain disordered dielectric superlattices," Phys. Rev. Lett. **58**, 2486 (1987)

46. J. Joannopoulos, R. Meade, and J. Winn, "Photonic Crystals," Princeton Press, New Jersey (1995)

47. J. C. Knight, T. A. Birks, P. J. Russell, and D. M. Atkins, "All-silica single-mode optical fiber with photonic crystal cladding," Optics Lett. **21**, 1547 (1996)

48. R. H. Ritchie, "Plasma losses by fast electrons in thin films," Phys. Rev. **106**, 874 (1957); H. Raether, "Surface Plasmons," Springer, Berlin Heidelberg New York (1988)

49. S. A. Maier, "Plasmonics: Fundamentals and Applications," Springer, Berlin Heidelberg New York (2007)

50. H. A. Atwater, "The promise of plasmonics," Verlag Scientific Am. **50**, 56 (2007)

51. V. G. Veselago, "The electrodynamics of substances with simultaneously negative values of permittivity and permeability," Soviet Phys. Uspekhi **10**, 509 (1968)

52. J. B. Pendry, D. Schurig, and D. R. Smith, "Controlling electromagnetic fields," Science **312**, 1780 (2006)

2

FUNDAMENTAL ELECTROMAGNETIC FIELD EQUATIONS

All large-scale electromagnetic wave phenomena are governed by the Maxwell equations and the appropriate boundary conditions. In this chapter we shall discuss the fundamental equations and relations dealing with electromagnetic waves [1–3].

2.1 Maxwell Equations

On the basis of the established experimental laws, Maxwell postulated that the electromagnetic field vectors are subject to the following equations:

$$\nabla \times \mathbf{E}(\mathbf{r}, t) = -\frac{\partial \mathbf{B}(\mathbf{r}, t)}{\partial t}, \tag{2.1}$$

$$\nabla \times \mathbf{H}(\mathbf{r}, t) = \mathbf{J}(r, t) + \frac{\partial \mathbf{D}(\mathbf{r}, t)}{\partial t}, \tag{2.2}$$

where

$\mathbf{E}(\mathbf{r}, t)$ = Electric field intensity ($\mathrm{V\,m^{-1}}$)

$\mathbf{H}(\mathbf{r}, t)$ = Magnetic field intensity ($\mathrm{A\,m^{-1}}$)

$\mathbf{D}(\mathbf{r}, t)$ = Electric displacement vector ($\mathrm{C\,m^{-2}}$)

$\mathbf{B}(\mathbf{r}, t)$ = Magnetic induction vector ($\mathrm{Wb\,m^{-2}}$)

$\mathbf{J}(\mathbf{r}, t)$ = Electric current density ($\mathrm{A\,m^{-1}}$)

These vectors are functions of space, $\mathbf{r}$ (in meters), and time, t (in seconds). The mks or Giorgi system of units will be used throughout. On a macroscopic scale, the conservation of charge law can be expressed as follows:

$$\nabla \cdot \mathbf{J}(\mathbf{r}, t) + \frac{\partial \rho}{\partial t} = 0, \tag{2.3}$$

here,

$$\rho(r, t) = \text{Electric charge density } (\mathrm{C\,m^{-3}}).$$

This is the equation of continuity. Faraday's Law, Ampere's Law, Gauss' Law, and Coulomb's Law are included or can be derived from the Maxwell equations and the equation of continuity. For example, (2.1) is a statement of Faraday's Law, while (2.2), without the displacement current term, $\partial \mathbf{D}(\mathbf{r}, t)/\partial t$, is a statement of Ampere's Law. Maxwell postulated the existence of the displacement current term in (2.2) to express the wave nature of the electromagnetic fields. Since the divergence of the curl of any vector vanishes identically, taking the divergence of (2.1) yields

$$\nabla \cdot \frac{\partial \mathbf{B}(\mathbf{r}, t)}{\partial t} = \frac{\partial}{\partial t}(\nabla \cdot \mathbf{B}(\mathbf{r}, t)) = 0 \tag{2.4}$$

or

$$\nabla \cdot \mathbf{B}(\mathbf{r}, t) = 0. \tag{2.5}$$

This is Gauss' Law. From (2.5), the field of the magnetic induction vector $\mathbf{B}(\mathbf{r}, t)$ is solenoidal. The divergence of (2.2) gives

$$\nabla \cdot \mathbf{J}(\mathbf{r}, t) + \nabla \cdot \frac{\partial \mathbf{D}(\mathbf{r}, t)}{\partial t} = 0 \tag{2.6}$$

and, from (2.3), one obtains

$$\frac{\partial}{\partial t}[\nabla \cdot (\mathbf{D}(\mathbf{r}, t) - \rho(\mathbf{r}, t)] = 0 \tag{2.7}$$

or

$$\nabla \cdot \mathbf{D}(\mathbf{r}, t) = \rho(\mathbf{r}, t). \tag{2.8}$$

This is Coulomb's Law. One notes that these divergence equations (2.5) and (2.8) are not independent relations of Maxwell equations (2.1) and (2.2) and the equation of continuity, (2.3). Limiting our investigation to linear phenomena, fields of arbitrary time variation can be constructed from harmonic solutions through the Fourier Transform Method, and there is no loss of generality with the assumption that the time-dependent variation of the fields may be factored out as follows:

$$\mathbf{E}(\mathbf{r}, t) = \mathrm{Re}[\mathbf{E}(\mathbf{r})\, e^{j\omega t}], \tag{2.9}$$

where ω is the harmonic frequency of the wave, Re means the real part of, and $\mathbf{E}(\mathbf{r})$, the electric field vector, is a spatially dependent, complex function. Similar time variations are assumed for the other field and source quantities, such as $\mathbf{D}(\mathbf{r}, t) = \mathrm{Re}[\mathbf{D}(\mathbf{r})\, e^{j\omega t}]$, $\mathbf{H}(\mathbf{r}, t) = \mathrm{Re}[\mathbf{H}(\mathbf{r})\, e^{j\omega t}], \ldots,$ etc. The time-harmonic Maxwell equations and the continuity equation now take the forms

$$\nabla \times \mathbf{E}(\mathbf{r}) = -j\omega \mathbf{B}(\mathbf{r}), \tag{2.10}$$

$$\nabla \times \mathbf{H}(\mathbf{r}) = \mathbf{J}(\mathbf{r}) + j\omega \mathbf{D}(\mathbf{r}), \tag{2.11}$$

$$\nabla \cdot \mathbf{J}(\mathbf{r}) = -j\omega \rho(\mathbf{r}). \tag{2.12}$$

The associated divergence equations are

$$\nabla \cdot \mathbf{B}(\mathbf{r}) = 0, \tag{2.13}$$

$$\nabla \cdot \mathbf{D}(\mathbf{r}) = \rho(\mathbf{r}). \tag{2.14}$$

The field vectors $\mathbf{E}(\mathbf{r})$, $\mathbf{D}(\mathbf{r})$, $\mathbf{H}(\mathbf{r})$, and $\mathbf{B}(\mathbf{r})$ are now spatially dependent complex functions. It is seen from (2.10) and (2.11) that given the source function $\mathbf{J}(\mathbf{r})$, there are four unknown quantities, $\mathbf{E}$, $\mathbf{B}$, $\mathbf{D}$, and $\mathbf{H}$, and two independent equations (2.10) and (2.11). Two additional independent equations relating the field quantities, $\mathbf{E}$, $\mathbf{B}$, $\mathbf{D}$, and $\mathbf{H}$ are needed in order that deterministic solutions for these quantities may be found. The needed equations are obtained from the constitutive relations.

2.2 The Constitutive Relations

The constitutive relations are derived from the description of the macroscopic properties of the medium in the immediate neighborhood of the specified field point. In general, we shall assume that, at any given point in a given medium, the vector $\mathbf{D}$ and $\mathbf{H}$ may be represented as a function of $\mathbf{E}$ and $\mathbf{B}$.

$$\mathbf{D} = F_1(\mathbf{E}, \mathbf{B}), \tag{2.15}$$

$$\mathbf{H} = F_2(\mathbf{E}, \mathbf{B}). \tag{2.16}$$

The functional dependencies of these functions are obtained from the macroscopic physical properties of the medium [4]. The behavior of a material medium in an electromagnetic field can be described in terms of distributions of electric and magnetic dipoles. The medium can be characterized by two polarization density functions: $\mathbf{P}$, the electric dipole moment per unit volume, and $\mathbf{M}$, the magnetic dipole moment per unit volume. The polarization may be induced under the action of the field from other sources, or it may be virtually permanent and independent of external fields. The permanent polarizations will be designated by $\mathbf{P}_0$ and $\mathbf{M}_0$. A few examples are given below.

2.2.1 Simple Medium (Linear and Isotropic)

A simple medium is taken to be (a) linear, where $\mathbf{D}$ is a linear function of $\mathbf{E}$ and $\mathbf{H}$ is a linear function of $\mathbf{B}$, and (b) isotropic, where $\mathbf{D}$ is parallel to $\mathbf{E}$ and $\mathbf{H}$ is parallel to $\mathbf{B}$. In this simple medium,

$$\mathbf{D} = \epsilon \mathbf{E}, \qquad \mathbf{H} = \frac{1}{\mu}\mathbf{B}. \tag{2.17}$$

The parameters ϵ and μ, which represent the macroscopic electromagnetic properties, are, respectively, the permittivity and permeability of the medium. For isotropic inhomogeneous media, ϵ and μ may be functions of positions. For free-space,

$$\epsilon = \epsilon_0, \qquad \mu = \mu_0, \tag{2.18}$$

where $\epsilon_0 = 8.854 \times 10^{-12}\,(\mathrm{F\,m^{-1}})$ and $\mu_0 = 4\pi \times 10^{-7}\,(\mathrm{H\,m^{-1}})$ are, respectively, the free-space permittivity and free-space permeability. The relationships between the field vectors and the polarization vectors are defined as follows:

$$\mathbf{P} + \mathbf{P}_0 = \mathbf{D} - \epsilon_0 \mathbf{E} = (\epsilon - \epsilon_0)\mathbf{E} = \chi_e \epsilon_0 \mathbf{E}, \tag{2.19}$$

$$\mathbf{M} + \mathbf{M}_0 = \frac{1}{\mu_0}\mathbf{B} - \mathbf{H} = \left(\frac{\mu}{\mu_0} - 1\right)\mathbf{H} = \chi_m \mathbf{H}, \tag{2.20}$$

where χ_e and χ_m are called the electric and magnetic susceptibilities. The electric and magnetic polarization vectors are zero in free-space. Strictly speaking, the relations (2.19) and (2.20) are definable only for time-periodic phenomena, since in general ϵ and μ are functions of the frequency. The frequency dependence of the constitutive parameters is known as the *dispersive property* of the medium. Hence, these relations are applicable to other than time-periodic, time-varying fields only when, over the significant part of the frequency spectrum covered by the Fourier components of the time dependence, the constitutive parameters ϵ and μ are sensibly independent of frequency.

2.2.2 Anisotropic Medium [5–7]

In an anisotropic material medium, the electromagnetic properties are functions of the field directions about a point. Thus, in general,

$$\mathbf{D} = \underline{\underline{\epsilon}} \cdot \mathbf{E} \qquad \underline{\underline{\epsilon}} = \begin{bmatrix} \epsilon_{11} & \epsilon_{12} & \epsilon_{13} \\ \epsilon_{21} & \epsilon_{22} & \epsilon_{23} \\ \epsilon_{31} & \epsilon_{32} & \epsilon_{33} \end{bmatrix}, \tag{2.21}$$

$$\mathbf{B} = \underline{\underline{\mu}} \cdot \mathbf{H} \qquad \underline{\underline{\mu}} = \begin{bmatrix} \mu_{11} & \mu_{12} & \mu_{13} \\ \mu_{21} & \mu_{22} & \mu_{23} \\ \mu_{31} & \mu_{32} & \mu_{33} \end{bmatrix}. \tag{2.22}$$

Here, ϵ_{ij} and μ_{ij} are elements of the permittivity matrix and the permeability matrix describing the anisotropic characteristics of the medium. For inhomogeneous and anisotropic medium, ϵ_{ij} and μ_{ij} are functions of positions. For anisotropic and dispersive medium, ϵ_{ij} and μ_{ij} are functions of the frequency. The electromagnetic properties of a few common anisotropic material media are characterized as follows:

(a) Magnetized Ferrite Medium

$$\underline{\underline{\epsilon}} = \begin{bmatrix} \epsilon_1 & 0 & 0 \\ 0 & \epsilon_1 & 0 \\ 0 & 0 & \epsilon_1 \end{bmatrix} = \text{a scalar} \qquad \underline{\underline{\mu}} = \begin{bmatrix} \mu_{11} & \mu_{12} & 0 \\ \mu_{21} & \mu_{22} & 0 \\ 0 & 0 & \mu_{33} \end{bmatrix} \tag{2.23}$$

with an impressed static magnetic field along the axial z-axis.

(b) Crystalline Medium

$$\underline{\underline{\epsilon}} = \begin{bmatrix} \epsilon_{11} & \epsilon_{12} & \epsilon_{13} \\ \epsilon_{21} & \epsilon_{22} & \epsilon_{23} \\ \epsilon_{31} & \epsilon_{32} & \epsilon_{33} \end{bmatrix} \qquad \underline{\underline{\mu}} = \begin{bmatrix} \mu_0 & 0 & 0 \\ 0 & \mu_0 & 0 \\ 0 & 0 & \mu_0 \end{bmatrix} = \text{a scalar}. \tag{2.24}$$

(c) Uniaxial Medium

$$\underline{\underline{\epsilon}} = \begin{bmatrix} \epsilon_1 & 0 & 0 \\ 0 & \epsilon_2 & 0 \\ 0 & 0 & \epsilon_3 \end{bmatrix} \qquad \underline{\underline{\mu}} = \begin{bmatrix} \mu_0 & 0 & 0 \\ 0 & \mu_0 & 0 \\ 0 & 0 & \mu_0 \end{bmatrix} = \text{a scalar}. \tag{2.25}$$

(d) Cold Plasma with Impressed Static Magnetic Field B_0

$$\underline{\underline{\epsilon}} = \begin{bmatrix} \epsilon_{11} & \epsilon_{12} & 0 \\ \epsilon_{21} & \epsilon_{22} & 0 \\ 0 & 0 & \epsilon_{33} \end{bmatrix} \qquad \underline{\underline{\mu}} = \begin{bmatrix} \mu_0 & 0 & 0 \\ 0 & \mu_0 & 0 \\ 0 & 0 & \mu_0 \end{bmatrix} = \text{a scalar.} \tag{2.26}$$

Here ϵ_{ij} is a function of frequency as follows:

$$\begin{aligned}
\epsilon_{11} &= \epsilon_0 \left[1 - \frac{\omega_{\mathrm{p}}^2 (\omega - j\nu)}{\omega \left[(\omega - j\nu)^2 - \omega_{\mathrm{c}}^2 \right]} \right], \\
\epsilon_{22} &= \epsilon_{11}, \\
\epsilon_{12} &= j\epsilon_0 \left[\frac{\omega_{\mathrm{p}}^2 \omega_{\mathrm{c}}}{\omega (\omega - j\nu + \omega_{\mathrm{c}}) (\omega - j\nu - \omega_{\mathrm{c}})} \right], \\
\epsilon_{21} &= -\epsilon_{12}, \\
\epsilon_{33} &= \epsilon_0 \left[1 - \frac{\omega_{\mathrm{p}}^2}{\omega (\omega - j\nu)} \right],
\end{aligned}$$

where $\omega_{\mathrm{p}} = \left(n_{\mathrm{e}} e^2 / m_{\mathrm{e}} \epsilon_0 \right)^{1/2}$ is the electron plasma frequency, n_{e} is the electron number density, e is the electronic charge, m_{e} is the electron mass, and $\omega_{\mathrm{c}} = eB_0/m_{\mathrm{e}}$ is the electron cyclotron frequency, where B_0 is the impressed static magnetic induction along the z-axis. The term ν is the collision frequency of the electrons with the heavier particles.

2.2.3 Left-Handed Medium (Metamaterial) [8–10]

A class of artificial media can be characterized as follows:

$$\mathbf{D} = -\epsilon \mathbf{E}, \qquad \mathbf{H} = -\frac{1}{\mu} \mathbf{B}. \tag{2.27}$$

A left-handed material medium (usually artificially made) is one with negative permittivity and negative permeability in the frequency range of interest. The index of refraction in such a metamaterial medium is also negative. The permittivity and the permeability of the medium are usually frequency dependent and lossy. In other words, they are complex quantities.

2.2.4 Conducting Medium

According to Ohm's law

$$\mathbf{J} = \sigma \mathbf{E}, \tag{2.28}$$

where σ, a scalar constant for most conducting materials, is called the conductivity, in units of mhos per meter. This is an added relation for the source-free Maxwell equations (2.10) and (2.11). In a conducting medium ($\sigma \neq 0$), there can be no permanent distribution of free charges, because any free charge will diffuse out of the conducting medium and reside on the surface. A more refined model that gives the equivalent dielectric representation of a conducting medium is given below. From the Drude theory [11], the frequency dependent conductivity has the form

$$\sigma = \frac{\omega_{\mathrm{p}}^2 \epsilon_0}{j\omega + \nu}, \tag{2.29}$$

where ω_{p} is the plasma frequency of the free electrons in the metal and ν is the collision frequency of the free electrons with the heavier elements. From the Maxwell equation formulation, the equivalent dielectric permittivity of a metal is given by

$$\epsilon_{\mathrm{m}} = \epsilon_0 \left(1 + \frac{\sigma}{j\omega\epsilon_0} \right). \tag{2.30}$$

From (2.29) and (2.30) and separating into the real and imaginary parts, (2.30) becomes

$$\epsilon_{\mathrm{m}} = \epsilon_0 \left(1 - \frac{\omega_{\mathrm{p}}^2}{\omega^2 + \nu^2} - j\frac{\nu}{\omega} \frac{\omega_{\mathrm{p}}^2}{\omega^2 + \nu^2} \right). \tag{2.31}$$

In good conducting metals, ω_{p} is of the order of 5×10^{15} rad s^{-1} and ν is of the order of 5×10^{13} rad s^{-1}. Below terahertz frequencies, it is seen that the conductivity is essentially a large real quantity and the displacement current is negligible, and it can be postulated that the tangential E field at the conductor is zero. However, as the frequency increases, the more exact expression for the metallic behavior must be used. As the frequency approaches the plasma frequency, the real part of the dielectric permittivity becomes a small negative quantity and slow wave guiding can be achieved (plasmon waves [12, 13]) on a small bounded conductor. When the frequency becomes larger than the plasma frequency, the metal acts like a dielectric, albeit a lossy one.

2.2.5 Dielectric Medium with Loss [11]

In general, a wave propagating in a dielectric undergoes losses, and the permittivity of the material can no longer be represented by a real value. These losses can be attributed to a number of causes, including conduction, relaxation phenomena, both in the dielectric as well as in impurities, molecular resonances, and molecular structure [14–16]. Because of losses, the dielectric must be represented by a complex value. The complex dielectric permittivity of that material can be written as follows:

$$\epsilon = \epsilon' - j\epsilon'', \tag{2.32}$$

where ϵ' and ϵ'' are the real and imaginary parts of the permittivity or dielectric constant. In a dielectric, the ratio $\epsilon''/\epsilon'(= \sigma/\omega\epsilon')$ is a direct measure of the ratio of the conduction current to the displacement current. For the principal case of interest, where there is only a small conductive loss, the permittivity is often written as

$$\epsilon = \epsilon'\left(1 - j\frac{\sigma}{\omega\epsilon'}\right) = \epsilon'(1 - j\tan\delta), \tag{2.33}$$

where $\tan\delta = \sigma/(\omega\epsilon')$, and the factor, $\tan\delta$, called the loss tangent, is commonly used to characterize the loss at microwave and millimeter wavelengths, even though the losses may be due to other than conduction. In general, ϵ' and ϵ'' are functions of frequency, although in many applications they may be considered to be constant over a limited frequency band of interest. The dispersive dielectric properties can also depend on temperature. Table 2.1 lists reported relative dielectric constants (ϵ'/ϵ_0) and loss tangents at room temperature for a number of materials of interest at millimeter wavelengths and shorter. The range of values reflects the different fabrication processes and material frequency dispersion as well as differences in the measurements by different authors [12, 14–16]. The complex permittivity is given by (2.32), and for the dielectric materials the imaginary part is also given as a loss tangent. For sapphire, O is for the ordinary ray and E is for the extraordinary ray.

2.2.6 Nonlinear Medium [17]

The constitutive relations for a nonlinear medium have the form

$$\mathbf{D} = \epsilon(\mathbf{E})\mathbf{E}, \tag{2.34}$$

$$\mathbf{B} = \mu(\mathbf{H})\mathbf{H}, \tag{2.35}$$

where $\epsilon(\mathbf{E})$ and $\mu(\mathbf{H})$ are functions of the field strengths. Substituting these constitutive relations into (2.1) and (2.2) gives a set of equations that are nonlinear.

It is important to understand the limitations imposed by describing a given medium by the macroscopic electromagnetic parameters $(\epsilon,\ \mu,\ \sigma)$ as given in Sects. 2.2.1–2.2.6. Microscopically speaking, any material medium is composed of a large number of atoms and/or molecules whose electromagnetic properties can only be accurately described by the principles of quantum electrodynamics. Macroscopically speaking, it is assumed that the ensemble electromagnetic

Table 2.1 Relative dielectric constant and loss tangent of selected materials

	f (GHz)	ϵ'/ϵ_0	$\tan\delta \times 10^4$
Crystalline material [14, 16]			
ZnS	100	8.4	20
KRS-5	94.75	30.5	190
KRS-6	94.75	28.5	230
$LiNbO_3$	94.75	6.7	80
Alumina	100–400	9.6–9.9	6.0–26.0
Sapphire O	100–900	9.4–9.6	2.1–17
Sapphire E	140–900	11.6–11.7	1.1–30
Quartz	94–1,000	3.5–4.6	1–8
Silicon	100–900	11.7–11.9	0.05–19
Germanium	800–1,200	15.9–16.1	1.3–14.4
Polymer Material [14, 15]			
Teflon	100–1,000	2.04–2.07	2–15
Polyethylene	100–1,000	2.27–2.31	1.3–14
Polypropylene	100–300	2.26–2.05	5.2–7.3
Rexolite	30–900	2.52–2.69	10–47
TPX	100–1,000	2.13	6–11
Metal [12]			
Aluminum	400	-3.3×10^4	4.8×10^5
Aluminum	4.7×10^5	−29.8	3.9×10^3
Silver	4.7×10^5	−19	279
Gold	5.4×10^5	−6	3.3×10^3

properties of the material medium can be characterized by a permittivity, ϵ, a permeability, μ, and a conductivity, σ. This means that the wavelength of the electromagnetic wave phenomenon of interest must be much larger than the size of individual atoms or molecules and their separations. Characterizing the electromagnetic properties of a given medium by its permittivity (electric polarizability), permeability (magnetic polarizability), and conductivity is the cornerstone of macroscopic electrodynamics [4]. The macroscopic behavior of the electromagnetic field in a given medium is therefore governed by Maxwell equations (2.10) and (2.11), the continuity equation (2.12), and the constitutive relations (2.15) and (2.16), and, if the medium is conductive, by (2.28). The Maxwell equations can then be reduced to two independent equations with two independent variables,

E and **H**. This macroscopic description of a material medium is still valid for any nanostructure [18, 19] as long as the wavelength of the electromagnetic wave phenomenon of interest is much larger than the physical dimensions of molecules but smaller than that of the nanostructure. If the material medium is macroscopically nonuniform, the medium is considered to be an inhomogeneous medium whose permittivity, permeability, and conductivity may vary spatially, that is, $\epsilon = \epsilon(\mathbf{r})$, $\mu = \mu(\mathbf{r})$, and $\sigma = \sigma(\mathbf{r})$, where **r** is the position vector.

2.3 Boundary Conditions, Radiation Condition, and Edge Condition

The solution of a given electromagnetic problem is uniquely determined if it satisfies the Maxwell equations, the associated constitutive relations, the appropriate boundary conditions, and/or the radiation condition, where appropriate, and/or the edge condition, where appropriate.

2.3.1 Boundary Conditions

Electromagnetic fields across a given boundary between two distinct media must satisfy a set of boundary conditions. Let P be a smooth surface separating two media, 1 and 2; let the unit vector normal to the boundary be **n**, pointing from medium 1 into medium 2. Consider the following three cases: (a) medium 1 and 2 are dielectrics; (b) medium 1 is a perfect conductor and medium 2 is a dielectric; and (c) medium 1 is an imperfect conductor and medium 2 is a dielectric.

Case a. Media 1 and 2 are dielectrics having constitutive parameters $(\epsilon_1, \mu_1, \sigma_1)$ *and* $(\epsilon_2, \mu_2, \sigma_2)$*, respectively*

To thoroughly understand the boundary conditions for the field quantities **E**, **D**, **H**, and **B**, let us consider the following scenarios:

(a) Scenario 1

Across the boundary, as shown in Fig. 2.1, let us introduce two small parallel surface areas of rectangular shape that are mirror images of each other.

The top rectangle, parallel to the interface, is located in medium 1, while the bottom rectangle, also parallel to the interface, is located in medium 2. The small rectangle has sides Δs_1 and Δs_2 .

The unit vectors $\mathbf{n}_1$ and $\mathbf{n}_2$ are normal to the rectangular surfaces. The vectors $\mathbf{e}_x$, $\mathbf{e}_y$, and $\mathbf{e}_z$ are the three unit vectors in the x, y, z directions, respectively. P is the plane that separates medium 1 from medium 2. The Maxwell equations are

$$\nabla \times \mathbf{E}(\mathbf{r}, t) = -\frac{\partial \mathbf{B}(\mathbf{r}, t)}{\partial t}, \tag{2.36}$$

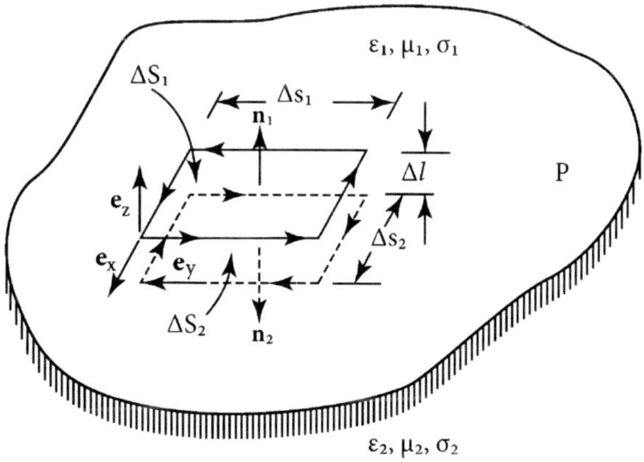

Figure 2.1. Geometry for scenario 1, showing the relationship between the tangential electric fields and the normal magnetic fluxes across a dielectric boundary, as well as the tangential magnetic fields and the normal displacement vectors. The rectangular area ΔS_1 is parallel to the rectangular area ΔS_2. The sides of the rectangle are Δs_1 and Δs_2 and the separation between the two rectangles is Δl. P is the plane that separates medium 1 and medium 2

$$\nabla \times \mathbf{H}(\mathbf{r}, t) = \mathbf{J}(\mathbf{r}, t) + \frac{\partial \mathbf{D}(\mathbf{r}, t)}{\partial t} \tag{2.37}$$

Integrating (2.36) over the rectangular area ΔS_1 in region 1 yields

$$\int_{\Delta S_1} (\nabla \times \mathbf{E}_1) \cdot \mathbf{n}_1 \, dS = -\frac{\partial}{\partial t} \left(\int_{\Delta S_1} \mathbf{B}_1 \cdot \mathbf{n}_1 \, dS \right) \tag{2.38}$$

and integrating (2.36) over the rectangular area ΔS_2 in region 2 yields

$$\int_{\Delta S_2} (\nabla \times \mathbf{E}_2) \cdot \mathbf{n}_2 \, dS = -\frac{\partial}{\partial t} \left(\int_{\Delta S_2} \mathbf{B}_2 \cdot \mathbf{n}_2 \, dS \right) . \tag{2.39}$$

Application of Stokes theorem to (2.38) and (2.39) and adding the resultant equations yield

$$\int_{c1} \mathbf{E}_1 \cdot d\mathbf{s}_1 + \int_{c2} \mathbf{E}_2 \cdot d\mathbf{s}_2 = - \left[\frac{\partial}{\partial t} \left(\int_{\Delta S_1} \mathbf{B}_1 \cdot \mathbf{n}_1 \, dS \right) + \right.$$

$$\left. \frac{\partial}{\partial t} \left(\int_{\Delta S_2} \mathbf{B}_2 \cdot \mathbf{n}_2 \, dS \right) \right] , \tag{2.40}$$

where c_1 and c_2 are, respectively, the circumferences of the rectangular areas $\triangle S_1$ and $\triangle S_2$. In rectangular coordinates, with $\Delta s_1 \to 0$ and $\Delta s_2 \to 0$, **E** has a constant value along each side. Allowing the separation Δl between the two parallel rectangular areas in medium 1 and medium 2 to approach zero, one has

$$(E_{1x} - E_{2x})2\triangle x + (E_{1y} - E_{2y})2\triangle y = -\frac{\partial}{\partial t}(B_{1z} - B_{2z})\triangle x\triangle y, \tag{2.41}$$

where $\Delta s_1 = \Delta x$, $\Delta s_2 = \Delta y$, $\mathbf{s}_1 = \mathbf{e}_x$, $\mathbf{s}_2 = \mathbf{e}_y$, $\mathbf{n}_1 = -\mathbf{n}_2 = \mathbf{n} = \mathbf{e}_z$, E_{1x} and E_{2x} are, respectively, the x-directed electric field tangential to the boundary in region 1 and region 2, E_{1y} and E_{2y} are, respectively, the y-directed electric field tangential to the boundary in region 1 and region 2, B_{1z} and B_{2z} are, respectively, the z-directed **B** field normal to the boundary surface in region 1 and region 2.

In a similar manner, one may derive the following relation from (2.37) for the tangential components of **H** and the normal component of **D** on the boundary surface:

$$(H_{1x} - H_{2x})2\triangle x + (H_{1y} - H_{2y})2\triangle y = \frac{\partial}{\partial t}(D_{1z} - D_{2z})\triangle x\triangle y + (J_{1z} - J_{2z})\triangle x\triangle y. \tag{2.42}$$

The significance of (2.41) and (2.42) will be discussed later.

(b) Scenario 2

Let us consider a rectangular path c crossing through the boundary as shown in Fig. 2.2.

Two sides with length Δs of this rectangle are parallel to the boundary P and the height is Δl. The unit vector $\mathbf{n}_0$ is normal to the rectangular area ΔS. The unit vectors $\mathbf{s}_1$and $\mathbf{s}_2$ are parallel to the boundary interface and are normal to $\mathbf{n}_0$ and

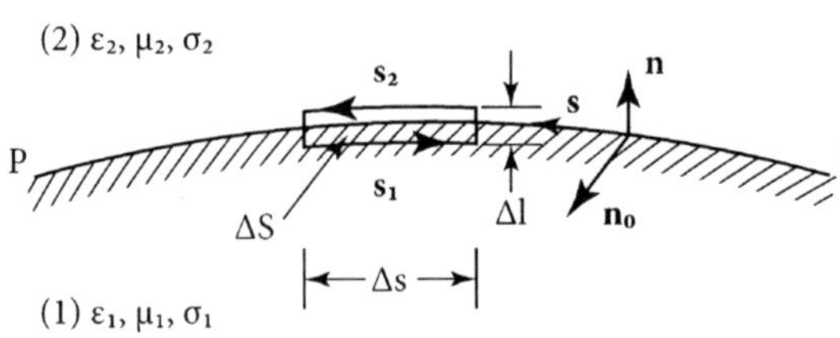

Figure 2.2. Geometry for scenario 2, showing the continuity of the tangential electric and magnetic fields across a dielectric boundary plane P

$\mathbf{n}$, where $\mathbf{n}$ is the unit vector normal to the boundary surface. Also, $\mathbf{s}_1 = -\mathbf{s}_2$, $\mathbf{s}_1 = \mathbf{n}_0 \times \mathbf{n}$, and $\mathbf{s}_2 = -\mathbf{n}_0 \times \mathbf{n}$. Integrating (2.36) over the rectangular area yields

$$\int_{\triangle S} (\nabla \times \mathbf{E}) \cdot \mathbf{n}_0 \, dS = -\frac{\partial}{\partial t} \left(\int_{\triangle S} \mathbf{B} \cdot \mathbf{n_0} \, dS \right). \tag{2.43}$$

Application of Stoke's theorem gives

$$\int_c \mathbf{E} \cdot d\mathbf{s} = -\frac{\partial}{\partial t} \left(\int_{\triangle S} \mathbf{B} \cdot \mathbf{n}_0 \, dS \right), \tag{2.44}$$

which for a very small rectangle with $\Delta s \rightarrow 0$ and $\Delta l \rightarrow 0$, $\mathbf{E}$ has a constant value along each side, giving

$$\begin{aligned} (\mathbf{s}_1 \cdot \mathbf{E}_1 + \mathbf{s}_2 \cdot \mathbf{E}_2)\Delta s &+ \text{contribution from ends} \\ &= -\frac{\partial \mathbf{B}}{\partial t} \cdot \mathbf{n}_0 \, dS, \\ &= -\frac{\partial \mathbf{B}}{\partial t} \cdot \mathbf{n}_0 \Delta s \Delta l. \end{aligned} \tag{2.45}$$

Taking the limit $\Delta s \rightarrow 0$ and $\Delta l \rightarrow 0$, one has

$$(\mathbf{n}_0 \times \mathbf{n} \cdot \mathbf{E}_1 - \mathbf{n}_0 \times \mathbf{n} \cdot \mathbf{E}_2) = -\lim_{\Delta l \rightarrow 0} \left(\frac{\partial \mathbf{B}}{\partial t} \cdot \mathbf{n}_0 \Delta l \right) \tag{2.46}$$

or

$$\mathbf{n}_0 \cdot [\mathbf{n} \times (\mathbf{E}_1 - \mathbf{E}_2)] = \lim_{\Delta l \rightarrow 0} \left(\mathbf{n}_0 \cdot \frac{\partial \mathbf{B}}{\partial t} \right) \Delta l. \tag{2.47}$$

The contribution from ends on the left side (2.45) is zero as $\Delta l \rightarrow 0$. So,

$$\mathbf{n} \times (\mathbf{E}_1 - \mathbf{E}_2) = 0. \tag{2.48}$$

In a similar manner, using (2.37), one may derive the relation

$$\mathbf{n} \times (\mathbf{H}_1 - \mathbf{H}_2) = \mathbf{J}_s, \tag{2.49}$$

where $\mathbf{J}_s$ is the surface current density at the boundary surface. It is shown that the tangential components of the electric and magnetic fields must be continuous across the dielectric boundary. That these boundary conditions on the tangential components of the electric and magnetic fields at the boundary of two dissimilar dielectric media are necessary and sufficient boundary conditions will be shown later.

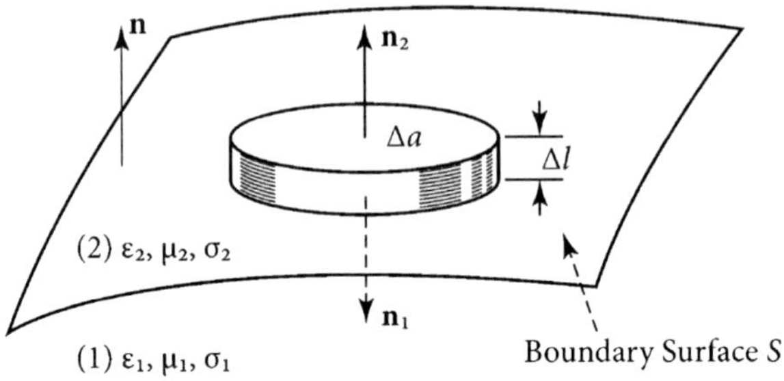

Figure 2.3. Geometry for scenario 3, showing the continuity of normal magnetic fluxes and the continuity of normal displacement vectors across a dielectric boundary

(c) Scenario 3

Let us consider a small right circular cylinder with an end area Δa and height Δl, situated across the boundary interface as shown in Fig. 2.3. The unit vectors $\mathbf{n}_1$ and $\mathbf{n}_2$ are normal to the end surfaces, which are normal to the boundary. The volume of this cylinder is ΔV. The boundary surface S separates medium 1 and medium 2. Integrating (2.5) over this enclosed volume gives

$$\int_{\Delta V} \boldsymbol{\nabla} \cdot \mathbf{B} \, dV = 0. \tag{2.50}$$

Applying the divergence theorem yields

$$\int \mathbf{B} \cdot \mathbf{n} \; da = 0, \tag{2.51}$$

where the area integration is over the walls and ends of the cylinder. For a very small cylinder with $\Delta a \rightarrow 0$ and $\Delta l \rightarrow 0$, $\mathbf{B}$ has a constant value over each end surface, and

$$(\mathbf{B}_1 \cdot \mathbf{n}_1 + \mathbf{B}_2 \cdot \mathbf{n}_2)\, \Delta a + \text{contribution of the side walls} = 0. \tag{2.52}$$

Here $\mathbf{B}_1$ is the B field in region 1 and $\mathbf{B}_2$ is the $\mathbf{B}$ field in region 2. Since $\mathbf{n}_2 = -\mathbf{n}_1 = -\mathbf{n}$, (2.52) becomes

$$(\mathbf{B}_1 - \mathbf{B}_2) \cdot \mathbf{n} = 0, \tag{2.53}$$

which is the needed boundary condition on $\mathbf{B}$. Similarly, using (2.8), one may derive the relation

$$(\mathbf{D}_1 - \mathbf{D}_2) \cdot \mathbf{n} = \rho_s, \tag{2.54}$$

where ρ_s is the surface charge density at the boundary surface. These equations indicate that across the boundary of dissimilar dielectric media, the normal components of **B** must be continuous and the normal components of **D** must be discontinuous by the surface charge density. That these boundary conditions are necessary but not sufficient boundary conditions will be shown in the following.

From the above discussion, according to the three scenarios, one may restate the following boundary conditions at the boundary of two dielectrics:

$$(E_{1x} - E_{2x})2\triangle x + (E_{1y} - E_{2y})2\triangle y = -\frac{\partial}{\partial t}(B_{1z} - B_{2z})\triangle x\triangle y, \qquad (2.41)$$

$$(H_{1x} - H_{2x})2\triangle x + (H_{1y} - H_{2y})2\triangle y = \frac{\partial}{\partial t}(D_{1z} - D_{2z})\triangle x\triangle y + (J_{1z} - J_{2z})\triangle x\triangle y, \qquad (2.42)$$

$$\mathbf{n} \times (\mathbf{E}_1 - \mathbf{E}_2) = 0, \qquad (2.48)$$

$$\mathbf{n} \times (\mathbf{H}_1 - \mathbf{H}_2) = \mathbf{J}_s, \qquad (2.49)$$

$$(\mathbf{B}_1 - \mathbf{B}_2) \cdot \mathbf{n} = 0, \qquad (2.53)$$

$$(\mathbf{D}_1 - \mathbf{D}_2) \cdot \mathbf{n} = \rho_s. \qquad (2.54)$$

The discussion on the necessary and sufficient boundary conditions at the boundary of two dielectrics was first given by Yeh in 1993 [20]. Referring to (2.41) and (2.42), for time-harmonic fields, we may replace $\partial/\partial t$ by $j\omega$, resulting in

$$(E_{1x} - E_{2x})2\triangle x + (E_{1y} - E_{2y})2\triangle y = -j\omega(B_{1z} - B_{2z})\triangle x\triangle y \qquad (2.55)$$

and

$$(H_{1x} - H_{2x})2\triangle x + (H_{1y} - H_{2y})2\triangle y = j\omega(D_{1z} - D_{2z})\triangle x\triangle y + (J_{1z} - J_{2z})\triangle x\triangle y. \qquad (2.56)$$

Here, ω is the frequency of the time-harmonic fields. All field symbols are now time-independent phasors. These equations are valid at the interface boundary. Equation (2.55) shows that if the boundary conditions on the tangential electric field are satisfied, that is, $E_{1x} = E_{2x}$ and $E_{1y} = E_{2y}$, then the left hand side of the equation is identically zero. Since, for time-varying fields, ω is nonzero, and $\triangle x \triangle y \neq 0$, then $B_{1z} - B_{2z} = 0$ or $B_{1z} = B_{2z}$, which is the boundary condition on the normal component of **B**. On the one hand, this is proof that satisfying the boundary conditions on the tangential components of electric field means that the boundary condition on the normal component of **B** is satisfied. On the other hand, if the boundary condition on the normal component of **B** is satisfied, that is, if the right hand side of (2.55) is zero, it only means that

$$(E_{1x} - E_{2x})2\triangle x + (E_{1y} - E_{2y})2\triangle y = 0. \tag{2.57}$$

It is not possible to conclude that the boundary conditions on the tangential components of **E** are satisfied. Equation (2.57) indicates that only the combined terms of $(E_{1x} - E_{2x})2\triangle x$ and $(E_{1y} - E_{2y})2\triangle y$ must be zero and not necessarily each term must be zero. Similar conclusion can be reached from (2.56), that is, satisfying the boundary conditions on the tangential components of magnetic field **H**, (2.49), means that the boundary condition on the normal component of **D**, (2.54), is satisfied, while the converse is not true.

To summarize, the necessary and sufficient boundary conditions on the time-varying electromagnetic fields across two distinct dielectric media are that the tangential electric fields must be continuous across the boundary and the tangential magnetic fields must be discontinuous by the surface current density, that is, (2.48) and (2.49). Satisfying the boundary conditions for the normal components of **D** *and* **B**, that is, *(2.53) and (2.54), does not guarantee that all the necessary boundary conditions are satisfied.*

Since the constitutive relations were never used to arrive at the above proof [20], the proof is equally applicable to isotropic or anisotropic, dispersive or nondispersive, moving or stationary, linear or nonlinear, and left-handed (metamaterial) or right-handed media.

Let us now investigate the special case of static fields. Here, ω is identically zero. In that case, (2.55) and (2.56) can be read as follows:

$$(E_{1x} - E_{2x})2\triangle x + (E_{1y} - E_{2y})2\triangle y = 0, \tag{2.55a}$$

$$(H_{1x} - H_{2x})2\triangle x + (H_{1y} - H_{2y})2\triangle y = (J_{1z} - J_{2z})\triangle x \triangle y. \tag{2.56a}$$

There is no connection between electric and magnetic fields. Hence, unlike the time-dependent case, satisfying the boundary conditions on the tangential electric or magnetic fields says nothing about the satisfaction of the boundary conditions on the normal component of $\mathbf{D}$ and $\mathbf{B}$. Furthermore, (2.55a) shows that, on the boundary, only the conditions $E_{1x} = E_{2x}$ and $E_{1y} = E_{2y}$ must be satisfied simultaneously. Indeed, for electrostatic problems, the complete boundary conditions require the satisfaction of the continuity of the tangential components of $\mathbf{E}$ across the boundary as well as the condition that the normal $\mathbf{D}$ is discontinuous by the surface charge density across the boundary. Using the same argument on (2.56a) for the magnetostatic case, one may reach similar conclusion, that is, the complete magnetostatic boundary conditions require the condition that the tangential components of $\mathbf{H}$ be discontinuous by the surface current density as well as the continuity of normal $\mathbf{B}$ across the boundary.

The necessary and sufficient boundary condition on the static electric and magnetic fields across two distinct dielectric media are that (a) for the electrostatic fields, the tangential component of $\mathbf{E}$ must be continuous and the normal component of $\mathbf{D}$ must be discontinuous by the surface charge density at the boundary, and, (b) for the magnetostatic fields, the tangential component of $\mathbf{H}$ must be discontinuous by the surface current density and the normal component of $\mathbf{B}$ must be continuous at the boundary.

Case b. Medium 2 is a perfect conductor and medium 1 is a dielectric

By definition, no electromagnetic field can exist within a perfect conductor (medium 2) and the potential on a perfect conductor is a constant. Therefore, no tangential electric field can exist on the surface of that perfect conductor, while, by Ampere's law, the tangential magnetic field must be equal to the surface current density and must be in a direction normal to that surface current. All electric field lines must terminate normally on the perfectly conducting surface. Hence the boundary conditions are

$$\mathbf{n} \times \mathbf{E}_1 = 0, \quad \mathbf{n} \times \mathbf{H}_1 = \mathbf{J}_s, \quad \mathbf{n} \cdot \mathbf{B}_1 = 0, \quad \mathbf{n} \cdot \mathbf{D}_1 = \rho_s, \tag{2.58}$$

where $\mathbf{J}_s$ is the total electric surface current density, which consists of the induced surface current density due to an incident electromagnetic field in medium 1 and any other impressed currents, and ρ_s is the surface charge density.

Case c. Medium 2 has a surface impedance Z_s and no field can penetrate into medium 2, and medium 1 is a dielectric

On the surface of medium 2, the surface impedance is defined as follows [21, 22]:

$$-\mathbf{n} \times \mathbf{E}_1 = Z_s \mathbf{H}_1. \tag{2.59}$$

Equation (2.59) becomes the boundary condition for the fields in medium 1. If medium 2 is a good conductor, in which the conductivity σ is large but finite, then

$Z_s = (1+j)(\omega\mu/2\sigma)^{1/2}$. The surface impedance boundary condition is valid only for time-harmonic fields. Note that the surface impedance is inductive for a planar imperfect conductor.

2.3.2 Radiation Condition [3]

The field associated with a finite distribution of sources or the field scattered from obstacles must satisfy conditions at infinity, which pertain to the finiteness of the energy radiated by the sources or scattered by obstacles as well as the assurance that the field at infinity represents an outgoing wave. In other words, the radiation condition not only requires that (a) for a finite three-dimensional (3D) source, the field intensities must vanish at infinity such that $\lim(r^2\mathbf{E})$ and $\lim(r^2\mathbf{H})$ are bounded as $r \to \infty$, where r is the 3D radial distance from the origin, (b) for a finite two-dimensional (2D) source, the field intensities must vanish at infinity such that $\lim(r\mathbf{E})$ and $\lim(r\mathbf{H})$ are bounded as $r \to \infty$, where r is 2D radial distance from the origin, but also (c) the electromagnetic wave represented by $\mathbf{E}$ and $\mathbf{H}$ must behave as an outgoing divergent traveling wave at great distances from the source. The finiteness of the energy radiated by the source is assured by condition (a) for a 3D source or by condition (b) for a 2D source. For a time-periodic field in a homogeneous medium the 3D radiation condition at infinity takes the form

$$\lim_{r\to\infty} r \left[\mathbf{H} - \left(\frac{\epsilon_0}{\mu_0}\right)^{1/2} (\mathbf{e}_r \times \mathbf{E}) \right] \to 0, \tag{2.60}$$

$$\lim_{r\to\infty} r\mathbf{E} \text{ is finite,} \tag{2.61}$$

where r is the radial distance from an arbitrary origin in the neighborhood of the sources and $\mathbf{e}_r$ is a unit vector directed from the origin in the radial direction.

2.3.3 Edge Condition [23, 24]

At sharp edges the field vectors may become infinite. But the order of this singularity is restricted by the Bouwkamp–Meixner edge condition. The energy density must be integrable over any finite domain even if this domain happens to include field singularities, that is, the energy in any finite region of space must be finite. For example, when applied to a perfectly conducting sharp edge, this condition states that the singular components of the electric and magnetic vectors are of order $\xi^{-1/2}$, where ξ is the distance from the edge, whereas the parallel components are always finite.

2.3.4 Uniqueness Theorem

A field in a lossy region is uniquely specified by the sources within the region plus the tangential components of $\mathbf{E}$ over the boundary, or the tangential components of $\mathbf{H}$ over the boundary, or the former over part of the boundary and the latter over the rest of the boundary. The uniqueness theorem for the lossy case can be carried over to the lossless case if we consider the field in a lossless medium to be the limit of the corresponding field in a lossy medium as the loss goes to zero.

2.4 Energy Relations: Poynting's Vector Theorem

Taking the scalar product of Maxwell equations (2.1) and (2.2) with $\mathbf{H}$ and $\mathbf{E}$, respectively, and subtracting the resultant equations gives the following energy relation for a simple medium:

$$\nabla \cdot \mathbf{S}(t) + \mathbf{E}(t) \cdot \mathbf{J}(t) = \frac{\partial \mathbf{B}}{\partial t} \cdot \mathbf{H} - \frac{\partial \mathbf{D}}{\partial t} \cdot \mathbf{E} = \frac{1}{2}\frac{\partial}{\partial t}(\mathbf{B} \cdot \mathbf{H} - \mathbf{D} \cdot \mathbf{E}), \quad (2.62)$$

where

$$\mathbf{S}(t) = \mathbf{E}(t) \times \mathbf{H}(t) \quad (2.63)$$

is defined as the instantaneous Poynting's vector representing the flow of energy associated with an electromagnetic field; the right hand side of (2.62) represents the time rate of change of the stored magnetic and electric energy; the factor $\mathbf{E}(t) \cdot \mathbf{J}(t)$ is the power supplied by the current source $\mathbf{J}$. Equation (2.62) is the differential form of Poynting's vector theorem.

Taking the volume integral of (2.62) gives the integral form of Poynting's vector theorem:

$$\int_A \mathbf{S}(t) \cdot \mathbf{n}\, dA + \int_V (\mathbf{E} \cdot \mathbf{J})\, dV = -\frac{\partial}{\partial t} \int_V \frac{1}{2}(\mathbf{B} \cdot \mathbf{H} - \mathbf{D} \cdot \mathbf{E})\, dV. \quad (2.64)$$

The first integral represents the electromagnetic energy flowing out or in per second from a volume V bounded by a surface A. The second integral represents power generated within the volume V, or if $\mathbf{J} = \sigma \mathbf{E}$, it represents power dissipated as Joules heat in the volume V. The third integral represents the time rate of change of electric and magnetic energies in the volume V.

For time-harmonic fields, the complex Poynting's vector is derived by taking the dot product of (2.10) and $\mathbf{H}(\mathbf{r})$and subtracting the dot product of the complex conjugate of (2.11) and $\mathbf{E}(\mathbf{r})$. Using the identity

$$\nabla \cdot [\mathbf{E}(\mathbf{r}) \times \mathbf{H}^*(\mathbf{r})] = \mathbf{H}^*(\mathbf{r}) \cdot [\nabla \times \mathbf{E}(\mathbf{r})] - \mathbf{E}(\mathbf{r}) \cdot [\nabla \times \mathbf{H}^*(\mathbf{r})],$$

one obtains

$$\nabla \cdot [\mathbf{E}(\mathbf{r}) \times \mathbf{H}^*(\mathbf{r})] = -\mathbf{E}(\mathbf{r}) \cdot \mathbf{J}^*(\mathbf{r}) - j\omega [\mathbf{B}(\mathbf{r}) \cdot \mathbf{H}^*(\mathbf{r}) - \mathbf{E}(\mathbf{r}) \cdot \mathbf{D}^*(\mathbf{r})] . \tag{2.65}$$

The complex Poynting's vector is defined as follows:

$$\mathbf{S}(\mathbf{r}) = \mathbf{E}(\mathbf{r}) \times \mathbf{H}^*(\mathbf{r}) \qquad \left(\mathrm{W\,m}^{-2}\right). \tag{2.66}$$

Integrating (2.65) over an enclosed volume V with a surface area A yields

$$P = \int_A \mathbf{S}(\mathbf{r}) \cdot \mathbf{n}\, dA = -\int_V \mathbf{E}(\mathbf{r}) \cdot \mathbf{J}^*(\mathbf{r})\, dV - j\omega \int_V [\mathbf{B} \cdot \mathbf{H}^* - \mathbf{E} \cdot \mathbf{D}^*]\, dV, \tag{2.67}$$

where P is defined as the complex power crossing the surface area A. The time-average power is

$$\overline{P} = \left\langle \int_A \mathbf{S}(\mathbf{r}, t) \cdot \mathbf{n}\, dA \right\rangle = \frac{1}{2} \mathrm{Re}(P) = \frac{1}{2} \mathrm{Re} \int_A \mathbf{S}(\mathbf{r}) \cdot \mathbf{n}\, dA \tag{2.68}$$

Following Kong [25], let us find the relationship between $\langle \mathbf{S}(\mathbf{r}, t) \rangle$ and $\mathbf{S}(\mathbf{r})$ in the harmonic case where the symbol $\langle \ \rangle$ means time-averaging. Recall the following relations for the harmonic case:

$$\mathbf{E}(\mathbf{r}, t) = \mathrm{Re}\left[\mathbf{E}(\mathbf{r})\, e^{j\omega t}\right] = \mathbf{E}_r(\mathbf{r}) \cos\omega t + \mathbf{E}_j(\mathbf{r}) \sin\omega t, \tag{2.69}$$

$$\mathbf{H}(\mathbf{r}, t) = \mathrm{Re}\left[\mathbf{H}(\mathbf{r})\, e^{j\omega t}\right] = \mathbf{H}_r(\mathbf{r}) \cos\omega t + \mathbf{H}_j(\mathbf{r}) \sin\omega t, \tag{2.70}$$

$$\mathbf{E}(\mathbf{r}) = \mathbf{E}_r(\mathbf{r}) + j\mathbf{E}_j(\mathbf{r}), \tag{2.71}$$

$$\mathbf{H}(\mathbf{r}) = \mathbf{H}_r(\mathbf{r}) + j\mathbf{H}_j(\mathbf{r}). \tag{2.72}$$

The subscripts r and j indicate the real and imaginary part, respectively. Now

$$\begin{aligned} \mathbf{S}(\mathbf{r}, t) &= \mathbf{E}(\mathbf{r}, t) \times \mathbf{H}(\mathbf{r}, t), \\ &= \mathbf{E}_r(\mathbf{r}) \times \mathbf{H}_r(\mathbf{r}) \cos^2\omega t + \mathbf{E}_j(\mathbf{r}) \times \mathbf{H}_j(\mathbf{r}) \sin^2\omega t \\ &\quad + [\mathbf{E}_r(\mathbf{r}) \times \mathbf{H}_j(\mathbf{r}) + \mathbf{E}_j(\mathbf{r}) \times \mathbf{H}_r(\mathbf{r})] \sin\omega t \cos\omega t. \end{aligned} \tag{2.73}$$

Taking the time average of (2.73) gives

$$
\begin{aligned}
\langle \mathbf{S}(\mathbf{r}, t) \rangle &= \frac{1}{2\pi} \int_0^{2\pi} \mathbf{S}(\mathbf{r},t)\, d(\omega t), \\
&= \frac{1}{2} \left[\mathbf{E}_r(\mathbf{r}) \times \mathbf{H}_r(\mathbf{r}) + \mathbf{E}_j(\mathbf{r}) \times \mathbf{H}_j(\mathbf{r}) \right], \\
&= \frac{1}{2} \operatorname{Re} \left[\mathbf{E}(\mathbf{r}) \times \mathbf{H}^*(\mathbf{r}) \right], \\
&= \frac{1}{2} \operatorname{Re} \left[\mathbf{S}(\mathbf{r}) \right].
\end{aligned}
\tag{2.74}
$$

This is the relationship we are seeking.

In the following, we shall summarize several useful formulas:

(a) The time-average power flow through a surface area A_n is

$$
\overline{P} = \frac{1}{2} \operatorname{Re} \int_{A_n} \mathbf{n} \cdot \left(\mathbf{E} \times \mathbf{H}^* \right) dA, \tag{2.75}
$$

where $\mathbf{n}$ is normal to the surface.

(b) The time-average power generated within the volume V is

$$
\overline{P}_{\mathrm{g}} = \frac{1}{2} \operatorname{Re} \int_V \left(\mathbf{J} \cdot \mathbf{E}^* \right) dA. \tag{2.76}
$$

(c) The time-average power dissipated as Joules heat within the volume V_{d} of dielectric material with loss σ is

$$
\overline{P}_{\mathrm{d}} = \frac{1}{2} \operatorname{Re} \int_{V_{\mathrm{d}}} \sigma \left(\mathbf{E} \cdot \mathbf{E}^* \right) dV, \tag{2.77}
$$

where $\mathbf{J} = \sigma \mathbf{E}$, $\sigma = \omega\epsilon \tan\delta$, ϵ is the dielectric constant, and $\tan\delta$ is the loss tangent of the dielectric material.

(d) The time-average power dissipated as Joules heat on the surface A_{c} of a conductor due to the surface resistance R_{s} is

$$
\overline{P}_{\mathrm{c}} = \frac{1}{2} R_{\mathrm{s}} \int_{A_{\mathrm{c}}} \mathbf{J}_{\mathrm{s}} \cdot \mathbf{J}_{\mathrm{s}}^*(\mathbf{r})\, dA, \tag{2.78}
$$

where $\mathbf{J}_s = \mathbf{n} \times \mathbf{H}$, $\mathbf{J}_s$ is the surface current on the conducting surface. The unit vector $\mathbf{n}$ is normal to the surface, $R_s = \sqrt{\omega\mu/2\sigma}$, σ is the conductivity of the conductor.

(e) The time-average electric energy density is

$$\overline{W}_e(\mathbf{r}) = \frac{1}{4}\mathbf{D}(\mathbf{r}) \cdot \mathbf{E}^*(\mathbf{r}) \qquad \left(\mathrm{J\,m^{-3}}\right) \tag{2.79}$$

and the time-average electric energy in a volume V is

$$U_e = \frac{1}{2}\int_V \left(\mathbf{D}(\mathbf{r}) \cdot \mathbf{E}^*(\mathbf{r})\right) dV \qquad (\mathrm{J}). \tag{2.80}$$

The time-average magnetic energy density is

$$\overline{W}_m(\mathbf{r}) = \frac{1}{4}\mathbf{B}(\mathbf{r}) \cdot \mathbf{H}^*(\mathbf{r}) \qquad \left(\mathrm{J\,m^{-3}}\right) \tag{2.81}$$

and the time-average magnetic energy in a volume V is

$$U_m = \frac{1}{2}\int_V \left(\mathbf{H}(\mathbf{r}) \cdot \mathbf{B}^*(\mathbf{r})\right) dV \quad (\mathrm{J}). \tag{2.82}$$

If the volume contains no sources or sinks, the stored electromagnetic energy oscillates between the electric and magnetic field. So, we may find the total stored energy by calculation of the energy stored in the electric (or magnetic) field at the instant when it is at a maximum,

$$U = (U_e)_{\max} = \frac{1}{2}\int_V \left(\mathbf{E}_{\max}(\mathbf{r}) \cdot \mathbf{D}^*_{\max}(\mathbf{r})\right) dV. \tag{2.83}$$

2.5 Classification of Fields

The propagation characteristics of guided waves are governed by the source-free Maxwell equations. Only when the excitation of the guided waves is under consideration will the source-present Maxwell equations be used. Let us now consider the source-free Maxwell equations. In a simple medium in which

$$\mathbf{D} = \epsilon\mathbf{E}, \tag{2.84}$$

$$\mathbf{B} = \mu\mathbf{H}, \tag{2.85}$$

the source-free time-harmonic Maxwell equations (2.10) and (2.11) become

$$\boldsymbol{\nabla} \times \mathbf{E}(\mathbf{r}) = -j\omega\mu\mathbf{H}(\mathbf{r}), \tag{2.86}$$

$$\boldsymbol{\nabla} \times \mathbf{H}(\mathbf{r}) = j\omega\epsilon\mathbf{E}(\mathbf{r}). \tag{2.87}$$

Taking the curl of (2.86) and substituting (2.87) into the resultant equation yields

$$\boldsymbol{\nabla} \times \boldsymbol{\nabla} \times \mathbf{E}(\mathbf{r}) - \omega^2\mu\epsilon\mathbf{E}(\mathbf{r}) = 0. \tag{2.88}$$

A similar equation for $\mathbf{H}$ can also be derived,

$$\boldsymbol{\nabla} \times \boldsymbol{\nabla} \times \mathbf{H}(\mathbf{r}) - \omega^2\mu\epsilon\mathbf{H}(\mathbf{r}) = 0. \tag{2.89}$$

Equation (2.88) or (2.89) is called the vector wave equation. Historically speaking, the scalar wave equation describing the behavior of sound waves is

$$\nabla^2 u(\mathbf{r}) + k^2 u(\mathbf{r}) = 0, \tag{2.90}$$

where $u(\mathbf{r})$ may represent the pressure wave and k is the sound wave number. The solutions of this scalar wave equation had been known for many years prior to the introduction of the vector wave equation describing the electromagnetic wave phenomenon. Therefore, the reduction of the vector wave equation to the well-studied scalar wave equation would be very desirable.

In a simple medium, the electromagnetic fields, governed by a linear vector wave equation, are linear. This means that the superposition theorem can be applied to the fields. The complete electromagnetic fields can be obtained by superposing partial fields, which are derived from a set of new potentials, called the Debye potentials. Using these potentials, it is possible to reduce the vector wave equation to a scalar wave equation in certain special but very useful situations.

2.5.1 The Debye Potentials

Following Debye, let us introduce two new potentials, $\Psi(\mathbf{r})$ and $\Phi(\mathbf{r})$, as follows:

$$\mathbf{E}(\mathbf{r}) = \boldsymbol{\nabla} \times (\mathbf{a}\Psi(\mathbf{r})) - (j/\omega\epsilon)\nabla \times \nabla \times (\mathbf{a}\Phi(\mathbf{r})), \tag{2.91}$$

$$\mathbf{H}(\mathbf{r}) = \boldsymbol{\nabla} \times (\mathbf{a}\Phi(\mathbf{r})) + (j/\omega\mu)\nabla \times \nabla \times (\mathbf{a}\Psi(\mathbf{r})), \tag{2.92}$$

where $\mathbf{a}$ is a unit vector or the position vector $\mathbf{r}$. For example, in spherical coordinates $\mathbf{a} = \mathbf{r}$, the radial position vector; in cylindrical coordinates $\mathbf{a} = \mathbf{e}_z$, the axial vector; in rectangular coordinates, $\mathbf{a} = \mathbf{e}_x$ or $\mathbf{e}_y$ or $\mathbf{e}_z$, the unit vector in x or y or

z direction, respectively. Since the divergence of a curl is identically zero, taking the divergence of (2.91) and (2.92) gives $\nabla \cdot \mathbf{E} = 0$ and $\nabla \cdot \mathbf{H} = 0$. This means that the representation of $\mathbf{E}$ and $\mathbf{H}$ by the Debye potentials has already satisfied the divergence equations for $\mathbf{E}$ and $\mathbf{H}$. The two scalar functions $\Psi(\mathbf{r})$ and $\Phi(\mathbf{r})$ are called the Debye potentials, which satisfy a pair of second-order partial differential equations. These partial differential equations are obtained by substituting (2.91) and (2.92) into (2.88) and (2.89) (the wave equations), respectively. One can show that in a homogeneous, isotropic medium, $\Psi(\mathbf{r})$ and $\Phi(\mathbf{r})$ satisfy the scalar Helmholtz equation,

$$\left(\nabla^2 + k^2\right) \Psi(\mathbf{r}) = 0, \tag{2.93a}$$

$$\left(\nabla^2 + k^2\right) \Phi(\mathbf{r}) = 0, \tag{2.93b}$$

with $k^2 = \omega^2 \mu \epsilon$. This means that by using the Debye potentials as introduced in (2.91) and (2.92), one may reduce the vector wave equations (2.88) and (2.89) to a set of scalar wave equations, (2.93). Since many of the solutions for the scalar Helmholtz equations are known, a number of solutions for the vector wave equations can be obtained by using (2.91) and (2.92). Since the vector wave equations for the homogeneous isotropic simple medium are linear, it follows that the superposition theorem can be used to obtain the total and complete wave fields from (2.91) and (2.92). In other words, (2.91) and (2.92) represent the complete electromagnetic field solutions in a simple medium using the Debye potentials, $\Psi(\mathbf{r})$ and $\Phi(\mathbf{r})$.

From a historical perspective, we note that the vector $\mathbf{a}\Phi(\mathbf{r})$ may be identified as the electric Hertz vector, while the vector $\mathbf{a}\Psi(\mathbf{r})$ may be identified as the magnetic Hertz vector. Here, $\mathbf{a}$ is a constant vector. When $\mathbf{a}$ is taken as a unit vector in the longitudinal z direction, the Hertz vectors are $\mathbf{e}_z\Phi(\mathbf{r})$ for the electric type and $\mathbf{e}_z\Psi(\mathbf{r})$ for the magnetic type. According to (2.91) and (2.92), all transverse components of the electric and magnetic fields can be generated from the knowledge of the axial Hertz vectors, $\mathbf{e}_z\Phi(\mathbf{r})$ and $\mathbf{e}_z\Psi(\mathbf{r})$. Therefore, it is necessary only to obtain solutions of $\Psi(\mathbf{r})$ and $\Phi(\mathbf{r})$ from the scalar wave equations (2.93). All electromagnetic field components can then be obtained.

2.5.2 Basic Wave Types

Knowing that the linear property of electromagnetic fields in a simple medium allows us to use the superposition theorem to obtain the complete field components from partial fields (i.e., different wave types), we can consider each wave type separately. Let us define the following basic wave types:

1. Transverse Electromagnetic Waves (TEM Waves). These waves contain neither an electric nor a magnetic field component in the direction of propagation. There is no E_z or H_z if $\mathbf{e}_z$ is the direction of propagation.

2. Transverse Magnetic Waves (TM or E waves). These waves contain an electric field component but not a magnetic field component in the direction of propagation. There is no H_z if $\mathbf{e}_z$ is the direction of propagation, that is, $\Psi(\mathbf{r}) = 0$.

3. Transverse Electric Waves (TE or H waves). These waves contain a magnetic field component but not an electric field component in the direction of propagation. There is no E_z if $\mathbf{e}_z$ is the direction of propagation, that is, $\Phi(\mathbf{r}) = 0$.

4. Hybrid waves (HE or EH waves). These waves contain all components of electric and magnetic fields. These hybrid waves are obtained by linear superposition of TE and TM waves, that is, $\Psi(\mathbf{r}) \neq 0$ and $\Phi(\mathbf{r}) \neq 0$.

The knowledge that the total electromagnetic fields may be separated into partial fields and wave types that satisfy the scalar wave equations prompted us to search for a direct way to obtain the scalar wave equations for the field components of guided waves propagating along a straight line path from the time-harmonic Maxwell equations without the use of the Debye potentials or the Hertz vectors. When the guided modes propagate along a perfectly straight line path, one may assume that every component of the electromagnetic wave may be represented in the form $f(u, v)e^{-j\beta z + j\omega t}$, in which z is chosen as the propagation direction and u, v are generalized orthogonal coordinates in the transverse plane. In the following, the factor $e^{j(\omega t - \beta z)}$ is assumed to be attached to all field components and will be suppressed. The symbol β is the propagation constant. The time-harmonic Maxwell equations in a simple homogeneous, isotropic medium (ϵ, μ) now read as

$$\frac{1}{h_2}\left[\frac{\partial E_z}{\partial v} + j\beta h_2 E_v\right] + j\omega\mu H_u = 0, \tag{2.94}$$

$$\frac{1}{h_1}\left[-j\beta h_1 E_u - \frac{\partial E_z}{\partial u}\right] + j\omega\mu H_v = 0, \tag{2.95}$$

$$\frac{1}{h_1 h_2}\left[\frac{\partial(h_2 E_v)}{\partial u} - \frac{\partial(h_1 E_u)}{\partial v}\right] + j\omega\mu H_z = 0, \tag{2.96}$$

$$\frac{1}{h_2}\left[\frac{\partial H_z}{\partial v} + j\beta h_2 H_v\right] - j\omega\epsilon E_u = 0, \tag{2.97}$$

$$\frac{1}{h_1}\left[-j\beta h_1 H_u - \frac{\partial H_z}{\partial u}\right] - j\omega\epsilon E_v = 0, \tag{2.98}$$

$$\frac{1}{h_1 h_2}\left[\frac{\partial(h_2 H_v)}{\partial u} - \frac{\partial(h_1 H_u)}{\partial v}\right] - j\omega\epsilon E_z = 0. \tag{2.99}$$

Here, $E_{u,v,z} = E_{u,v,z}(u,v)e^{-j\beta z}$ and $H_{u,v,z} = H_{u,v,z}(u,v)e^{-j\beta z}$ and (h_1, h_2) are the transverse coefficients of any orthogonal system of curvilinear coordinates with a longitudinal z-coordinate. Combining (2.95) and (2.97) and solving for E_u and H_v in terms of $\partial E_z/\partial u$ and $\partial H_z/\partial v$, we obtain (2.100) and (2.103). Combining (2.94) and (2.98) and solving for E_v and H_u in terms of $\partial E_z/\partial v$ and $\partial H_z/\partial u$, we obtain (2.101) and (2.102). The results are

$$E_u = \frac{1}{p^2}\left(-\frac{j\beta}{h_1}\frac{\partial E_z}{\partial u} - \frac{j\omega\mu}{h_2}\frac{\partial H_z}{\partial v}\right), \tag{2.100}$$

$$E_v = \frac{1}{p^2}\left(-\frac{j\beta}{h_2}\frac{\partial E_z}{\partial v} + \frac{j\omega\mu}{h_1}\frac{\partial H_z}{\partial u}\right), \tag{2.101}$$

$$H_u = \frac{1}{p^2}\left(\frac{j\omega\epsilon}{h_2}\frac{\partial E_z}{\partial v} - \frac{j\beta}{h_1}\frac{\partial H_z}{\partial u}\right), \tag{2.102}$$

$$H_v = \frac{1}{p^2}\left(-\frac{j\omega\epsilon}{h_1}\frac{\partial E_z}{\partial u} - \frac{j\beta}{h_2}\frac{\partial H_z}{\partial v}\right), \tag{2.103}$$

with $p^2 = \omega^2\mu\epsilon - \beta^2$. Substituting (2.100) and (2.101) into (2.96) yields

$$\left[\frac{1}{h_1 h_2}\left(\frac{\partial}{\partial u}\frac{h_2}{h_1}\frac{\partial}{\partial u} + \frac{\partial}{\partial v}\frac{h_1}{h_2}\frac{\partial}{\partial v}\right) + p^2\right] E_z = 0 \tag{2.104}$$

and substituting (2.102) and 2.103) into (2.99) yields

$$\left[\frac{1}{h_1 h_2}\left(\frac{\partial}{\partial u}\frac{h_2}{h_1}\frac{\partial}{\partial u} + \frac{\partial}{\partial v}\frac{h_1}{h_2}\frac{\partial}{\partial v}\right) + p^2\right] H_z = 0. \tag{2.105}$$

These are scalar equations for E_z and H_z. Using the vector notation, we can write (2.104) and (2.105) as

$$[\nabla_t^2 + p^2]\begin{bmatrix} E_z \\ H_z \end{bmatrix} = 0. \tag{2.106}$$

The transverse Laplacian ∇_t^2 in curvilinear coordinates is given by

$$\nabla_t^2 = \frac{1}{h_1 h_2}\left(\frac{\partial}{\partial u}\frac{h_2}{h_1}\frac{\partial}{\partial u} + \frac{\partial}{\partial v}\frac{h_1}{h_2}\frac{\partial}{\partial v}\right).$$

According to (2.100)–(2.103), all transverse field components, E_u, E_v, H_u, and H_v, of a wave propagating in a simple medium (ϵ, μ) along a straight line path in the z-direction can be obtained from the longitudinal axial fields (E_z, H_z). And, the longitudinal fields are governed by the scalar equations (2.104) and (2.105). Recall that by the Debye formulation all field components may be derived from two separate potentials, $\Phi(\mathbf{r})$ and $\Psi(\mathbf{r})$. The relationship between the Hertz vectors and the Debye potentials has been discussed earlier. One may sensibly inquire whether there is a one-to-one correspondence between the longitudinal components of a propagating wave (E_z, H_z) and the Debye potentials (Φ, Ψ). From the Debye field expressions (2.91) and (2.92) with $\mathbf{a} = \mathbf{e}_z$, the following expressions are obtained:

$$E_z(u, v) = -\frac{jp^2}{\omega\epsilon}\Phi(u, v), \tag{2.107}$$

$$H_z(u, v) = \frac{jp^2}{\omega\mu}\Psi(u, v), \tag{2.108}$$

where the scalar wave equations (2.93) for Φ and Ψ have been used. The one-to-one correspondence is thus established. Since the field components $E_{u,v,z}$ and $H_{u,v,z}$ are much more recognizable, they will be often used in the subsequent text.

We recall that the TM or E modes (Transverse Magnetic Waves) refer to waves having $H_z = 0$, TE or H modes (Transverse Electric Waves) refer to waves having $E_z = 0$, HE or EH modes (Hybrid Waves) refer to waves having all field components, and the TEM mode (Transverse Electromagnetic Waves) refer to waves having $E_z = H_z = 0$. It is seen from (2.100)–(2.105) that when E_z and H_z are both zero, the only nonzero solution for the transverse components occurs when $p^2 = 0$ or $\beta = \omega\sqrt{\mu\epsilon}$. This means all TEM waves must propagate with a propagation constant of $\omega\sqrt{\mu\epsilon}$. Substituting $E_z = 0$ and $H_z = 0$ in (2.94)–(2.99) yields

$$H_u = -\sqrt{\frac{\epsilon}{\mu}}E_v, \tag{2.109}$$

$$H_v = \sqrt{\frac{\epsilon}{\mu}}E_u, \tag{2.110}$$

$$\frac{\partial(h_2 E_v)}{\partial u} - \frac{\partial(h_1 E_u)}{\partial v} = 0, \tag{2.111}$$

$$\frac{\partial(h_2 H_v)}{\partial u} - \frac{\partial(h_1 H_u)}{\partial v} = 0. \tag{2.112}$$

Let

$$\mathbf{E} = \nabla_t V = \frac{1}{h_1}\frac{\partial V}{\partial u}\mathbf{e}_u + \frac{1}{h_2}\frac{\partial V}{\partial v}\mathbf{e}_v, \tag{2.113}$$

where V is a scalar function, ∇_t is the transverse del operator, and $\mathbf{e}_u$ and $\mathbf{e}_v$ are unit vectors in the u and v directions. Substituting (2.109), (2.110), and (2.113) in (2.111) yields

$$\frac{\partial}{\partial u}\left(\frac{h_2}{h_1}\frac{\partial V}{\partial u}\right)+\frac{\partial}{\partial v}\left(\frac{h_1}{h_2}\frac{\partial V}{\partial v}\right)=0 \tag{2.114}$$

or

$$\nabla_t^2 V = 0.$$

To summarize, the field components in a TEM wave are

$$E_t\left(u,v,z\right)=E_u\mathbf{e}_u+E_v\mathbf{e}_v=\left(\nabla_t V\right)e^{\pm j\beta z}, \tag{2.115}$$

$$H_t\left(u,v,z\right)=H_u\mathbf{e}_u+H_v\mathbf{e}_v=\left(\mathbf{e}_z\times\nabla_t V\right)e^{\pm j\beta z}, \tag{2.116}$$

where V is a two-dimensional solution of the Laplace equation

$$\nabla_t^2 V = 0. \tag{2.117}$$

As an example, let us obtain the field components of the TEM wave in rectangular coordinates (x, y, z). From (2.109) and (2.110) we find that the nonzero field expressions for the two independent sets of TEM waves are

$$E_x = E_{0x}\, e^{\pm j\beta z}, \tag{2.118}$$

$$H_y = \sqrt{\frac{\epsilon}{\mu}}E_{0x}\, e^{\pm j\beta z}, \tag{2.119}$$

and

$$E_y = E_{0y}\, e^{\pm j\beta z}, \tag{2.120}$$

$$H_x = -\sqrt{\frac{\epsilon}{\mu}}E_{0y}\, e^{\pm j\beta z}. \tag{2.121}$$

Here E_{0x} and E_{0y} are two independent constants, $\beta=\omega\sqrt{\mu\epsilon}$, $\partial\beta/\partial\omega=\sqrt{\mu\epsilon}$, and $\partial^2\beta/\partial\omega^2=0$. Other examples of TEM waves in other coordinate systems may be found in the literature [26]. It will be shown later that pure TEM waves cannot exist along dielectric waveguides.

2.5.3 Separation of Variables

It is possible to reduce the partial differential equations (2.104) or (2.105) to ordinary differential equations through the separation of variables in four coordinate systems with a longitudinal z-coordinate [2]. Analytic solutions for these ordinary differential equations can be found [27]. From (2.104) or (2.105), one has

$$\begin{bmatrix} E_z \\ H_z \end{bmatrix} = U(u)V(v), \tag{2.122}$$

where $U(u)$ and $V(v)$ satisfy two second-order total differential equations. Expansions for all field components of TE or H modes (with $E_z = 0,\ H_z \neq 0$) or TM or E modes (with $E_z \neq 0,\ H_z = 0$) can be generated from (2.100)–(2.103) using the appropriate solutions given by (2.122). Expressions for all field components of the hybrid modes (with $E_z \neq 0,\ H_z \neq 0$) are obtained using linear superposition of the field components for the TE and TM modes. The appropriate choice for the kind of solutions and the type of modes depends on the boundary conditions that must be satisfied for a given physical problem.

Let us now consider the solutions in the four coordinate systems as follows.

2.5.3.1 Rectangular Coordinates (*x, y, z*)

The variables and the metrical coefficients are

$$u = x \quad v = y, \tag{2.123}$$

$$h_1 = 1 \quad h_2 = 1. \tag{2.124}$$

The functions U and V are

$$U(x) = e^{\pm jk_x x}, \tag{2.125}$$

$$V(y) = e^{\pm jk_y y}, \tag{2.126}$$

and

$$p^2 - k_x^2 - k_y^2 = 0, \tag{2.127}$$

$$p^2 = \omega^2 \mu\epsilon - \beta^2. \tag{2.128}$$

Here, k_x and k_y are the separation constants. The complete expressions for the axial field components in rectangular coordinates are

$$\begin{bmatrix} E_z(x,y,z) \\ H_z(x,y,z) \end{bmatrix} = \left[e^{\pm jk_x x}\right]\left[e^{\pm jk_y y}\right]\left[e^{\pm j\beta z}\right]. \tag{2.129}$$

All transverse components may be derived from (2.100)–(2.103) as follows

$$E_x = \frac{1}{p^2}\left(-j\beta\frac{\partial E_z}{\partial x} - j\omega\mu\frac{\partial H_z}{\partial y}\right), \tag{2.130}$$

$$E_y = \frac{1}{p^2}\left(-j\beta\frac{\partial E_z}{\partial y} + j\omega\mu\frac{\partial H_z}{\partial x}\right), \tag{2.131}$$

$$H_x = \frac{1}{p^2}\left(j\omega\epsilon\frac{\partial E_z}{\partial y} - j\beta\frac{\partial H_z}{\partial x}\right), \tag{2.132}$$

$$H_y = \frac{1}{p^2}\left(-j\omega\epsilon\frac{\partial E_z}{\partial x} - j\beta\frac{\partial H_z}{\partial y}\right), \tag{2.133}$$

$$p^2 = \omega^2\mu\epsilon - \beta^2, \qquad p^2 - k_x^2 - k_y^2 = 0. \tag{2.134}$$

2.5.3.2 Circular Cylinder Coordinates (*r*, θ, *z*)

The variables and the metrical coefficients are

$$u = r \qquad v = \theta, \tag{2.135}$$

$$h_1 = 1 \quad h_2 = r. \tag{2.136}$$

The functions U and V are

$$U(r) = Z_\nu^{(1),(2)}(pr), \tag{2.137}$$

$$V(\theta) = e^{\pm j\nu\theta}, \tag{2.138}$$

and

$$\beta^2 = k^2 - p^2 \qquad k^2 = \omega^2\mu\epsilon. \tag{2.139}$$

Here $Z_\nu^{(1),(2)}$ are two linearly independent solutions to the Bessel differential equation of order ν and argument *pr* [28]. The complete expressions for the axial field components in the circular cylindrical coordinates are

$$\begin{bmatrix} E_z(r,\theta,z) \\ H_z(r,\theta,z) \end{bmatrix} = \left[Z_\nu^{(1),(2)}(pr)\right]\left[e^{\pm j\nu\theta}\right]\left[e^{\pm j\beta z}\right]. \tag{2.140}$$

All transverse components may be derived from (2.100)–(2.103) as follows:

$$E_r = \frac{1}{p^2}\left(-j\beta\frac{\partial E_z}{\partial r} - j\omega\mu\frac{1}{r}\frac{\partial H_z}{\partial \theta}\right), \tag{2.141}$$

$$E_\theta = \frac{1}{p^2}\left(-j\beta\frac{1}{r}\frac{\partial E_z}{\partial \theta} + j\omega\mu\frac{\partial H_z}{\partial r}\right), \tag{2.142}$$

$$H_r = \frac{1}{p^2}\left(j\omega\epsilon\frac{1}{r}\frac{\partial E_z}{\partial \theta} - j\beta\frac{\partial H_z}{\partial r}\right), \tag{2.143}$$

$$H_\theta = \frac{1}{p^2}\left(-j\omega\epsilon\frac{\partial E_z}{\partial r} - j\beta\frac{1}{r}\frac{\partial H_z}{\partial \theta}\right), \tag{2.144}$$

$$\beta^2 = k^2 - p^2 \qquad k^2 = \omega^2\mu\epsilon. \tag{2.145}$$

2.5.3.3 Elliptical Cylinder Coordinates (ξ, η, z)

The variables and the metrical coefficients are

$$u = \cosh\xi \qquad v = \cos\eta, \tag{2.146}$$

$$h_1 = \frac{q(\cosh^2\xi - \cos^2\eta)^{1/2}}{\sinh\xi} \qquad h_2 = \frac{q(\cosh^2\xi - \cos^2\eta)^{1/2}}{\sin\eta}, \tag{2.147}$$

$$2q = \text{focal distance of the ellipse.} \tag{2.148}$$

In terms of the rectangular coordinates (x, y), the elliptical coordinates (ξ, η) are defined by the following relations: $x = q\cosh\xi\cos\eta$, $y = q\sinh\xi\sin\eta$, with the conditions $(0 \le \xi \le \infty,\ 0 \le \eta \le 2\pi)$. The scalar wave equation becomes

$$\left[\frac{1}{q^2(\cosh^2\xi - \cos^2\eta)}\left(\frac{\partial^2}{\partial\xi^2} + \frac{\partial^2}{\partial\eta^2}\right) + p^2\right]\begin{bmatrix}E_z\\ H_z\end{bmatrix} = 0. \tag{2.149}$$

The functions U and V are

$$U(\xi) = \begin{bmatrix} Ce_n(\xi, \gamma^2),\ Fey_n(\xi, \gamma^2)\\ Se_n(\xi, \gamma^2),\ Gey_n(\xi, \gamma^2)\end{bmatrix}, \tag{2.150}$$

$$V(\eta) = \begin{bmatrix} ce_n(\eta, \gamma^2)\\ se_n(\eta, \gamma^2)\end{bmatrix}, \tag{2.151}$$

$$\gamma = \frac{q}{2}p, \qquad p^2 = \omega^2\mu\epsilon - \beta^2. \tag{2.152}$$

Here $Ce_n(\xi,\gamma^2)$, $Fey_n(\xi,\gamma^2)$ and $Se_n(\xi,\gamma^2)$, $Gey_n(\xi,\gamma^2)$ are the even and odd radial Mathieu functions, n is an integer representing the order of the function, $ce_n(\eta,\gamma^2)$ and $se_n(\eta,\gamma^2)$ are the even and odd angular Mathieu functions [29]. The complete expressions for the axial field components in the elliptical cylindrical coordinates are

$$\begin{bmatrix} E_z(\xi,\eta,z) \\ H_z(\xi,\eta,z) \end{bmatrix} = \begin{bmatrix} Ce_n(\xi,\gamma^2), & Fey_n(\xi,\gamma^2) \\ Se_n(\xi,\gamma^2), & Gey_n(\xi,\gamma^2) \end{bmatrix} \begin{bmatrix} ce_n(\eta,\gamma^2) \\ se_n(\eta,\gamma^2) \end{bmatrix} e^{\pm j\beta z}. \tag{2.153}$$

All transverse components may be derived from (2.100)–(2.103) as follows:

$$E_\xi = \frac{1}{p^2 s}\left(-j\beta\frac{\partial E_z}{\partial \xi} - j\omega\mu\frac{\partial H_z}{\partial \eta}\right), \tag{2.154}$$

$$E_\eta = \frac{1}{p^2 s}\left(-j\beta\frac{\partial E_z}{\partial \eta} + j\omega\mu\frac{\partial H_z}{\partial \xi}\right), \tag{2.155}$$

$$H_\xi = \frac{1}{p^2 s}\left(j\omega\epsilon\frac{\partial E_z}{\partial \eta} - j\beta\frac{\partial H_z}{\partial \xi}\right), \tag{2.156}$$

$$H_\eta = -\frac{1}{p^2 s}\left(j\omega\epsilon\frac{\partial E_z}{\partial \xi} + j\beta\frac{\partial H_z}{\partial \eta}\right), \tag{2.157}$$

$$p^2 = \omega^2\mu\epsilon - \beta^2, \qquad \gamma = \frac{q}{2}p, \tag{2.158}$$

$$s = q(\cosh^2\xi - \cos^2\eta)^{1/2}. \tag{2.159}$$

2.5.3.4 Parabolic Cylinder Coordinates (ξ, η, z)

The variables and the metrical coefficients are

$$u = \xi, \qquad v = \eta, \tag{2.160}$$

$$h_1 = h_2 = \left(\xi^2 + \eta^2\right)^{1/2}. \tag{2.161}$$

In terms of the rectangular coordinates x, y, the parabolic coordinates of ξ, η are defined by the following relations: $x = \frac{1}{2}\left(\xi^2 - \eta^2\right), y = \xi\eta$, with $(-\infty < \xi < \infty)$ and $(-\infty < \eta < \infty)$. The scalar wave equation becomes

$$\left[\frac{1}{\xi^2+\eta^2}\left(\frac{\partial^2}{\partial\xi^2} + \frac{\partial^2}{\partial\eta^2}\right) + p^2\right]\begin{bmatrix} E_z \\ H_z \end{bmatrix} = 0, \tag{2.162}$$

$$p^2 = \omega^2\mu\epsilon - \beta^2. \tag{2.163}$$

The functions $U(\xi)$ and $V(\eta)$ are

$$U(\xi) = U_m^{(1),(2)}(\xi), \tag{2.164}$$

$$V(\eta) = V_m^{(1),(2)}(\eta), \tag{2.165}$$

where $U(\xi)$ and $V(\eta)$, respectively, satisfy the following differential equations, called the Weber's equations of the confluent hypergeometric type [30]:

$$\left[\frac{\partial^2}{\partial \xi^2} + \left(p^2\xi^2 + m\right)\right] U(\xi) = 0 \tag{2.166}$$

and

$$\left[\frac{\partial^2}{\partial \eta^2} + \left(p^2\eta^2 - m\right)\right] V(\eta) = 0. \tag{2.167}$$

Here, m is an integer. The complete expressions for the axial field components in the parabolic cylinder coordinates are

$$\begin{bmatrix} E_z \\ H_z \end{bmatrix} = \left[U_m^{(1),(2)}(\xi)\right] \left[V_m^{(1),(2)}(\eta)\right] e^{\pm j\beta z}. \tag{2.168}$$

All transverse components may be derived from (2.100)–(2.103) as follows:

$$E_\xi = \frac{1}{p^2 s}\left(-j\beta\frac{\partial E_z}{\partial \xi} - j\omega\mu\frac{\partial H_z}{\partial \eta}\right), \tag{2.169}$$

$$E_\eta = \frac{1}{p^2 s}\left(-j\beta\frac{\partial E_z}{\partial \eta} + j\omega\mu\frac{\partial H_z}{\partial \xi}\right), \tag{2.170}$$

$$H_\xi = \frac{1}{p^2 s}\left(j\omega\epsilon\frac{\partial E_z}{\partial \eta} - j\beta\frac{\partial H_z}{\partial \xi}\right), \tag{2.171}$$

$$H_\eta = \frac{1}{p^2 s}\left(-j\omega\epsilon\frac{\partial E_z}{\partial \xi} - j\beta\frac{\partial H_z}{\partial \eta}\right), \tag{2.172}$$

with

$$p^2 = \omega^2\mu\epsilon - \beta^2 \qquad s^2 = \left(\xi^2 + \eta^2\right)^{1/2}. \tag{2.173}$$

In all the above cases the complete electromagnetic fields are given by the sum of all fields for TE, TM, HE, and TEM waves. The existence of these waves depends on the excitation condition, the boundary conditions, and the allowed eigen solutions.

2.6 Polarization of Waves

Consider a plane wave in free space propagating in the z-direction and having the following components:

$$\mathbf{E} = \mathbf{e}_x E_1 \, e^{j(\omega t - kz)} + \mathbf{e}_y E_2 \, e^{j(\omega t - kz)}, \tag{2.174}$$

$$\mathbf{B} = -\mathbf{e}_x E_2 \sqrt{\mu_0 \epsilon_0} \, e^{j(\omega t - kz)} + \mathbf{e}_y E_1 \sqrt{\mu_0 \epsilon_0} \, e^{j(\omega t - kz)}, \tag{2.175}$$

with $k = \omega \sqrt{\mu_0 \epsilon_0}$. Note that (E_x, B_y) and (E_y, B_x) are linearly independent fields and E_1 and E_2 are complex constants.

2.6.1 Linearly Polarized Waves

For linearly polarized waves, E_1 and E_2 have the same phase. In this case, $\mathbf{E}$, at any point in space, oscillates along a directional line, which makes a constant angle ϕ with the x-axis, this angle being given by $\phi = \arctan(E_2/E_1)$.

2.6.2 Circularly Polarized Waves

For circularly polarized waves, E_1 and E_2 have the same magnitude but have a phase difference of $90°$. Hence,

$$\mathbf{E} = \mathrm{Re}\left[(\mathbf{e}_x \pm j\mathbf{e}_y) \, E_1 \, e^{j(\omega t - kz)}\right]$$

$$= E_1 \left[\mathbf{e}_x \cos(\omega t - kz) \pm \mathbf{e}_y \sin(\omega t - kz)\right]. \tag{2.176}$$

Here, at any point in space, $\mathbf{E}$ does not oscillate. Its magnitude is constant and its direction rotates at the angular velocity ω. When $E_2 = jE_1$, the wave is said to be right-handed circularly polarized, and when $E_2 = -jE_1$, the wave is said to be left-handed circularly polarized.

2.6.3 Elliptically Polarized Waves

In elliptically polarized waves, E_1 and E_2 have arbitrary relative amplitudes and phases. If the magnitude of $\mathbf{E}$ at any point in space is depicted as a vector, the tip of the vector rotates through an ellipse at the angular frequency ω.

2.7 Phase Velocity and Group Velocity

For a guided wave propagating in the z-direction, its surface of constant phase is normal to the z-direction. The phase velocity of the propagating wave is defined as the velocity with which its surface of constant phase travels in the z-direction and is given by

$$v_{\mathrm{p}} = \frac{\omega}{\beta}, \tag{2.177}$$

where β is the propagation constant of that wave and is usually a function of the frequency, ω.

For a TEM wave, $\beta = \omega/c$, where c is the velocity of propagation of this wave in a given medium. If c is a constant, not a function of frequency, then it is called dispersionless propagation.

For a TE, TM, or HE wave, the propagation constant β is not a linear function of ω. Therefore, its phase velocity is a function of ω. This is called dispersive propagation.

The group velocity of a propagating wave along a guide is usually defined as the velocity at which a wave packet is propagating along this guide. The frequency spectrum of the wave packet is assumed to be narrow and centered about ω_0. The propagation constant of this wave packet can be approximated by the following expression:

$$\beta(\omega) \approx \beta_0 + (\omega - \omega_0) \left.\frac{\partial \beta}{\partial \omega}\right|_{\omega=\omega_0}, \tag{2.178}$$

where β_0 is the β at $\omega = \omega_0$ and $\partial\beta/\partial\omega|_{\omega=\omega_0}$ is $\partial\beta/\partial\omega$ at $\omega = \omega_0$. Let us represent the general expression for a wave packet as follows:

$$A(z,t) = \int_{-\infty}^{\infty} A(\omega)\, e^{-j\beta z} e^{j\omega t}\, d\omega, \tag{2.179}$$

where β is given by (2.178). If the spectrum is narrow and centered at ω_0, we can evaluate the above integral over a narrow interval from $\omega_0 - \Delta\omega$ to $\omega_0 + \Delta\omega$, resulting in

$$A(z,t) = 2A(\omega_0) \frac{\sin\left[\Delta\omega \left(t - z \left.\dfrac{\partial \beta}{\partial \omega}\right|_{\omega=\omega_0}\right)\right]}{\left(t - z \left.\dfrac{\partial \beta}{\partial \omega}\right|_{\omega=\omega_0}\right)}\, e^{-j\beta_0 z} e^{j\omega t}. \tag{2.180}$$

The above expression has a maximum at

$$t - z \left.\frac{\partial \beta}{\partial \omega}\right|_{\omega=\omega_0} = 0. \tag{2.181}$$

The velocity dz/dt with which this maximum moves down the guide is called the group velocity, v_{g}. It is

$$v_{\mathrm{g}} = \frac{dz}{dt} = \left.\frac{d\omega}{d\beta}\right|_{\omega=\omega_0}. \tag{2.182}$$

This group velocity is the velocity at which this wave packet (or signal) propagates down the guide.

2.8 The Impedance Concept [31]

Conventional transmission line analysis on distributed circuits predates the development of field analysis of waveguiding structures. The familiarity and simplicity of the circuit based transmission line theory therefore provide the impetus to seek an equivalent transmission line analogy for the more complex waveguiding structures whenever possible. For example, the guidance on a conventional transmission line is described by a characteristic impedance Z_{c} and a propagation constant β. For a hollow metallic waveguide, supporting either a TE or TM mode, one may also define a characteristic impedance of a given mode in terms of the ratio of the transverse electric field and the transverse magnetic field and a propagation constant as follows:

$$Z_{\mathrm{c}}^{(\mathrm{TE})} = \left|\frac{E_t^{(\mathrm{TE})}}{H_t^{(\mathrm{TE})}}\right| = \frac{\omega\mu_0}{\beta^{(\mathrm{TE})}} \tag{2.183}$$

and

$$Z_{\mathrm{c}}^{(\mathrm{TM})} = \left|\frac{E_t^{(\mathrm{TM})}}{H_t^{(\mathrm{TM})}}\right| = \frac{\beta^{(\mathrm{TM})}}{\omega\epsilon_0}. \tag{2.184}$$

Here $\beta^{(\mathrm{TE})}$ and $\beta^{(\mathrm{TM})}$ are the propagation constants of the TE and TM modes. For a coaxial cable supporting the dominant TEM mode, the characteristic impedance is

$$Z_{\mathrm{c}}^{(\mathrm{TEM})} = \sqrt{\frac{\mu_0}{\epsilon_0}}. \tag{2.185}$$

The situation is more complicated for a dielectric waveguide. For a dielectric waveguide supporting a TE, TM, or HE mode, the characteristic impedance defined as the ratio of the transverse electric field and the transverse magnetic field is no longer a constant, but is a function of the transverse spatial coordinates. For example, for an HE mode, using (2.100)–(2.103), we find

$$\left|\frac{E_u}{H_v}\right| = \left|\frac{\left(-\frac{j\beta}{h_1}\frac{\partial E_z}{\partial u} - \frac{j\omega\mu}{h_2}\frac{\partial H_z}{\partial v}\right)}{\left(-\frac{j\omega\epsilon}{h_1}\frac{\partial E_z}{\partial u} - \frac{j\beta}{h_2}\frac{\partial H_z}{\partial v}\right)}\right|, \tag{2.186}$$

$$\left|\frac{E_v}{H_u}\right| = \left|\frac{\left(-\frac{j\beta}{h_2}\frac{\partial E_z}{\partial v} + \frac{j\omega\mu}{h_1}\frac{\partial H_z}{\partial u}\right)}{\left(\frac{j\omega\epsilon}{h_2}\frac{\partial E_z}{\partial v} - \frac{j\beta}{h_1}\frac{\partial H_z}{\partial u}\right)}\right|. \tag{2.187}$$

Obviously, these expressions are functions of (u, v), the transverse coordinates. Even for the TE or TM mode, different regions of the dielectric waveguide will possess different characteristic impedances. It is therefore not possible to obtain an equivalent one-dimensional transmission line analogy for a dielectric waveguide. Additionally, the presence of any discontinuities, inhomogeneity, or curvature of the dielectric waveguide structure will induce radiated waves, which could not be accounted for in the transmission line model. At best, one may obtain an approximate value for the characteristic impedance using an average value weighted by the amount of power carried by the various cross-sectional regions of the guide. Nevertheless, one may still make use of the impedance concept to gain a first order understanding of problems involving dielectric waveguide structures and to obtain preliminary guidance in the design of dielectric waveguide circuits.

2.9 Validity of the Scalar Wave Approach

It has been universally accepted that the full set of Maxwell equations resulting in the vector-wave equation must be used to treat propagation in waveguides. This requirement confines the analytical treatment to only a few simple structures. The advent of optical-fiber guides as viable communication links as well as the dawn of small high-bandwidth integrated optical circuits demand that the analytical horizon be expanded. It is recognized that many more problems can be solved if only the scalar-wave equation need be considered. Here we give an in depth investigation on the conditions under which the scalar-wave equation can be used instead of the vector-wave equation. Concrete examples on the validity of the scalar wave approach in providing accurate results for the graded index fiber guide are given.

Starting with the vector-wave equation for the electric field vector $\mathbf{E}$ in the fiber structure,

$$\nabla \times \nabla \times \mathbf{E} - \omega^2 \mu_0 \epsilon \mathbf{E} = 0, \tag{2.188}$$

where ω is the frequency of the wave, μ_0 is the permeability, and $\epsilon(r)$ is the inhomogeneous permittivity of the structure; and making use of the vector identity

$$\nabla \times \nabla \times \mathbf{E} = \nabla(\nabla \cdot \mathbf{E}) - \nabla^2 \mathbf{E} \tag{2.189}$$

and the relation

$$\nabla \cdot \mathbf{E} = -\frac{1}{\epsilon} \nabla \epsilon \cdot \mathbf{E}, \tag{2.190}$$

we obtain

$$\nabla^2 \mathbf{E} + \omega^2 \mu_0 \epsilon \mathbf{E} - \nabla \left(\frac{1}{\epsilon} \nabla \epsilon \cdot \mathbf{E} \right) = 0. \tag{2.191}$$

Rewriting (2.191) gives

$$\nabla^2 \mathbf{E} + \omega^2 \mu_0 \epsilon_0 \left\{ \frac{\epsilon}{\epsilon_0} \mathbf{E} - \left[\frac{1}{\omega^2 \mu_0 \epsilon_0} \nabla \left(\frac{1}{\epsilon} \nabla \epsilon \cdot \mathbf{E} \right) \right] \right\} = 0. \tag{2.192}$$

The relative importance of the terms within the curly brackets can be determined from the following:

$$\frac{\epsilon}{\epsilon_0} \mathbf{E} = O \left(\frac{\epsilon}{\epsilon_0} \right) \mathbf{E}, \tag{2.193}$$

$$\frac{1}{\omega^2 \mu_0 \epsilon_0} \nabla \left(\frac{1}{\epsilon} \nabla \epsilon \cdot \mathbf{E} \right) = \frac{1}{k_0^2} O \left(\frac{\nabla \epsilon \cdot \nabla \mathbf{E}}{\epsilon} \right) = O \left(\frac{\epsilon}{\epsilon_0} \frac{1}{k_0 l} \mathbf{E} \right), \tag{2.194}$$

where O means "order of magnitude" and l is the smaller of the distances over which ϵ/ϵ_0 and $\mathbf{E}$ change appreciably. For single-mode fiber structures, the values of ϵ/ϵ_0 and $k_0 l$ are typically in the range

$$\frac{\epsilon}{\epsilon_0} = O(2), \tag{2.195}$$

$$k_0 l \simeq \left(\frac{2\pi}{\lambda} \right) l = O \left(10^2 \text{ or } 10^3 \right),$$

$$l \simeq O(10 - 100\,\mu), \quad \lambda \simeq O(1\,\mu). \tag{2.196}$$

It follows that the second term within the curly brackets in (2.192) is several orders of magnitude smaller than the first term, $(\epsilon/\epsilon_0)\,\mathbf{E}$. It is therefore justifiable to neglect the second term and write (2.192) in the form

$$\nabla^2 \mathbf{E} + k_0^2 \left(\frac{\epsilon}{\epsilon_0} \right) \mathbf{E} = 0. \tag{2.197}$$

The physical significance of replacing (2.192) by (2.197) is the following. By discarding the term $\nabla\left(\epsilon^{-1}\nabla\epsilon\cdot\mathbf{E}\right)$, we are neglecting any depolarization effects that may occur. This means that the wave retains the linear polarization it has at the source, which is evidenced by writing $\mathbf{E}(\mathbf{r})$ in the form

$$\mathbf{E}(\mathbf{r}) = \mathbf{e}_{\mathrm{p}} u(\mathbf{r}), \tag{2.198}$$

where $\mathbf{e}_{\mathrm{p}}$ is a unit vector in the direction of the initial polarization of the wave. Substituting (2.198) into (2.197), we find that $u(\mathbf{r})$ satisfies the scalar wave equation

$$\nabla^2 u + k_0^2\left(\frac{\epsilon}{\epsilon_0}\right)u = 0. \tag{2.199}$$

This equation with the boundary condition on the initial surface, and the radiation condition at infinity, completely specifies $u(\mathbf{r})$, from which we can then obtain the electromagnetic field vector $\mathbf{E}$ and $\mathbf{H}$.

To verify the fact that the scalar-wave approach may be used to obtain accurate results for the case of wave propagation along single-mode inhomogeneous fiber structures, we shall now consider the special case of the dominant mode propagating in a fiber with a parabolic index profile. This case was chosen because exact vector-wave solutions exist for this problem. By comparing our results with the exact results, one may verify the applicability of this scalar-wave approach to single-mode problems.

Vector-wave solutions exist for radially inhomogeneous fibers [32, 33]. Two methods can usually be used to obtain the propagation characteristics of dominant modes on these types of fibers: (1) the radially inhomogeneous cylinder is subdivided into a number of concentric layers and the problem is solved by matching the solution for each homogeneous layer at the subdivided boundaries [32]; (2) the problem for the radially inhomogeneous cylinder is formulated in terms of a set of four coupled first-order differential equations for the transverse field components, and direct numerical integration is then performed to obtain the propagation constants of the lower-order modes [33]. Both methods have been used to obtain the dispersion characteristics of the dominant HE_{11} mode for a parabolic index profile fiber. Results are shown in Fig. 2.4.

The normalized propagation constant $\beta c/n_2'\omega$ is plotted against the normalized frequency $n_2'\omega a/c$ in Fig. 2.4 for the following index profile:

$$n = n_2' b_0^{1/2}\left(1 - \frac{1}{2}\frac{b_2}{b_0}\rho^2\right) \qquad \text{for} \quad \rho \le 1, \tag{2.200}$$

$$n = n_2' b_0^{1/2}\left(1 - \frac{1}{2}\frac{b_2}{b_0}\right) \qquad \text{for} \quad \rho \ge 1, \tag{2.201}$$

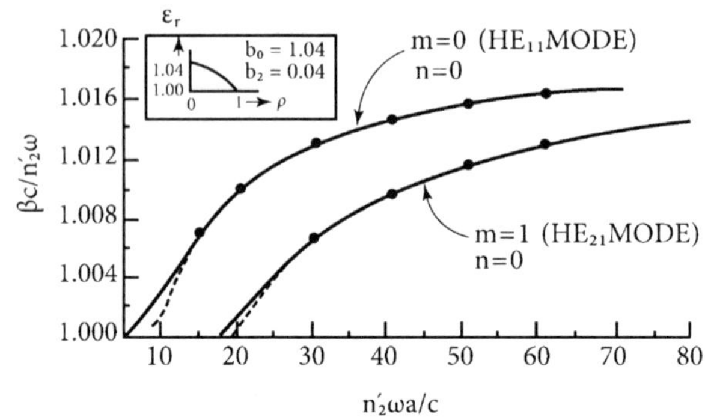

Figure 2.4. Comparison of the scalar-wave results with vector-wave results for the dispersion characteristics of HE$_{11}$ and HE$_{21}$ modes. The data points and the dashed curves are scalar-wave results and the solid curves are exact vector wave results. The core index variations are given by $n_1 = n_2'\epsilon_r^{1/2} = n_2'\left(b_0 - b_2\rho^2\right)^{1/2}$, where $\rho = r/$a, n_2' is approximately the index of the cladding, and a is the core radius [34]

with $\rho = r/a$, a is the radius of the inhomogeneous core, ω is the frequency of the wave, c is the speed of light in vacuum, and n_2', b_0, and b_2 are given constants. Data given in Fig. 2.4 will be used to check the accuracy of the scalar-wave results.

The scalar-wave-solution for the propagating wave in a parabolic index profile medium can be written in the form [35]

$$u = u_0(x, y)\, \exp\left\{-j\left[\frac{2\pi n_0}{\lambda} - (m+n+1)\left(\frac{n_2}{n_0}\right)^{1/2}\right]z\right\}, \tag{2.202}$$

where the propagating field is assumed to be linearly polarized, $u_0(x, y)$ is a real Hermite–Gaussian function, $\lambda = 2\pi c/\omega$, m and n are integers $(0, 1, 2, \ldots)$, and the index profile is given in the standard Gaussian beam notation

$$n = n_0\left[1 - \frac{1}{2}\frac{n_2}{n_0}a^2\rho^2\right]. \tag{2.203}$$

The values of n_2 and n_0 in this expression are related to n_2', b_0, and b_2 given previously by the relationships

$$n_0 = n_2' b_0^{1/2}, \tag{2.204}$$

$$\frac{n_2}{n_0}a^2 = \frac{b_2}{b_0}. \tag{2.205}$$

The propagation constant β for the field can be obtained readily from (2.202)

$$\beta = \frac{2\pi n_0}{\lambda} - (m+n+1)\left(\frac{n_2}{n_0}\right)^{1/2} \tag{2.206}$$

or

$$\frac{\beta c}{n_2'\omega} = \sqrt{b_0} - (m+n+1)\left(1-\frac{1}{b_0}\right)^{1/2}\left(\frac{c}{n_2'\omega a}\right). \tag{2.207}$$

This is the analytic result for the normalized propagation constant as a function of the normalized frequency for various modes $(m, n = 0, 1, 2, \ldots)$ in a parabolic index guide based on the scalar-wave equation. Numerical results calculated according to this equation are also displayed in Fig. 2.4 by the dashed curves [34]. It is clearly seen that excellent agreement between the scalar-wave results and the vector-wave results is obtained. Only when the operating frequencies are near the cutoff frequency of the modes does any difference exist. In fact, the difference is *not* caused by the inadequacy or inaccuracy of the scalar-wave approach, but rather by the difference in our choice of index profiles for the vector and the scalar cases. In the vector case, the index profile is no longer parabolic for $\rho \geq 1$, while in the scalar case it is always parabolic. By extending the parabolic index profile beyond $\rho = 1$ for the vector case, we can show (according to our computations) that the difference becomes very small indeed. It appears, therefore, that if the conditions described by (2.195) and (2.196) are satisfied, excellent results may be obtained using the scalar-wave equation.

In conclusion, we can state that the scalar-wave approach can be used with confidence to solve single-mode problems dealing with fiber or Integrated Optical Circuits (IOC) structures with general inhomogeneous index profiles, provided that the following conditions are satisfied:

(a) The depolarization effects are negligible, or

$$\left|\frac{1}{\omega^2\mu_0\epsilon_0}\nabla\left(\frac{1}{\epsilon}\nabla\epsilon\cdot\mathbf{E}\right)\right| \ll \left|\frac{\epsilon}{\epsilon_0}\mathbf{E}\right|, \tag{2.208}$$

where ϵ is the permittivity, $\mathbf{E}$ is the electric field vector, μ_0 and ϵ_0 are, respectively, the free-space permeability and permittivity, and ω is the frequency of the wave. Equation (2.208) implies that the index profile of the fiber varies little over distances of the order of the wavelength.

(b) The paraxial ray approximation may be used, or the factor $\left[\partial^2 A(x,y,z)/\partial z^2\right]$ is negligible compared to the factor

$$\left\{2jkn_0\frac{\partial}{\partial z} + \nabla_t^2 + k^2\left[n^2(x,y,z) - n_0^2\right]\right\} A(x,y,z), \tag{2.209}$$

where

$$u = \exp(jkn_0 z)\, A(x, y, z). \tag{2.210}$$

This condition means that the complex amplitude $A(x, y, z)$ varies much more rapidly transverse to the direction of propagation than it does along the direction of propagation, which is satisfied for fields propagating at small angles to the z-axis.

References

1. J. C. Maxwell, "A Treatise on Electricity and Magnetism," Dover, New York (1954)
2. J. A. Stratton, "Electromagnetic Theory," McGraw-Hill, New York (1941)
3. A. Sommerfeld, "Electromagnetics," Academic, New York (1949)
4. R. P. Feynman, R. B. Leighton, and M. Sands, "Feynman Lectures on Physics, The Definitive and Extended Edition, 2/E," Addison-Wesley, Reading, MA (2006)
5. J. F. Nye, "Physical Properties of Crystals," Oxford Science Publications, Oxford (1985)
6. B. Lax and K. J. Button, "Microwave Ferrites and Ferromagnetics," McGraw-Hill, New York (1962)
7. L. Spitzer, "Physics of Fully Ionized Gases," 2nd edn., Interscience Publications, Wiley, New York (1962)
8. V. G. Veselago, "The electrodynamics of substances with simultaneously negative values of ϵ and μ," Sov. Phys. Uspekhi **10**, 509 (1968)
9. J. B. Pendry, "Negative refraction makes a perfect lens" Phys. Rev. Lett. **85**, 3966 (2000)
10. P. W. Milonni, "Fast Light, Slow Light and Left-Handed Light," Institute of Physics Publishing (IOP), London (2005)
11. A. R. von Hippel, "Dielectric Materials and Applications," MIT Press, Cambridge, MA (1954)
12. H. Raether, "Surface Plasmons on Smooth and Rough Surfaces and on Gratings," Springer Tracts in Modern Physics, Vol. III, Springer Berlin Heidelberg New York (1988); P. Stoller, V. Jacobsen, and V. Sandoghdar, "Measurement of the complex dielectric constant of a single gold nanoparticle," Opt. Lett. **31**, 247402476 (2006)
13. W. L. Barnes, A. Dereux, and T. W. Ebbesen, "Surface plasmon wavelength optics," Nature **424**, 824 (2003)
14. M. N. Afsar and K. J. Button, "Millimeter-wave dielectric measurement of materials," Proc. IEEE **73**, 131 (1985); J. W. Lamb, "Miscellaneous data on materials for millimetre and submillimetre optics," Int. J. Infrared MM Waves **17**, 1997 (1996)
15. J. R. Birch, J. D. Dromey, and J Lisurf, "The optical constants of some common low-loss polymers between 4 and 40 cm^{-1}," Infrared Phys. **21**, 225 (1981); M. N. Afsar,

"Precision dielectric measurements of non-polar polymers in millimeter wavelength range," IEEE Trans. Microw. Theory Tech. **MTT-33**, 1410 (1985)

16. J. R. Birch and T. J. Parker, "Infrared and Millimeter Waves," Vol. 2, K. J. Button, ed., Academic, New York (1979); W. B. Bridges, M. B. Kline, and E. Schwarz, "Low loss flexible dielectric waveguides for millimeter wave transmission and its application to devices," IEEE Trans. Microw. Theory Tech. **MTT-30**, 286 (1982)
17. N. Bloembergen, "Non-Linear Optics," W. A. Benjamin, New York (1965)
18. I. Brodie and J. J. Murray, "The Physics of Micro/Nano Fabrication," Plenum, New York (1992)
19. J. A. Schwarz, C. I. Contescu, and K. Putyera, eds., "Dekker Encyclopedia of Nanoscience and Nanotechnology," Marcel Dekker, New York (2004)
20. C. Yeh, "Boundary conditions in electromagnetics," Phys. Rev. E **48**, 1426 (1993)
21. M. A. Leontovich, "Investigation of Propagation of Radio Waves, Part II," Printing House of the Academy of Sciences, Moscow (1948)
22. T. B. A. Senior, "Impedance boundary condition for imperfectly conducting surface," Appl. Sci. Res. B **8**, 418 (1960)
23. C. J. Bouwkamp, "A note on singularities at sharp edges in electromagnetic theory," Physica **12**, 467 (1960)
24. J. Meixner, "The edge condition in the theory of electromagnetic waves at perfectly conducting plane screens," Ann. Physik **6**, 2 (1949)
25. J. A. Kong, "Electromagnetic Wave Theory," Wiley, New York (1986)
26. C. Yeh, "Dynamic fields," American Institute of Physics Handbook, 3rd edn., 5b-9 (1972)
27. J. Mathews and R. L. Walker, "Mathematical Methods of Physics," 3rd edn., W. A. Benjamin, New York (1970)
28. G. N. Watson, "A treatise on the theory of Bessel functions," Cambridge University Press (1944)
29. N. W. McLachlan, "Theory and Application of Mathieu Functions," Oxford Press, Oxford, England (1947)
30. E. T. Whittaker and G. N. Watson, "A Course of Modern Analysis," Cambridge University Press (1927)
31. S. A. Schelkunoff, "Electromagnetic Waves," Van Nostrand, New York (1943)
32. C. Yeh and G. Lindgren, "Computing the propagation characteristics of radially stratified fibers-An efficient method," Appl. Opt. **16**, 483 (1977)
33. J. G. Dil and H. Blok, "Propagation of electromagnetic surface waves in a radially inhomogeneous optical waveguide," Opto-Electron. **5**, 415 (1973)
34. C. Yeh, L. Casperson, and W. P. Brown, "Scalar-wave approach for single-mode inhomogeneous fiber problems," Appl. Phys. Lett. **34**, 460 (1979)
35. L. Casperson, "Gaussian light beams in inhomogeneous media," Appl. Opt. **12**, 2434 (1973)

3

PROPAGATION CHARACTERISTICS OF GUIDED WAVES ALONG A DIELECTRIC GUIDE

In this chapter the general propagation characteristics of guided waves along a typical dielectric waveguide will be presented. Problems associated with pulse degradation, attenuation, excitation–launching–coupling of guided waves, design of dielectric waveguide systems in the presence of noise, as well as radiation losses caused by bends and corners will be discussed. The discussion, emphasizing the fundamental aspects of the surface wave propagation characteristics, is generally applicable to all guided waves on conventional surface wave structures.

3.1 Typical Surface Waveguide Structures

Any guided wave structure on which the guided field can extend to infinity is considered to be an open surface wave structure. A few typical examples are illustrated in Fig. 3.1.

There are basically two types of surface waveguides. One type is the field-penetrable kind, such as the various dielectric structures that include solid dielectrics [1], plasmon [2], artificial [1], and photonic–periodic structures [3–9], shown in Fig. 3.1a. The guided field is allowed to penetrate significantly into the core material and extend well beyond the exterior of the core surface. The other type is the field-impenetrable kind, such as the various surface impedance conducting structures, including the Sommerfeld–Goubau line [10], shown in Fig. 3.1b. The guided field does not penetrate much beyond the surface of the guiding structure. The guided wave simply adheres to the guiding surface. Figure 3.1c shows a number of unconventional surface waveguides having structures made of metamaterial [11] or moving material [12] or anisotropic material [13]. The structures shown in Fig. 3.1 are capable of supporting a finite number of guided modes as

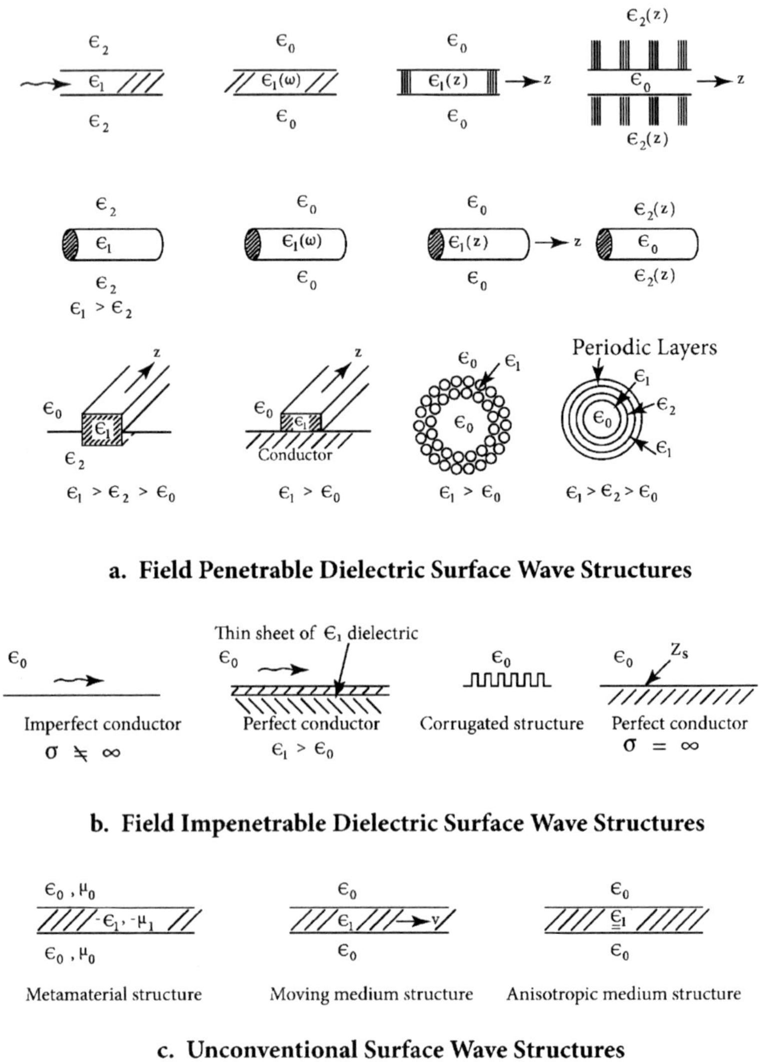

a. Field Penetrable Dielectric Surface Wave Structures

b. Field Impenetrable Dielectric Surface Wave Structures

c. Unconventional Surface Wave Structures

Figure 3.1. Examples of surface wave structures

well as a continuous spectrum of unguided radiated waves. Hence, it is impossible to excite the desired guided modes along a surface waveguide without simultaneously exciting the unwanted radiated waves. Furthermore, deviations from perfect waveguide geometry or from a straight line path not only convert power among the guided modes but also scatter power into the continuous spectrum of radiated waves. The radiated wave carries power away from the guiding structure, and therefore the radiated power is considered totally lost. As far as terminology is concerned, surface waveguides encompass dielectric waveguides, optical fibers, integrated optical circuits (IOC), plasmon guides, Sommerfeld–Goubou wires, photonic waveguides, metamaterial waveguides, or any guiding structure whose guided field can extend to infinity.

3.2 Formal Approach to the Surface Waveguide Problems

To obtain the formal analytical solution to the problem of wave guidance along a straight line dielectric (or surface wave) structure, the following items must be taken into consideration:

1. For guided modes propagating along a perfectly straight line path, we may assume that all components of the electromagnetic wave can be represented in the form

$$f(u, v)\, e^{-j\beta z} e^{j\omega t}, \tag{3.1}$$

in which z is chosen as the propagation direction, (u, v) are generalized orthogonal coordinates in the transverse plane, β is the propagation constant, and ω is the angular frequency of the wave.

2. Solutions for the longitudinal fields (E_z, H_z) must satisfy the scalar wave equation

$$\left[\nabla_t^2 + (\omega^2 \mu\epsilon - \beta^2)\right] \begin{bmatrix} E_z \\ H_z \end{bmatrix} = 0, \tag{3.2}$$

where ∇_t is the transverse del operator, (μ, ϵ) are the material parameters for the medium in which the fields reside. For inhomogeneous, anisotropic, or nonlinear medium, the above equation must be modified. (See subsequent chapters)

3. Field solutions for a given guiding structure must satisfy the appropriate boundary conditions given below:

- Tangential components of **E** and **H** must be continuous across the boundary of an interface between different material media
- Tangential components of **E** must be zero at a perfectly conducting surface
- The ratio of the tangential component of **E** and the tangential component of **H** is equal to the surface impedance of an imperfect conductor
- The radiation condition must be satisfied at infinity

Satisfying the boundary conditions will yield a transcendental equation in ω and β for the guided modes. For a given frequency, the transcendental equation may yield a finite number of real roots for β. These discrete roots represent the eigenvalues of the surface wave modes. Corresponding to these eigenvalues are the eigenfunctions of the guided modes. It should be pointed out that these finite number of eigen solutions for the guided modes for a given frequency cannot and do not represent the complete solution for the problem. The radiation solution must be included to form the complete solution. Furthermore, for a given frequency ω, there may be a case where no real root of β exists. In that case, a guided mode simply does not exist for that situation; only a radiated wave can be present. This distinctive feature for the guided surface wave problem is very different from the case of guided waves in an enclosed metallic waveguide, where an infinite number of evanescent modes exist even when no propagating mode (with β = real value) is present [14].

4. In general, the transverse electric fields for the electromagnetic wave on an open structure satisfy certain orthogonality relations and can be expressed as follows [1]:

$$\mathbf{E}_{\mathrm{t}}(x,y,z) = C\mathbf{E}_{\mathrm{t}}^{\mathrm{radiated}}(x,y,z) + \sum_p A_p \mathbf{E}_{\mathrm{t}p}^{\mathrm{guided}}(x,y)\, e^{-j\beta_p z}, \tag{3.3}$$

where the subscript t signifies the transverse fields (transverse to the direction of propagation of the guided wave), the index p represents the summation index for the number of eigen surface wave modes that can exist for a given frequency, C is the amplitude coefficient for the radiated wave, A_p are the amplitude coefficients for the surface wave modes, and β_p are the propagation constants for the surface wave modes. It can also be shown that in any z plane, the following relations apply:

$$\int_{-\infty}^{\infty} \left(\mathbf{E}_{\mathrm{t}p} \cdot \mathbf{E}_{\mathrm{t}q}^{*}\right) dxdy = 0 \qquad for\ p \neq q, \tag{3.4}$$

$$\int_{-\infty}^{\infty} \mathbf{e}_z \cdot \left(\mathbf{E}_{\mathrm{t}p} \times \mathbf{H}_{\mathrm{t}q}^{*}\right) dxdy = 0 \qquad for\ p \neq q, \tag{3.5}$$

where $\mathbf{e}_z$ is a unit vector in the z-direction, the asterisk denotes the complex conjugate value, p signifies the pth guided mode or the radiated wave, and q also signifies the qth guided mode or the radiated wave. Equations (3.4) and (3.5) are valid for any two modes regardless whether mode p and mode q are the guided modes, whether both are radiated waves, or whether one of the modes is a guided mode while the other one is a radiated wave. These are called the orthogonality relations for the surface wave fields. Strictly speaking, these relations are valid only for a lossless medium. Using these relations, any arbitrary field distribution in a given z-plane of this surface wave problem can be expressed in terms of the sum of orthogonal wave functions representing either guided modes or the radiated wave. The expansion coefficients, C or A_p, can be obtained with the help of the orthogonality relations, (3.4) and (3.5). The presence of these orthogonality relations also means that the modal power for each guided mode and the radiated power are additive and independent.

3.3 The ω-β Diagram: Dispersion Relations

How the propagation constant, β, of a given guided mode varies as a function of frequency is obtained by solving the governing transcendental equation. The ω-β diagram is simply a way of displaying the behavior of β vs. ω. From this diagram one may obtain β, $\partial\beta/\partial\omega$, and $\partial^2\beta/\partial\omega^2$, as a function of frequency. These values specify the dispersion characteristics of a given mode. For example, the phase velocity is ω/β for that mode, the group velocity is $\partial\beta/\partial\omega$ for that mode, and the pulse broadening factor for that mode is related to $\partial^2\beta/\partial\omega^2$. More will be said about these quantities later.

Intuitively speaking, we may reason that the propagation constant of a guided mode is most influenced by the distribution of the guided power of that mode. Let us consider the case shown in Fig. 3.2. When most of the guided power is within the core region of the guiding structure, that is, at high normalized frequencies, the situation must approach the case of wave propagation in an infinite medium made with the core material. In that case, the TEM wave in an infinite medium will have a propagation constant $\beta = k_1 - \omega\sqrt{\mu_0\epsilon_1}$, where ϵ_1 is the dielectric constant of the medium. When most of the guided power is outside the core region, that is, at low normalized frequencies or near cutoff, the situation must approach the case of TEM wave propagation in an infinite medium with dielectric constant ϵ_2. In that case, its propagation constant $\beta = k_2 = \omega\sqrt{\mu_0\epsilon_2}$. For intermediate guiding case, we anticipate the propagation constant of a guided mode to lie somewhere in between. Indeed, this is the case shown in Fig. 3.2, where typical ω-β diagrams (dispersion characteristics) of several lower-order modes on a simple di electric waveguide structure consisting of a core region with dielectric constant ϵ_1 surrounded by a cladding region with dielectric constant ϵ_2, where $\epsilon_1 > \epsilon_2$, are given. Most commonly used dielectric waveguides such as optical fibers, microwave, millimeter, and terahertz dielectric waveguides, thin film waveguides, and optical integrated circuits belong to this family.

From Fig. 3.2 we note that all guided propagating mode solutions lie between the lines $\beta = \omega n_1/c$ and $\beta = \omega n_2/c$, where $n_1 \left(= \sqrt{\epsilon_1/\epsilon_0}\right)$ is the index of refraction of the core, $n_2 \left(= \sqrt{\epsilon_2/\epsilon_0}\right)$ is the index of refraction of the cladding, ϵ_0 is the permittivity of free space, and c is the speed of light in a vacuum. In other words, the phase velocity ($v_{\text{ph}} = \omega/\beta$), as well as the group velocity ($v_{\text{g}} = \partial\beta/\partial\omega$), is bounded between c/n_1 and c/n_2. Those modes, whose phase velocities are closer

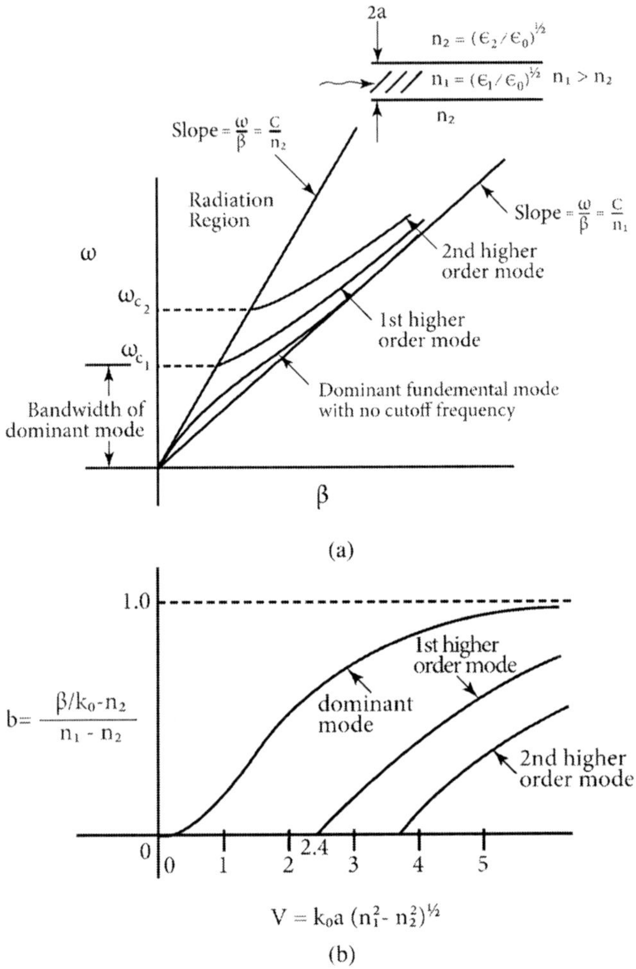

Figure 3.2. The top figure shows the ω-β diagram for a typical dielectric waveguide. The bottom figure shows another version of ω-β diagram for a typical circular fiber waveguide with radius a, with $n_1 \approx n_2$. Here b is defined by (3.6b), and is approximately equal to $(\beta/k_0 - n_2)/(n_1 - n_2)$. The core index is $n_1 = \sqrt{\epsilon_1/\epsilon_0}$ and the cladding index is $n_2 = \sqrt{\epsilon_2/\epsilon_0}$

to c/n_1 are tightly bound to the guiding structure since most of the guided power is confined to the core region, whereas the modes whose phase velocities are closer to c/n_2 are loosely coupled to the core since a significant amount of the guided power is distributed in the cladding. The loosely bound modes can easily be disturbed by imperfections of the guiding structure and converted into radiation modes and lost.

At a fixed frequency ω, only a finite number of guided eigen modes whose cutoff frequencies are less than ω can be supported by this dielectric waveguide. This means that below the cutoff frequency of a given guided mode that mode simply ceases to exist. Thus, if at a given frequency only a finite number of modes can exist on the structure, any exciting source can, at most, excite those guided modes plus the radiated wave. A fundamental mode (the lowest-order mode) that possesses no cutoff frequency can exist in this dielectric structure according to Fig. 3.2. For example, the fundamental mode in a circular optical fiber is the hybrid HE_{11} mode. Two versions of ω-β diagrams are shown in Fig. 3.2. The traditional ω vs. β diagram is given in the top plot in Fig. 3.2. If we introduce the following constants,

$$V = k_0 a(n_1^2 - n_2^2)^{1/2}, \tag{3.6a}$$

$$b = \frac{(\beta/k_0)^2 - n_2^2}{n_1^2 - n_2^2} \approx \frac{\beta/k_0 - n_2}{n_1 - n_2} \quad \text{when } n_1 \approx n_2, \tag{3.6b}$$

where a is the radius of the circular core, the ω-β diagram may also be expressed by the bottom plot in Fig. 3.2. Figure 3.3 shows the variations of b, $d(Vb)/dV$, and $Vd^2(Vb)/dV^2$ as a function of V for a typical dominant mode [15]. Sketches of the electric field line configuration and the intensity distribution across the fiber are shown in Fig. 3.4.

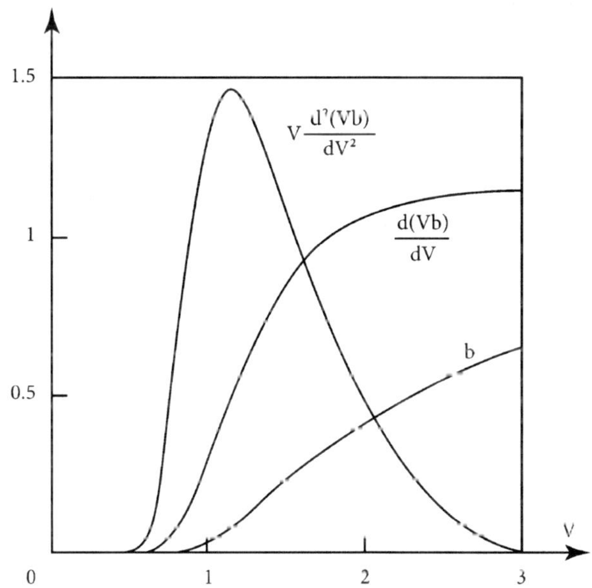

Figure 3.3. Typical variations of b, $d(Vb)/dV$, and $Vd^2(Vb)/dV^2$ as a function of V for a dominant mode [15]

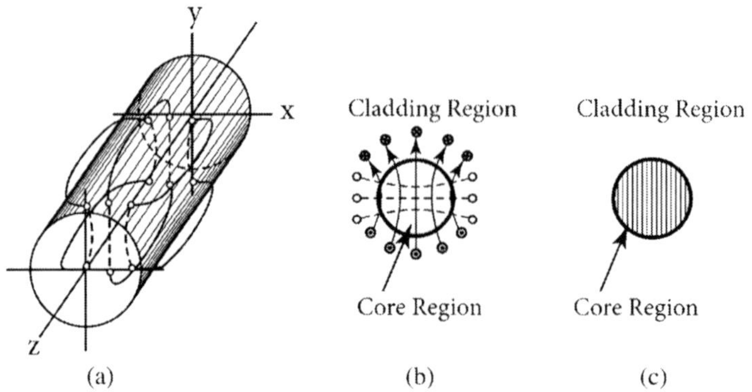

Figure 3.4. For the HE_{11} mode along a circular dielectric cylinder. (**a**) Sketch of the electric lines of force; (**b**) Cross-sectional view of the electric field lines (*solid lines*) and the magnetic field lines (*dashed lines*); and (**c**) The intensity distribution

As one may note, the intensity distribution for the HE_{11} mode is quite uniform throughout the core region. Consequently, efficient launching of this mode can be accomplished by simply illuminating one end of the guide with a plane wave source.

This ω-β diagram also shows that single-mode operation for a dielectric waveguide can be obtained by the appropriate choice of core size, the frequency of operation, and the dielectric constants for the core and cladding material. For a circular dielectric waveguide, the single mode condition is met if $V < 2.4$. The dispersive nature of a given guided mode can be seen from the ω-β diagram, that is, how $\partial^2\beta/\partial\omega^2$ varies as a function of frequency. More will be said later. What is shown in this section is an example of the ω vs. β behavior of a typical guided mode in a dielectric structure whose core dielectric constant is larger than the cladding dielectric constant and both relative dielectric constants are positive and greater than unity. The ω vs. β behavior may be significantly different than those described if the core or cladding material of the guiding structure is significantly altered. For example, if the core material is a plasma and the cladding material is glass [2], the ω-β diagram will be very different than that shown in Figs. 3.2 and 3.3. Some of these guiding structures will be studied in subsequent chapters.

3.4 Geometrical Optics Approach

Although the emphasis of this book is on the propagation characteristics of the dominant single mode or the lowest order modes in a dielectric waveguide structure, because of its historical significance and conceptual simplicity, the geometrical optics approach to the solution of the dielectric waveguide problem will be introduced here. Strictly speaking, dielectric waveguide structures with dimensions comparable or smaller than the free-space wavelength should be analyzed using the exact wave/modal approach rather than the ray optics approach. However, for large-core dielectric waveguide structures, which may support many surface waveguide modes (i.e., a multimode guide), an excellent physical insight can be gained from the following consideration. According to the simplified geometrical optics theory, rays within the core of a dielectric waveguide are totally internally reflected if $\theta \leq \theta_c$ (see Fig. 3.5), where θ is the incident angle of the ray at the boundary surface and θ_c satisfies the relationship

$$\cos\,\theta_c = \frac{n_2}{n_1}, \tag{3.7}$$

where n_1 and n_2 are the indices of refraction in the core and cladding, respectively, with $n_1 > n_2$.

These trapped rays follow zig zag paths in the core material as shown in Fig. 3.5. There is an evanescent (nonradiating) field in the cladding, which decays exponentially away from the core. In this simplified picture, modes can be understood as propagating along different zig zag paths characterized by discrete grazing angles. Hence, there exists a time delay per unit guide length for different propagating modes. The time delay per unit length between the largest and the shortest path length, which is the straight line path through the middle of the core region, is [16]

$$\tau_{d1} = \frac{1}{c}\frac{n_1}{n_2}\left(n_1 - n_2\right), \tag{3.8}$$

where c is the velocity of light in vacuum. For example, a multimode dielectric waveguide with an index difference of 1% or so would have a time spread of 33 ns km^{-1}. Hence, the usable bandwidth of a 1 km guide is about 15 MHz. This delay contributes significantly to the distortion of a pulse in a multimode dielectric

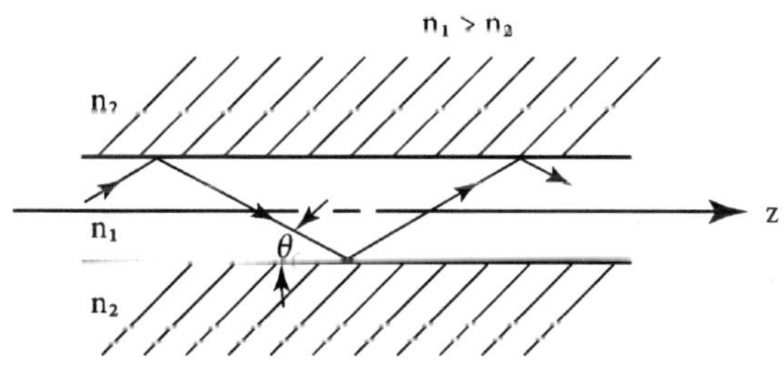

Figure 3.5. Ray path of a guided wave inside the core region of a dielectric waveguide. Rays are totally reflected if $\theta \leq \theta_c$, where θ is the incident angle of the ray as shown and θ_c is given by $\cos\theta_c = n_2/n_1$

(or fiber) guide. To minimize this delay, a graded index profile for the core may be used. It is noted that no index profile can achieve equal transit time along all possible ray trajectories from one guide cross-section to another. The parabolic index profile

$$n(r) = n_1\left(1 - D\frac{r^2}{a^2}\right), \tag{3.9}$$

where D is a constant <1, n_1 is the given index of refraction, and a is the radius of the core, is close to optimal for the core region of a circular cylindrical dielectric waveguide (or fiber) [17]. It has been assumed that the surrounding cladding has an index of $n_1(1-D)$. The time delay per unit length between the longest path length, which corresponds to the helical path at the interface of the core and cladding surface, and the shortest path, which corresponds to the axial path through the center of the core, is [17]

$$\tau_{\mathrm{d2}} = \frac{n_1}{c}\left[\frac{(1-D)^{3/2}}{(1-3D)^{1/2}} - 1\right]. \tag{3.10}$$

For example, if $n_1 = 1.5$, and $D = 1\%$, a time delay of $0.77\,\mathrm{ns\,km^{-1}}$ is possible. This corresponds to a usable 0.65 GHz bandwidth for a 1 km fiber. The limitation of the simplified ray-optics approach can be highlighted by the following example. Given a point source in the cladding region of a dielectric waveguide as shown in Fig. 3.6, the simplified ray-optics picture shows that all rays from the source will pass through the core region of the waveguide when $n_1 > n_2$ and no rays will be trapped in the core region. Therefore, no guided modes can be excited by this source. But according to the more exact wave approach, some guided modes may be excited because the evanescent, decaying fields in the cladding region can be excited by the source, resulting in the excitation of guided wave modes possessing these evanescent, decaying fields.

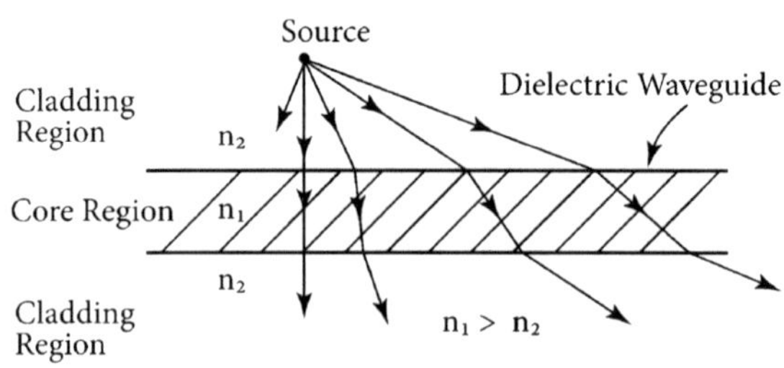

Figure 3.6. According to simplified ray-optics theory, all rays originating in the cladding region will pass through the core region of the dielectric waveguide. No rays will be trapped within the core region

3.5 Attenuation Constant

An imperfect dielectric medium with a complex permittivity, or an imperfect conductor with a finite surface conductivity, will contribute to the attenuation of guided waves. In the absence of mode coupling due to any causes, each guided mode can exist independently of any other guided, evanescent, or radiated modes. Then this attenuation is manifested in making the propagation constant of each guided mode, β_p, a complex quantity, that is,

$$\beta_p = \beta_p^r - j\alpha_p. \tag{3.11}$$

Here, ω/β_p^r is the phase velocity of the pth propagating mode with frequency, ω, and α_p is the attenuation constant for the pth mode. This complex propagation constant is obtained by solving the appropriate boundary value problem in which the lossy nature of the dielectric material and/or the imperfect nature of the conducting surface is taken into consideration.

If the dielectric materials were slightly lossy and if the conducting surfaces were highly conductive, then the perturbation technique may be used to find the power loss and the attenuation constant [18–20]. According to the perturbation technique, the problem is first solved assuming lossless dielectrics and/or perfectly conducting surfaces. Furthermore, the orthogonality relations for the lossless case, (3.4) and (3.5), are assumed to be valid for the slighly lossy case. The electromagnetic fields so obtained, called the unperturbed fields, are then used in the perturbation formula to calculate the attenuation constant and the power loss. In other words, the a priori assumption is that the electromagnetic fields of the modes are not affected by the losses in the dielectric or imperfect conducting surface. From (2.77) the time-average power loss in a lossy dielectric of volume V is

$$\overline{P}_{\text{dielectric loss}} = \frac{1}{2}\sigma_{\text{d}} \int_V [\mathbf{E}(\mathbf{r}) \cdot \mathbf{E}^*(\mathbf{r})]\, dV, \tag{3.12}$$

where σ_{d} (mhos/meter) is the conductivity of the lossy dielectric and $\mathbf{E}(\mathbf{r})$ is the unperturbed electric field in the dielectric. The parameters, ϵ (the complex dielectric constant of the material), ϵ' (the real part of ϵ), ϵ'' (the imaginary part of ϵ), $\tan\delta$ (the loss tangent), and σ_{d} (the conductivity of the lossy dielectric), are related as follows:

$$\begin{aligned} \epsilon &= \epsilon' - j\epsilon'', \\ \tan\delta &= \frac{\epsilon''}{\epsilon'}, \\ \epsilon'' &= \frac{\sigma_{\text{d}}}{\omega}, \end{aligned} \tag{3.13}$$

where ω is the frequency. From (2.78) the power loss over a surface area A due to the imperfect conducting surface, S, is

$$\begin{aligned}
\overline{P}_{\text{conduction loss}} &= \frac{1}{2} R_s \int_S [\mathbf{J}_s(\mathbf{r}) \cdot \mathbf{J}_s^*(\mathbf{r})] \, dA, \\
&= \frac{1}{2} R_s \int_S [(\mathbf{n} \times \mathbf{H}_s(\mathbf{r})) \cdot (\mathbf{n} \times \mathbf{H}_s^*(\mathbf{r}))] \, dA, \\
&= \frac{1}{2} R_s \int_S \mathbf{n} \cdot [\mathbf{H}_s(\mathbf{r}) \times (\mathbf{n} \times \mathbf{H}_s^*(\mathbf{r}))] \, dA, \\
&= \frac{1}{2} R_s \int_S \mathbf{n} \cdot [(\mathbf{H}_s(\mathbf{r}) \cdot \mathbf{H}_s^*(\mathbf{r}))\mathbf{n} - (\mathbf{H}_s(\mathbf{r}) \cdot \mathbf{n})\mathbf{H}_s^*(\mathbf{r})] \, dA, \\
&= \frac{1}{2} R_s \int_S [\mathbf{H}_s(\mathbf{r}) \cdot \mathbf{H}_s^*(\mathbf{r})] \, dA, \qquad (3.14)
\end{aligned}$$

where R_s is the surface resistivity of the imperfectly conducting surface and $\mathbf{H}_s(\mathbf{r})$ is the unperturbed tangential magnetic field at the conducting surface. $\mathbf{H}_s(\mathbf{r}) \cdot \mathbf{n}$ is zero on the conducting surface.

In the rest of this section, we shall deal exclusively with cases under the low-loss approximation condition.

3.5.1 Single Mode Case

When a single mode, say the mth mode, is propagating along a guiding structure, the time average power carried by this mode can be expressed as follows:

$$\overline{P}^{(m)}(z) = \overline{P}_0^{(m)} \, e^{-2\alpha^{(m)} z}, \qquad (3.15)$$

where $\overline{P}_0^{(m)}$ is the initial input power of the mth mode, z is the distance along the guide, and $\alpha^{(m)}$is the attenuation constant of the single mode along the guide. The rate of decrease in $\overline{P}^{(m)}(z)$ vs. z equals the time average power dissipated per unit length, $\overline{P}_{\text{d}}^{(m)}$, or,

$$\overline{P}_{\text{d}}^{(m)} = -\frac{\partial \overline{P}^{(m)}}{\partial z} = 2\alpha^{(m)} \overline{P}^{(m)}. \qquad (3.16)$$

Thus, the attenuation constant is given by

$$\alpha^{(m)} = \frac{\overline{P}_{\text{d}}^{(m)}}{2\overline{P}^{(m)}} \quad (\text{Np m}^{-1}). \qquad (3.17)$$

Here, the power dissipated per unit length represents the total power loss per unit length due to the imperfection of both the dielectric and the conducting material.

$$\overline{P}_{\mathrm{d}}^{(m)} = \overline{P}_{\text{dielectric loss}}^{(m)} + \overline{P}_{\text{conduction loss}}^{(m)}, \tag{3.18}$$

where $\overline{P}_{\text{dielectric loss}}^{(m)}$ and $\overline{P}_{\text{conduction loss}}^{(m)}$ for the mth mode have been given by (3.12) and (3.14), respectively. The time average power carried by mode m is

$$\overline{P}^{(m)} = \frac{1}{2}\mathrm{Re}\int_A [\mathbf{E}(\mathbf{r}) \times \mathbf{H}^*(\mathbf{r})] \cdot \mathbf{e}_z \, dA, \tag{3.19}$$

where A is the cross-sectional area of the guide. The single mode attenuation constant due to the loss in the dielectric material is

$$\alpha_{\text{dielectric}}^{(m)} = \frac{\frac{1}{2}\sigma_{\mathrm{d}} \int_A \left[\mathbf{E}^{(m)}(\mathbf{r}) \cdot \mathbf{E}^{(m)*}(\mathbf{r})\right] dA}{\mathrm{Re}\int_A \left[\mathbf{E}_{\mathrm{t}}^{(m)}(\mathbf{r}) \times \mathbf{H}_{\mathrm{t}}^{(m)*}(\mathbf{r})\right] \cdot \mathbf{e}_z \, dA} \quad (\mathrm{Np\,m^{-1}}) \tag{3.20}$$

and for the imperfect conductor

$$\alpha_{\text{conductor}}^{(m)} = \frac{\frac{1}{2}R_{\mathrm{s}} \oint_c \left[\mathbf{H}_{\mathrm{s}}^{(m)}(\mathbf{r}) \cdot \mathbf{H}_{\mathrm{s}}^{(m)*}(\mathbf{r})\right] dc}{\mathrm{Re}\int_A \left[\mathbf{E}_{\mathrm{t}}^{(m)}(\mathbf{r}) \times \mathbf{H}_{\mathrm{t}}^{(m)*}(\mathbf{r})\right] \cdot \mathbf{e}_z \, dA} \quad (\mathrm{Np\,m^{-1}}). \tag{3.21}$$

Note that the attenuation constant $\alpha^{(m)}$ for single unidirectional propagation mode is a constant, not a function of the propagation distance. Here, A is the cross-sectional area of the dielectric waveguide and c is the contour path along the conducting surface that is normal to z, the direction of propagation. The subscript s means along the conducting surface, and the subscript t means tangential to the cross-sectional area and normal to $\mathbf{e}_z$. In other words, along the waveguide, uniform power loss is produced for this mode. Typical attenuation characteristics of a few surface wave modes on a dielectric rod in free space are shown in Fig. 3.7 [21].

It is seen that at high normalized frequencies most of the guided power is contained within the core region, and the waveguide loss must approach that of a plane wave in an infinite medium with dielectric loss. At lower normalized frequencies or near cutoff, most of the guided power is contained in the lossless free space cladding region and the waveguide loss must approach zero.

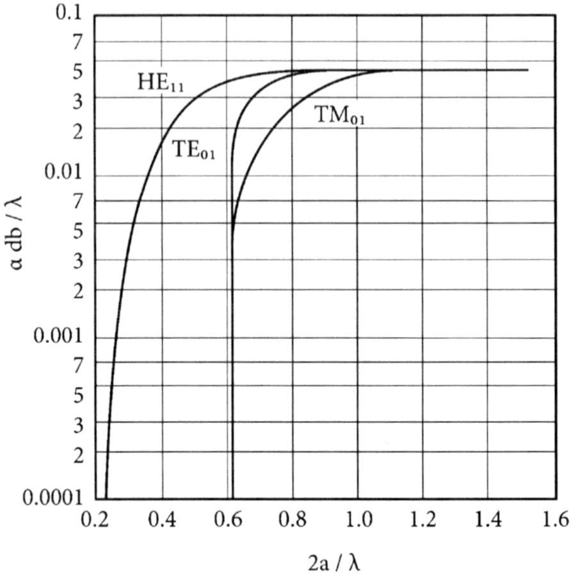

Figure 3.7. Attenuation for several lower order surface wave modes along a polystyrene rod of radius a ($\epsilon_1 = 2.56$ and $\tan\delta$ =0.001). Note that the attenuation α is in dB/λ, where λ is the free space wavelength [21]

3.5.2 Multimode Case

When more than a single mode are propagating simultaneously along a guiding structure, the above formulas for a single unidirectional propagating mode can no longer be used. Because of the nonzero contributions of the cross-product terms, the power loss is no longer uniform along the guide. It is *incorrect* to state that the total power loss for the guided wave containing multiple modes can be expressed as the sum of all the power loss for each of these multiple modes propagating independently. In other words, the following formula is incorrect.

$$\overline{P}_{\text{total}}(z) = \sum_m \overline{P}_0^{(m)} e^{-2\alpha^{(m)} z} \qquad \text{(incorrect)}, \tag{3.22}$$

where $P_0^{(m)}$ is the initial input power for the mth mode and $\alpha^{(m)}$ is the attenuation constant for the mth mode, propagating alone. It should be pointed out that the orthogonality condition for the guided modes dictates that the total power is additive under the low-loss approximation, that is,

$$\overline{P}_{\text{total}}(z) = \sum_m \overline{P}^{(m)}(z) \qquad \text{(correct)}, \tag{3.23}$$

where $\overline{P}^{(m)}(z)$ is the total power for the mth mode at z. This means that *when multiple modes are propagating simultaneously*, the relation

$$\overline{P}^{(m)}(z) = \overline{P}_0^{(m)}\, e^{-2\alpha^{(m)}z} \qquad \text{(incorrect)} \tag{3.24}$$

for *each* mode is not necessarily correct. In the following, the necessary conditions under which this formula is approximately correct will be shown.

The correct way of finding the total power loss in a multimode situation was first shown by Yeh [22] and subsequently verified by experiment [23]. For the sake of clarity, let us describe the case where only two modes are propagating simultaneously in a hollow uniform metallic tube with a slightly imperfect conducting wall [22, 23]. This approach can be extended to treat the case with many modes (more than two) propagating simultaneously, or to the case of multimode propagation in a dielectric waveguide with slight loss. Let us consider the specific case of two propagating TE modes in a conducting tube. The field components of these two modes are

$$\mathbf{E}^{(\mathrm{TE})}(x,y,z) = A_1\mathbf{E}_{1t}(x,y)\, e^{-j\beta_1 z} + A_2\mathbf{E}_{2t}(x,y)\, e^{-j\beta_2 z}, \tag{3.25}$$

$$\mathbf{H}^{(\mathrm{TE})}(x,y,z) = A_1\left[\mathbf{H}_{1t}(x,y) + H_{1z}(x,y)\mathbf{e}_z\right] e^{-j\beta_1 z}$$
$$+ A_2\left[\mathbf{H}_{2t}(x,y) + H_{2z}(x,y)\mathbf{e}_z\right] e^{-j\beta_2 z}. \tag{3.26}$$

Here A_1 and A_2 are known amplitude constants and β_1 and β_2 are the two different propagation constants for the two propagating modes. The power loss due to the imperfect conducting surface S can be calculated by substituting (3.26) in (3.14) and carrying out the integration in z:

$$\overline{P}_{\mathrm{L}}(l) = \text{Part 1} + \text{Part 2}, \tag{3.27}$$

where

$$\text{Part 1} = \frac{1}{2}R_{\mathrm{s}}l\left[|A_1|^2 I_1 + |A_2|^2 I_2\right], \tag{3.28}$$

$$\text{Part 2} = \frac{1}{2}R_{\mathrm{s}}l\left[A_1A_2^*I_{12}\left(\frac{\exp(-j(\beta_1-\beta_2)l)-1}{-j(\beta_1-\beta_2)l}\right)\right.$$
$$\left. + A_2A_1^*I_{21}\left(\frac{\exp(-j(\beta_2-\beta_1)l)-1}{-j(\beta_2-\beta_1)l}\right)\right], \tag{3.29}$$

$$I_{1,2} = \oint_c \left[|H_{1,2c}|^2 + |H_{1,2z}|^2\right] dc, \tag{3.30}$$

$$I_{12} = \oint_c [H_{1c}H_{2c}^* + H_{1z}H_{2z}^*] \, dc, \tag{3.31}$$

$$I_{21} = \oint_c [H_{2c}H_{1c}^* + H_{2z}H_{1z}^*] \, dc. \tag{3.32}$$

The subscript c represents the component of the transverse field that is tangential to the contour around the surface of the tube and is normal to the z-axis. The symbol l represents the length of the waveguide of interest. It is seen that Part 1 represents the power loss contributed by each mode as if it were propagating as a single mode. [See (3.14) without the cross product terms]. Part 2, because of the cross-product terms of (3.14), provides the necessary missing component to calculate the correct power loss for the two copropagating modes. The importance of Part 2 diminishes when l, the length of the guide increases, as well as when a larger difference for the propagation constants occurs. So, for a very long waveguide, the incorrect expression, (3.22) may be used and the contributions from the cross-product terms or the coupling effects of copropagating modes may be ignored. Experiments using a beam-waveguide system was performed to verify the validity of the above analysis [23]. Results are shown in Fig. 3.8. Excellent agreement was found verifying the approach. This approach may be used to derive the accurate formula for the power loss for multimodes propagating simultaneously in a dielectric waveguide, using (3.12) with the cross product terms. Similar conclusions as those given above will be reached.

3.6 Signal Dispersion and Distortion

One of the most important parameters characterizing a communication line made with a dielectric waveguide is its information-carrying capacity, which is directly related to the bandwidth of the line. The bandwidth of a dielectric waveguide is limited, because signals are distorted as they pass through the dielectric waveguide. The major contributing factors are (1) *Differences between the group velocities of different modes.* In a multimode dielectric waveguide, the energy of an incident pulse is distributed among many modes; since each mode propagates at its own velocity, the resultant output signal, which is the sum of all signals carried by different modes, is necessarily broadened. (2) *Dispersion of the dielectric waveguide.* The propagation constant of each mode is a function of frequency. When it is not linearly proportional to the frequency, dispersion occurs. This is caused by the mode dispersion properties of the waveguide and by the dispersive and/or the nonlinear characteristics of the index of refraction of the dielectric material. Since the input

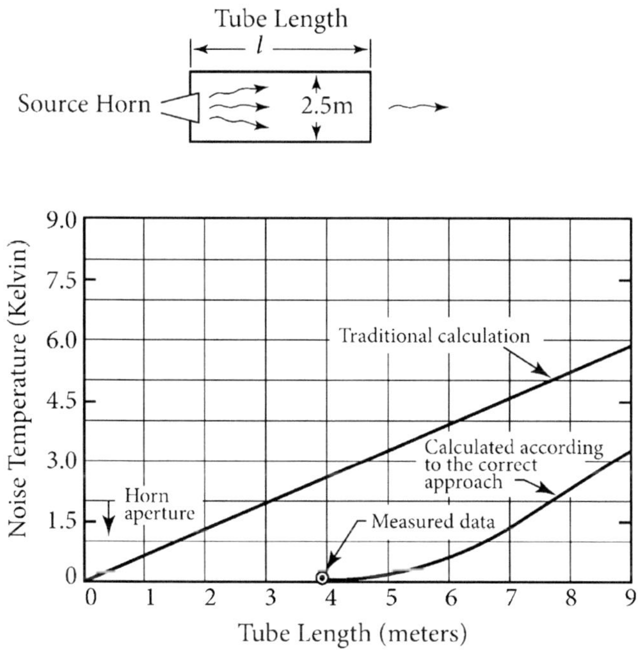

Figure 3.8. Illustration on the importance of using the correct formula to calculate power loss for multimoded waveguides. Here, the noise temperature, defined as $(P_L/P_T)T_0$, where P_L is the total power loss, P_T is the total input power, and T_0 is the ambient temperature in K (for room temperature, $T_0 = 293.1$ K), is plotted against tube length for a 22.5 dBi gain horn. Because of the large size of the tube guide, more than 45 modes are generated simultaneously. It is seen that the measured data agree very well with the calculated results according to the approach given in the text for the multimode case [23]

pulse has a certain spectral width, the output pulse is necessarily distorted even under single-mode operation. (3) *Mode conversion and nonuniform attenuation of various propagating modes*. The expected output pulse width computed according to the first two effects may be significantly altered due to the existence of mode conversion, which may be caused by the imperfection of the guide or by the presence of bends along the guide, and due to the nonuniform attenuation of different modes. An illustration of the effect of signal degradation due to the above causes on the output pulse is given in Fig. 3.9. Here, a series of input Gaussian pulses are broadened as they propagate along the waveguide.

In Sect. 3.3 on the discussion of the ω-β behavior of a given guided mode, the factor $\partial^2\beta/\partial\omega^2$ is noted as the pulse broadening factor. The derivation on this fact is now given [24]. Assuming that the tangential fields of the pth mode at the input end of the dielectric waveguide is $\mathbf{G}_{tp}^{(i)}(t, r, \theta) = \mathbf{E}_{tp}(r, \theta)f(t)$, where (r, θ) are the

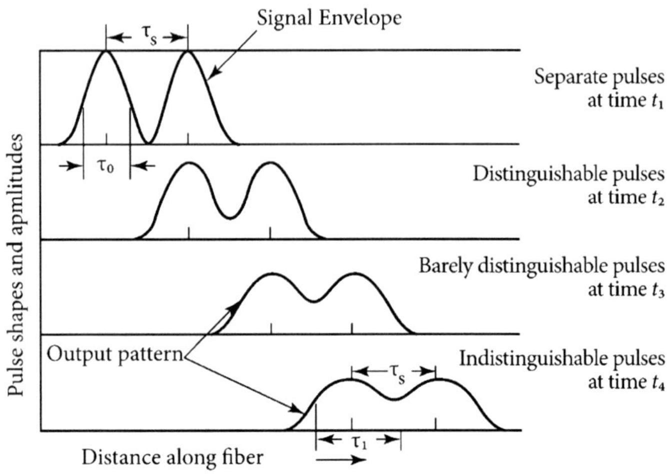

Figure 3.9. A sketch on pulse degradation effects on an input series of signal pulses. These degradation effects cause two neighboring pulses to overlap each other and to deform so that they become indistinguishable as separate pulses, thereby losing their information carrying capability. This would limit the width of the spacing between the nearest pulses; thus limiting the signal bandwidth

transverse coordinates in the circular cylindrical coordinate system, the tangential electric field of the pth mode at the output end of the guide is

$$\mathbf{G}_{tp}^{(\mathrm{o})}(t, r, \theta) = \mathbf{E}_{tp}(r, \theta)\frac{1}{2\pi}\int_{-\infty}^{\infty} e^{-j\beta_p l} F(\omega)\, e^{j\omega t}\, d\omega, \tag{3.33}$$

$$\begin{aligned} F(\omega) &= \int_{-\infty}^{\infty} f(t)\, e^{-j\omega t}\, dt, \\ &= \int_{-\infty}^{\infty} g(t)\, e^{j\omega_0 t} e^{-j\omega t}\, dt = G(\omega - \omega_0), \end{aligned} \tag{3.34}$$

where $f(t) = g(t)\, e^{j\omega_0 t}$, $g(t)$ is the information carrying envelope wave modulating a rapidly oscillating carrier wave of frequency ω_0, and l is the length of the guide. The superscript (o) means the output and the superscript (i) means the input. The dispersion characteristics of the dielectric waveguide are included in $\beta_p(\omega)$, which gives the propagation constant of the pth mode as a function of frequency. If a normalized Gaussian pulse

$$g(t) = \frac{1}{(\tau_0 \pi^{1/2})^{1/2}} \exp\left(-\frac{t^2}{2\tau_0^2}\right), \qquad \tau_0 \gg \frac{1}{\omega_0}, \tag{3.35}$$

is chosen as the input envelope whose spectral extent is narrow compared to the carrier frequency ω_0 and if we Taylor-expand β_p about ω_0, that is,

$$\beta_p(\omega) = \beta_p(\omega_0) + \left(\frac{\partial \beta_p}{\partial \omega}\right)_{\omega=\omega_0} (\omega - \omega_0) + \frac{1}{2}\left(\frac{\partial^2 \beta_p}{\partial \omega^2}\right)_{\omega=\omega_0} (\omega - \omega_0)^2 + \cdots,$$

then (3.33) may be evaluated analytically to yield

$$\mathbf{G}_{tp}^{(o)}(t, r, \theta) = \mathbf{E}_{tp}(r, \theta) \exp\left[j\omega_0 (t - t_0)\right] q(t), \tag{3.36}$$

$$q(t) \approx \frac{1}{(\tau_0 \pi^{1/2})^{1/2}} \frac{1}{(1 + jb_0)^{1/2}} \exp\left[-\frac{(t - t_d)^2}{2\tau_0^2 (1 + jb_0)}\right], \tag{3.37}$$

$$|q(t)|_p^2 \sim \exp\left[-\frac{(t - t_d)^2}{\tau_0^2 (1 + b_0^2)}\right], \tag{3.38}$$

where

$$\beta_{p0} = \beta_p(\omega_0), \quad \beta'_{p0} = \left(\frac{\partial \beta_p}{\partial \omega}\right)_{\omega=\omega_0}, \quad \beta''_{p0} = \left(\frac{\partial^2 \beta_p}{\partial \omega^2}\right)_{\omega=\omega_0},$$

$$t_0 = \frac{\beta_{p0} l}{\omega_0}, \quad t_d = \frac{l}{\beta'_{p0}}, \quad b_0 = \frac{\beta''_{p0} l}{\tau_0^2}.$$

The approximation sign in (3.38) means that $\beta_p(\omega)$ is adequately represented by the first three terms of the Taylor series expansion. According to (3.38) we note that apart from a time delay $t_d = l/\beta'_{p0}$, for the pth mode, the Gaussian pulse of the pth mode is broadened from a pulse width of τ_0 to τ_1, where

$$\tau_1 = \tau_0 \left[1 + \left(\frac{\beta''_{p0} l}{\tau_0^2}\right)^2\right]^{1/2}. \tag{3.39}$$

Using the mode orthogonality relations (3.4) and (3.5), we can show that the total output power is equal to the sum of the output power of each mode. The distribution of energy among all propagating modes at the input is represented by the different amplitude coefficients of these modes. So for the multimode waveguide, the different time delays for various modes may be the principal cause for pulse broadening, while for single-mode guides the finite value of β''_{p0} is the major cause of pulse distortion. From (3.39), we notice that the pulse broadening for a single mode is caused by the existence of $\beta''_p(\omega_0)$, which represents the curvature of the ω-β curve. In other words, if the ω-β curve is a straight line within the bandwidth of interest, then the signal can be transmitted through the given single mode waveguide without dispersion, that is, without pulse broadening. A waveguide that supports a single dominant propagating mode with a straight line ω-β curve is called a dispersionless transmission line [18].

A sketch of $c\beta_p/\omega$ as a function of c/ω (c is the speed of light in vacuum) is given in Fig. 3.10. It can be seen that material dispersion is the dominant feature that contributes to the curvature $\left(\beta''_{p0}\right)$ of the curves. Hence, the principal cause of pulse broadening for a single-mode dielectric waveguide is the material dispersion of the dielectrics. Typical dispersion $c\omega\beta''_{p0}$ as a function of frequency for fused silica and Schott glasses is shown in Fig. 3.11. It was pointed out by Smith and Snitzer [25] that by properly choosing the dielectric material with low dispersion, we may minimize the dispersion in a dielectric guide. In the literature, a parameter D is frequently used to represent the dispersion factor. It is related to β''_{p0} by the following relation [26]:

$$D = \frac{d}{d\lambda}\left(\frac{1}{v_g}\right) = -\frac{2\pi c}{\lambda^2}\beta''_{p0}, \tag{3.40}$$

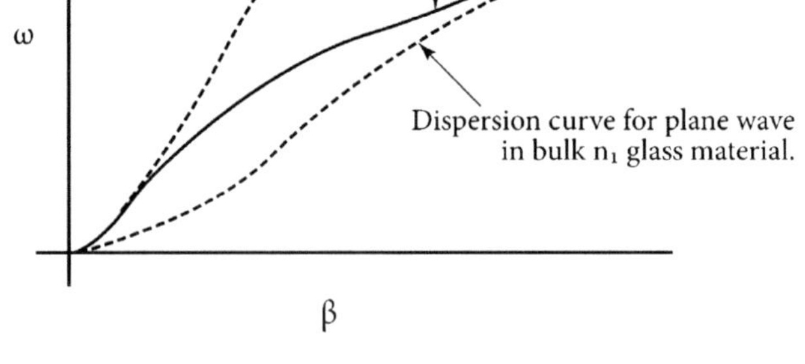

Figure 3.10. A sketch of the ω-β diagram for the dominant mode with waveguide and material dispersion

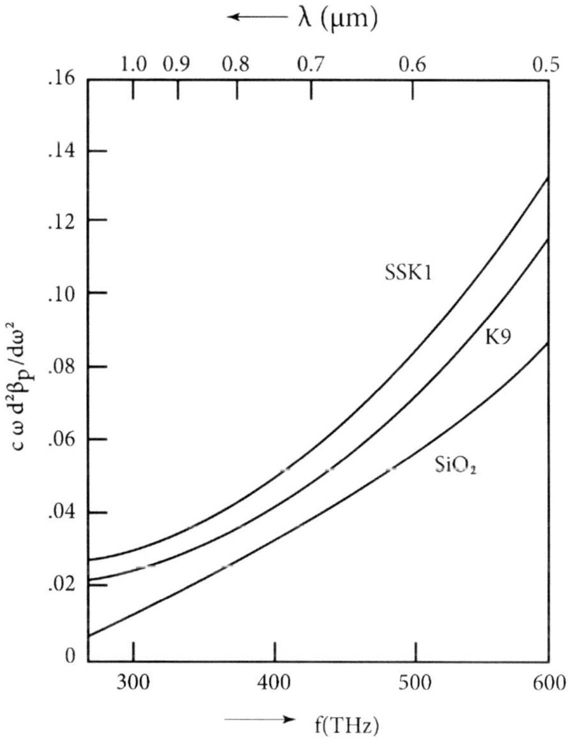

Figure 3.11. Dispersion coefficient $c\omega\, d^2\beta_p/d\omega^2$ as a function of frequency for fused silica (SiO_2) and the Schott glasses K9 and SSK1. β_p is the propagation constant of the pth mode. Waveguide dispersion is assumed to be negligible [16]

where λ is the wavelength, v_g is the group velocity, and the units of D is ps km^{-1} nm^{-1}. A typical D vs. λ curve for the standard dispersion-shifted, and dispersion-flattened fiber is shown in Fig. 3.12.

For a typical multimode dielectric guide made of glass whose $\delta = 1 - \left(n_2^2/n_1^2\right)$ is very small, the trade-off point for determining the major contributor to pulse broadening normally occurs at $V \simeq 20$, with V defined by (3.6). In other words, for $V < 20$, material dispersion provides the dominant cause for pulse broadening; for $V > 20$, group velocity delays of various modes with uniform energy distribution among all propagating modes and uniform attenuation for all modes provides the dominant cause. It can also be shown that the bandwidth of a single-mode dielectric waveguide is approximately proportional to $1/\sqrt{l}$ and that of a multimode waveguide to $1/l$, where l is the length of the guide [27]. Typically, the bandwidth of a single mode dielectric waveguide limited by the material (glass) dispersion factor is about 20–40 GHz km^{-1}, while the bandwidth of a multimode dielectric waveguide is about 35–45 MHz km^{-1} (for $V \simeq 50$).

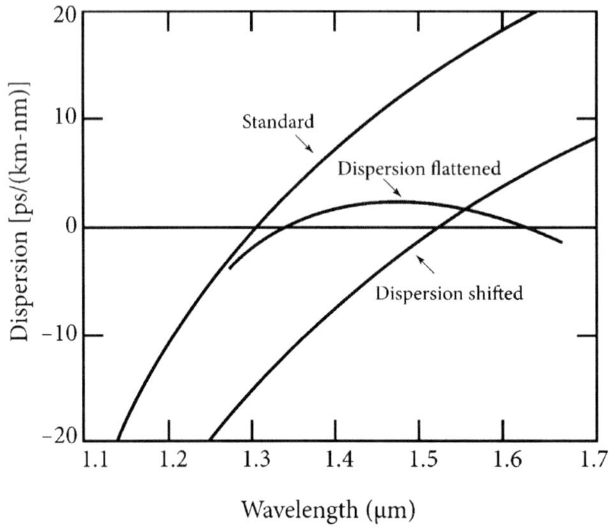

Figure 3.12. Typical wavelength dependence of the dispersion parameter D for standard dispersion-shifted and dispersion-flattened fibers. [26]

Analysis for the pulse dispersion characteristics in a lens-like medium, whose dielectric constant varies in the radial direction with the variation

$$\epsilon(r) = \epsilon(0) + [1 - (gr)^2 + b(gr)^4], \tag{3.41}$$

where $\epsilon(0)$ is the dielectric constant on the axis and b and g are constants, has also been carried out using a scalar wave approximation [28]. As expected the group velocity differences can be minimized with the use of a dielectric waveguide having the appropriate radial refractive index variation, such as the "Selfoc" fiber [28]. Hence a typical multimode Selfoc fiber may yield a bandwidth of 300–500 MHz km^{-1}.

For a single-mode dielectric waveguide, under special circumstances, the presence of material and dispersion effects tends to broaden a propagating pulse, and the presence of nonlinear material effects tends to narrow the pulse such that there can exist a specially tailored propagating pulse that can retain its pulse shape indefinitely while propagating along the dielectric waveguide [29]. This pulse is called a soliton pulse. More detailed discussion will be given later.

3.7 α and Q

A low-loss dielectric waveguide can be used to construct a cavity resonator by short-circuiting both ends of the dielectric waveguide with perfectly conducting surfaces or by joining the input end of the guide with the output end. A relation exists between α, the attenuation constant of the low-loss waveguide for a given single-mode, and Q, the quality factor of the resonator, which is defined as $Q =$ ($\omega\times$energy stored)/(average power dissipated). This provides a convenient way of measuring the small attenuation constant of a low-loss dielectric waveguide. The equation relating α and Q will now be derived.

In 1944, Davidson and Simmonds [30] derived a relation between the Q of a cavity composed of a uniform transmission line with short-circuiting ends and the attenuation constant, α, of such a transmission line. Later, in 1950, Barlow and Cullen [31] rederived this relation. These authors showed that this relation is quite general and is applicable to uniform metal tube waveguides with arbitrary cross-section. Since then, one of the standard techniques for the measurement of the attenuation constant has been the use of the cavity method. This method offers an excellent way of measuring the attenuation constant of a waveguide when the loss is very small. Later, this method was generalized and applied to open waveguide structures, such as the single wire transmission line and the dielectric waveguide by various authors [10, 21].

However, it is noted that the formula by the early authors [30, 31] was derived under the assumption that there exists a single equivalent transmission line for the mode under consideration. This assumption is true for a pure TE, TM, or TEM mode, but it is not clear that such a single equivalent transmission line exists for a hybrid wave, which is the dominant mode on a typical cylindrical dielectric waveguide. This suspicion originates from the fact that on a cylindrical dielectric waveguide, (1) the TE and TM waves are intimately coupled to each other and (2) the characteristic impedance defined by Schelkunoff [32] is not constant with respect to the transverse coordinates (see Sect. 2.7). It is, therefore, very difficult to conceive the possibility that there exists a single equivalent transmission line for this hybrid wave; at best the hybrid wave may be represented by a set of equivalent transmission lines coupled tightly with one another. Hence, the formula by Davidson et al. [30] may not be applicable to a hybrid wave.

A more general relation between Q and α was obtained by Yeh [33] in 1962 without using the transmission line equivalent circuit, provided that α is very small compared with the phase constant β and the loss contributed by short circuiting the end plates is negligible compared with the total loss of the waveguide section under consideration. The propagation constant of a guided wave with a small attenuation constant at ω_0 is

$$\Gamma(\omega_0) = \alpha(\omega_0) - j\beta(\omega_0) \tag{3.42}$$

At resonance, the following relation is true:

$$\Gamma(\omega_0) + \frac{\partial\Gamma}{\partial\omega}\Delta\omega \approx -j\beta(\omega_0). \tag{3.43}$$

Combining (3.42) and (3.43) gives

$$\alpha(\omega_0) = j\frac{\partial\beta}{\partial\omega}\Delta\omega. \tag{3.44}$$

Since the group velocity v_g and Q are given by the relations

$$v_g = \frac{\partial\omega}{\partial\beta},$$

$$Q = \frac{\omega_0}{j2\Delta\omega},$$

one arrives at the relation

$$\alpha = \frac{\omega_0}{2Qv_g} = \frac{v_p}{v_g}\frac{\beta}{2Q}, \tag{3.45}$$

where $v_{\rm p}(=\omega_0/\beta)$ is the phase velocity of the wave. This is the general relation that is sought.

Substituting the values of $v_{\rm p}/v_{\rm g}$ for a TE, TM, or TEM wave in (3.45), one gets the relation derived by Davidson et al. [30]. For a TM or TE wave in a metal waveguide,

$$\frac{v_{\rm p}}{v_{\rm g}} = \frac{1}{1-\left(\dfrac{\lambda}{\lambda_{\rm c}}\right)^2}, \tag{3.46}$$

where $\lambda_{\rm c}$ is the cutoff wavelength of the wave under consideration. Thus

$$\alpha_{\rm TE,TM} = \frac{1}{1-\left(\dfrac{\lambda}{\lambda_{\rm c}}\right)^2}\frac{\beta}{2Q}. \tag{3.47}$$

For a TEM wave, $v_{\rm p}/v_{\rm g} = 1$. Hence,

$$\alpha_{\rm TEM} = \frac{\beta}{2Q}. \tag{3.48}$$

For a hybrid wave, v_p and v_g are not simply related. They must be obtained graphically or numerically from the ω-β relation. However, for a dominant hybrid wave on a dielectric cylinder at very low frequencies or at very high frequencies, the relation (3.48) is a good approximation since, at these frequencies, $v_{\rm p} \approx v_{\rm g}$.

Using this technique, the ultra low-loss behavior of a dielectric ribbon waveguide can be measured. The results are illustrated in Fig. 3.13.

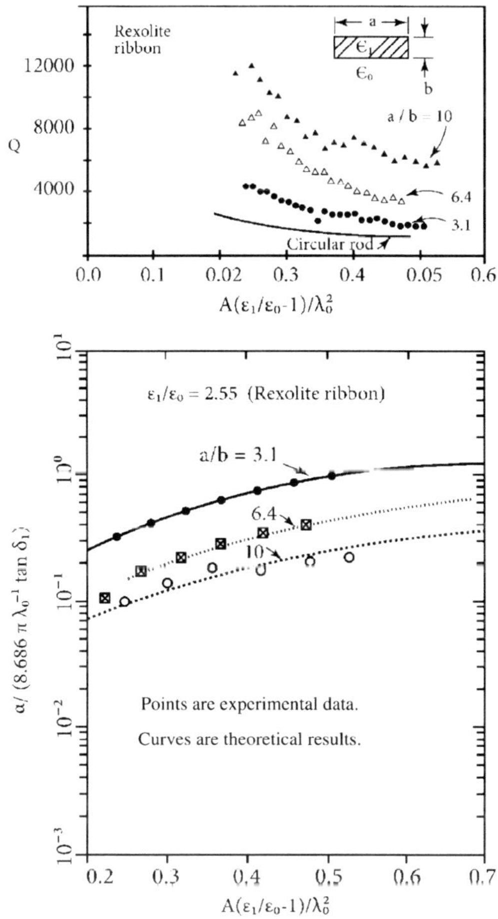

Figure 3.13. The top graph shows measured Q as a function of frequency for the four Rexolite ribbon waveguides. The relative dielectric constant ϵ_1/ϵ_0 and loss tangent of Rexolite are, respectively, 2.55 and 0.9×10^{-3}. Only the low-loss dominant HE_{11} mode is supported by the structure. A is the cross-sectional area of the ribbon and λ_0 is the free-space wavelength. The bottom graph shows the comparison between the calculated loss factor and measured data for the dielectric ribbon waveguide with three different aspect ratios. α is the attenuation constant in dB/m, $\tan\delta_1$is the loss tangent of Rexolite [33]

3.8 Excitation of Modes on a Dielectric Waveguide

3.8.1 Excitation Through Direct Incidence

One of the simplest and rather efficient ways of exciting guided modes on a dielectric waveguide is to illuminate an incident wave or beam directly on the end of the guide [34] (see Fig. 3.14).

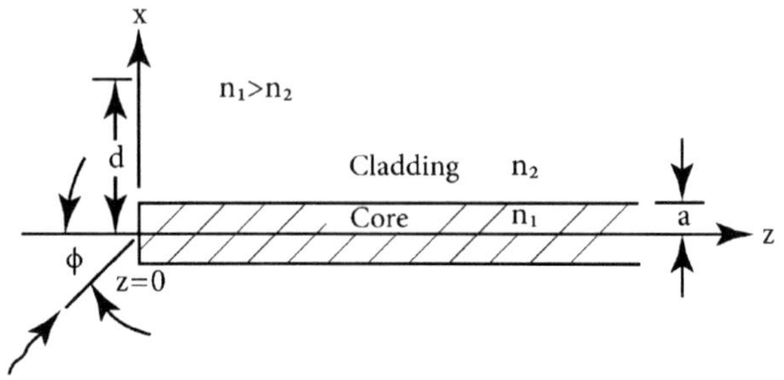

Figure 3.14. End on excitation of a dielectric rod by an incident wave or beam. d is the radius of the illuminated aperture

Since any arbitrary field distribution of the dielectric waveguide can be expressed in terms of the orthonormal modes (which includes guided modes as well as the radiation mode) of the guide, the transverse electric fields at $z = 0$ (one end of the semi-infinite dielectric rod) can be represented as follows:

$$\mathbf{E}_{\mathrm{t}}(x, y) = \sum_{p} A_p \mathbf{E}_{\mathrm{t}p}^{\mathrm{guided}}(x, y), \tag{3.49}$$

where $\mathbf{E}_{\mathrm{t}}$ and $\mathbf{E}_{\mathrm{t}p}$ are, respectively, the transverse electric fields (incident plus scattered fields) and the transverse orthonormal modal electric fields, the subscript p indicates the pth mode, and A_p is the amplitude coefficient of the pth mode, which can be obtained by using the orthogonality relation (3.4). Assuming that (1) no reflected wave exists at $z = 0$, that is, $\mathbf{E}_{\mathrm{t}} = \mathbf{E}_{\mathrm{t}}^{(\mathrm{inc})}$; (2) the angle of incidence ϕ is so small that ϕ^2 may be neglected; (3) the factor $\delta = 1 - \epsilon_2/\epsilon_1 = 1 - \left(n_2^2/n_1^2\right)$ is small, then the excited modal power P_p is given by

$$P_p = \frac{1}{2}\left|A_p\right|^2 = \frac{1}{2}\frac{\epsilon_1}{\mu_0}\,\mathrm{Re}\left|\int_{\mathrm{ap}}\left(\mathbf{E}_{\mathrm{t}}^{(\mathrm{inc})}\cdot\mathbf{E}_{\mathrm{t}p}\right)dA\right|, \tag{3.50}$$

where $\mathbf{E}_{\mathrm{t}}^{(\mathrm{inc})}$ is the incident transverse electric field at the aperture (ap), which is the illuminated portion at the end of the dielectric rod, and the transverse orthonormal fields $\mathbf{E}_{\mathrm{t}p}$ are normalized such that

$$\left(\frac{\epsilon_1}{\mu_0}\right)^{1/2}\int_A \left|\mathbf{E}_{\mathrm{t}p}\right|^2 dA = 1, \tag{3.51}$$

with A being the total cross-sectional area of the circular dielectric rod. As a specific example, for the weakly guiding case where $\delta \ll 1$ [15], the transverse orthonormal electric fields for guided modes on a circular dielectric rod are the following:

For HE_{nm} and EH_{nm} modes

$$\mathbf{E}_{tp} = \frac{1}{\sqrt{N_p}} \left\{ \pm \mathbf{e}_x \begin{bmatrix} \sin(n \mp 1)\theta \\ \text{or} \\ \cos(n \mp 1)\theta \end{bmatrix} + \mathbf{e}_y \begin{bmatrix} \cos(n \mp 1)\theta \\ \text{or} \\ -\sin(n \mp 1)\theta \end{bmatrix} \right\} f_n(r). \tag{3.52}$$

For TM_{0m} modes

$$\mathbf{E}_{tp} = \frac{1}{\sqrt{N_p}} \left(-\mathbf{e}_x \cos\theta + \mathbf{e}_y \sin\theta\right) f_0(r). \tag{3.53}$$

For TE_{0m} modes

$$\mathbf{E}_{tp} = \frac{1}{\sqrt{N_p}} \left(-\mathbf{e}_x \sin\theta + \mathbf{e}_y \cos\theta\right) f_0(r), \tag{3.54}$$

$$f_n(r) = \begin{bmatrix} J_{n\mp1}(pr)/J_{n\mp1}(pa) & r \le a \\ K_{n\mp1}(qr)/K_{n\mp1}(qa) & r \ge a \end{bmatrix},$$

$$N_p = a^2\pi \left(\frac{\epsilon_1}{\mu_0}\right)^{1/2} \left(\frac{V}{pa}\right)^2 \left[\frac{K_n(qa)K_{n\mp2}(qa)}{K^2_{n\mp1}(qa)}\right], \tag{3.55}$$

where $V = k_0 a \left(n_1^2 - n_2^2\right)^{1/2}$, $k_0 a = \omega\sqrt{\mu_0\epsilon_0}a$, $p^2 = k_0^2 n_1^2 - \beta^2$, $q^2 = \beta^2 - k_0^2 n_2^2$, β = propagation constant, and a = radius of the rod. The circular cylindrical coordinates (r, θ, z) have been used and J and K refer to the Bessel function and the modified Bessel function of the second kind, respectively. The upper sign is for the HE_{nm} modes and the lower for the EH_{nm} modes. Two types of incident fields are of special interest.

3.8.1.1 Incident Plane Wave

The propagation vector of the incident plane wave, which is assumed to be in the x-z plane, makes an angle ϕ with the positive z-axis. The electric vector of the incident wave takes the form (at $z = 0$)

$$\mathbf{E}^{\mathrm{inc}} = E_0 \exp\left[-j\omega\sqrt{\mu_0\epsilon_0}\, \sin\phi\right] \mathbf{e},$$

$$\mathbf{E}^{\mathrm{inc}} \simeq E_0 \exp\left[-j\omega\sqrt{\mu_0\epsilon_0}\phi\right] \mathbf{e} = E_0 \exp\left[-j\omega\sqrt{\mu_0\epsilon_0}\,\phi r\cos\theta\right] \mathbf{e}, \tag{3.56}$$

where $\phi \ll 1$ and $\mathbf{e}$ is a unit vector in the direction of $\mathbf{E}^{(\mathrm{inc})}$ and E_0 is the normalized amplitude constant given as

$$E_0 = \left[\frac{2\sqrt{\mu_0/\epsilon_0}}{\pi a^2} \right]^{1/2} \tag{3.57}$$

for unit incident power, a is the core radius of the circular dielectric rod. Substituting (3.54) and (3.56) in (3.50) gives

$$P_p = \left(\frac{2pa}{Vd} \right)^2 \frac{K_{n\mp1}^2(qa)}{K_n(qa)K_{n\mp2}(qa)} (I_1 + I_2), \tag{3.58}$$

with

$$I_1 = \frac{1}{J_{n\mp1}(pa)} \int_0^1 J_{n\mp1}(pa\chi) J_{n\mp1}(\omega\sqrt{\mu_0\epsilon_1}\, a\phi\chi)\chi\, d\chi,$$

$$I_2 = \frac{1}{K_{n\mp1}(pa)} \int_0^{d/a} K_{n\mp1}(qa\chi) K_{n\mp1}(\omega\sqrt{\mu_0\epsilon_1}\, a\phi\chi)\chi\, d\chi,$$

where d is the radius of the illuminated aperture. The polarization vector $\mathbf{e}$ for the incident plane wave is taken to be either $\mathbf{e}_x$ or $\mathbf{e}_y$. When $\mathbf{e} = \mathbf{e}_x$, $\overline{P}_p = 0$ for TE$_{0m}$ modes and when $\mathbf{e} = \mathbf{e}_y$, $\overline{P}_p = 0$ for TM$_{0m}$ modes. In other words no TE$_{0m}$ modes are excited when $\mathbf{e} = \mathbf{e}_x$ and no TM$_{0m}$ modes are excited when $\mathbf{e} = \mathbf{e}_y$. Numerical results for $\overline{P}_p$ as a function of the angle of incidence ϕ for several lower-order modes are shown in Fig. 3.15. Some interesting results may be summarized as follows:

(1) At normal incidence only the HE$_{1m}$ modes are excited. Almost 80% of the incident power is transmitted by the HE$_{11}$ mode and 15% by the HE$_{12}$ mode if only the core area is illuminated and if the frequency is far from cutoff.

(2) It is almost impossible to excite only one mode without exciting other modes using this end-on technique.

3.8.1.2 Incident Gaussian Beam

The electric field of a focused Gaussian laser beam normally incident upon the end of a dielectric rod takes the form (at $z = 0$) [36]

$$\mathbf{E}^{\mathrm{inc}} = E_0 \exp\left[-\left(\frac{r}{w} \right)^2 \right] \mathbf{e}_x, \tag{3.59}$$

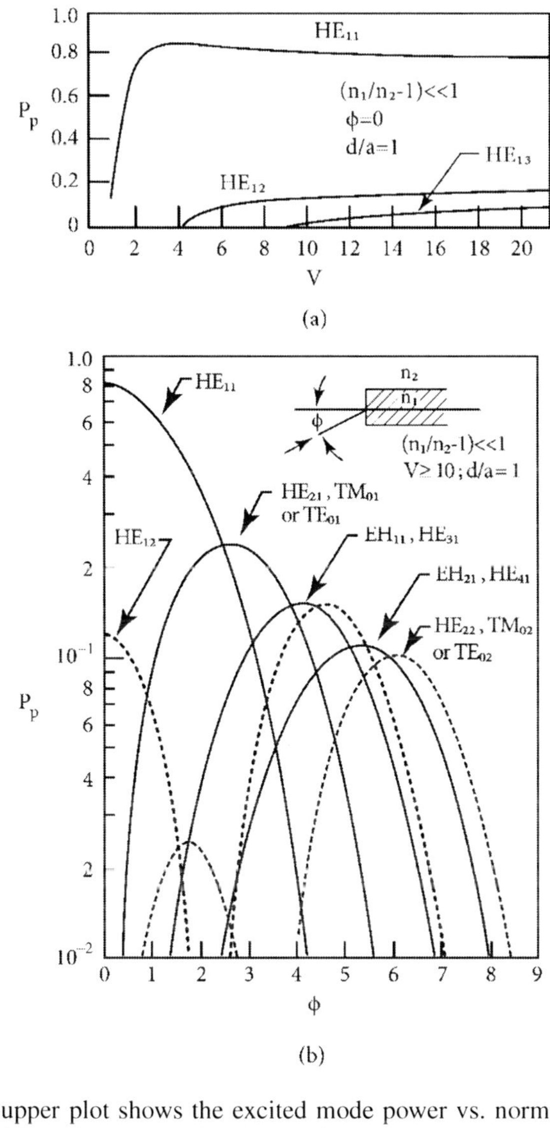

Figure 3.15. The upper plot shows the excited mode power vs. normalized frequency V for a normally incident plane wave. The lower plot shows the excited mode power vs. the incident angle, ϕ. Only the core area is illuminated and the incident power is normalized to unity [35]

where w is a constant and E_0 is the normalized amplitude constant given as

$$E_0 = \left[\frac{4\sqrt{\mu_0/\epsilon_0}}{\pi w^2} \right]^{1/2} \tag{3.60}$$

for unit incident power. Substituting (3.52) and (3.59) in (3.50) gives

$$P_{\mathrm{HE}_{11}} = 2\left\{\frac{qa}{V}\frac{J_0(pa)}{J_1(pa)}\frac{2a}{w}\left[\int_0^1 \frac{J_0(pa\chi)}{J_0(pa)}\exp\left(-\chi^2a^2/w^2\right)\chi\,d\chi\right.\right.$$
$$\left.\left. + \int_1^\infty \frac{K_0(qa\chi)}{K_0(qa)}\exp\left(-\chi^2a^2/w^2\right)\chi\,d\chi\right]\right\}^2 . \qquad (3.61)$$

Numerical computation of (3.61) shows that the excitation of the HE_{11} mode by a normally incident Gaussian beam is extremely efficient (see Fig. 3.16). For $V \geq 2$, more than 90% of the incident power may be coupled to the HE_{11} mode. It has also been shown by Marcuse [36] that slight offset of the incident Gaussian beam from the axis of the dielectric rod would not affect greatly the excitation efficiency of the HE_{11} mode, and the direction in which the beam is offset with respect to the polarization of the input field was found to be unimportant. He also found that by tilting the incident Gaussian beam with respect to the rod axis, more higher-order modes are generated and that tilts of input field are more serious, as far as the excitation of HE_{11} modes is concerned, for small values of the ratio of core radius to wavelength.

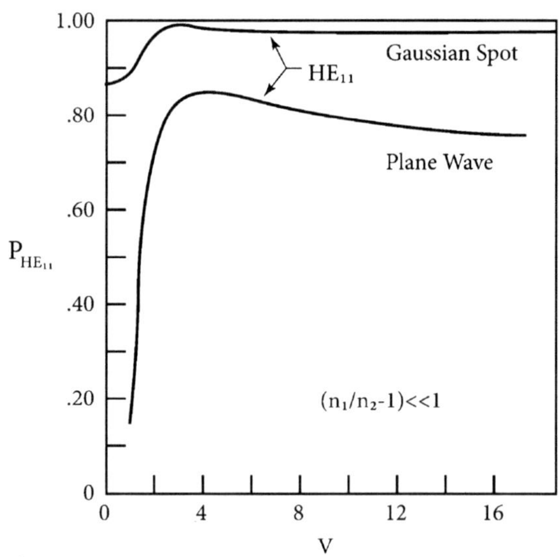

Figure 3.16. Excited HE_{11} mode power vs. normalized frequency V for normally incident plane wave and Gaussian beam. The spot size is assumed to fill the core region. The incident power is normalized to unity [36]

It has also been shown that when a semi-infinite dielectric rod is illuminated by an incoherent source, as $V \to \infty$ all possible modes are excited with approximately the same power [37]. When the index of refraction of the dielectric rod is only slightly greater than that of its cladding, the trapped modes (the guided modes) account for half of the total incident power.

Guided modes of a dielectric rod can also be excited by other less convenient or less efficient means. For example, launching of guided modes on a dielectric rod may be done with a prism coupler or a grating coupler, although these devices are best suited on thin film waveguides or optical circuits channel waveguides [38]. Cross-talk between neighboring rods as well as scattering by obstacles may also cause the excitation of guided modes.

3.8.2 Excitation Through Efficient Transitions [39]

Three conditions must be satisfied to achieve efficient coupling of power into and out of a dielectric waveguide through the use of a transition/coupler. (1) Impedance matching: The wave impedance of an incident wave (mode) must be as close as possible to that of the guided mode on the dielectric waveguide. Minimizing Fresnel loss belongs to this category. This means that transition between guides with highly dissimilar impedances must be very gradual. (2) Field matching: The transverse-field configuration (pattern) of the incident wave (mode) must be as close as possible to that of the guided mode on the dielectric waveguide. The presence of dissimilar transverse field configurations will induce radiated waves or higher order modes. (3) Phase velocity matching: The phase velocity of the incident wave (mode) must be as close as possible to that of the guided mode on the dielectric waveguide. This means that the transition between guides with very different phase velocities must be very gentle. It is perhaps worthwhile to mention that there exists another way of transferring power from one waveguide to another by the coupled-mode approach [40]. According to the coupled-mode perturbation theory, power can be transferred from one guide to another if the modes on these guides possess a similar phase velocity and if the proximity of the guides does not significantly affect the mode pattern of the guided wave on each guide. This concept has been successfully used to design traveling wave tubes, optical fiber couplers, and integrated optical planar waveguide couplers. The same concept can certainly be used here to design waveguide couplers.

In the following we shall discuss the specific example of the excitation of the dominant mode on a dielectric waveguide:

1. Transition between a rectangular metallic waveguide and a low dielectric constant rod.

A convenient way of transferring power from a metallic waveguide to a low dielectric constant (e.g., polymer) rod is by the use of a transition horn [41]. The cross-section of the polymer waveguide is normally larger than that of the metallic waveguide. To provide a good impedance match and a good field pattern match, the polymer is inserted into the metallic waveguide, fully filling its cross-section, and the inserted end of the polymer guide is gently tapered to a point. Most of the guided power is already contained within the polymer at the mouth of the waveguide. Mismatching of the transverse field for the polymer-filled metallic waveguide and that of a the polymer rod will excite a radiated wave, which will detract from the guided power. To remedy this situation, a transition horn is placed at the end of a polymer-filled metallic waveguide (Fig. 3.17). In this way a very gentle transition of the field inside the polymer-filled metallic waveguide to that of the open structure polymer rod occurs. The guided power inside a polymer-filled waveguide is fully launched onto the polymer rod, generating very little lost radiated power. A coupling efficiency of as high as 98% has been achieved for this transition [41].

2. Transition from a metallic waveguide to a high dielectric constant rod.

There are two salient points: (1) The transverse electric field pattern for the high dielectric constant rod has a significant dip in the core relative to the field outside, while that for the rectangular metallic waveguide does not. (2) Because of the severe discontinuity of the transverse electric field at the surface of a high dielectric constant rod, tapering the thickness of the rod can cause significant perturbation to the guided wave, thus producing a large radiation loss. This means that the conventional way of making the transition for a rectangular metallic waveguide to a polymer rod cannot be used for the high dielectric constant rod. Here, to minimize the impedance mismatch, the high dielectric constant rod should not be inserted into the rectangular metallic waveguide. Instead, it will be placed partially into the mouth of the horn and only its width is tapered in the transition region (see Fig. 3.18). Note that tapering the width of the rod would not cause a large disturbance to the transverse electric field due to the tangential field continuity condition.

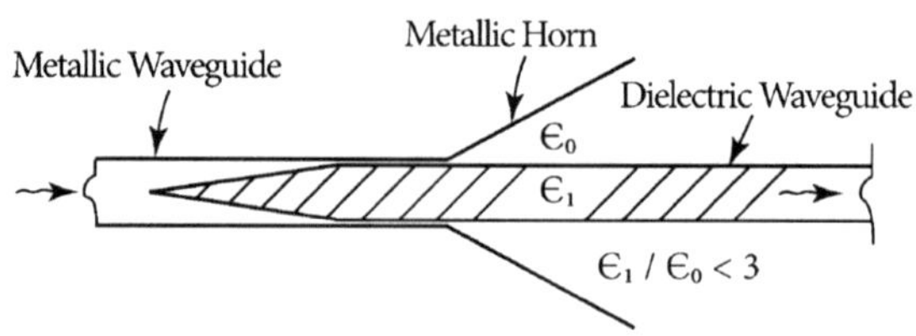

Figure 3.17. Rectangular metal waveguide to low dielectric constant waveguide transition

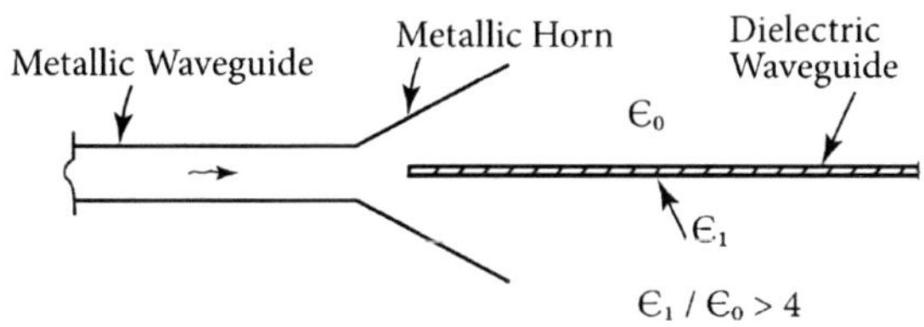

Figure 3.18. Rectangular metal waveguide to high dielectric constant waveguide transition

Let us consider the special case of a small high dielectric constant waveguide supporting the dominant mode whose fields reside mostly outside the core region of the guide, and whose transverse fields are polarized mostly in the same direction as the incident field from the horn. Despite the field mismatch in the region inside the high dielectric constant guide, strong coupling of the electric field on the surface of the high dielectric constant guide with the relatively uniform transverse electric field of the horn and the good matching of the phase velocities for these fields can provide an excellent condition for the efficient launching of the desired guided wave on the high dielectric constant guide. Using this transition method, a measured value of 0.35 dB mismatch loss was attained at $K_{\rm a}$ band [42].

Finally, we wish to add that the most elegant way of treating the excitation problem is through the use of Green's functions. That mathematical approach has been treated in many textbooks [19]. We have chosen to provide a more physical and intuitive approach for the excitation problem.

3.9 Coupled Mode Theory

A dielectric waveguide of a given cross-section will support a finite number of guided modes and a continuous spectrum of radiation modes. Any irregularity in a circular waveguide due to diameter variation, bending, interface irregularities, or the presence of scattering centers will produce coupling of one mode to the others. Mode coupling within a dielectric waveguide may produce the desirable effect of reducing the delay distortion that results from uncoupled multimode operation or may produce the undesirable effect of signal contamination. Coupling between guided modes and the continuum radiation modes is usually not desirable unless the guide is intended to serve as a radiator. Mode coupling may also occur between parallel guides in a bundle of dielectric waveguides. This kind of coupling may be used to advantage by constructing devices such as directional couplers. Exact treatment of the mode coupling is difficult due to the complexity of the problem. If, however, the disturbance or perturbation is small, the coupled mode theory approach may be used [40, 43]. This is accomplished by expanding the fields of the complicated (perturbed) system in terms of a complete set of known modes for a simpler system. These known modes do not individually satisfy the boundary conditions for the complicated system and hence couple to each other. The modal coefficients are found by solving a set of coupled first-order differential equations with a given set of initial conditions [40, 43].

The electromagnetic fields of the pth mode for the simpler system are assumed to be known:

$$\mathbf{E}_p(x, y, z) = \mathbf{E}_p(x, y)\, e^{-j\beta_p z}, \tag{3.62}$$

$$\mathbf{H}_p(x, y, z) = \mathbf{H}_p(x, y)\, e^{-j\beta_p z}, \tag{3.63}$$

where β_p is the pth modal propagation constant. The positive (negative) values of p represent modes propagating in the positive (negative) z-direction, so that $\beta_{-p} = -\beta_p$, $\mathbf{E}_{-p}(x, y) = \mathbf{E}_p(x, y)$, $\mathbf{H}_{-p}(x, y) = -\mathbf{H}_p(x, y)$. The modal functions satisfy the following orthogonality relation:

$$\int_{A_\infty} \mathbf{e}_z \cdot \left[\mathbf{E}_p(x, y) \times \mathbf{H}_q^*(x, y)\right]\, dA = \pm\delta_{pq}, \tag{3.64}$$

where A_∞ represents integration over the infinite cross section, the negative sign is for negative p and q, and the asterisk indicates complex conjugation. The symbol δ_{pq} is the Kronecker delta function for discrete values of p and q; it is zero if one of the indices labels a guided mode while the other labels a radiated mode, and it becomes a Dirac delta function if both indices label radiation modes.

The completeness of the known transverse modal fields of the uniform system $[\mathbf{E}_{\mathrm{t}p}(x, y), \mathbf{H}_{\mathrm{t}p}(x, y)]$ enables one to expand the transverse fields of the perturbed system $[\mathbf{E}_t'(x, y, z), \mathbf{H}_t'(x, y, z)]$ as follows:

$$\mathbf{E}_{\mathrm{t}}' = \sum_p a_p(z)\mathbf{E}_{\mathrm{t}p}(x, y) \tag{3.65}$$

$$\mathbf{H}_{\mathrm{t}}' = \sum_p a_p(z)\mathbf{H}_{\mathrm{t}p}(x, y) \tag{3.66}$$

where the subscript t represents the transverse vector, the summation is understood to extend to an integral for the continuous modes, and $a_p(z)$ are the unknown amplitude coefficients, which are to be determined. We shall assume that the radiated fields are negligibly small. Here we shall provide an example of coupling between parallel uniform dielectric waveguides. Given i identical parallel dielectric waveguides, it is known from coupled mode theory that power transfer occurs only between the mode of one waveguide with the other modes of the other guides when their propagation constants are similar, that is, $\beta_p^{(1)} \simeq \beta_p^{(2)} \simeq \beta_p^{(3)} \simeq ....\beta_p^{(i)}$, where $\beta_p^{(i)}$ is the propagation constant of the pth mode of the ith waveguide. Only forward modes of one waveguide couple to forward modes of the others. Hence, the coupled mode equations are

$$\frac{da_p^{(i)}}{dz} = -j\beta_p^{(i)} a_p^{(i)} + j\sum_{s \neq i} a_p^{(s)} c_{pp}^{(i)(s)}, \tag{3.67}$$

$$c_{pp}^{(i)(s)} = \frac{\omega}{2} \int_{A^{(s)}} \left[\epsilon^{(s)} - \epsilon\right] \left[\mathbf{E}_p^{(i)}(x,y) \cdot \mathbf{E}_p^{(s)*}(x,y)\right] dA, \tag{3.68}$$

where $a_p^{(i)}(z)$ is the amplitude coefficient of the pth mode in the ith guide, $c_{pp}^{(i)(s)}$ is the coupling coefficient between the s and i guides for the pth mode, $\epsilon^{(s)} - \epsilon$ is the difference between the dielectric constant of guide s and its surrounding medium.

As a specific example, let us consider the case of the coupling between two parallel identical dielectric waveguides with i=1,2. The solution of (3.67) is

$$a_p^{(1)}(z) = \left[a_p^{(1)}(0)\cos(\Delta\beta z) + ja_p^{(2)}(0)\sin(\Delta\beta z)\right] e^{-j\beta z}, \tag{3.69}$$

$$a_p^{(2)}(z) = \left[a_p^{(2)}(0)\cos(\Delta\beta z) + ja_p^{(1)}(0)\sin(\Delta\beta z)\right] e^{-j\beta z}, \tag{3.70}$$

where $\beta_p^{(1)} \simeq \beta_p^{(2)} \simeq \beta$, $\Delta\beta = c_{pp}^{(1)(2)} = c_{pp}^{(2)(1)}$, and $a_p^{(1)}(0)$ and $a_p^{(2)}(0)$ are the given initial conditions. If the coupling coefficient $c_{pp}^{(1)(2)}$ is real and if we assume $a_p^{(2)}(0) = 0$, the above solution shows that, after a sufficiently large distance, the periodic exchange of power disappears and the power in both guides equalizes. Figure 3.19 gives the HE_{11} mode coupling coefficient for two identical circular dielectric waveguides as a function of the distance between their center for various V. The results are calculated according to (3.68).

3.10 Bends and Corners for Dielectric Waveguides

Since the dielectric waveguide is an open structure, a portion of the guided wave may escape as a radiated wave when the waveguide deviates from a straight path [44]. If most of the guided power is contained within the core region of the guide, it is expected that most of the guided power will remain trapped within the core region when a gentle bend occurs.

To illustrate the curvature effect on the transmission of guided power in an optical fiber, let us introduce the results of an experiment [45]. A typical glass fiber was tightly wound ten times around a post with a predetermined radius of

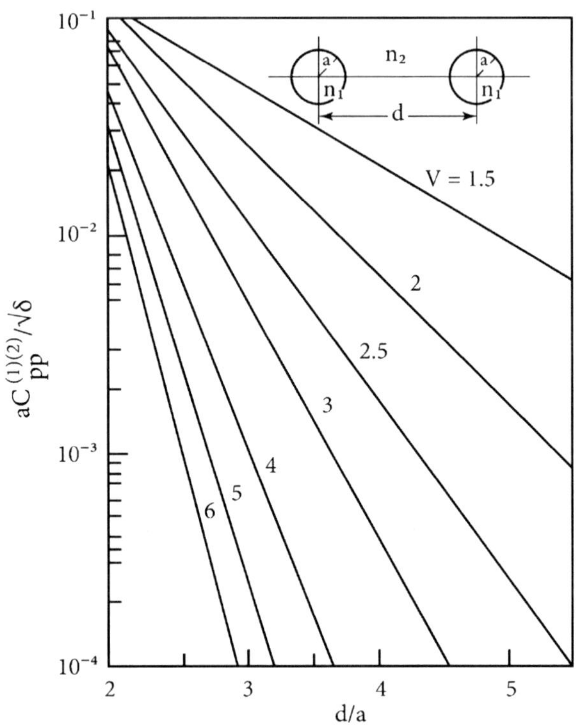

Figure 3.19. Coupling coefficient for HE_{11} mode for two identical fibers as a function of the distance between their centers for several values of normalized frequency, V. $\delta = 1 - (n_2/n_1)^2$ [43]

curvature. By using posts of different radii, one can adjust the bending radius of the fiber. Results of the measurements are shown in Fig. 3.20. It can be seen that leakage from the fiber is very significant when the radius is less than a certain value. This also means that gentle curvature can be tolerated by the fiber transmission line in order to retain most of the guided power.

Let us now discuss another more practical case of how to minimize the radiation loss on a dielectric ribbon waveguide made with a high dielectric constant core material [39]. Since the highest concentration of guided field resides just outside the surface of the high dielectric constant waveguide, the ribbon waveguide is a true surface waveguide. As such, the external field can easily "slide" off the guiding structure and transform into the radiated field when the ribbon guide bends.

The loss is clearly seen in Fig. 3.21 (left picture), where the behavior of the dominant electric field is plotted as the ribbon guide turns a $90°$ corner with a radius of curvature of $4\lambda_0$. Almost all the guided field is radiated. It is known that for the strongly guided case where most of the guided surface wave power is

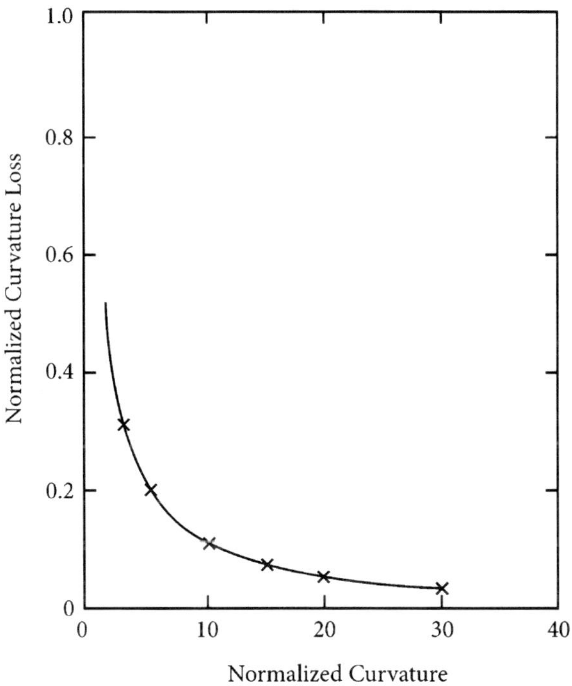

Figure 3.20. Measured normalized curvature loss vs. normalized curvature. Crosses are measured values [45]

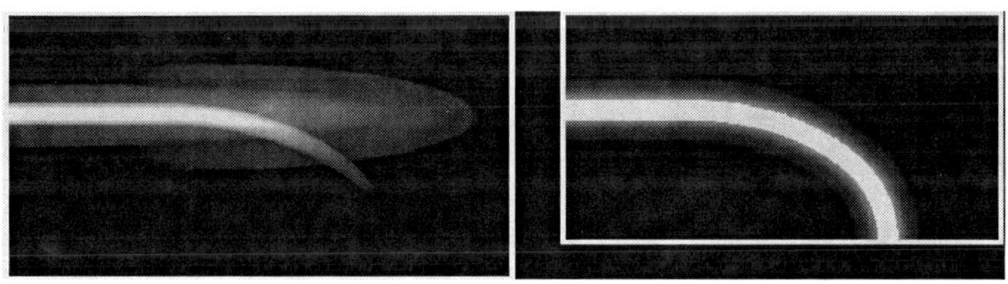

Figure 3.21. A 90° alumina ribbon bend. The ribbon is $0.6\lambda_0$ wide with a $1\lambda_0$ input and output straight section and a $4\lambda_0$ inside radius bend. The left view shows the envelope of the transverse E field for the TM-like mode just outside the high dielectric constant ribbon, as it propagates outside the bend. The lighter shades indicate regions of higher field intensity Almost all the initial guided wave is lost to radiation. The right view shows the envelope of the transverse E field for the TM-like mode just inside the Teflon–air boundary, for the high dielectric constant ribbon bend shown in Fig. 3.20a with a Teflon coat ($0.26\lambda_0$ thick) on either side of the alumina. Almost all the guided wave is transmitted through the bend. The radiation loss is less than 0.4 dB [39]

confined within the core region of the surface waveguide, very little power will be radiated when the guide curves with a radius of curvature >3–4 wavelengths. So, to develop a low-loss bend, it is necessary first to transform the weakly guided low-loss high dielectric constant ribbon to a strongly guided polymer-coated high dielectric constant ribbon. In other words, the low-loss bends can be made with a polymer-coated high dielectric constant ribbon. Computer simulation of the guided field around a 90° bend having a radius of curvature of $4\lambda_0$ on a polymer coated high dielectric constant ribbon was carried out.

This is displayed in Fig. 3.21 (right picture). It is seen that since most of the guided field is already confined within the polymer-coated guide prior to turning the corner, very little guided field is transformed into a radiated field when the polymer-coated guide turns a corner. Computer simulation shows that the total 90° cornering loss was less than 0.4 dB.

3.11 Systems and Noise

Information carrying capability of a dielectric (fiber) waveguide system is limited by the dispersion (dispersion-limited) and attenuation (loss-limited) characteristics of the power carrying propagating mode(s) [26]. The intrinsic noise power in the detector/receiver [46] ultimately sets a lower limit to the useful signal power transported by the waveguide, while the signal flattening dispersive behavior of the guided mode waveguide limits the bandwidth of the transmitted signal. The maximum length of a waveguide link, between repeaters, is either dispersion limited or loss limited. The waveguide system is characterized by the data rate (bandwidth) and the required bit error rate (BER) or signal-to-noise ratio (SNR). The relation between BER and SNR is shown in Fig. 3.22.

For example, for an optical signal, with a data rate of 10^9 bits/s with a BER of 10^{-9}, the SNR at the receiver must be approximately 36. For an ideal receiver, which is shot noise limited [46], we have

$$\mathrm{SNR} = \frac{m^2 R_0 P_\mathrm{r}}{4qB} \tag{3.71}$$

where m = modulation index (assumed to be 1), $R_0 = 0.5\,\mathrm{A\,W^{-1}}$for many materials, P_r = received power, q = electronic charge = 1.6×10^{-19} C, and B is the bandwidth.

If $B = 1$ GHz, SNR $= 36$, then $P_\mathrm{r} = 46$ nW (-43 dB m). If the transmitting source is a laser providing an output power of $P_0 = 1$ mW (0 dB m) the operating budget for this fiber is 43 dB. This means that a total of 43 dB loss can be spent, including source–fiber and fiber–detector interface losses,[1] and the attenuation and

[1] There are basically three types of interface losses: (a) Fresnel reflective losses due to mismatch

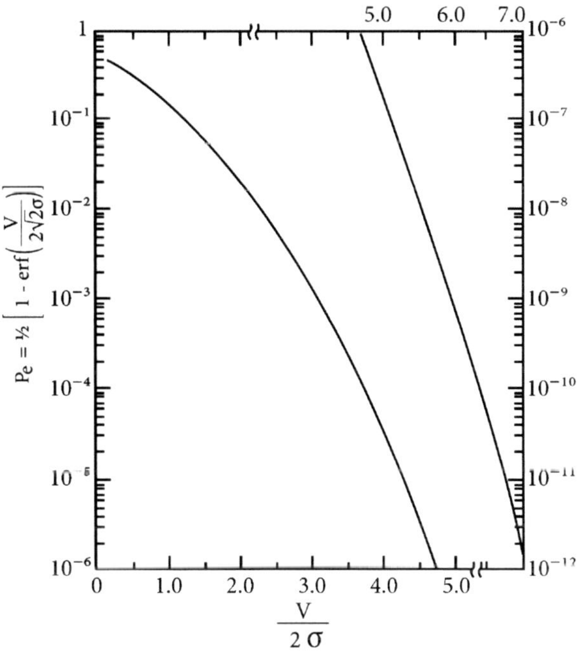

Figure 3.22. Bit error rate as a function of signal-to-noise ratio. Signal-to-noise ratio (SNR) $= (S/N)_{\mathrm{dB}} = 20\log(V/\sigma)$. P_e is the bit error rate (BER)

other losses for the fiber between repeater stations. Given the total interface loss of 6 dB and the total fiber loss of 0.2 dB km^{-1}, the distance between repeater stations can be as long as 187 km. These are theoretical values. The practical values are somewhat less.

In general, there are two types of photodetectors: the pin photodetector and the avalanche photodetector. The pin photodetector is commonly used when the incoming optical signal is relatively large. For this case, the quantum noise dominates the circuit noise. When the incoming optical signal is weak, the avalanche photodetector is commonly used. For this case, the gain of the avalanche photodetector can be used to increase the SNR. (There exists an optimum gain at which the quantum noise is comparable to circuit noise and at which SNR reaches a maximum. Beyond this gain the SNR decreases). The performance (SNR) of pin and avalanche photodetectors as a function of received optical power for bandwidths of 5 and 25 MHz is shown in Fig. 3.23.

of indices of refraction at the interface (roughly amounts to about 0.5 dB loss per interface). (b) Area mismatch losses. Signal is lost if the area of an incident beam is larger than the core area of the dielectric waveguide. (c) Angular mismatch losses. The signal is incident at an angle greater than the acceptance angle of the dielectric waveguide (~8 dB loss).

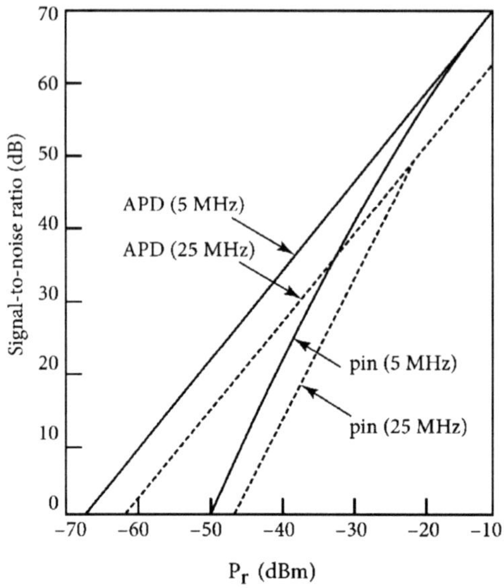

Figure 3.23. The signal-to-noise ratio for pin and avalanche photodiodes as a function of received optical power for bandwidths of 5 and 25 MHz [46]

Let us now consider the minimum optical power that must be received in order to obtain the required BER for a given data rate. Shown in Fig. 3.24 is a plot of minimum received power (P_r) as a function of data rate for a given BER for several types of photodetectors.

For example, for a 10^{-11} BER using the InGaAs APD and operating at 1,550 nm, at a data rate of 50 Mb s^{-1}, $P_r = -52$ dB m (6.3 nW) and at a data rate of 1 Gb s^{-1}, $P_r = -38$ dB m (150 nW). It is also seen that the required minimum received power is considerably higher for pin diodes.

A typical fiber link is shown in Fig. 3.25. It is seen that signal processing (modulation–demodulation, switching, multiplexing, amplification, signal reconstruction, etc.) are all done electronically using, perhaps, ICs, or LSI circuits. More sophisticated future generation systems must await the successful developments in the integrated optics area.

There exist several multiplexing schemes that may be used to take full advantage of the large bandwidth capability of the optical fiber as well as its small physical dimensions:

1. Wavelength Division Multiplexing

The outputs of several lasers emitting at different wavelengths are optically combined to form a single beam and transmitted over the same optical fiber. Frequency selective switches must be used to select the wanted signal from the output beam.

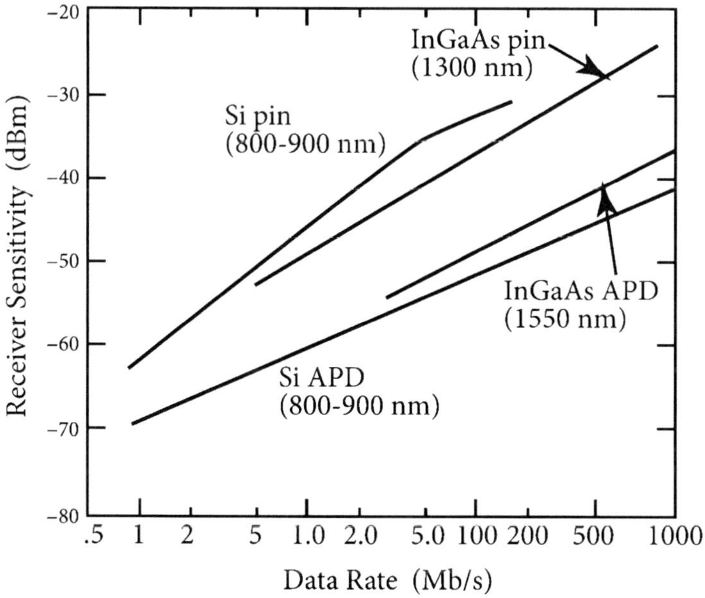

Figure 3.24. Receiver sensistivities as a function of bit rate. The Si pin, Si APD, and InGaAs pin curves are for a 10^{-9} BER. The InGaAs APD curve is for a 10^{-11} BER [46]

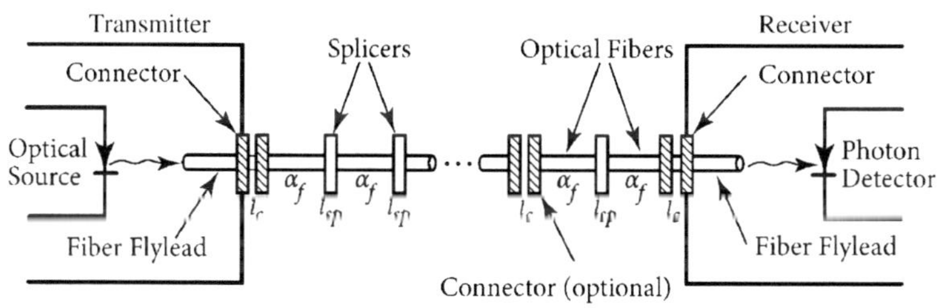

Figure 3.25. Typical point-to-point fiber link. The losses occur at connectors (l_c), at splices (l_{sp}), and in the fiber (α_f) [46]

2. Time Division Multiplexing

A time-multiplexing system would provide a common single-mode optical fiber line interconnecting a large number of terminals. The source would be a single high-power, mode-locked, cw laser. Each terminal would have its own "time window" within the cycle of the system. The receiving function (beam-splitting and detection) and transmitting function (modulation) would be accomplished by optical circuitry through which the carrier would pass at each terminal. Each terminal would have the capability of transmitting or receiving during its own or any other time window.

3. Space Division Multiplexing

A group of single-mode fibers may be bundled together to form a single transmission line. Each parallel channel is associated with a single fiber with its own source, modulator, detector, and demodulator.

References

1. R. E. Collin, "Field Theory of Guided Waves," McGraw-Hill , New York (1960)
2. S. A. Maier, "Plasmonics: Fundamentals and Applications," Springer, Berlin Heidelberg New York (2007)
3. L. Brillouin, "Wave Propagation in Periodic Structures," 2nd edn., Dover, New York (1953)
4. T. Tamir, H. C. Wang, and A. A. Oliner, "Wave propagation in sinusoidally stratified dielectric media (TE Waves)", IEEE Trans. Microw. Theory Tech. **MTT-12**, 323 (1964)
5. C. Yeh, K. F. Casey, and Z. A. Kaprielian, "Transverse magnetic wave propagation in sinusoidally stratified dielectric media," IEEE Trans. Microw. Theory Tech. **MTT-13**, 297 (1965)
6. C. Elachi and C. Yeh, "Periodic structures in integrated optics," J. Appl. Phys. **44**, 3146 (1973)
7. S. John, "Strong localization of photons in certain disordered dielectric superlattices," Phys. Rev. Lett. **58**, 2486 (1987)
8. E. Yablonovitch, "Inhibited spontaneous emission in solid-state physics and electronics," Phys. Rev. Lett. **58**, 2059 (1987)
9. J. Joannopoulos, R. Meade, and J. Winn, "Photonic Crystals," Princeton Press, New Jersey (1995)
10. G. Goubau, "Surface waves and their application to transmission lines," J. Appl. Phys. **21**, 119 (1950)
11. O. N. Singh and A. Lakhtakia, eds., "Electromagnetic Fields in Unconventional Materials and Structures," Wiley, New York (2000)
12. C. Yeh, "Propagation along moving dielectric waveguides," J. Opt. Soc. Am. **58**, 767 (1968)
13. B. Lax and K. J. Button, "Microwave Ferrites and Ferrimagnetics," McGraw-Hill, New York (1962)
14. F. E. Borgnis and C. H. Papas, "Handbuch der Physik," Vol. **16**, Springer, Berlin Heidelberg New york (1958)

15. D. Gloge, "Weakly guiding fibers," Appl. Opt. **10**, 2252 (1971); D. Gloge, "Dispersion in weakly guiding fibers," Appl. Opt. **10**, 2442 (1971)

16. D. Gloge, "Optical waveguide transmission," Proc. IEEE **58**, 1513 (1970)

17. S. Kawakami and J. Nishizawa, "An optical waveguide with optimum distribution of the refractive index with reference to waveform distortion," IEEE Trans. Microw. Theory Tech. **MTT-16**, 814 (1968)

18. S. Ramo, J. R. Whinnery, and T. Van Duzer, "Fields and Waves in Communication Electronics," Wiley, New York (1967)

19. A. Ishimaru, "Electromagnetic Wave Propagation, Radiation, and Scattering," Prentice Hall, New York (1991); R. F. Harrington, "Time-Harmonic Electromagnetic Field," McGraw-Hill, New York (1961); R. E. Collin, "Field Theory of Guided Waves," McGraw-Hill, New York (1960)

20. C. Yeh, "Dynamic Fields," Am. Inst. of Phys. Handbook, 3rd edn., D. E. Gray, ed., McGraw-Hill, New York (1972)

21. W. M. Elsasser, "Attenuation in a dielectric circular rod", J. Appl. Phys. **20**, 1193 (1949); C. H. Chandler, "An investigation of dielectric rod as waveguides," J. Appl. Phys. **20**, 1188 (1949)

22. C. Yeh, Private Communication, JPL (1992)

23. W. A. Imbriale, T. Y. Otoshi, and C. Yeh, "Power loss for multimode waveguides and its application to beam-waveguide systems," IEEE Trans. Microw. Theory Tech. **MTT-46**, 523 (1998)

24. F. P. Kapron and D. B. Keck, "Pulse transmission through a dielectric optical waveguide," Appl. Opt. **10**, 1519 (1971)

25. L. Smith and E. Snitzer, "Dispersion-minimization in dielectric waveguide," Appl. Opt. **12**, 1592 (1973)

26. G. P. Agrawal, "Fibre-Optic Communication Systems," Wiley, New York (2002)

27. H. F. Taylor, "Transfer of information on naval vessels via fiber optics transmission lines," NELC-TR-1763, Naval Electronics Laboratory Center, San Diego, CA (1971)

28. W. A. Gambling and H. Matsumura, "Pulse dispersion in a lenslike medium," Opto-Electronics **5**, 429 (1973)

29. A. Hasegawa and Y. Kodama, "Solitons in Optical Communications," Clarendon Press, Oxford (1995)

30. C. F. Davidson and J. C. Simmonds, "Cylindrical cavity resonator," Wireless Eng. **21**, 420 (1944)

31. H. M. Barlow and A. L. Cullen, "Microwave Measurements," Constable, London (1950)

32. S. A. Schelkunoff, "The impedance concept and its application to problems of reflection, refraction, shielding and power absorption," Bell Syst. Tech. J. **17**, 17 (1938)

33. C. Yeh, "A relation between $\alpha$and Q," Proc. IRE **50**, 2145 (1962); F. I. Shimabukuro and C. Yeh, "Attenuation measurement of very low loss dielectric waveguides by the cavity resonator method applicable in the millimeter/submillimeter wavelength range," IEEE Trans. Microw. Theory Tech. **MTT-36**, 1160 (1988)
34. N. S. Kapany, J. J. Burke, and T. Sawatari, "A technique for launching an arbitrary mode on an optical dielectric waveguide," J. Opt. Soc. Am. **60**, 1178 (1970)
35. A. W. Snyder, C. Pask, and D. J. Mitchell, "Light acceptance property of an optical fiber," J. Opt. Soc. Am. **63**, 59 (1973); A. W. Snyder, "Asymptotic expression for eigenfunctions and eigenvalues of a dielectric or optical waveguide," IEEE Trans. Microw. Theory Tech. **MTT-17**, 1130 (1969)
36. J. R. Stern, M. Peace, and R. B. Dyott, "Launching into optical-fibre waveguide," Electron. Lett. **6**, 160–162 (1970); D. Marcuse, "Excitation of the dominant mode of a round fiber by a Gaussian beam," Bell Syst. Tech. J. **49**, 1695 (1970)
37. A. W. Snyder and C. Pask, "Incoherent illumination of an optical fiber," J. Opt. Soc. Am. **63**, 806 (1973)
38. D. Marcuse, "Integrated Optics," IEEE Press, New York (1973)
39. C. Yeh, F. Shimabukuro, and P. H. Siegel,"Low-loss terahertz ribbon waveguides," Appl. Opt. **44**, 5937(2005)
40. A. W. Snyder, "Coupled-mode theory for optical fibers," J. Opt. Soc. Am. **62**, 1267 (1972)
41. W. Schlosser and H. G. Unger, " Advances in Microwaves," L. Young ed., Academic Press, New York (1966)
42. C. Yeh, F. Shimabukuro, P. Stanton, V, Jamnejad, W. Imbriale, and A. F. Manshadi, "Communication at millimetre-submillimtre wavelengths using ceramic ribbon," Nature **404**, 584 (2000)
43. D. Marcuse, "Coupled mode theory of round optical fibers," Bell Syst. Tech. J. **52**, 817 (1973)
44. E. G. Neumann and H. D. Rudolph, "Losses from corners in dielectric-rod or optical-fiber waveguides," Appl. Phys. **8**, 107 (1975); E. G. Neumann and H. D. Rudolph, "Radiation from bends in dielectric-rod transmission lines," IEEE Trans. Microw. Theory Tech. **MTT-23**, 142 (1975)
45. C. Yeh and A. Johnston, "How does one induce leakage in an optical fiber link," paper presented in AGARD Conference on Optical fibers, integrated optics, and their military applications, Jan 26– May 26, Preprint no: 219, Harford House, London (1977)
46. G. Keiser, "Optical fiber communications," 2nd edn., McGraw-Hill, New York (1991)

4

PLANAR DIELECTRIC WAVEGUIDES

Because of the simplicity of the planar geometry and the tractability of the analytic solutions associated with this geometry, the very first canonical solution for a given physical problem is usually obtained analytically for a planar structure. This also implies that solutions for the planar structure, which can yield great insight into the problems, are of fundamental importance [1, 2].

Here, we not only shall present the analysis of a canonical slab (planar) dielectric waveguide leading to the detailed theoretical solutions, but also provide detailed analyses on other planar structures – a leaky slab dielectric waveguide [3], a multilayered (or inhomogeneous) dielectric waveguide [4], and coupled planar dielectric waveguides [5]. We shall also discuss the very first, historically speaking, surface wave guiding structure – the Sommerfeld–Zenneck surface impedance guiding structure [6].

4.1 Fundamental Equations

From Sect. 2.5, using rectangular coordinates (x, y, z), one obtains

$$E_x = \frac{1}{p^2}\left(-j\beta\frac{\partial E_z}{\partial x} - j\omega\mu\frac{\partial H_z}{\partial y}\right), \tag{4.1}$$

$$E_y = \frac{1}{p^2}\left(-j\beta\frac{\partial E_z}{\partial y} + j\omega\mu\frac{\partial H_z}{\partial x}\right), \tag{4.2}$$

$$H_x = \frac{1}{p^2}\left(j\omega\epsilon\frac{\partial E_z}{\partial y} - j\beta\frac{\partial H_z}{\partial x}\right), \tag{4.3}$$

$$H_y = \frac{1}{p^2}\left(-j\omega\epsilon\frac{\partial E_z}{\partial x} - j\beta\frac{\partial H_z}{\partial y}\right), \tag{4.4}$$

with

$$\left[\frac{\partial^2}{\partial x^2} + \frac{\partial^2}{\partial y^2} + p^2\right]\left[\begin{array}{c} E_z \\ H_z \end{array}\right] = 0, \tag{4.5}$$

$$p^2 = \omega^2\mu\epsilon - \beta^2. \tag{4.6}$$

Here β is the propagation constant of the fields and ϵ and μ are the constitutive parameters of the medium in which fields reside. The factor $e^{-j\beta z+j\omega t}$ is attached to all field components and is suppressed. For the waves guided by any planar structures, the propagation constant β must be identical in all regions of the guiding structure. This is a necessary, but not sufficient condition, for the existence of a given guided mode. Other conditions, such as the boundary conditions, the radiation condition, and/or the edge conditions, must also be satisfied by the modal fields. As discussed in Chap. 2, the modal fields may be classified into four types:

$$\text{TEM } [(E_x, H_y) \text{ } or \text{ } (E_y, H_x)] \text{ with } E_z \text{ and } H_z = 0,$$

$$\text{TE } [(H_z, E_x, E_y, H_x, H_y)] \text{ with } E_z = 0,$$

$$\text{TM } [(E_z, E_x, E_y, H_x, H_y)] \text{ with } H_z = 0,$$

$$\text{HE } [(E_x, E_y, E_z, H_x, H_y, H_z)] \, .$$

4.2 Dielectric Slab Waveguide

One of the simplest planar structure is a dielectric slab structure shown in Fig. 4.1. The slab is taken to be infinite in extent in both the z and y directions. The constitutive parameters for the three regions are (ϵ_0, μ_0), (ϵ_1, μ_0), and (ϵ_2, μ_0) with

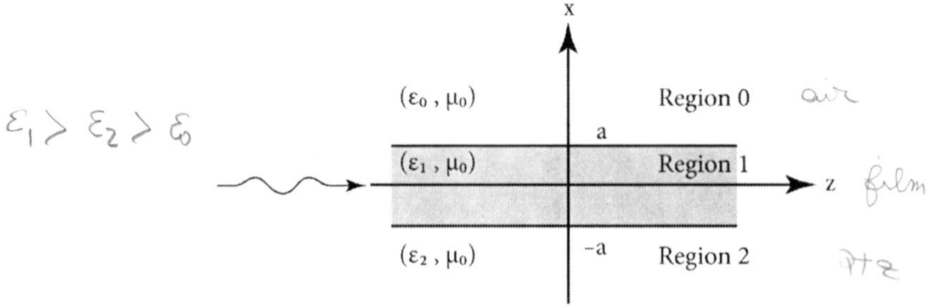

Figure 4.1. The geometry of a dielectric slab waveguide with thickness $2a$

$\epsilon_1 > \epsilon_2 > \epsilon_0$. The slab has a thickness $2a$. We are interested in wave propagation in the $+z$-direction, which can be supported by the structure. According to Sect. 2.5, this planar structure can support two types of modes: TM modes with $H_z = 0$ and TE modes with $E_z = 0$. The complete fields are obtained by a linear combination of these modes. It will be shown that each set of TM modes or each set of TE modes can satisfy all the necessary boundary conditions and the radiation condition.

Because of the symmetry of the planar structure, all field components are independent of the y-coordinate, that is, $\partial/\partial y = 0$.

4.2.1 The TM Surface Wave Modes

Let us consider the case of a TM wave propagating along the structure. The required TM fields in the three regions are, using (4.1)–(4.5),

Region 1 $(-a \leq x \leq a)$

$$E_z^{(1)} = A_1 \sin p_1 x + B_1 \cos p_1 x, \tag{4.7}$$

$$E_x^{(1)} = \frac{1}{p_1^2}(-j\beta)\frac{\partial E_z^{(1)}}{\partial x} = -\frac{j\beta}{p_1}\left[A_1 \cos p_1 x - B_1 \sin p_1 x\right], \tag{4.8}$$

$$H_y^{(1)} = \frac{1}{p_1^2}(-j\omega\epsilon_1)\frac{\partial E_z^{(1)}}{\partial x} = -\frac{j\omega\epsilon_1}{p_1}\left[A_1 \cos p_1 x - B_1 \sin p_1 x\right], \tag{4.9}$$

with

$$p_1^2 = \omega^2\mu\epsilon_1 - \beta^2. \tag{4.10}$$

Region 0 $(x \geq a)$

$$E_z^{(0)} = A_0\, e^{-q_0 x}, \tag{4.11}$$

$$E_x^{(1)} = -\frac{1}{q_0^2}(-j\beta)\frac{\partial E_z^{(0)}}{\partial x} = -\frac{j\beta}{q_0} A_0\, e^{-q_0 x}, \tag{4.12}$$

$$H_y^{(1)} = -\frac{1}{q_0^2}(-j\omega\epsilon_0)\frac{\partial E_z^{(0)}}{\partial x} = -\frac{j\omega\epsilon_0}{q_0} A_0\, e^{-q_0 x}, \tag{4.13}$$

with

$$q_0^2 = \beta^2 - \omega^2\mu\epsilon_0. \tag{4.14}$$

Region 2 $(x \leq -a)$

$$E_z^{(2)} = A_2\, e^{q_2 x}, \tag{4.15}$$

$$E_x^{(2)} = -\frac{1}{q_2^2}(-j\beta)\frac{\partial E_z^{(2)}}{\partial x} = \frac{j\beta}{q_2} A_2\, e^{q_2 x}, \tag{4.16}$$

$$H_y^{(2)} = -\frac{1}{q_2^2}(-j\omega\epsilon_2)\frac{\partial E_z^{(2)}}{\partial x} = \frac{j\omega\epsilon_2}{q_2} A_0\, e^{q_2 x}, \tag{4.17}$$

with

$$q_2^2 = \beta^2 - \omega^2 \mu \epsilon_2. \tag{4.18}$$

Here, A_1, B_1, A_0, and B_2 are arbitrary constants. Appropriate functions for the above fields are chosen so that they may satisfy the radiation condition at $x = \pm\infty$. In other words, all fields must vanish at $x = \pm\infty$.

Matching the tangential fields at $x = \pm a$ yields

$$A_1 \sin p_1 a + B_1 \cos p_1 a = A_0\, e^{-q_0 a}, \tag{4.19}$$

$$\frac{\epsilon_1}{p_1}\left[A_1 \cos p_1 a - B_1 \sin p_1 a\right] = \frac{\epsilon_0}{q_0} A_0\, e^{-q_0 a}, \tag{4.20}$$

$$-A_1 \sin p_1 a + B_1 \cos p_1 a = A_2\, e^{-q_2 a}, \tag{4.21}$$

$$-\frac{\epsilon_1}{p_1}\left[A_1 \cos p_1 a + B_1 \sin p_1 a\right] = \frac{\epsilon_2}{q_2} A_2\, e^{-q_0 a}, \tag{4.22}$$

or

$$\begin{bmatrix} \sin p_1 a & \cos p_1 a & -e^{-q_0 a} & 0 \\ s_1 \cos p_1 a & -s_1 \sin p_1 a & -e^{-q_0 a} & 0 \\ -\sin p_1 a & \cos p_1 a & 0 & -e^{-q_2 a} \\ -s_2 \cos p_1 a & -s_2 \sin p_1 a & 0 & -e^{-q_2 a} \end{bmatrix} \begin{bmatrix} A_1 \\ B_1 \\ A_0 \\ A_2 \end{bmatrix} = 0, \tag{4.23}$$

with

$$s_1 = \frac{q_0}{p_1}\frac{\epsilon_1}{\epsilon_0}, \tag{4.24}$$

$$s_2 = \frac{q_2}{p_1}\frac{\epsilon_1}{\epsilon_2}. \tag{4.25}$$

Setting the determinant of the above simultaneous linear equation to zero and simplifying yields the dispersion relation:

$$\begin{aligned}&(\sin p_1 a - s_1 \cos p_1 a)(\cos p_1 a + s_2 \sin p_1 a)\\ &\quad + (\sin p_1 a - s_2 \cos p_1 a)(\cos p_1 a + s_1 \sin p_1 a) = 0.\end{aligned} \tag{4.26}$$

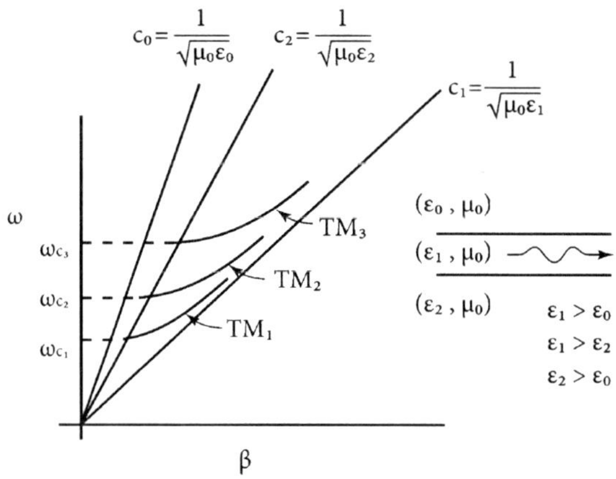

Figure 4.2. A sketch of the ω-β diagram for TM modes. These modes possess nonzero cutoff frequencies. The modes can only exist between the c_1 and c_2 lines. Outside these boundaries the modes are not allowed to exist

This equation is a function of βa, $k_0 a$, ϵ_1/ϵ_0, and ϵ_2/ϵ_0, or

$$F(\beta a, k_0 a, \epsilon_1/\epsilon_0, \epsilon_2/\epsilon_0) = 0, \tag{4.27}$$

with $k_0 = \omega\sqrt{\mu_0\epsilon_0}$. So, given $(k_0 a, \epsilon_1/\epsilon_0, \epsilon_2/\epsilon_0)$ we may solve for βa, the normalized propagation constant. Therefore, the ω-β diagram may be generated from the solutions to this equation. A sketch of the ω-β diagrams for various TM modes are shown in Fig. 4.2

Equation (4.23) shows that the arbitrary constants A_1, B_1, A_0, and B_2 are related to each other as follows:

$$\frac{A_1}{A_0} = f_1, \tag{4.28}$$

$$\frac{B_1}{A_0} = f_2, \tag{4.29}$$

$$\frac{A_2}{A_0} = f_3. \tag{4.30}$$

The constants f_1, f_2, and f_3 can be found from (4.23). In other words, the only unknown coefficient is A_0, which is determined by the excitation condition.

4.2.1.1 Cutoff Conditions for TM Modes

"Cutoff" refers to the cutoff frequency of a mode at which that mode ceases to exist. Another way to interpret the cutoff frequency for a guided surface wave mode is that at or below the cutoff frequency the structure can no longer support that mode. When a mode ceases to exist, $q_0 a$ or $q_2 a$ must be zero. For the case $\epsilon_2 > \epsilon_0$, then $q_0 a > q_2 a$, and so $q_2 a$ will approach zero before $q_0 a$. The cutoff condition for a given mode on this structure shown in Fig. 4.2 is

$$q_2 a = 0. \tag{4.31}$$

At this condition the fields in region 2 can no longer decay to zero at $x = -\infty$. To satisfy the radiation condition, that mode must not exist. Using (4.31), we have $s_2 = 0$. Equation (4.26) becomes, at cutoff,

$$(\sin p_1 a - s_1 \cos p_1 a) \cos p_1 a + \sin p_1 a (\cos p_1 a + s_1 \sin p_1 a) = 0 \quad \text{(at cutoff)}. \tag{4.32}$$

From (4.18), using (4.31), we find

$$\beta^2 a^2 - \omega^2 \mu_0 \epsilon_2 a^2 = q_2^2 a^2 = 0 \quad \text{(at cutoff)}.$$

So,

$$\beta a \text{ (at cutoff)} = \frac{\epsilon_2}{\epsilon_0} k_0 a \quad \text{(at cutoff)}. \tag{4.33}$$

Substituting (4.33) in (4.10) and (4.14), we get

$$q_0 a \text{ (at cutoff)} = \left(\frac{\epsilon_2}{\epsilon_0} - 1\right) k_0 a \quad \text{(at cutoff)}, \tag{4.34}$$

$$p_1 a \text{ (at cutoff)} = \left(\frac{\epsilon_1}{\epsilon_0} - \frac{\epsilon_2}{\epsilon_0}\right) k_0 a \quad \text{(at cutoff)}. \tag{4.35}$$

The cutoff frequency $k_0 a$ (at cutoff) can be found by solving (4.32) using (4.34) and (4.35). To learn if there is a zero cutoff frequency, let us set $p_1 a = 0$ in (4.32). We have

$$-s_1 = 0 \quad \text{(at cutoff)}. \tag{4.36}$$

Since $s_1 = \frac{q_0}{p_1}\frac{\epsilon_1}{\epsilon_0} = \frac{(\epsilon_2/\epsilon_0 - 1)}{(\epsilon_1/\epsilon_0 - \epsilon_2/\epsilon_0)}\frac{\epsilon_1}{\epsilon_0}$ (at cutoff), which is a nonzero positive number, (4.36) cannot be satisfied. This means that there is no zero cutoff frequency for any TM mode on this structure. All TM modes on this structure must possess nonzero cutoff frequencies, which can be found by solving (4.32).

4.2.1.2 Distribution of Guided Power

It is of interest to learn how the guided power is distributed within the guiding structure. According to Poynting vector theorem discussed in Chap. 2, the complex Poynting vector representing the power flow is given by

$$\mathbf{S} = \mathbf{E} \times \mathbf{H}^* \qquad (\mathrm{W\,m^{-2}}), \tag{4.37}$$

where $\mathbf{E}$ and $\mathbf{H}$ are the complex field vectors. The time average power flow through a surface area A is

$$\overline{P} = \frac{1}{2}\int_A (\mathbf{S}\cdot\mathbf{n})\; dA \qquad (\mathrm{W}). \tag{4.38}$$

The distribution of the guided power is given by

$$S_z = \mathbf{S}\cdot\mathbf{e}_z,$$

$$S_z = E_x^{(0)} H_y^{(0)*} = A_0^2 \frac{\omega\epsilon_0\beta}{q_0^2}\, e^{-2q_0 x} \qquad (x \geq a)\,, \tag{4.39}$$

$$S_z = E_x^{(1)} H_y^{(1)*} = A_0^2 \frac{\omega\epsilon_1\beta}{p_1^2}\, (f_1\cos p_1 x - f_2 \sin p_1 x)^2 \quad (-a \leq x \leq a)\,, \tag{4.40}$$

$$S_z = E_x^{(2)} H_y^{(2)*} = A_0^2 \frac{\omega\epsilon_2\beta}{q_2^2}\, e^{2q_2 x} \qquad (x \leq -a)\,, \tag{4.41}$$

where A_0 is the unknown coefficient. The time average power flow carried in each region is

$$\overline{P}^{(0)} = \frac{1}{2} A_0^2 \frac{\omega\epsilon_0\beta}{q_0^2} \int_a^\infty e^{-2q_0 x}\, dx \qquad \text{(region 0)}, \tag{4.42}$$

$$\overline{P}^{(1)} = \frac{1}{2} A_0^2 \frac{\omega\epsilon_1\beta}{p_1^2} \int_{a}^{a} (f_1\cos p_1 x - f_2 \sin p_1 x)^2\, dx \qquad \text{(region 1)}, \tag{4.43}$$

$$\overline{P}^{(2)} = \frac{1}{2} A_0^2 \frac{\omega\epsilon_2\beta}{q_2^2} \int_{-\infty}^{-a} f_3\, e^{2q_2 x}\, dx \qquad \text{(region 2)}. \tag{4.44}$$

The total time average power is

$$\overline{P} = \overline{P}^{(0)} + \overline{P}^{(1)} + \overline{P}^{(2)}. \tag{4.45}$$

4.2.1.3 Attenuation

If the material media are slightly lossy, that is, the dielectric constants are complex containing small imaginary parts, the guided modes will suffer attenuation. For single mode propagation, we may make use of the perturbation formula discussed in Chap. 3. It is

$$\alpha_{\mathrm{d}} = \frac{1}{2}\sigma_{\mathrm{d}} \frac{\displaystyle\int_V \mathbf{E}\cdot\mathbf{E}^*\,dV}{\displaystyle\int_A (\mathbf{E}\times\mathbf{H}^*)\cdot\mathbf{e}_z\,dA}, \tag{4.46}$$

where σ_{d} is the conductivity of the medium, $\sigma_{\mathrm{d}} = \omega\epsilon''$, $\epsilon = \epsilon' - j\epsilon''$, $\tan\delta = \epsilon''/\epsilon'$, and $\mathbf{E}$ and $\mathbf{H}$ are the guided fields for that single mode. Assuming that the dielectric loss only exists in region 1, that is, $\epsilon_1 = \epsilon_1' - j\epsilon_1''$ and $\sigma_{\mathrm{d}} = \sigma_{\mathrm{d}1}$, then

$$\begin{aligned}\int_V \mathbf{E}\cdot\mathbf{E}^*\,dV &= \int_{-a}^{a}\left(E_x^{(1)2} + E_z^{(1)2}\right)dx,\\ &= A_0^2\int_{-a}^{a}[(\beta^2/p_1^2)\times(f_1\cos p_1x - f_2\sin p_1x)^2\\ &\quad + (f_1\sin p_1x + f_2\cos p_1x)^2]\,dx,\\ &= A_0^2\,g_1,\end{aligned} \tag{4.47}$$

$$\begin{aligned}\int_A (\mathbf{E}\times\mathbf{H}^*)\cdot\mathbf{e}_z\,dA &= A_0^2\left\{\frac{\omega\epsilon_2\beta}{q_2^2}\int_a^{\infty} e^{-2q_2x}\,dx + \frac{\omega\epsilon_1\beta}{p_1^2}\right.\\ &\quad\left.\int_{-a}^{a}(f_1\cos p_1x - f_2\sin p_1x)^2\,dx + \frac{\omega\epsilon_2\beta}{q_2^2}\int_{-\infty}^{-a} e^{2q_2x}\,dx\right\},\\ &= A_0^2\,g_2.\end{aligned} \tag{4.48}$$

Substituting (4.47) and (4.48) into (4.46) gives

$$\alpha_{\mathrm{d}} = \frac{1}{2}\sigma_{\mathrm{d}1}\frac{g_1}{g_2}, \tag{4.49}$$

Here g_1 and g_2 have been defined in (4.47) and (4.48).

Let us investigate the limiting case of a large slab where most of the guided fields are contained within the slab region. In this case, the fields are mostly transverse to the direction of propagation, that is,

$$\mathbf{E} = E_x \mathbf{e}_x,$$
$$\mathbf{H} = H_y \mathbf{e}_y.$$

So, (4.47) becomes

$$\int_V \mathbf{E} \cdot \mathbf{E}^* \, dV = A_0^2 \frac{\beta^2}{p_1^2} \int_{-a}^{a} (f_1 \cos p_1 x - f_2 \sin p_1 x)^2 \, dx \tag{4.50}$$

and (4.48) becomes

$$\int_A (\mathbf{E} \times \mathbf{H}^*) \cdot \mathbf{e}_z \, dA = A_0^2 \frac{\omega \epsilon_1 \beta}{p_1^2} \int_{-a}^{a} (f_1 \cos p_1 x - f_2 \sin p_1 x)^2 \, dx. \tag{4.51}$$

Substituting (4.50) and (4.51) in (4.46) yields

$$\alpha_\mathrm{d} = \frac{1}{2} \sigma_{\mathrm{d}1} \frac{\left(A_0^2 \dfrac{\beta^2}{p_1^2} \right)}{\left(A_0^2 \dfrac{\omega \epsilon_1 \beta}{p_1^2} \right)} = \frac{1}{2} \sigma_{\mathrm{d}1} \frac{\beta}{\omega \epsilon_1} = \frac{1}{2} \sigma_{\mathrm{d}1} \tan \delta \, \omega \sqrt{\mu_0 \epsilon_1}. \tag{4.52}$$

We observe that the propagation constant β is $\sqrt{\mu_0 \epsilon_1}$ and the integrals in (4.50) and (4.51) cancel. As expected, (4.52) is the attenuation constant of a plane wave in a slightly lossy dielectric medium.

4.2.2 The TE Surface Wave Mode

The TE surface wave modes may be treated in a similar manner as the case for the TM modes. For the TE case, the nonzero field components are (H_z, E_x, E_y). The appropriate expressions are the following:

Region 1 ($a \leq x \leq a$)

$$H_z^{(1)} = C_1 \sin p_1 x + D_1 \cos p_1 x, \tag{4.53}$$

$$H_x^{(1)} = -\frac{j\beta}{p_1} (C_1 \cos p_1 x - D_1 \sin p_1 x), \tag{4.54}$$

$$E_y^{(1)} = \frac{j\omega\mu_0}{p_1} (C_1 \cos p_1 x - D_1 \sin p_1 x), \tag{4.55}$$

with

$$p_1^2 = \omega^2 \mu_0 \epsilon_1 - \beta^2. \tag{4.56}$$

Region 0 $(x \geq a)$

$$H_z^{(0)} = C_0\, e^{-q_0 x}, \tag{4.57}$$

$$H_x^{(0)} = -\frac{j\beta}{q_0} C_0\, e^{-q_0 x}, \tag{4.58}$$

$$E_y^{(0)} = \frac{j\omega\mu_0}{q_0} C_0\, e^{-q_0 x}, \tag{4.59}$$

with

$$q_0^2 = \beta^2 - \omega^2 \mu_0 \epsilon_0. \tag{4.60}$$

Region 2 $(x \leq -a)$

$$H_z^{(2)} = C_2\, e^{q_2 x}, \tag{4.61}$$

$$H_x^{(2)} = \frac{j\beta}{q_2} C_2\, e^{q_2 x}, \tag{4.62}$$

$$E_y^{(2)} = -\frac{j\omega\mu_0}{q_2} C_2\, e^{q_2 x}, \tag{4.63}$$

with

$$q_2^2 = \beta^2 - \omega^2 \mu_0 \epsilon_2, \tag{4.64}$$

where C_0, C_1, D_1, and C_2 are arbitrary constants. Satisfying the boundary conditions yields

$$\begin{bmatrix} \sin p_1 a & \cos p_1 a & -e^{-q_0 a} & 0 \\ t_1 \cos p_1 a & -t_1 \sin p_1 a & -e^{-q_0 a} & 0 \\ -\sin p_1 a & \cos p_1 a & 0 & -e^{-q_2 a} \\ -t_2 \cos p_1 a & -t_2 \sin p_1 a & 0 & -e^{-q_2 a} \end{bmatrix} \begin{bmatrix} C_1 \\ D_1 \\ C_0 \\ C_2 \end{bmatrix} = 0, \tag{4.65}$$

with

$$t_1 = \frac{q_0}{p_1}, \tag{4.66}$$

$$t_2 = \frac{q_2}{p_1}. \tag{4.67}$$

The dispersion relation is

$$\begin{aligned} &(\sin p_1 a - t_1 \cos p_1 a)(\cos p_1 a + t_2 \sin p_1 a) \\ &\quad + (\sin p_1 a - t_2 \cos p_1 a)(\cos p_1 a + t_1 \sin p_1 a) = 0. \end{aligned} \tag{4.68}$$

From (4.65) we can derive

$$\frac{C_1}{A_0} = h_1,$$

$$\frac{D_1}{C_0} = h_2,$$

$$\frac{C_2}{C_0} = h_3.$$

It can be shown in a similar manner that there is no zero cutoff frequency for any TE mode. The cutoff frequencies for various TE modes must be found from the following equation:

$$(\sin p_1 a - t_1 \cos p_1 a) \cos p_1 a + \sin p_1 a \, (\cos p_1 a + t_1 \sin p_1 a) = 0. \qquad (4.69)$$

Since the power distribution and the attenuation constant for the TE modes can be derived in a similar manner as in the case for the TM modes, they will not be repeated here.

4.2.3 Special Cases and Numerical Examples

Let us treat the special cases of wave guidance along a symmetric slab where $\epsilon_2 = \epsilon_0$. From (4.26), we have, for $s_2 = s_1$,

$$2(\sin p_1 a - s_1 \cos p_1 a) \, (\cos p_1 a + s_1 \sin p_1 a) = 0 \qquad \text{(TM case)}, \qquad (4.70)$$

or,

$$(\sin p_1 a - s_1 \cos p_1 a) = 0 \qquad \text{(Defined as TM even case)}, \qquad (4.71)$$

$$(\cos p_1 a + s_1 \sin p_1 a) = 0 \qquad \text{(Defined as TM odd case)}, \qquad (4.72)$$

$$s_1 = \frac{q_0}{p_1} \frac{\epsilon_1}{\epsilon_0}$$

From (4.68). we have, for $t_2 = t_1$,

$$2(\sin p_1 a - t_1 \cos p_1 a) \, (\cos p_1 a + t_2 \sin p_1 a) = 0 \qquad \text{(TE case)}, \qquad (4.73)$$

or

$$(\sin p_1 a - t_1 \cos p_1 a) = 0 \qquad \text{(Defined as TE even case)}, \tag{4.74}$$

$$(\cos p_1 a + t_1 \sin p_1 a) = 0 \qquad \text{(Defined as TE odd case)}, \tag{4.75}$$

$$t_1 = \frac{q_0}{p_1}.$$

At cutoff, $q_0 = 0$. So from (4.71), (4.72), (4.74), and (4.75), respectively,

$$\sin p_1 a = 0 \qquad \text{(cutoff for TM even case)}, \tag{4.76}$$

$$\cos p_1 a = 0 \qquad \text{(cutoff for TM odd case)}, \tag{4.77}$$

$$\sin p_1 a = 0 \qquad \text{(cutoff for TE even case)}, \tag{4.78}$$

$$\cos p_1 a = 0 \qquad \text{(cutoff for TE odd case)}. \tag{4.79}$$

The cutoff for TM (even) and TE (even) modes is given by

$$\sin p_1 a = 0 \qquad p_1 a = 0,\ \pi,\ 2\pi, \ldots. \tag{4.80}$$

The cutoff for TM (odd) and TE (odd) modes is given by

$$\cos p_1 a = 0 \qquad p_1 a = \frac{\pi}{2}, \frac{3\pi}{2}, \frac{5\pi}{2}, \ldots. \tag{4.81}$$

These are the cutoff conditions for modes on a symmetric slab. Combining (4.10) and (4.14) and eliminating β gives

$$p_1^2 + q_0^2 = k_0^2 \left(\frac{\epsilon_1}{\epsilon_0} - 1 \right). \tag{4.82}$$

At cutoff, $q_0 = 0$, the cutoff wave number $k_0 = \omega\sqrt{\mu_0 \epsilon_0}$ is given by

$$k_0 a \ \text{(at cutoff)} = \frac{p_1 a}{\sqrt{\dfrac{\epsilon_1}{\epsilon_0} - 1}}. \tag{4.83}$$

Putting (4.80) and (4.81) in (4.83) gives

$$V\text{ (cutoff)} = \left(\frac{\epsilon_1}{\epsilon_0} - 1\right)^{1/2} k_0 a \text{ (at cutoff)} = 0,\ \pi,\ 2\pi, \ldots \tag{4.84}$$

(for TM even and TE even modes),

$$V\text{ (cutoff)} = \left(\frac{\epsilon_1}{\epsilon_0} - 1\right)^{1/2} k_0 a \text{ (at cutoff)} = \frac{\pi}{2}, \frac{3\pi}{2}, \frac{5\pi}{2}, \ldots \tag{4.85}$$

(for TM odd and TE odd modes).

Let us define the following normalized parameters:

$$V = k_0 a \left(\frac{\epsilon_1}{\epsilon_0} - 1\right)^{1/2}, \tag{4.86}$$

$$U = \left[(k_0 a)\left(\frac{\epsilon_1}{\epsilon_0}\right) - \beta a\right]^{1/2}. \tag{4.87}$$

The orderly appearances of these cutoff values prompted us to use them to label mode numbers. Let us assign the lowest order mode to the lowest cutoff value, the next order mode to the second lowest cutoff value, and so forth, that is, TE_0, TM_0 with V (cutoff) $= 0$; TE_1, TM_1 with V (cutoff) $= \pi/2$; TE_2, TM_2 with V (cutoff) $= \pi$, etc. Using this notation, we can state that the two dominant modes having zero cutoff frequency are the TE_0 and TM_0 modes. The ω-β diagram can now be represented by the V-U diagram. Figure 4.3 is a plot of the U vs. V curves for various lowest order modes on a symmetric dielectric waveguide with $\epsilon_1/\epsilon_0 = 2.77$. It is seen that the TE_0 and TM_0 modes are the dominant modes with no cutoff frequency. Another way to represent the ω-β diagram is to plot (β/k_0) vs. V, as shown in the left plot of Fig. 4.4 with $\epsilon_1/\epsilon_0 = 2.25$. Plots of normalized power distribution vs. V for the dominant TE_0 and TM_0 modes on a slab with $\epsilon_1/\epsilon_0 = 2.25$ are shown in the right plot of Fig. 4.4.

Finally, plots of the attenuation factor $(\epsilon_1/\epsilon_0)\,R$ as a function of normalized frequency $2a/\lambda_0$ for the dominant modes are displayed in Fig. 4.5 for various values of ϵ_1/ϵ_0. The attenuation factor is related to α_d by the following formula:

$$\alpha_\text{d} = \frac{\pi \tan\delta_1}{\lambda_0}\frac{\epsilon_1}{\epsilon_0}R. \tag{4.88}$$

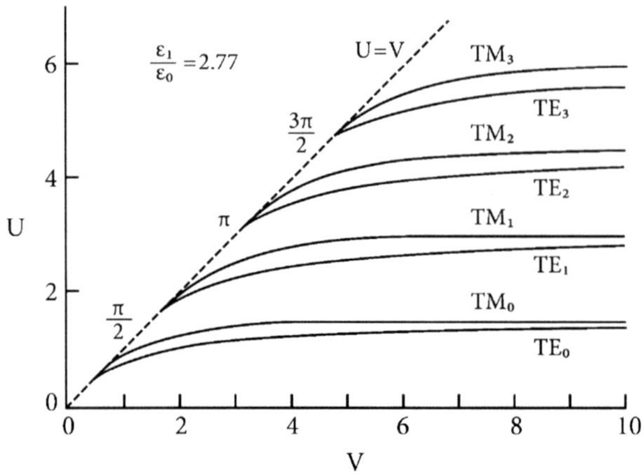

Figure 4.3. Normalized propagation constant U vs. normalized frequency V for several lower order modes [2]

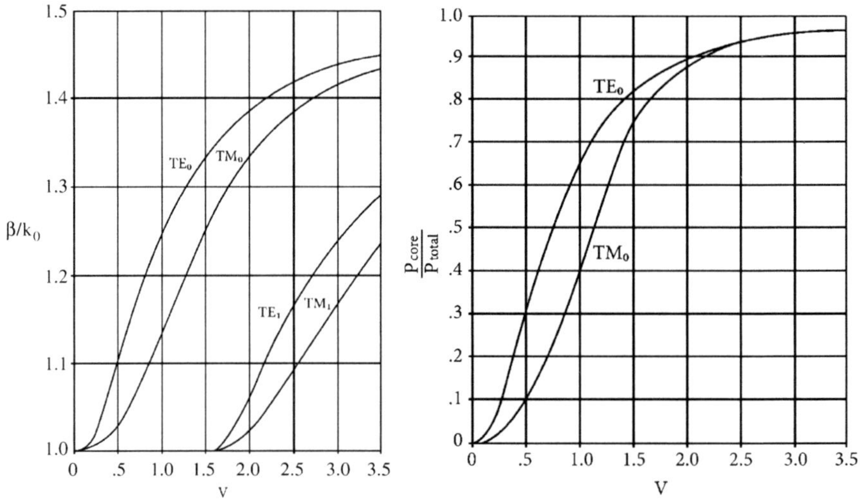

Figure 4.4. (*Left*) Normalized propagation constant β/k_0 vs. V for the TE_0, TM_0, TE_1, and TM_1 mode on a dielectric slab; (*right*) P_{core}/P_{total} vs. V for the dominant modes with $\epsilon_1/\epsilon_0 = 2.25$ [7]

4.3 Leaky Wave in a Heteroepitaxial Film Slab Waveguide [3]

Here we shall present an example of guidance of a surface wave with leakage. In this case, the guiding slab structure $(0 < x < 2a)$ with dielectric constant ϵ_2 and thickness $2a$ is situated on top of a substrate $(0 > x)$ whose dielectric constant is ϵ_3 with $\epsilon_3 > \epsilon_2$. The region above the slab structure $(x > 2a)$ has a dielectric constant

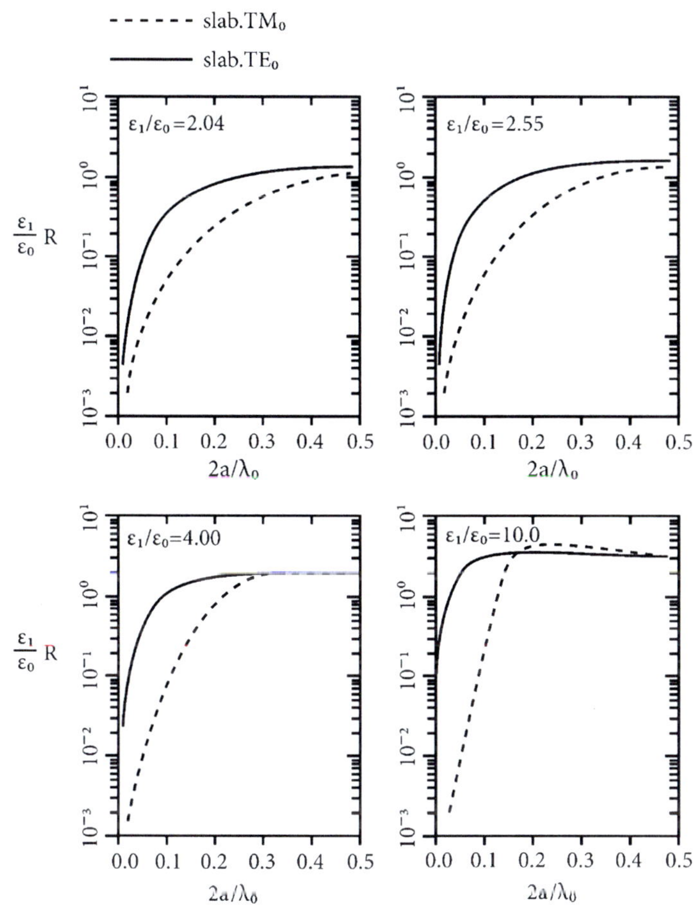

Figure 4.5. Configuration loss factor $(\epsilon_1/\epsilon_0)\, R$ as a function of normalized frequency for a dielectric slab of thickness $2a$ supporting the dominant TE_0 and dominant TM_0 modes [8]

of ϵ_1 with $\epsilon_2 > \epsilon_1$. The direction of propagation is z and there is no y-dependence for all fields. For surface wave guidance in the slab without leakage, the condition $\epsilon_2 > \epsilon_1$ or ϵ_3 must be satisfied. However, in many important physical situations, such as the deposition of ZnS ($\epsilon/\epsilon_0 = 5.48$) or ZnSe ($\epsilon/\epsilon_0 = 6.66$) on GaAs ($\epsilon/\epsilon_0 \approx 15$), the dielectric constant of the substrate is larger than that of the layer. Hence, the ordinary surface guided wave cannot exist in this structure. The purpose of this section is to present the investigation of the problem of wave propagation in a structure when $\epsilon_1 < \epsilon_2 < \epsilon_3$. It will be shown that when the thickness of the layer is large compared with the wavelength of the guided wave, low-loss leaky modes may exist.

4.3.1 Leaky Modes along an Asymmetric Dielectric Waveguide

Two types of leaky modes may exist along an asymmetric dielectric guide: TE and TM leaky modes. The TE leaky modes have a single component of electric field E_y and magnetic field components H_x and H_z, while the TM leaky modes have a single component of magnetic field H_y and electric field components E_x and E_z. Because of the asymmetry of the three-layer structure, in general, the modes may not be separated into even and odd modes. Strictly speaking, no surface mode can exist on this structure, because no solution can be found for purely real values of the propagation constant. This example is given to show the presence of leaky modes on this structure.

The electric field component of the TE leaky modes, which propagate in the $+z$ direction, takes the form

$$E_y = A\, e^{-j(p_1 x + \gamma z - \omega t)} \qquad (x \geq 2a)\,, \tag{4.89}$$

$$E_y = (B \cos p_2 x + C \sin p_2 x)\, e^{-j(\gamma z - \omega t)} \qquad (2a \geq x \geq 0)\,, \tag{4.90}$$

$$E_y = D\, e^{j(-p_3 x + \gamma z - \omega t)} \qquad (x \leq 0)\,, \tag{4.91}$$

where the transverse wave numbers p_1, p_2, and p_3 are defined in the following relations:

$$p_1^2 = \frac{\epsilon_1}{\epsilon_2} k_2^2 - \beta^2, \tag{4.92}$$

$$p_2^2 = k_2^2 - \gamma^2, \tag{4.93}$$

$$p_3^2 = \frac{\epsilon_3}{\epsilon_2} k_2^2 - \gamma^2, \tag{4.94}$$

with $k_2^2 = \omega^2 \mu_0 \epsilon_2$. Since $\epsilon_1 < \epsilon_2 < \epsilon_3$, no TE surface-wave guided modes are possible. All TE modes are leaky in nature, that is, the propagation constant for these modes are complex with $\mathrm{Re}(\gamma) > 0$ and $\mathrm{Im}(\gamma) > 0$. A, B, C, and D are arbitrary constants. The tangential component of the magnetic field can be found from the Maxwell equations:

$$H_z = A \frac{p_1}{\omega \mu_0}\, e^{-j(p_1 x + \gamma z - \omega t)} \qquad x \geq 2a, \tag{4.95}$$

$$H_z = \frac{j}{\omega \mu_0} (-B p_2 \sin p_2 x + C p_2 \cos p_2 x)\, e^{-j(\gamma z - \omega t)} \quad 2a \geq x \geq 0, \tag{4.96}$$

$$H_z = -D\frac{p_3}{\omega\mu_0}\, e^{-j(-p_3 x+\gamma z-\omega t)} \qquad 0 \le x. \tag{4.97}$$

Matching the tangential electric and magnetic fields at the boundaries $x = 0$ and $x = 2a$, we obtain

$$A\, e^{-2jp_1 a} - B\cos(2p_2 a) - C\sin(2p_2 a) = 0, \tag{4.98}$$

$$Ap_1\, e^{-2jp_1 a} + jBp_2 \sin 2p_2 a - jp_2 C\cos(2p_2 a) = 0, \tag{4.99}$$

$$B - D = 0, \tag{4.100}$$

$$-jp_2 C - p_3 D = 0. \tag{4.101}$$

Setting the determinant of the linear simultaneous algebraic equations to zero gives the characteristic equation for the TE leaky modes:

$$\tan(2p_2 a) = -j\frac{p_1 p_2 + p_3 p_2}{p_2^2 + p_1 p_3}. \tag{4.102}$$

The analysis for the TM leaky modes is similar to the one for the TE leaky modes. H_y and E_z, the tangential electric and magnetic fields, are matched at the boundaries $x = 0$ and $x = 2a$. The determinant for the resulting linear simultaneous equations is set to zero and gives the characteristic equation for the TM leaky modes:

$$\tan(2p_2 a) = -j\frac{\epsilon_2\epsilon_3 p_1 p_2 + \epsilon_1\epsilon_2 p_3 p_2}{\epsilon_1\epsilon_3 p_2^2 + \epsilon_2^2 p_1 p_3}. \tag{4.103}$$

We note that the characteristic equation for the TE and TM leaky modes possess the same form except the right-hand side is different.

4.3.2 Approximate Solutions of the Characteristic Equations

It can be shown that when $\epsilon_1 < \epsilon_2 < \epsilon_3$, no real roots for γ, the propagation constant may be found from the characteristic equations, that is, no surface wave modes may exist on such structures. All roots for γ will be complex. In other words, waves "guided" along this structure will be attenuated in the direction of propagation. We are interested in the relatively low-loss modes. It is expected that when $k_2 a \to \infty$, the low-loss wave will approach the case of a plane wave propagating in a medium with dielectric constant ϵ_2 and permeability μ_0. Hence, making the initial approximation

$$\gamma \approx k_2, \tag{4.104}$$

we have from (4.92)–(4.94) that

$$p_1 \approx -jk_2\sqrt{1-\epsilon_1/\epsilon_2}, \tag{4.105}$$

$$p_3 \approx k_2\sqrt{\epsilon_3/\epsilon_2 - 1}, \tag{4.106}$$

and

$$k_2 a \gg p_2 a. \tag{4.107}$$

Using these approximations, (4.102) for the TE leaky mode becomes

$$\tan\left(2p_2 a\right) \approx -j\left(\frac{p_2}{p_1} + \frac{p_2}{p_3}\right) \tag{4.108}$$

and (4.103) for the TM leaky modes becomes

$$\tan\left(2p_2 a\right) \approx -j\left(\frac{\epsilon_1 p_2}{\epsilon_2 p_1} + \frac{\epsilon_3 p_2}{\epsilon_2 p_3}\right). \tag{4.109}$$

The right hand sides of (4.108) and (4.109) are complex numbers much smaller in magnitude than unity. Hence, applying the perturbation technique we obtain for the TE modes

$$2p_2 a \approx (n+1)\,\pi\left(1 + \frac{\theta_r}{2ka} - j\frac{\theta_i}{2ka}\right), \tag{4.110}$$

where $n = 0, 1, 2, \ldots$ is the mode order and

$$\theta_i = \sqrt{\epsilon_3/\epsilon_2 - 1}, \tag{4.111}$$

$$\theta_r = \sqrt{1-\epsilon_1/\epsilon_2}. \tag{4.112}$$

The corresponding characteristic equation for the TM modes is

$$2p_2 a \approx (n+1)\,\pi\left[1 + \frac{\epsilon_1}{\epsilon_2}\frac{\theta_r}{2ka} - j\frac{\epsilon_3}{\epsilon_2}\frac{\theta_i}{2ka}\right]. \tag{4.113}$$

Substitution of (4.110)–(4.113) in (4.93) gives, for the TE modes,

$$\begin{aligned}\gamma_{\mathrm{TE}} &= k_2\sqrt{1-\left(\frac{p_2}{k_2}\right)^2} \approx k_2\left[1-\frac{1}{2}\left(\frac{p_2}{k_2}\right)^2\right],\\ &\approx k_2\left[1-\frac{(n+1)^2\,\pi^2}{8k_2^2a^2}\left(1+\frac{\theta_r}{k_2 a} - j\frac{\theta i}{k_2 a}\right)\right],\end{aligned} \tag{4.114}$$

and, for the TM modes,

$$\gamma_{\text{TM}} = k_2 \left[1 - \frac{(n+1)^2 \pi^2}{8k_2^2 a^2} \left(1 + \frac{\epsilon_1}{\epsilon_2} \frac{\theta_r}{k_2 a} - j \frac{\epsilon_3}{\epsilon_2} \frac{\theta_i}{ka} \right) \right]. \tag{4.115}$$

Remembering that $ka \gg 1$, the phase constants and rates of power attenuation for the TE and TM modes are, to first order,

$$\beta_{\text{TE}} = \text{Re}(\gamma_{\text{TE}}) \approx k_2 \left[1 - \pi^2 \frac{(n+1)^2}{8k_2^2 a^2} \right], \tag{4.116}$$

$$\alpha_{\text{TE}} = 2\,\text{Im}(\gamma_{\text{TE}}) \approx k_2 \pi^2 \frac{(n+1)^2 \theta_i}{4k_2^3 a^3}, \tag{4.117}$$

$$\beta_{\text{TM}} = \text{Re}(\beta_{\text{TM}}) \approx k_2 \left[1 - \pi^2 \frac{(n+1)^2}{8k_2^2 a^2} \right], \tag{4.118}$$

$$\alpha_{\text{TM}} = 2\,\text{Im}(\gamma_{\text{TM}}) \approx k_2 \frac{\pi^2 (n+1)^2 \theta_i}{4k_2^3 a^3} \frac{\epsilon_3}{\epsilon_2}. \tag{4.119}$$

Upon comparison of (4.116) and (4.118) and of (4.117) and (4.119), we note that the frequency dependence and the guide-width dependence of the attenuation constants of the TE and TM modes of the same mode order are identical. Only the magnitudes are different: $\alpha_{\text{TM}} = \alpha_{\text{TE}} (\epsilon_3/\epsilon_2)$. Also noted is the fact that the phase constants for these modes are identical. Rearranging (4.117) and (4.119) gives

$$\Delta^{\text{TE}} = \frac{\epsilon_2}{\epsilon_0} \frac{\alpha_{\text{TE}} a}{(n+1)^2 \theta_i} = 4.34 \frac{\pi^2}{4k_0^2 a^2} \quad \text{(dB)}, \tag{4.120}$$

$$\Delta^{\text{TM}} = \frac{\epsilon_2}{\epsilon_0} \frac{\alpha_{\text{TM}} a}{(n+1)^2 \theta_i} \frac{\epsilon_2}{\epsilon_3} = 4.34 \frac{\pi^2}{4k_0^2 a^2} \quad \text{(dB)}, \tag{4.121}$$

where $k_0 = 2\pi/\lambda_0$. The term λ_0 is the free-space wavelength and Δ^{TE} and Δ^{TM} are called, respectively, the normalized attenuation constants for the TE and TM modes. To illustrate how the normalized attenuation constant varies as a function of $k_0 a$, Fig. 4.6 is introduced. It can be seen that substantial attenuation results for moderate values of $k_0 a$. As expected, when the thickness of the slab becomes very large, the guided-wave approaches the plane-wave case and the attenuation constant becomes very small.

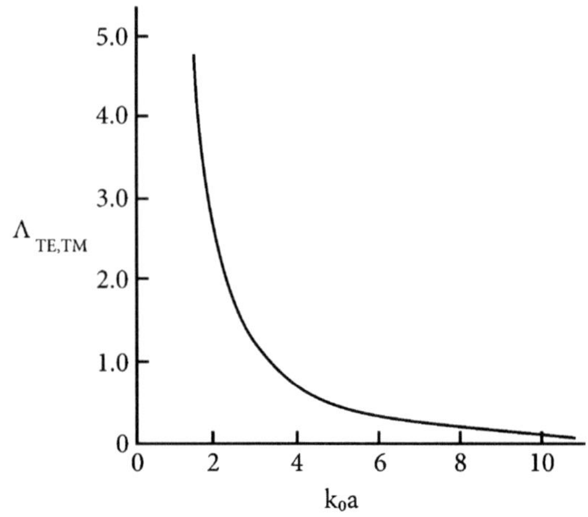

Figure 4.6. Normalized attenuation constant as a function of $k_0 a$ [3]

4.4 Multilayered Dielectric Slab Waveguides [4]

Analysis will now be presented for the case of surface wave propagation along a dielectric slab waveguide consisting of multiple parallel layers of dielectrics with different dielectric constants as shown in Fig. 4.7. This many-layered structure may correspond to a variety of practical structures of interest in microwave and optical integrated circuits. Furthermore, it may also approximate the case of an inhomogeneous slab with spatially varying permittivity $\epsilon(x)$ and spatially varying permeability $\mu(x)$. The continuously varying $\epsilon(x)$ and $\mu(x)$ may be approximated by many dielectric layers of constant permittivity and constant permeability.

We note that only TE or TM modes may be supported by this structure and that it is not necessary to use linear combination of these modes to satisfy the boundary conditions. Without loss of generality we may assume that the expressions for the field components of all modes are multiplied by the factor $e^{-j\beta z+j\omega t}$, which will be suppressed throughout.

Dividing the layered structure into $m+1$ regions, as shown in Fig. 4.7, we may write the expressions for the TM fields tangential to the interface in these regions using (4.1)–(4.6) as follows:

Region 1

$$\begin{bmatrix} E_z^{(1)} \\ \eta H_y^{(1)} \end{bmatrix} = \begin{bmatrix} a_1^{(1)}(x) & 0 \\ 0 & a_4^{(1)}(x) \end{bmatrix} \begin{bmatrix} A_1 \\ B_1 \end{bmatrix}. \tag{4.122}$$

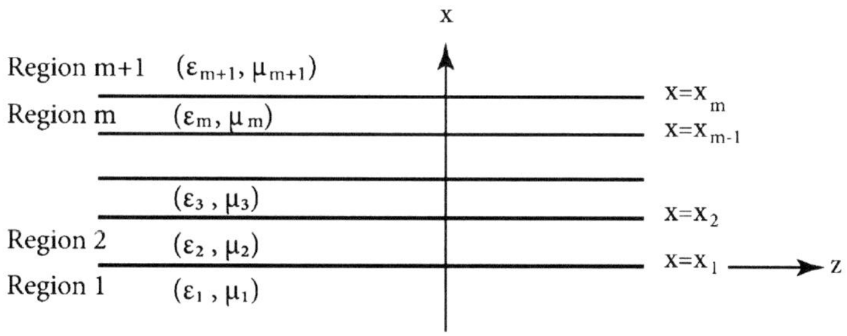

Figure 4.7. The geometry of the layered slab

Region m $(m > 1)$

$$\begin{bmatrix} E_z^{(m)} \\ \eta H_y^{(m)} \end{bmatrix} = \begin{bmatrix} b_1^{(m)}(x) & b_3^{(m)}(x) \\ b_2^{(m)}(x) & b_4^{(m)}(x) \end{bmatrix} \begin{bmatrix} A_m \\ B_m \end{bmatrix}. \tag{4.123}$$

Region $m+1$

$$\begin{bmatrix} E_z^{(m+1)} \\ \eta H_y^{(m+1)} \end{bmatrix} = \begin{bmatrix} c_1^{(m+1)}(x) & 0 \\ 0 & c_4^{(m+1)}(x) \end{bmatrix} \begin{bmatrix} A_{m+1} \\ B_{m+1} \end{bmatrix}, \tag{4.124}$$

where

$$\begin{aligned}
& a_1^{(1)}(x) = e^{q_1 x} && a_4^{(1)}(x) = -\omega\epsilon_1\,(\eta/q_1)\,e^{q_1 x} \\
& b_1^{(m)}(x) = \sin(p_m x) && b_2^{(m)}(x) = -j\omega\epsilon_m\,(\eta/p_m)\cos(p_m x) \\
& b_3^{(m)}(x) = \cos(p_m x) && b_4^{(m)}(x) = j\omega\epsilon_m\,(\eta/p_m)\sin(p_m x) \\
& c_1^{(m+1)}(x) = e^{-q_{m+1}x} && c_4^{(m+1)}(x) = \omega\epsilon_{m+1}\,(\eta/q_{m+1})\,e^{-q_{m+1}x} \\
& q_1^2 = \beta^2 - k_1^2 && q_{m+1}^2 = \beta^2 - k_{m+1}^2 \\
& p_m^2 = k_m^2 - \beta^2 && k_m^2 = \omega^2\mu_m\epsilon_m \\
& \eta = \sqrt{\mu_0/\epsilon_0} && \\
& (m = 2, 3, 4, \ldots) &&
\end{aligned} \tag{4.125}$$

and $A_1,\ B_1,\ A_2,\ B_2, \ldots, A_{m+1},\ B_{m+1}$ are arbitrary constants. Matching the tangential electric and magnetic fields at the boundary surfaces, that is, at $x = x_1, x_2, \ldots, x_m$, gives

$$\begin{bmatrix} a_1^{(1)}(x_1) & 0 \\ 0 & a_4^{(1)}(x_1) \end{bmatrix} \begin{bmatrix} A_1 \\ B_1 \end{bmatrix} = \begin{bmatrix} b_1^{(2)}(x_1) & b_3^{(2)}(x_1) \\ b_2^{(2)}(x_1) & b_4^{(2)}(x_1) \end{bmatrix} \begin{bmatrix} A_2 \\ B_2 \end{bmatrix}, \tag{4.126}$$

$$\begin{bmatrix} b_1^{(2)}(x_2) & b_3^{(2)}(x_2) \\ b_2^{(2)}(x_2) & b_4^{(2)}(x_2) \end{bmatrix} \begin{bmatrix} A_2 \\ B_2 \end{bmatrix} = \begin{bmatrix} b_1^{(3)}(x_2) & b_3^{(3)}(x_2) \\ b_2^{(3)}(x_2) & b_4^{(3)}(x_2) \end{bmatrix} \begin{bmatrix} A_3 \\ B_3 \end{bmatrix}, \tag{4.127}$$

$$\cdots\cdots\cdots\cdots\cdots\cdots\cdots\cdots\cdots\cdots\cdots\cdots$$

$$\cdots\cdots\cdots\cdots\cdots\cdots\cdots\cdots\cdots\cdots\cdots\cdots$$

$$\begin{bmatrix} b_1^{(m-1)}(x_{m-1}) & b_3^{(m-1)}(x_{m-1}) \\ b_2^{(m-1)}(x_{m-1}) & b_4^{(m-1)}(x_{m-1}) \end{bmatrix} \begin{bmatrix} A_{m-1} \\ B_{m-1} \end{bmatrix}$$
$$= \begin{bmatrix} b_1^{(m)}(x_{m-1}) & b_3^{(m)}(x_{m-1}) \\ b_2^{(m)}(x_{m-1}) & b_4^{(m)}(x_{m-1}) \end{bmatrix} \begin{bmatrix} A_m \\ B_m \end{bmatrix}, \tag{4.128}$$

$$\begin{bmatrix} b_1^{(m)}(x_m) & b_3^{(m)}(x_m) \\ b_2^{(m)}(x_m) & b_4^{(m)}(x_m) \end{bmatrix} \begin{bmatrix} A_m \\ B_m \end{bmatrix}$$
$$= \begin{bmatrix} c_1^{(m+1)}(x_m) & 0 \\ 0) & c_4^{(m+1)}(x_m) \end{bmatrix} \begin{bmatrix} A_{m+1} \\ B_{m+1} \end{bmatrix} \tag{4.129}$$

$$(m \geq 2).$$

It is clear that if we set the determinant of the above simultaneous equations to zero, one obtains in a straightforward manner, the dispersion relation for the layered slab problem. However, we note that the size of the resultant determinant depends directly on the number of layers that are present. The unattractive numerical problems associated with a very large size matrix are well-known. Here, we shall introduce a scheme to avoid working with this large size matrix [4].

Let us express the array $\begin{bmatrix} A_2 \\ B_2 \end{bmatrix}$ in terms of $\begin{bmatrix} A_3 \\ B_3 \end{bmatrix}$ and express the array $\begin{bmatrix} A_3 \\ B_3 \end{bmatrix}$ in terms of $\begin{bmatrix} A_4 \\ B_4 \end{bmatrix}$, etc. The following chain equation can be obtained.

$$M_1(x_1)\begin{bmatrix} A_1 \\ B_1 \end{bmatrix} = M_2(x_1)M_2^{-1}(x_2)M_3(x_2)M_3^{-1}(x_3)M_4(x_3) \cdots M_m^{-1}(x_m)M_{m+1}(x_m)\begin{bmatrix} A_{m+1} \\ B_{m+1} \end{bmatrix}, \quad (4.130)$$

where

$$M_1(x_{1)} = \begin{bmatrix} a_1^{(1)}(x_1) & 0 \\ 0 & a_4^{(1)}(x_1) \end{bmatrix},$$

$$M_m(x_{m-1}) = \begin{vmatrix} b_1^{(m)}(x_{m-1}) & b_3^{(m)}(x_{m-1}) \\ b_2^{(m)}(x_{m-1}) & b_4^{(m)}(x_{m-1}) \end{vmatrix},$$

$$M_m^{-1}(x_m) = \begin{bmatrix} b_1^{(m)}(x_m) & b_3^{(m)}(x_m) \\ b_2^{(m)}(x_m) & b_4^{(m)}(x_m) \end{bmatrix}^{-1},$$

$$M_{m+1}(x_m) = \begin{bmatrix} c_1^{(m+1)}(x_m) & 0 \\ 0) & c_4^{(m+1)}(x_m) \end{bmatrix}. \quad (4.131)$$

Here, $[M]^{-1}$ means the inverse matrix of M. Hence we are now dealing only with matrices of the size 2×2. This rearrangement eliminates the need to work with large matrices. The size of our present matrix is independent of the number of layers for the slab. This technique was first introduced by Yeh and Lindgren and subsequently popularized by Yeh et al. [4]. Rewriting (4.130) gives

$$\begin{bmatrix} a_1^{(1)}(x_1) & -(N_{11}+N_{12}) \\ a_2^{(1)}(x_1) & -(N_{21}+N_{22}) \end{bmatrix} \begin{bmatrix} A_1 \\ B_{m+1} \end{bmatrix} = 0, \tag{4.132}$$

where

$$N = \begin{bmatrix} N_{11} & N_{12} \\ N_{21} & N_{22} \end{bmatrix} = M_2(x_1)M_2^{-1}(x_2)\cdots M_m^{-1}(x_m)M_{m+1}(x_m). \tag{4.133}$$

Setting the determinant of (4.132) to zero, we obtain the dispersion relation from which the propagation constants of various modes along a layered slab may be found. The arbitrary constants for the field in each region can be expressed in terms of a single arbitrary constant A_1. Only 2×2 matrix operations need be used. For example, $(A_2,\ B_2)$ can be found from (4.126) and $(A_3,\ B_3)$ can be found from (4.127), and so forth.

The TE modes may be treated in the similar manner. Details will not be given here. The same technique may be used to solve the problem of guided waves on radially stratified or radially inhomogeneous dielectric cylinders. This will be discussed in the Chap. 5.

4.5 Coupling Between Two Parallel Dielectric Slab Waveguides [5]

Not only is this subject of interest in fiber or integrated optics applications, but also the detailed analytic solution of this problem can provide an in-depth understanding on how guided power may be interchanged between parallel dielectric waveguides and can provide a check on the region of validity for which the coupled mode theory [2] will yield accurate results.

The geometry of the problem is shown in Fig. 4.8. Two identical dielectric slab waveguides are located in the regions $d/2 \le |x| \le a + d/2$. The permittivity of the slab guides is ϵ_1 and that of the regions outside the guides is ϵ_2; furthermore, $\epsilon_1 > \epsilon_2$. The permeability is assumed to be μ_0 everywhere. A perfectly conducting

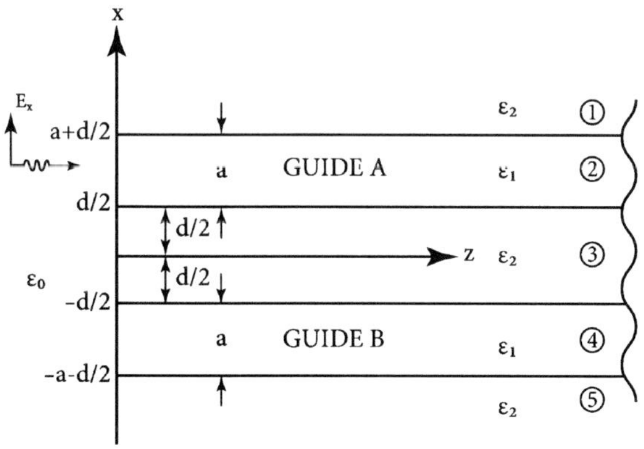

Figure 4.8. Geometry of the canonical problem

plane covers the surface $z = 0$ except for the region $d/2 \le |x| \le a + d/2$. In that aperture, the tangential electric field is $E_0\mathbf{e}_x$. All electromagnetic field quantities are independent of the coordinate y and the field components of all modes are multiplied by the factor $e^{-j\beta z+j\omega t}$, which will be suppressed throughout.

We are concerned with the excitation by the aperture field of the guided modes on this structure; the guided field will be the dominant portion of the total field for large z. Only E_x, E_z, and H_y will be excited. The appropriate solutions for the guided wave are the following:

$$\begin{aligned}
H_{yn}^{(1)}(x) &= A_n\, e^{[q_n(x+a+d/2)]} && (-\infty \le x \le -a - d/2)\,, \\
H_{yn}^{(2)}(x) &= B_n \sin p_n(x + d/2) + C_n \cos p_n(x + d/2) && (-a - d/2 \le x \le -d/2)\,, \\
H_{yn}^{(3)}(x) &= D_n \sinh q_n x + E_n \cosh q_n x && (-d/2 \le x \le d/2)\,, \\
H_{yn}^{(4)}(x) &= F_n \sin p_n(x - d/2) + G_n \cos p_n(x - d/2) && (d/2 \le x \le a + d/2)\,, \\
H_{yn}^{(5)}(x) &= H_n\, e^{[-q_n(x-a-d/2)]} && (a + d/2 \le x \le \infty)\,,
\end{aligned} \tag{4.134}$$

in which $p_n^2 = k_1^2 - \beta_n^2$, $q_n^2 = \beta_n^2 - k_2^2$, and the coefficients $A_n \cdots H_n$ are to be determined. The n subscripts refer to the nth guided-wave mode.

The characteristic equation from which the values of β_n may be determined is found by forcing the tangential fields components E_z and H_y to be continuous at the four dielectric interfaces. By virtue of the symmetry of the configuration about $x = 0$, the electromagnetic field may be separated into two parts whose axial electric field E_z have either even or odd symmetry about $x = 0$. For the even modes, $H_n = -A_n$, $G_n = -C_n$, $F_n = B_n$, and $E_n = 0$. For the odd modes $H_n = A_n$, $G_n = C_n$, $F_n = -B_n$, and $D_n = 0$. Thus, considering the two cases separately, one matches the boundary conditions at $x = d/2$ and $x = a + d/2$, and obtains

$$\left[\tan p_n a - \frac{2rp_n q_n}{r^2p_n^2 - q_n^2 + (r^2p_n^2 + q_n^2)\, e^{-q_n d}}\right] \times \left[\tan p_n a - \frac{2qp_n q_n}{r^2p_n^2 - q_n^2 - (r^2p_n^2 + q_n^2)\, e^{-q_n d}}\right] = 0, \tag{4.135}$$

in which $r = \epsilon_2/\epsilon_1$. The roots of the first factor in square brackets yield the even modes; those of the second factor, the odd modes. We may also evaluate the coefficients $A_n \cdots H_n$ in terms of any single one of them, say A_n. The relationship is expressed as $B_n = b_n A_n, \ldots, H_n = h_n A_n$, in which $b_n \cdots h_n$ are given by

(a) Even modes

$$\begin{aligned}
b_n^{\mathrm{e}} &= f_n^{\mathrm{e}} = (1/\Delta_n^{\mathrm{e}})\,(q_n/\epsilon_2)\cosh(q_n d/2),\\
c_n^{\mathrm{e}} &= -g_n^{\mathrm{e}} = -\,(1/\Delta_n^{\mathrm{e}})\,(p_n/\epsilon_1)\sinh(q_n d/2),\\
d_n^{\mathrm{e}} &= (1/\Delta_n^{\mathrm{e}})\,(p_n/\epsilon_1)\,,\\
e_n^{\mathrm{e}} &= 0,\\
h_n^{\mathrm{e}} &= -1,\\
\Delta_n^{\mathrm{e}} &= -\,(q_n/\epsilon_2)\cosh(q_n d/2)\sin p_n a - (p_n/\epsilon_1)\sinh(q_n d/2)\cos p_n a.
\end{aligned} \tag{4.136}$$

(b) Odd modes

$$\begin{aligned}
b_n^{\mathrm{o}} &= -f_n^{\mathrm{o}} = (1/\Delta_n^{\mathrm{o}})\,(q_n/\epsilon_2)\sinh(q_n d/2),\\
c_n^{\mathrm{o}} &= g_n^{\mathrm{o}} = -\,(1/\Delta_n^{\mathrm{o}})\,(p_n/\epsilon_1)\cosh(q_n d/2),\\
d_n^{\mathrm{o}} &= 0,
\end{aligned}$$

$$e_n^{\mathrm{o}} = -\left(1/\Delta_n^{\mathrm{o}}\right)\left(p_n/\epsilon_1\right),$$

$$h_n^{\mathrm{o}} = 1,$$

$$\Delta_n^{\mathrm{o}} = -\left(q_n/\epsilon_2\right)\sinh(q_n d/2)\sin p_n a - \left(p_n/\epsilon_1\right)\cosh(q_n d/2)\cos p_n a. \tag{4.137}$$

We now consider the boundary condition at $z = 0$, that is, $E_x(x,0) = E_0$ for $d/2 \leq |x| \leq a + d/2$ and $E_x(x,0) = 0$ elsewhere. Equating the excitation field to the total E_x field for $z \geq 0$ at $z = 0$ yields

$$E_0\left[U(x - d/2) - U(x - a - d/2)\right] = \sum_n -\beta_n H_{yn}(x)/\omega\epsilon(x) + \text{(radiated field)}, \tag{4.138}$$

where U is the unit step function. The sum is taken over the guided modes for which β_n is a proper root, that is, a root for which $\mathrm{Re}(q_n) \geq 0$, of (4.135). By virtue of the orthogonality relation

$$\int_{-\infty}^{\infty} \frac{1}{\epsilon(x)} H_y(x,\beta)\cdot H_y^*(x,\beta')dx = 0 \qquad (\beta \neq \beta'), \tag{4.139}$$

we readily obtain

$$\begin{aligned}
-\frac{\omega}{\beta_n}E_0 \int_{d/2}^{a+d/2} & H_{yn}^{(4)*}(x)\,dx \\
&= \frac{1}{\epsilon_2}\int_{-\infty}^{-a-d/2}\left|H_{yn}^{(1)}(x)\right|^2 dx + \frac{1}{\epsilon_1}\int_{-a-d/2}^{-d/2}\left|H_{yn}^{(2)}(x)\right|^2 dx \\
&+ \frac{1}{\epsilon_2}\int_{-d/2}^{d/2}\left|H_{yn}^{(3)}(x)\right|^2 dx + \frac{1}{\epsilon_1}\int_{d/2}^{a+d/2}\left|H_{yn}^{(4)}(x)\right|^2 dx \\
&+ \frac{1}{\epsilon_2}\int_{a+d/2}^{\infty}\left|H_{yn}^{(5)}(x)\right|^2 dx + \text{(radiated fields)}.
\end{aligned} \tag{4.140}$$

The integrals in (4.140) are easily evaluated and yield a relation between E_0 and A_n. We find that

$$A_n = -\frac{\omega E_0}{p_n\beta_n}\left[f_n^*\left(1 - \cos p_n a\right) + g_n^*\sin p_n a\right]\frac{1}{K}, \tag{4.141}$$

where

$$K = \frac{1}{q_n \epsilon_2} + \frac{1}{p_n \epsilon_1} S + \frac{p_n^2}{2\Delta_n^2 q_n \epsilon_1^2 \epsilon_2} (\sinh q_n d \pm q_n d),$$

$$S = |g_n|^2 \left(p_n a + \frac{1}{2} \sin(2p_n a) \right) + |f_n|^2 \left(p_n a - \frac{1}{2} \sin(2p_n a) \right)$$

$$+ (f_n g_n^* + f_n^* g_n) \sin^2 p_n a,$$

in which the lower ($-$) sign is taken for the even modes, and the upper ($+$) sign for the odd modes. Δ_n^2 is either Δ_n^{e2} or Δ_n^{o2} as appropriate, as are f_n and g_n. Here we have assumed that the contribution of the exciting field to the radiated field is negligible as compared with the guided field.

We now consider the special case which is of particular interest for the purposes of this illustration, that in which only a single mode is possible on each of the dielectric slabs in isolation. We shall further require that the lowest-order even and odd modes will propagate on the dual-guide structure. Thus,

$$2a/rd < p_1 a \tan p_1 a < \infty,$$

where $0 < p_1 a < \pi/2$. We now evaluate the z component of the Poynting vector in each of the slab waveguides under the assumption that only these lowest-order modes exist; their propagation constants are denoted β_e and β_o, and the associated fields $H_y(x)$ are denoted $H_{y\mathrm{e}}(x)$ and $H_{y\mathrm{o}}(x)$, respectively. In the "upper" guide $(d/2 < x < a + d/2)$ or "guide A," we have

$$P_{z\mathrm{A}}(x,z) = \frac{1}{2\omega} \mathrm{Re} \left\{ \frac{1}{\epsilon_1} \left[\beta_\mathrm{e} \left| H_{ye}^{(4)}(x) \right|^2 + \beta_\mathrm{o} \left| H_{yo}^{(4)}(x) \right|^2 \right.\right.$$

$$\left.\left. + \beta_\mathrm{e} H_{ye}^{(4)}(x) H_{yo}^{(4)*}(x) e^{-j(\beta_\mathrm{e} - \beta_\mathrm{o}^*)z} + \beta_\mathrm{o} H_{yo}^{(4)}(x) H_{ye}^{(4)*}(x) e^{-j(\beta_\mathrm{e}^* - \beta_\mathrm{o})z} \right] \right\}. \tag{4.142}$$

In the "lower" guide or "guide B,"

$$P_{z\mathrm{B}}(x,z) = \frac{1}{2\omega} \mathrm{Re} \left\{ \frac{1}{\epsilon_1} \left[\beta_\mathrm{e} \left| H_{ye}^{(2)}(x) \right|^2 + \beta_\mathrm{o} \left| H_{yo}^{(2)}(x) \right|^2 \right.\right.$$

$$\left.\left. + \beta_\mathrm{e} H_{ye}^{(2)}(x) H_{yo}^{(2)*}(x) e^{-j(\beta_\mathrm{e} - \beta_\mathrm{o}^*)z} + \beta_\mathrm{o} H_{yo}^{(2)}(x) H_{ye}^{(2)*}(x) e^{-j(\beta_\mathrm{e}^* - \beta_\mathrm{o})z} \right] \right\}. \tag{4.143}$$

We now calculate the total power per unit width carried in each of the slab waveguides by integrating $P_{z\text{A}}$ from $x = d/2$ to $x = a + d/2$ and $P_{z\text{B}}$ from $x = -a - d/2$ to $x = -d/2$. (The power carried outside the slab waveguides is not included in the calculation. Thus P_A and P_B will not reduce to the total power carried in a single slab waveguide mode in the limit $d \to \infty$.) Denoting these powers by P_A and P_B, we obtain, assuming that ϵ_1, ϵ_2, β_e, and β_o are purely real,

$$P_\text{A}(z) = \int_{d/2}^{a+d/2} P_{z\text{A}}(x, z)\, dx,$$

$$P_\text{B}(z) = \int_{-a-d/2}^{-d/2} P_{z\text{B}}(x, z)\, dx.$$

Now we are in a position to compare exactly how the input power is transferred from one guide to the other. The only assumption is that no radiated power is present. We must first determine the propagation constants of the guided modes along the slabs from (4.135). Results for the two lowest-order modes are displayed in Fig. 4.9 for the symmetric antisymmetric modes. In these figures, the normalized propagation constant β/k_0 is plotted against the normalized thickness $k_0 a$ of the slab for various values of the normalized separation $k_0 d$ of the slabs. The constant k_0, the free-space wave number, is defined as $\omega\sqrt{\mu_0 \epsilon_0}$. It can be seen that there exists no cutoff frequency for the lowest-order symmetric mode, while a cutoff frequency does exist for the lowest-order antisymmetric mode. Of course, all higher-order symmetric and antisymmetric modes possess cutoff frequencies.

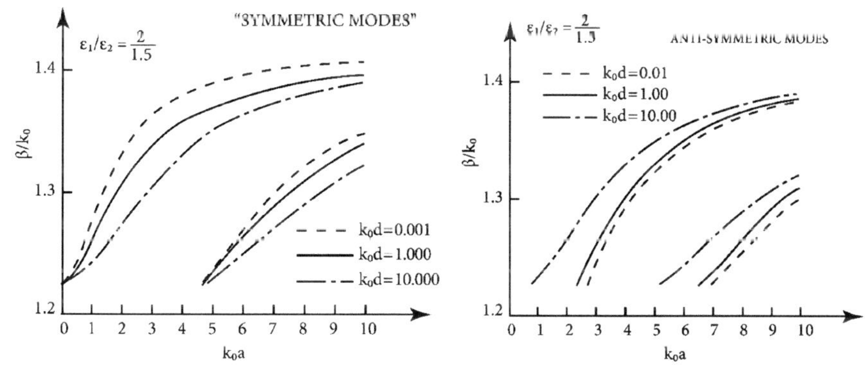

Figure 4.9. (*Left*) Normalized propagation constant as a function of normalized frequencies for symmetric modes. The lowest order mode $n = 1$ has zero cutoff frequency (*Right*) Normalized propagation constant as a function of normalized frequencies for antisymmetric modes. All antisymmetric modes have cutoff frequencies [5]

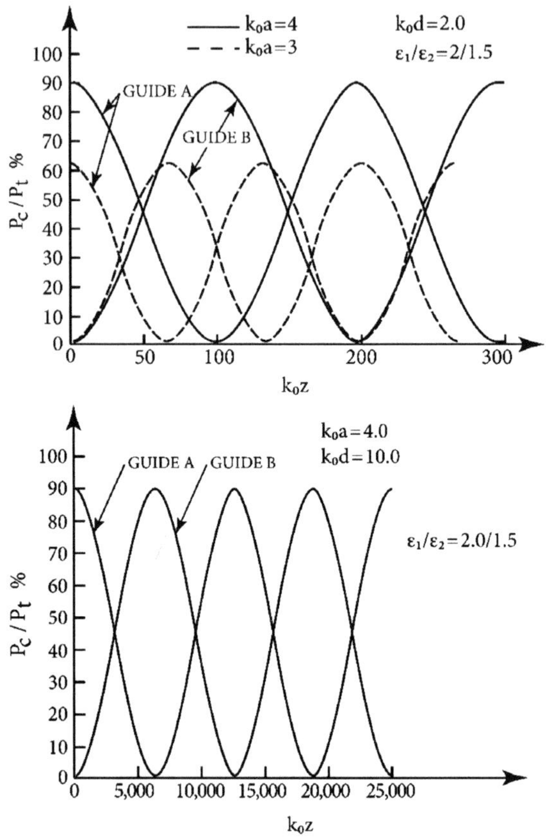

Figure 4.10. (*Top*) Normalized power in the core region of guides A and B as a function of normalized longitudinal distance. The coupling length (defined as the length for which complete exchange of power in the cores of guide A and guide B occurs) is longer for less tightly bounded fields. (*Bottom*) Normalized power in the core region of guides A and B as a function of normalized longitudinal distance. P_c is the power in the core region of guide A or guide B as appropriate. P_t is the total guided power [5]

To illustrate how the guided power exchanges between the two guiding structures, Fig. 4.10 is introduced. The operating frequency is so selected that only the dominant symmetric mode and the lowest-order antisymmetric mode may exist along the guiding structures. It is interesting to note that although the initial exciting field exists only at the entrance of the core region of guide A, according to our computed results there exists a small amount of guided power in guide B. The reason for this is that to satisfy the initial given field configuration at the entrance of the guiding structure, the radiation mode as well as the guided modes must be taken into account. Since we have a priori ignored the radiation mode in our calculation, we can only satisfy the given field approximately. The quantity P_B at $z = 0$

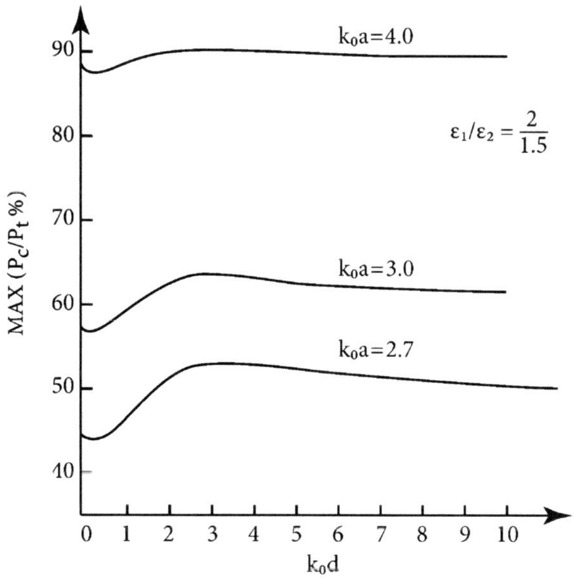

Figure 4.11. Maximum normalized power in the core region of guide A as a function of separation distance of the two guides [5]

represents the power calculated from the field that extends from guide A into the core region of guide B. It can also be seen from these figures that guided power exchanges from one guide to the other in a periodic fashion as expected.

The distance for which maximum guided power is transferred from one guide to the other is called the coupling length. It is noted that the coupling length becomes shorter as the separation distance between the guides is shortened. To correlate the maximum power contained in the core region of guide A with the normalized separation distance k_0d,, we have performed the computation at the entrance of the coupled guide. Results are shown in Fig. 4.11. It can be seen that the normalized maximum power in the core region of guide A varies in a rather unusual fashion for small separation distances. At large k_0d, the maximum power contained in the core region approaches that for the case of an isolated slab guide, as expected. To further understand the behavior of the transverse fields in the guides when the separation distance is small, we have plotted in Fig. 4.12 the quantity $|E_x|$ vs. the transverse distance. The complexity of the evolution of the transverse electric field as the separation distances are changed indicates the complex nature of the curves in Fig. 4.11 when the separation distances are small.

Recall that the primary purpose of this illustration is to determine how accurate the coupled-mode theory is in its treatment of the coupled dielectric waveguide problem. We have carried out the cases above according to the coupled-mode

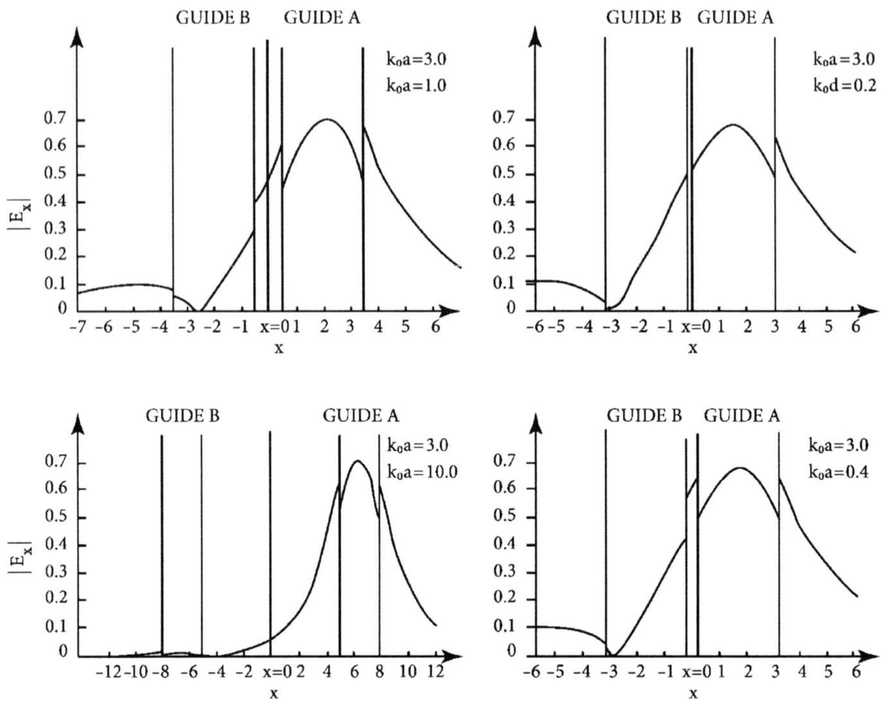

Figure 4.12. Transverse electric field distribution across two coupled guides [5]

theory described by Marcuse [9] and by the NELC researchers [8]. The "exact" normal-mode results are then compared with those obtained according to the various coupled-mode theories. Displayed in Fig. 4.13 are the curves for the coupling length as a function of the normalized separation distance between two parallel dielectric guides as shown for two different k_0a values in Fig. 4.8. We note that as the separation distance is increased, the agreement between the results based on coupled-mode theories and our normal-mode results becomes better, and that closer agreement is obtained for larger k_0a values or when more power is confined within the core of the guide. This is because large k_0a corresponds to more tightly bound field; hence, the coupling field is weaker and the coupled-mode theory, which is based on the perturbation concept, tends to be more accurate. It is worthwhile to point out that for certain combinations of k_0a and k_0d values, such as for $k_0a = 1$ and $k_0d \leq 7$, only the dominant symmetric mode exists, and so according to the exact normal-mode theory, no back and forth exchange of propagating power takes place between guide A and guide B. On the other hand, the approximate coupled-mode theory continues to predict the power exchange phenomenon. Another way of expressing the differences for the results based on

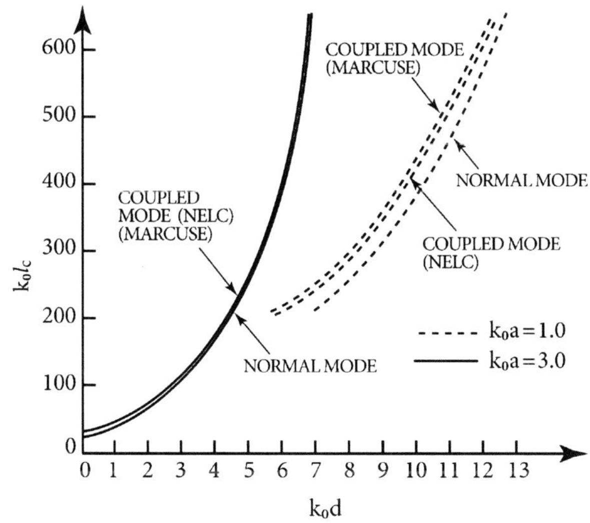

Figure 4.13. Normalized coupling length as a function of the normalized separation distance. Note that the coupling length ceases to exist for $k_0 d < 7.0$ when $k_0 a = 1.0$ according to the normal-mode theory [5]

different theories is shown in Fig. 4.13 where the percent differences between different coupled-mode theories and the normal-mode theory are plotted against the normalized separation distances.

It can be seen that the coupled-mode theory is surprisingly good (within 20%) in predicting the coupling length of two parallel dielectric slab guides even when the separation distances is relatively small and the confinement of guided power is relatively weak. Extrapolating our present results to other geometries involving optical fibers or integrated optical guides, it is inferred that the coupling distances predicted according to the coupled-mode theory are accurate to within 20% of the actual values if the symmetric and antisymmetric modes are above cutoff. Finally, it should be noted that when the separation distance is small, the transverse field configurations of the coupled guides (see Fig. 4.14) are significantly different than those assumed in coupled-mode theory.

4.6 The Sommerfeld–Zenneck Surface Impedance Waveguide [6]

Historically speaking, the very first solution to the surface wave problem was found for the Sommerfeld–Zenneck structure, which was a plane interface separating a dielectric and a good (not perfect) conductor. It has been shown in Sect. 2.3.1 that a good conductor may effectively be represented by a surface impedance $Z_s^{\text{conductor}}$,

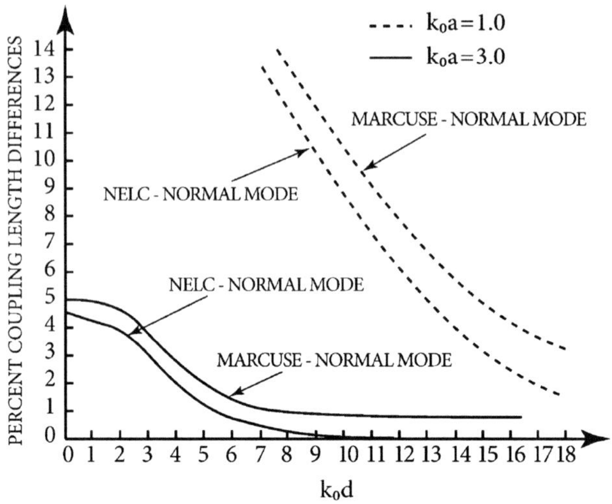

Figure 4.14. Percent coupling length differences between normal-mode theory and coupled-mode theory as a function of normalized separation distance [5]

where

$$Z_{\mathrm{S}}^{\text{conductor}} = (1+j)(\omega\mu_1/2\sigma_1)^{1/2}. \tag{4.144}$$

Here μ_1 is the permeability of the conductor and σ_1 is the conductivity. Thus the problem is reduced to finding the surface wave solution along the interface between free-space and a conductor with a surface impedance Z_{S}. In this case, no field can exist in the conducting half-space. Only external fields in the free-space region (region 0) can exist. From (4.1)–(4.5), the field expressions tangential to the interface that can satisfy the radiation condition are given as follows. For the TM mode

$$E_z^{(0)} = A_0\, e^{-q_0 x}, \tag{4.145}$$

$$H_y^{(0)} = A_0 \left(\frac{-j\omega\epsilon_0}{q_0}\right) e^{-q_0 x}, \tag{4.146}$$

and for the TE mode,

$$H_z^{(0)} = B_0\, e^{-q_0 x}, \tag{4.147}$$

$$E_y^{(0)} = A_0 \left(\frac{j\omega\mu_0}{q_0}\right) e^{-q_0 x}, \tag{4.148}$$

where $q_0^2 = \beta^2 - \omega^2 \mu_0 \epsilon_0$. Using the definition given by (2.59) on the surface $x = 0$, one has, for the TM mode,

$$-\mathbf{e}_x \times E_z^{(0)} \mathbf{e}_z = Z_s H_y^{(0)} \mathbf{e}_y \,|_{\text{at } x=0} \tag{4.149}$$

and for the TE mode,

$$-\mathbf{e}_x \times E_y^{(0)} \mathbf{e}_y = Z_s H_z^{(0)} \mathbf{e}_z \,|_{\text{at } x=0} \,, \tag{4.150}$$

or

$$\frac{E_z^{(0)}}{H_y^{(0)}} = Z_s \,|_{\text{at } x=0} \qquad \text{TM}, \tag{4.151}$$

$$\frac{E_y^{(0)}}{H_z^{(0)}} = -Z_s \,|_{\text{at } x=0} \quad \text{TE}. \tag{4.152}$$

Using (4.144)–(4.148), one obtains the dispersion relations

$$\frac{q_0}{j\omega\epsilon_0} = -Z_s \qquad \text{TM}, \tag{4.153}$$

$$\frac{j\omega\mu_0}{q_0} = -Z_s \qquad \text{TE}. \tag{4.154}$$

Since Z_s is a complex quantity, the solution for β must also be a complex quantity. This means the surface wave will propagate as well as decay along the z-axis. Thus the existence of the Sommerfeld–Zenneck wave is established.

Let us now investigate the case where the surface impedance is purely reactive, that is, $Z_s = \pm jX_s$. The upper + sign indicates inductive reactance while the lower – sign indicates capacitive reactance. Equations (4.153) and (4.154) become

$$q_0^{(\text{TM})} = \pm X_s \omega \epsilon_0, \tag{4.155}$$

$$q_0^{(\text{TE})} = \mp \frac{\omega\mu_0}{X_s}, \tag{4.156}$$

with $q_0^{(\text{TE}),(\text{TM})} = \left[\beta^{(\text{TE}),(\text{TM})2} - \omega^2\mu_0\epsilon_0\right]^{1/2}$. For propagating modes, $q_0^{(\text{TM})}$ or $q_0^{(\text{TE})}$ must be positive and real, so that the permitted reactance for the TM mode is inductive (the upper sign), and that for the TE mode is capacitive (the lower sign).

Two other questions of interest are (a) Can TEM modes be supported by a slab dielectric waveguide? and (b) Can a dielectric half-space support a surface wave?

It is known from the discussion of TEM modes in Chap. 2 that

$$\beta_{0,1}^{(\mathrm{TEM})} = k_{0,1} = \omega\sqrt{\mu_{0,1}\epsilon_{0,1}} = \frac{\omega}{c_{0,1}}, \tag{4.157}$$

where the subscripts $(0, 1)$ represent, respectively, the medium outside and inside a slab, c_0 is the speed of light outside the slab, and c_1 is the speed of light inside the slab. This means that the fields for the TEM wave always propagate at the speed of light for the medium in which they apply. But the fields for a surface wave mode must possess a single propagation constant no matter in which region the fields are located. In other words, $\beta_0^{(\mathrm{TEM})} = \beta_1^{(\mathrm{TEM})}$. However, this condition is not allowed because $k_0 \neq k_1$. This contradiction can only be resolved if the TEM mode is not allowed to exist in this open waveguide structure.

For the dielectric half-space problem, according to Sect. 2.5, the field components tangential to the interface at $x = 0$ in the upper $(x > 0)$ and lower $(x < 0)$ regions are, respectively,

$$\text{TM} \quad (x > 0) \quad E_z^{(0)} = A_0\, e^{-q_0 x}, \qquad H_y^{(0)} = A_0 \left(\frac{-j\omega\epsilon_0}{q_0}\right) e^{-q_0 x},$$

$$\text{TE} \quad (x > 0) \quad H_z^{(0)} = B_0\, e^{-q_0 x}, \qquad E_y^0 = B_0 \left(\frac{j\omega\mu}{q_0}\right) e^{-q_0 x},$$

and

$$\text{TM} \quad (x < 0) \quad E_z^{(1)} = A_1\, e^{q_1 x}, \qquad H_y^{(0)} = A_1 \left(\frac{j\omega\epsilon_1}{q_1}\right) e^{q_1 x},$$

$$\text{TE} \quad (x < 0) \quad H_z^{(1)} = B_1\, e^{q_1 x}, \qquad E_y^{(0)} = B_1 \left(\frac{-j\omega\mu}{q_1}\right) e^{q_1 x},$$

with $q_{0,1}^2 = \beta^2 - \omega^2\mu\epsilon_{0,1}$. Satisfying the boundary conditions at $x = 0$ yields

$$\text{TM:} \quad \epsilon_0 q_1 = -\epsilon_1 q_0 \qquad\qquad \text{TE:} \quad q_1 = -q_0.$$

Since q_0, q_1, ϵ_0, and ϵ_1 are all real and positive, the above equations cannot be satisfied. Thus, both TM and TE modes are not allowed to exist. Hence, the dielectric half-space cannot support a surface wave.

So, the conclusion is that a TEM mode cannot exist on any surface wave structure and no surface wave can exist on the interface of a lossless dielectric half-space.

References

1. R. E. Collin, "Field Theory of Guided Waves," McGraw-Hill, New York (1960)
2. A. W. Snyder and J. D. Love, "Optical Waveguide Theory," Chapman and Hall, London (1983)
3. D. B. Hall and C. Yeh, "Leaky waves in a heteroepitaxial film," J. Appl. Phys. **44**, 2271 (1973)
4. C. Yeh and G. Lindgren, "Computing the propagation characteristics of radially stratified fibers: An efficient method," Appl. Opt. **16**, 483 (1977); P. Yeh, A. Yariv, and C. S. Hong, "Electromagnetic propagation in periodic media. I. General theory," J. Opt. Soc. Am. **67**, 423 (1977)
5. C. Yeh, F. Manshadi, K. F. Casey, and A. Johnston, "Accuracy of directional coupler theory in fibers or integrated optics applications," J. Opt. Soc. Am. **68**, 1079 (1978)
6. J. A. Stratton, "Electromagnetic Theory," McGraw-Hill, New York (1941)
7. R. B. Dyott, "Elliptical Fiber Waveguides," Artech House, Boston (1995)
8. D. B. Hall, "Frequency selective coupling between planar waveguides," NELC Technical Note TN-2583, Naval Electronics Lab Center, San Diego, CA (1974)
9. D. Marcuse, "Light Transmission Optics," Van Nostrand-Reinhold, New York (1972)

5

CIRCULAR DIELECTRIC WAVEGUIDES

The first analytic solution for guiding a circularly symmetric TM wave along a solid lossless dielectric cylinder was obtained by Hondros and Debye in 1910 [1]. Until 1936, there was no complete mathematical analysis for this problem which was carried out by Carson et al. [2]. They were the first to show that to satisfy all the boundary conditions, a hybrid wave (i.e., the coexistence of longitudinal electric and magnetic fields) must be present. In other words, asymmetric TE and TM modes were inextricably coupled to each other along a circular dielectric rod. They also showed that (1) pure TE and TM waves could only exist in the circularly symmetric case, and (2) there existed one and only one mode, namely the lowest order hybrid wave called the HE_{11} mode, which possessed no cutoff frequency and could propagate at all frequencies. All other circularly symmetric or nonsymmetric modes had cutoff frequencies. The dispersion relations of these modes were also obtained in their paper, but no numerical results were given.

In 1945, Mallach [3] published his results on the use of the dielectric rod as a directive radiator. He showed experimentally that the radiation pattern obtained by the use of the asymmetric HE_{11} mode produced only one lobe in the principal direction of radiation. Immediately after Mallach's paper, Wegener [4] presented a dissertation in which the asymmetric HE_{11} mode together with the lowest order circularly symmetric TE and TM modes was analyzed in detail. Not only were the numerical results of the propagation constants for these waves obtained, but also their attenuation characteristics. Apparently he was unaware of Carson et al.'s work. A few experimental points were also included in his work to substantiate his theoretical results. Elsasser [5] in 1949, independent of Wegener's work, published his computation on the attenuation properties of these three lowest order waves by the perturbation method. In a companion paper, Chandler [6] verified experimentally Elsasser's results considering the dominant HE_{11} mode. He found that the guiding effect was retained even when the rod was only a fraction of a wavelength in diameter. Since the greater part of the guided energy was outside the dielectric, very little loss was observed. For the first time the cavity resonator technique was introduced to measure the attenuation constant of the HE_{11} mode. The resonator technique was very suitable for investigating very low loss uniform waveguides. It should be noted, however, that the formula relating the Q of the resonator and the attenuation constant α in Chandler's paper is only applicable for very small $2a/\lambda_0$, where a is the radius of the rod and λ_0 is the free space wavelength. A more general relation between α and Q was derived by Yeh [7] and given in Sect. 3.7.

King [8] in 1952 proposed the so-called "dielectric image line" as a practical surface guiding device. The dielectric image line was made up of a semicircular dielectric rod mounted on a conducting sheet and was designed specifically for the dominant HE_{11} mode. He indicated that the conducting sheet not only could act as a supporting device but also as a polarization anchor for this dominant mode. A detailed study on the attenuation and the radial field decay was reported by King and Schlesinger [9] in 1957. Again, the cavity resonator method, used by Chandler, was used for the attenuation constant measurement.

In 1961, Snitzer [10] rederived the analytic results originally given by Carson et al. and applied them specifically to the optical fiber problem. The purpose of this chapter is to provide the analytic approach and results for the problem of wave propagation along a solid circular cylindrical waveguide and its variations, layered radially inhomogeneous dielectric cylinders, hollow cylinders, etc.

5.1 Fundamental Equations

From Sect. 2.5, using circular cylindrical coordinates (r, θ, z), one obtains

$$E_r = \frac{1}{p^2}\left(-j\beta\frac{\partial E_z}{\partial r} - \frac{j\omega\mu_0}{r}\frac{\partial H_z}{\partial \theta}\right), \tag{5.1}$$

$$E_\theta = \frac{1}{p^2}\left(-\frac{j\beta}{r}\frac{\partial E_z}{\partial \theta} + j\omega\mu_0\frac{\partial H_z}{\partial r}\right), \tag{5.2}$$

$$H_r = \frac{1}{p^2}\left(\frac{j\omega\epsilon}{r}\frac{\partial E_z}{\partial \theta} - j\beta\frac{\partial H_z}{\partial r}\right), \tag{5.3}$$

$$H_\theta = \frac{1}{p^2}\left(-j\omega\epsilon\frac{\partial E_z}{\partial r} - \frac{j\beta}{r}\frac{\partial H_z}{\partial \theta}\right), \tag{5.4}$$

$$\left[\frac{1}{r}\left(\frac{\partial}{\partial r}r\frac{\partial}{\partial r} + \frac{\partial}{\partial \theta}\frac{1}{r}\frac{\partial}{\partial \theta}\right) + p^2\right]\left[\begin{array}{c} E_z \\ H_z \end{array}\right] = 0, \tag{5.5}$$

$$p^2 = \omega^2\mu_0\epsilon - \beta^2. \tag{5.6}$$

Here, β is the propagation constant of the fields and (ϵ, μ_0) are the constitutive parameters of the medium in which the fields reside. All fields vary as $e^{j(\omega t - \beta z)}$. This factor has been suppressed.

In general, the cylindrical dielectric waveguide can support a family of circularly symmetric TE_{0m} or TM_{0m} modes (whose fields are independent of the azimuthal coordinates) and a family of hybrid HE_{nm} or EH_{nm} modes. In this chapter, the subscripts n and m denote, respectively, the number of cyclic variations with the azimuthal coordinates and the mth root of the characteristic equation, which is obtained by satisfying the boundary conditions. The symbol HE refers to the modes with the ratio $(\mu_0\omega/\beta)/(H_z/E_z) = -1$ far from the cutoff frequency, while the symbol EH refers to the modes with the ratio $(\mu_0\omega/\beta)/(H_z/E_z) = +1$ far from the cutoff frequency.

5.2 Modes on Uniform Solid Core Circular Dielectric Cylinder

Shown in Fig. 5.1 is the geometry of the problem. A circular dielectric cylinder of radius a with permittivity ϵ_1 and permeability μ_0 is immersed in an infinite medium with permittivity ϵ_2 and permeability μ_0. Here, $\epsilon_1 > \epsilon_2$. Solutions of (5.5) for the axial fields (E_z, H_z) in terms of the radial Bessel functions and the angular circular functions have been given in (2.140). We must select the appropriate ones that will satisfy the required boundary conditions. In region 1 $(r < a)$, all field components must be finite.

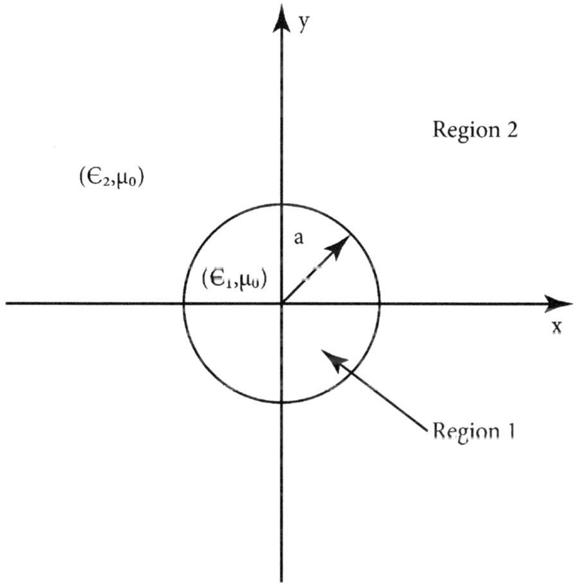

Figure 5.1. Geometry of a circular dielectric cylinder immersed in an infinite medium

Thus,

$$E_z^{(1)}(r,\theta) = \sum_{n=-\infty}^{\infty} A_n J_n(p_1 r) e^{jn\theta}, \tag{5.7}$$

$$H_z^{(1)}(r,\theta) = \sum_{n=-\infty}^{\infty} B_n J_n(p_1 r) e^{jn\theta}, \tag{5.8}$$

where the coefficients A_n and B_n are arbitrary constants. The transverse fields can be obtained from (5.1)–(5.4) and using (5.7) and (5.8) with $p^2 = p_1^2$. One has

$$E_r^{(1)}(r,\theta) = \frac{1}{p_1^2} \sum_{n=-\infty}^{\infty} \left[-j\beta p_1 A_n J_n'(p_1 r) + \frac{\omega\mu_0 n}{r} B_n J_n(p_1 r) \right] e^{jn\theta}, \tag{5.9}$$

$$E_\theta^{(1)}(r,\theta) = \frac{1}{p_1^2} \sum_{n=-\infty}^{\infty} \left[\frac{\beta n}{r} A_n J_n(p_1 r) + j\omega\mu_0 p_1 B_n J_n'(p_1 r) \right] e^{jn\theta}, \tag{5.10}$$

$$H_r^{(1)}(r,\theta) = \frac{1}{p_1^2} \sum_{n=-\infty}^{\infty} \left[-\frac{\omega\epsilon_1 n}{r} A_n J_n(p_1 r) - j\beta p_1 B_n J_n'(p_1 r) \right] e^{jn\theta}, \tag{5.11}$$

$$H_\theta^{(1)}(r,\theta) = \frac{1}{p_1^2} \sum_{n=-\infty}^{\infty} \left[-j\omega\epsilon_1 p_1 A_n J_n'(p_1 r) + \frac{\beta n}{r} B_n J_n(p_1 r) \right] e^{jn\theta}. \tag{5.12}$$

The prime signifies the derivative of the function with respect to its argument, $J_n(p_1 r)$ is the Bessel functions of the first kind of order n and argument $p_1 r$, and

$$p_1^2 = k_1^2 - \beta^2, \tag{5.13}$$

$$k_1^2 = \omega^2 \mu_0 \epsilon_1. \tag{5.14}$$

In region 2 ($r > a$), all field components must decay to zero as $r \to \infty$. Thus

$$E_z^{(2)}(r,\theta) = \sum_{n=-\infty}^{\infty} C_n K_n(q_2 r) e^{jn\theta}, \tag{5.15}$$

$$H_z^{(2)}(r,\theta) = \sum_{n=-\infty}^{\infty} D_n K_n(q_2 r) e^{jn\theta}, \tag{5.16}$$

where the coefficients C_n and D_n are arbitrary constants. The modified Bessel function $K_n(q_2 r)$ is chosen since as $q_2 r \to \infty$, $K_n(q_2 r) \to 0$, as required by the radiation condition at ∞. The transverse fields can be obtained from (5.1)–(5.4) using (5.15) and (5.16) with $p^2 = -q_2^2$. We have

$$E_r^{(2)}(r,\theta) = -\frac{1}{q_2^2}\sum_{n=-\infty}^{\infty}\left[-j\beta q_2 C_n K_n'(q_2 r) + \frac{\omega\mu_0 n}{r} D_n K_n(q_2 r)\right] e^{jn\theta}, \tag{5.17}$$

$$E_\theta^{(2)}(r,\theta) = -\frac{1}{q_2^2}\sum_{n=-\infty}^{\infty}\left[\frac{\beta n}{r} C_n K_n(q_2 r) + j\omega\mu_0 q_2 D_n K_n'(q_2 r)\right] e^{jn\theta}, \tag{5.18}$$

$$H_r^{(2)}(r,\theta) = -\frac{1}{q_2^2}\sum_{n=-\infty}^{\infty}\left[-\frac{\omega\epsilon_2 n}{r} C_n K_n(q_2 r) - j\beta q_2 D_n K_n'(q_2 r)\right] e^{jn\theta}, \tag{5.19}$$

$$H_\theta^{(2)}(r,\theta) = -\frac{1}{q_2^2}\sum_{n=-\infty}^{\infty}\left[-j\omega\epsilon_2 q_2 C_n K_n'(q_2 r) + \frac{\beta n}{r} D_n K_n(q_2 r)\right] e^{jn\theta}, \tag{5.20}$$

where the prime signifies the derivative of the function with respect to its argument, $K_n(q_2 r)$ is the modified Bessel function of the second kind of order n and argument $q_2 r$, and

$$q_2^2 = \beta^2 - k_2^2, \tag{5.21}$$

$$k_2^2 = \omega^2 \mu_0 \epsilon_2. \tag{5.22}$$

5.2.1 Dispersion Relations

The propagation constant β of the guided wave on this dielectric cylinder will be obtained from the dispersion relation, which is derived by enforcing the satisfaction of the boundary conditions by the fields, as follows: At the boundary surface $r = a$, the tangential fields must be continuous, that is,

$$E_z^{(1)}(a,\theta) = E_z^{(2)}(a,\theta), \tag{5.23}$$

$$H_z^{(1)}(a,\theta) = H_z^{(2)}(a,\theta), \tag{5.24}$$

$$E_\theta^{(1)}(a,\theta) = E_\theta^{(2)}(a,\theta), \tag{5.25}$$

$$H_\theta^{(1)}(a,\theta) = H_\theta^{(2)}(a,\theta). \tag{5.26}$$

Thus,

$$\sum_{n=-\infty}^{\infty} C_n K_n(q_2 a) e^{jn\theta} = \sum_{n=-\infty}^{\infty} A_n J_n(p_1 a) e^{jn\theta}, \tag{5.27}$$

$$\sum_{n=-\infty}^{\infty} D_n K_n(q_2 a) e^{jn\theta} = \sum_{n=-\infty}^{\infty} B_n J_n(p_1 a) e^{jn\theta}, \tag{5.28}$$

$$\begin{aligned} -\frac{1}{q_2^2} \sum_{n=-\infty}^{\infty} & \left[\frac{\beta n}{a} C_n K_n(q_2 a) + j\omega\mu_0 q_2 D_n K_n'(q_2 a) \right] e^{jn\theta} \\ &= \frac{1}{p_1^2} \sum_{n=-\infty}^{\infty} \left[\frac{\beta n}{a} A_n J_n(p_1 a) + j\omega\mu_0 p_1 B_n J_n'(p_1 a) \right] e^{jn\theta}, \end{aligned} \tag{5.29}$$

$$\begin{aligned} -\frac{1}{q_2^2} \sum_{n=-\infty}^{\infty} & \left[-j\omega\epsilon_2 q_2 C_n K_n'(q_2 a) + \frac{\beta n}{a} D_n K_n(q_2 a) \right] e^{jn\theta} \\ &= \frac{1}{p_1^2} \sum_{n=-\infty}^{\infty} \left[-j\omega\epsilon_1 p_1 A_n J_n'(p_1 a) + \frac{\beta n}{a} B_n J_n(p_1 a) \right] e^{jn\theta}. \end{aligned} \tag{5.30}$$

Using the orthogonality relations for the trigonometric functions, the θ dependence and the summation may be eliminated, resulting in a set of simultaneous linear equations for the unknown coefficients A_n, B_n, C_n, and D_n, as follows:

$$[M] \begin{bmatrix} A_n \\ B_n \\ C_n \\ D_n \end{bmatrix} = 0, \tag{5.31}$$

where

$$[M] = \begin{bmatrix} J_n(p_1 a) & 0 & -K_n(q_2 a) & 0 \\ 0 & J_n(p_1 a) & 0 & -K_n(q_2 a) \\ \frac{\beta n}{a p_1^2} J_n(p_1 a) & \frac{j\omega\mu_0}{p_1} J_n'(p_1 a) & \frac{\beta n}{a q_2^2} K_n(q_2 a) & \frac{j\omega\mu_0}{q_2} K_n'(q_2 a) \\ -\frac{j\omega\epsilon_1}{p_1} J_n'(p_1 a) & \frac{\beta n}{a p_1^2} J_n(p_1 a) & -\frac{j\omega\epsilon_2}{q_2} K_n'(q_2 a) & \frac{\beta n}{a q_2^2} K_n(q_2 a) \end{bmatrix}. \tag{5.32}$$

We note that the unknown coefficients B_n, C_n, and D_n can be expressed in terms of the arbitrary coefficient A_n, which can be determined from the excitation condition. So, from (5.31)

$$B_n = f_n^{(1)} A_n, \tag{5.33}$$

$$C_n = f_n^{(2)} A_n, \tag{5.34}$$

$$D_n = f_n^{(3)} A_n. \tag{5.35}$$

The functions $f_n^{(1),(2),(3)}$ may be found from (5.31).

For a nontrivial solution of (5.31), the determinant of the simultaneous equations, (5.31), must be set to zero, that is, $|M| = 0$. The determinant of this equation is purely real and dimensionless. Evaluation of this determinant yields the following eigenvalue equation for β:

$$\left[\frac{J_n'(p_1 a)}{p_1 J_n(p_1 a)} + \frac{K_n'(q_2 a)}{q_2 K_n(q_2 a)}\right]\left[\frac{k_1^2 J_n'(p_1 a)}{p_1 J_n(p_1 a)} + \frac{k_2^2 K_n'(q_2 a)}{q_2 K_n(q_2 a)}\right]$$
$$= \left(\frac{\beta n}{a}\right)^2 \left(\frac{1}{p_1^2} + \frac{1}{q_2^2}\right)^2. \tag{5.36}$$

Equation (5.36) is called the dispersion relation. It is a function of βa, ϵ_{r1}, ϵ_{r2}, $k_0 a$, and n, or $f(\beta a, \epsilon_{r1}, \epsilon_{r2}, k_0 a, n) = 0$. Here, $\epsilon_{ri} = \epsilon_i/\epsilon_0$. Given ϵ_{r1}, ϵ_{r2}, $k_0 a$, and n, we can solve for βa that will satisfy the above equation.

The fact that for each angular integer n there exists a distinct dispersion relation for β is worth nothing. This means the required boundary conditions may be satisfied by each order of angular integer n separately. (This is not the case for the elliptical dielectric cylinder case, as will be demonstrated in a later chapter.) It is further noted that for the $n = 0$ case there are two distinct dispersion relations:

$$\text{TE}_{0m} \text{ Modes:} \qquad \frac{J_0'(p_1 a)}{p_1 a J_0(p_1 a)} + \frac{K_0'(q_2 a)}{q_2 a K_0(q_2 a)} = 0 \tag{5.37}$$

and

$$\text{TM}_{0m} \text{ Modes:} \qquad \frac{k_1^2 J_0'(p_1 a)}{p_1 a J_0(p_1 a)} + \frac{k_2^2 K_0'(q_2 a)}{q_2 a K_0(q_2 a)} = 0. \tag{5.38}$$

These representations show that for the circular symmetric case, that is, $n = 0$ case, the boundary conditions may be satisfied by a pure TE wave or by a pure TM wave. Only for the $n \neq 0$ case will hybrid waves be needed to satisfy all the boundary conditions.

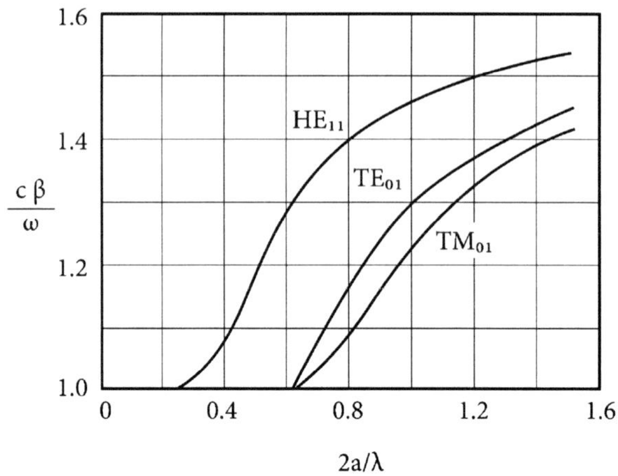

Figure 5.2. Velocity ratio of c/v for polystyrene rod ($\epsilon_{r1} = 2.56$) embedded in free space. c is the velocity of light in free space. a is the radius of the rod and λ_0 is the free space wavelength [5]

Let us elaborate on the integer m. For each of the above dispersion relations, (5.36)–(5.38), there exists m number of roots that would satisfy that relation. For example, for a given $k_0 a$, that is, for a given normalized frequency, ϵ_{r1}, ϵ_{r2}, and n, (5.36) will yield m number of distinct real roots for βa. The finite number of roots are consecutively labeled according to the lowest value, m=1, to the highest value.

Typical numerical solutions for the dispersion relations (5.36)–(5.38) for a dielectric rod in free-space are shown in Fig. 5.2, where the normalized phase velocities as a function of frequency are displayed for three lowest order modes [5]. Figure 5.3 shows the normalized guide wavelength λ/λ_0 as a function of normalized cross-sectional area $(2a/\lambda_0)^2$ for the dominant HE_{11} mode for various relative core dielectric constants. The cylinder is immersed in free space. Here, λ_0 is the free-space wavelength.

5.2.2 Cutoff Conditions

Referring to the external fields (5.15)–(5.20), we note that the argument of the modified Bessel function must be real in order to satisfy the radiation condition. Thus,

$$q_2 \geq 0. \tag{5.39}$$

From (5.21), using (5.39) we have

$$\beta \geq k_2. \tag{5.40}$$

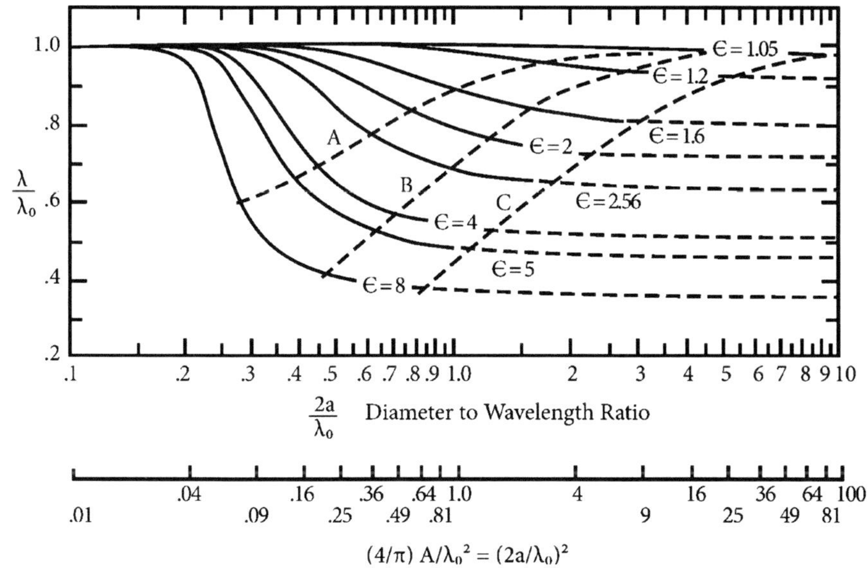

Figure 5.3. Normalized guide wavelength λ/λ_0 as a function of normalized cross-sectional area $(2a/\lambda_0)^2$ for the dominant HE_{11} mode for various relative core dielectric constants

In order that the dispersion relation, (5.36), may yield real roots, the radial function describing the internal fields ($r \leq a$) must be real. Thus, $p_1 a > 0$. From (5.13), we have

$$\beta \leq k_1. \tag{5.41}$$

Combining (5.40) and (5.41) yields

$$k_2 \leq \beta \leq k_1. \tag{5.42}$$

At cutoff, the mode is no longer bound to the core of the dielectric waveguide. Therefore, its fields no longer decay on the outside of the core, or

$$q_2 \text{ (at cutoff)} = 0. \tag{5.43}$$

Combining (5.13) and (5.21) and eliminating β^2 yields $p_1^2 = k_1^2 - k_2^2 - q_2^2$. Substituting (5.43) into the previous expression gives

$$k_0 \text{ (at cutoff)} = \frac{p_1}{(\epsilon_{r1} - \epsilon_{r2})^{1/2}}$$

or

$$\omega \text{ (at cutoff)} = \frac{p_1 c}{(\epsilon_{r1} - \epsilon_{r2})^{1/2}}, \tag{5.44}$$

where c is the speed of light in vacuum and p_1a corresponds to the root of the dispersion relation with $q_2a = 0$. Physically, it means that at or below this cutoff frequency, the dielectric structure can no longer support such a mode and thereby ceases to be a binding medium for that mode. Simple cutoff conditions can be obtained for the circular dielectric cylinder structure by setting $q_2a = 0$ in (5.36)–(5.38) [10, 11].

For TE_{0m} and TM_{0m} modes

$$J_0(p_1a) = 0. \tag{5.45}$$

For HE_{1m} or EH_{1m} modes

$$J_1(p_1a) = 0. \tag{5.46}$$

For HE_{nm} modes with $n \geq 2$

$$\left(\frac{\epsilon_1}{\epsilon_2} + 1\right) J_{n-1}(p_1a) = \frac{p_1a}{n-1} J_n(p_1a). \tag{5.47}$$

For EH_{nm} modes with $n \geq 2$

$$J_n(p_1a) = 0. \tag{5.48}$$

It is understood that $m = 1, 2, 3, \ldots$. The roots of these equations provide the values for p_1a for various modes at cutoff. Knowing the cutoff value of p_1a for a particular mode, one can calculate the cutoff frequency of that mode from (5.44). Only one mode, the HE_{11} mode, has zero cutoff frequency since the first root of (5.46) with $m = n = 1$ is $p_1a = 0$. The HE_{11} mode is called the dominant mode. The second lowest cutoff value occurs when $p_1a = 2.405$ for the TE_{01} or TM_{01} mode. Hence, the frequency range of single-guided mode (HE_{11} mode) operation is given by

$$0 < \omega < 2.405\,(c/a)\,(\epsilon_{r1} - \epsilon_{r2})^{-1/2}. \tag{5.49}$$

Table 5.1 gives the cutoff values of p_1a for various lower-order modes.

We recall from Chap. 3 that the following normalized parameters are frequently used in the literature to represent normalized frequency and normalized propagation constant:

$$V = a\left(p_1^2 + q_2^2\right)^{1/2} = k_0a(n_1^2 - n_2^2)^{1/2},$$

$$b = \frac{q_2^2a^2}{V^2} = \frac{(\beta/k_0)^2 - n_2^2}{n_1^2 - n_2^2},$$

where $n_1 = \sqrt{\epsilon_1/\epsilon_0}$, $n_2 = \sqrt{\epsilon_2/\epsilon_0}$, $k_0^2 = \omega^2\mu_0\epsilon_0$, λ_0 is the free-space wavelength, and a is the core radius. For single mode operation, $V \leq 2.405$. At cutoff $b = 0$.

Table 5.1 Cutoff values p_1a of several lower-order modes[a]

$n\backslash m$	1	2	3	4	5	6	Modes	
0	2.405	5.520	8.654	11.792	14.931	18.072	TE, TM	
1	0	3.832	7.016	10.173	13.324	16.471	HE	
1	3.832	7.016	10.173	13.324	16.471	19.616	EH	Independent
2	5.136	8.417	11.620	14.796	17.960	21.117	EH	of
3	6.380	9.761	13.015	16.223	19.409	22.583	EH	$\sqrt{\epsilon_{r1}}$ and $\sqrt{\epsilon_{r2}}$
4	7.588	11.065	14.373	17.616	20.827	24.019	EH	
5	8.772	12.339	15.700	18.980	22.218	25.450	EH	
2	2.413	5.524	8.657	11.793	14.932	18.072	HE	
3	3.842	7.021	10.177	13.327	16.473	19.618	HE	$\sqrt{\epsilon_{r1}} = 1.515$
4	5.147	8.424	11.625	14.800	17.963	21.120	HE	$\sqrt{\epsilon_{r2}} = 1.500$
5	6.393	9.769	13.021	16.228	19.413	22.586	HE	
2	2.421	5.527	8.658	11.795	14.934	18.073	HE	
3	3.853	7.027	10.181	13.330	16.476	19.620	HE	$\sqrt{\epsilon_{r1}} = 1.530$
4	5.159	8.432	11.630	14.804	17.967	21.123	HE	
5	6.405	9.777	13.027	16.233	19.418	22.590	EH	

[a]Note that the cutoff values for TE_{0m}, TM_{0m}, HE_{1m}, and EH_{nm} are independent of the core index n_1 or the cladding index n_2

5.2.3 Attenuation

If the material for the surface waveguide is imperfect, then there exists an attenuation constant α for all field components (i.e., all field components vary as $e^{-j\beta z}e^{-\alpha z}e^{j\omega t}$). There are two ways to calculate the attenuation constant: (a) an exact way where the problem is treated as an exact boundary-value problem; and (b) a perturbation approach where all fields are assumed to remain unchanged by the small loss factor.

5.2.3.1 The Exact Approach

As a boundary-value problem, all fields are derived from the wave equation with lossy or complex constitutive parameters, that is,

$$\epsilon = \epsilon^{\mathrm{r}} - j\epsilon^{\mathrm{i}},$$
$$\mu = \mu^{\mathrm{r}} - j\mu^{\mathrm{i}},$$

where the superscript r means the real part of and the superscript i means the imaginary part of. For all nonmagnetic materials, $\mu = \mu_0$. The primary effect of including the complex constitutive parameters in the wave equation is that the argument of its solutions will be complex. Therefore, the resultant dispersion relation will contain functions whose arguments will be complex quantities. For example, the dispersion relation (5.36) for a lossy circular dielectric rod in free-space is

$$\left[\frac{J_n'(p_1^{\mathrm{c}}a)}{p_1^{\mathrm{c}}J_n(p_1^{\mathrm{c}}a)}+\frac{K_n'(q_0^{\mathrm{c}}a)}{q_0^{\mathrm{c}}K_n(q_0^{\mathrm{c}}a)}\right]\left[\frac{k_1^{\mathrm{c}2}J_n'(p_1^{\mathrm{c}}a)}{p_1^{\mathrm{c}}J_n(p_1^{\mathrm{c}}a)}+\frac{k_0^2K_n'(q_0^{\mathrm{c}}a)}{q_0^{\mathrm{c}}K_n(q_0^{\mathrm{c}}a)}\right]$$

$$=\left(\frac{(\beta-j\alpha)\,n}{a}\right)^2\left(\frac{1}{p_1^{\mathrm{c}2}}+\frac{1}{q_0^{\mathrm{c}2}}\right)^2, \tag{5.50}$$

with

$$p_1^{\mathrm{c}2}=k_1^{\mathrm{c}2}-(\beta-j\alpha)^2, \tag{5.51}$$

$$k_1^{\mathrm{c}2}=\omega^2\mu_0\left(\epsilon_1^r-j\epsilon_1^i\right), \tag{5.52}$$

$$q_0^{\mathrm{c}2}=(\beta-j\alpha)^2-k_0^2, \tag{5.53}$$

$$k_0^2=\omega^2\mu_0\epsilon_0. \tag{5.54}$$

Note that even though the modified Bessel functions apply to the lossless free-space region, their arguments must still be complex to satisfy (5.53)

The complex roots of (5.50) will provide for β and α. Here, no assumption has been made on the smallness of the imaginary part of the dielectric constant. Therefore the solution is exact.

The above derivation for β and α of a single mode is still valid only when a single mode is propagating alone in the dielectric waveguide. When more than one mode is copropagating in the waveguide, the power loss must be calculated in the manner described in Sect. 3.5.

5.2.3.2 The Perturbation Approach

If the imaginary part of the dielectric constant for the dielectric rod is very small and if the power loss per wavelength along the rod is small, compared to the power carried by the rod, a perturbation approach may be used to find the attenuation constant of the wave supported by the rod.

$$\alpha_{\mathrm{d}}=4.343\left(\sigma_{\mathrm{d}1}R_1+\sigma_{\mathrm{d}2}R_2\right)\qquad(\mathrm{dB\,m^{-1}}), \tag{5.55}$$

$$R_1=L_1/S, \tag{5.56}$$

$$R_2 = L_2/S, \tag{5.57}$$

$$L_1 = \int_{A_1} [\mathbf{E}_1(\mathbf{r}) \cdot \mathbf{E}_1^*(\mathbf{r})]\, dA, \tag{5.58}$$

$$L_2 = \int_{A_2} [\mathbf{E}_2(\mathbf{r}) \cdot \mathbf{E}_2^*(\mathbf{r})]\, dA, \tag{5.59}$$

$$S = \mathrm{Re}\left\{\int_{A_1} [\mathbf{E}_1(\mathbf{r}) \times \mathbf{H}_1^*(\mathbf{r})] \cdot \mathbf{e}_z\, dA + \int_{A_2} [\mathbf{E}_2(\mathbf{r}) \times \mathbf{H}_2^*(\mathbf{r})] \cdot \mathbf{e}_z\, dA\right\}, \tag{5.60}$$

where σ_{d1} and σ_{d2} are, respectively, the conductivity of the core and cladding dielectrics ($\sigma = \omega\epsilon \tan\delta$), A_1 and A_2 are, respectively, the core cross-sectional area and the cladding cross-sectional area. The total time-average power carried by mode m along the guide is $S/2$ and the subscripts 1,2 refer to the core and cladding regions. The fields $\mathbf{E}_{1,2}(\mathbf{r})$ and $\mathbf{H}_{1,2}(\mathbf{r})$ are the unperturbed fields. As pointed out in Sect. 3.5, this perturbation formula (5.55) is valid only for a single propagating mode.

Since for most cases we are only interested in single mode propagation along slightly lossy guides, the perturbation formula will be used. Let us consider the specific case of a dielectric rod in lossless free-space. So, $\sigma_{\mathrm{d2}} = 0$. The attenuation formula becomes

$$\alpha_{\mathrm{d}} = 4.343\, \sigma_{\mathrm{d1}} L_1/S, \tag{5.61}$$

$$L_1 = \int_{A_1} [\mathbf{E}_1(\mathbf{r}) \cdot \mathbf{E}_1^*(\mathbf{r})]\, dA = \int_0^{2\pi}\int_0^a \left[\mathbf{E}_r^{(1)} \cdot \mathbf{E}_r^{(1)*} + \mathbf{E}_\theta^{(1)} \cdot \mathbf{E}_\theta^{(1)*}\right] r dr d\theta, \tag{5.62}$$

$$\begin{aligned} S &= \mathrm{Re}\left\{\int_{A_1} [\mathbf{E}_1(\mathbf{r}) \times \mathbf{H}_1^*(\mathbf{r})] \cdot \mathbf{e}_z\, dA + \int_{A_2} [\mathbf{E}_2(\mathbf{r}) \times \mathbf{H}_2^*(\mathbf{r})] \cdot \mathbf{e}_z\, dA\right\}, \\ &= \mathrm{Re}\int_0^{2\pi}\int_0^a \left[E_r^{(1)} H_\theta^{(1)*} - E_\theta^{(1)} H_r^{(1)*}\right] r dr d\theta \\ &\quad + \mathrm{Re}\int_0^{2\pi}\int_a^\infty \left[E_r^{(2)} H_\theta^{(2)*} - E_\theta^{(2)} H_r^{(2)*}\right] r dr d\theta. \end{aligned} \tag{5.63}$$

Evaluation of the above integral can be done in a straightforward manner and will not be given here. The attenuation characteristics of a typical dielectric rod is shown in Fig. 3.7. There the calculated attenuation constants, α_{d}, for three lower order modes are plotted against the normalized frequency. It is seen that near the cutoff frequency, most of the modal fields are outside the core region, which is the region without loss. Therefore, very little loss is experienced by the mode, or $\alpha \to 0$. As the frequency increases from cutoff, more and more fields migrate toward the core region, increasing the attenuation until it reaches an asymptote where all the modal power is carried within the core region. The modal fields degenerate into a plane wave field propagating in an infinitely large core region. The attenuation constant for a plane wave in the core region is

$$\alpha_{\text{plane wave in core}} = 4.343\, \omega \sqrt{\mu_0 \epsilon_0} \tan \delta_1 \sqrt{\epsilon_{r1}} \qquad (\text{dB m}^{-1}), \tag{5.64}$$

where $\epsilon_{r1} = \epsilon_1/\epsilon_0$.

If conductors are included as part of the surface wave structures (such as dielectric coated wires), then in addition to the dielectric loss there is an attenuation constant α_{c} associated with the loss due to the finite conductivity, that is,

$$\alpha_{\text{c}} = 4.343\, R_{\text{s}} L_3 / S \qquad (\text{dB m}^{-1}), \tag{5.65}$$

$$L_3 = \oint_c [\mathbf{H}_{\text{s}}(\mathbf{r}) \cdot \mathbf{H}_{\text{s}}^*(\mathbf{r})]\, dc, \tag{5.66}$$

where S has been given by (5.60), $\mathbf{H}_{\text{s}}$ is the magnetic field tangential to the conducting surface, and c is the contour path along the conducting surface that is normal to z, the axial direction of the guide. The total attenuation is

$$\alpha_{\text{T}} = \alpha_{\text{d}} + \alpha_{\text{c}}. \tag{5.67}$$

5.2.4 Field Configurations

One of the most distinguishing features of a dielectric waveguide as compared with other types of waveguides containing metallic structures is that all electric and magnetic field lines for each mode must form closed loops and can extend beyond the core region of the guide. A sketch of the HE_{11} mode has been shown in Fig. 3.4.

The cross-sectional views of the transverse field vectors for four lowest-order modes in a circular dielectric waveguide are shown in Fig. 5.4. Since the normal component of the displacement vector $\mathbf{D}$ must be continuous across the dielectric boundary, the normal component of the electric field must undergo a large discontinuity at the core-free space boundary when the core dielectric material has a large dielectric constant value relative to free space (say, $\epsilon_1/\epsilon_2 > 9$). This means that the conventional view of expecting higher power density to occur within the core region of a dielectric waveguide is not necessarily correct.

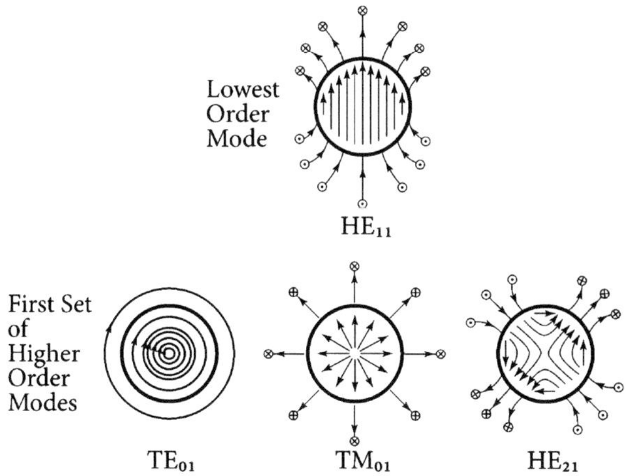

Figure 5.4. Cross-sectional views of the transverse electric field vectors for the four lowest-order modes in a step-index fiber

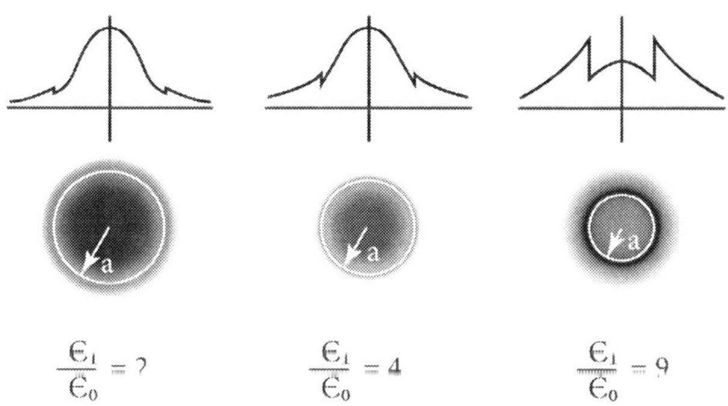

Figure 5.5. A sketch of the power density distributions for the dominant HE$_{11}$ mode for different core dielectric constants. $V = (\omega a/c)\,(\epsilon_1/\epsilon_0 - 1)^{1/2} = 2.0$

Shown in Fig. 5.5 are the cross-sectional power density distributions of the HE$_{11}$ mode for three circular dielectric waveguides with three different core dielectric constants and with the same normalized frequency, $V = 2$. For small values of ϵ_1/ϵ_0 (< 2), the power density distribution for the dominant mode peaks in the core center and decays rapidly from a relatively small value outside the core boundary to zero. For large values of ϵ_1/ϵ_0 (> 8), the power density distribution for the dominant mode may even peak just outside the core boundary due to the large discontinuity of the normal **E** field away from the core boundary. This doughnut shape dominant mode power distribution indicates that the way to excite the dominant mode on a circular dielectric waveguide must be different for different values of ϵ_1/ϵ_0.

5.3 The Sommerfeld–Goubau Wire

When a perfectly conducting wire is coated with a thin sheet of low-loss dielectric material as shown in Fig. 5.6, improved guidance of surface waves is observed [12, 13]. Using (5.1)–(5.6), we find that the axial fields in region 1 must take the following forms:

$$E_z^{(1)}(r,\theta) = \sum_{n=-\infty}^{\infty} \left[A_n^{(1)} J_n(p_1 r) + A_n^{(2)} Y_n(p_1 r)\right] e^{jn\theta}, \tag{5.68}$$

$$H_z^{(1)}(r,\theta) = \sum_{n=-\infty}^{\infty} \left[B_n^{(1)} J_n(p_1 r) + B_n^{(2)} Y_n(p_1 r)\right] e^{jn\theta}, \tag{5.69}$$

where $A_n^{(1)}$, $A_n^{(2)}$, $B_n^{(1)}$, and $B_n^{(2)}$ are arbitrary constants. The axial fields in region 0 must be

$$E_z^{(0)}(r,\theta) = \sum_{n=-\infty}^{\infty} C_n K_n(q_0 r)\, e^{jn\theta}, \tag{5.70}$$

$$H_z^{(0)}(r,\theta) = \sum_{n=-\infty}^{\infty} D_n K_n(q_0 r)\, e^{jn\theta}, \tag{5.71}$$

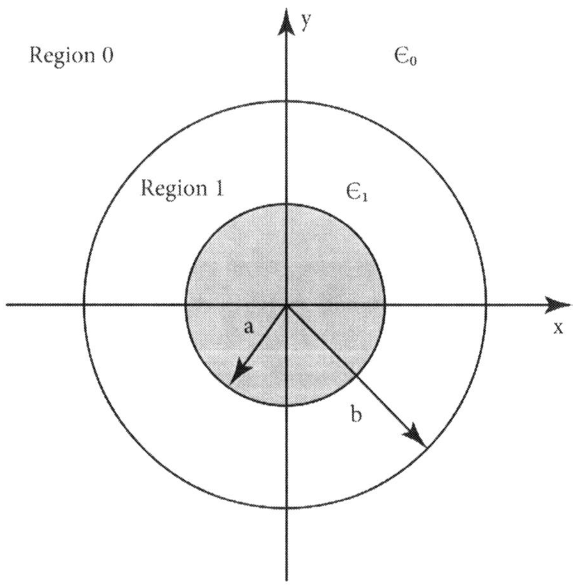

Figure 5.6. Geometry of the Sommerfeld–Goubau wire. The central cylinder is a solid conductor

where C_n and D_n are arbitrary constants. All transverse fields in these regions can be found using (5.1)–(5.4), with the appropriate substitution of p by p_1 and ϵ by ϵ_1 for region 1 and p^2 by -q_0^2 and ϵ by ϵ_0 for region 0. All fields are multiplied by $e^{-j\beta z+j\omega t}$, which is suppressed.

On the conducting cylinder, $r = a$, the tangential fields must be zero:

$$E_z^{(1)}(a,\theta) = 0, \tag{5.72}$$

$$E_\theta^{(1)}(a,\theta) = 0. \tag{5.73}$$

They yield

$$A_n^{(2)} = -A_n^{(1)} \frac{J_n(p_1 a)}{Y_n(p_1 a)}, \tag{5.74}$$

$$B_n^{(2)} = -B_n^{(1)} \frac{J_n'(p_1 a)}{Y_n'(p_1 a)}. \tag{5.75}$$

The tangential fields at $r = b$ must be continuous:

$$E_z^{(1)}(b,\theta) = E_z^{(0)}(b,\theta), \tag{5.76}$$

$$E_\theta^{(1)}(b,\theta) = E_\theta^{(0)}(b,\theta), \tag{5.77}$$

$$H_z^{(1)}(b,\theta) = H_z^{(0)}(b,\theta), \tag{5.78}$$

$$H_\theta^{(1)}(b,\theta) = H_\theta^{(0)}(b,\theta). \tag{5.79}$$

Satisfying these conditions gives

$$\begin{bmatrix} m_{11} & 0 & m_{13} & 0 \\ 0 & m_{22} & 0 & m_{24} \\ m_{31} & m_{32} & m_{33} & m_{34} \\ m_{41} & m_{42} & m_{43} & m_{44} \end{bmatrix} \begin{bmatrix} A_n^{(1)} \\ B_n^{(1)} \\ C_n \\ D_n \end{bmatrix} = 0, \tag{5.80}$$

with

$$m_{11} = J_n(p_1 b) - \frac{J_n(p_1 a) Y_n(p_1 b)}{Y_n(p_1 a)},$$

$$m_{13} = -K_n(q_0 b),$$

$$
\begin{aligned}
m_{22} &= J_n(p_1 b) - \frac{J_n'(p_1 a) Y_n(p_1 b)}{Y_n'(p_1 a)},\\
m_{24} &= -K_n(q_0 b),\\
m_{31} &= \frac{\beta n}{b p_1^2} m_{11},\\
m_{32} &= \frac{j\omega\mu_0}{p_1}\left[J_n'(p_1 b) - \frac{J_n'(p_1 a) Y_n'(p_1 b)}{Y_n'(p_1 a)}\right],\\
m_{33} &= \frac{\beta n}{b q_0^2} K_n(q_0 b),\\
m_{34} &= \frac{j\omega\mu_0}{q_0} K_n'(q_0 b),\\
m_{41} &= -\frac{j\omega\epsilon_1}{p_1^2}\left[J_n'(p_1 b) - \frac{J_n(p_1 a) Y_n'(p_1 b)}{Y_n(p_1 a)}\right],\\
m_{42} &= \frac{\beta n}{b p_1^2} m_{22},\\
m_{43} &= -\frac{j\omega\epsilon_0}{q_0} K_n'(q_0 b),\\
m_{44} &= \frac{\beta n}{b q_0^2} K_n(q_0 b),
\end{aligned}
\tag{5.81}
$$

with

$$p_1^2 = \omega^2 \mu_0 \epsilon_1 - \beta^2,$$

$$q_0^2 = \beta^2 - \omega^2 \mu_0 \epsilon_0.$$

Setting the determinant of (5.80) to zero yields the dispersion relation from which the ω vs. β behavior for various modes can be obtained.

One may show that there are two dominant modes that have zero cutoff frequencies: the HE_{11} mode and the TM_{01} mode. For imperfect conductors and dielectrics, the attenuation of the mode can be found by using (5.61), (5.65), and (5.67).

5.4 Modes on Radially Inhomogeneous Core Circular Dielectric Cylinder

It is known that radially inhomogeneous dielectric fibers may be designed to maximize the information capacities (i.e., to minimize the dispersive behavior) of the dielectric fiber transmission system [11]. Here, we introduce an efficient analytical means to calculate the propagation characteristics of various modes in such inhomogeneous fibers. By subdividing a radially inhomogeneous column into concentric layers, one may obtain field solutions in each homogeneous layer. The fields in each layer are expanded in appropriate eigenfunctions and the expansion coefficients determined by matching boundary conditions. This approach will yield a prohibitively large number of simultaneous equations to be solved as the number of layers increases. In the following, we show that this obstacle may be overcome by appropriately manipulating these simultaneous equations, resulting in dealing with only 4×4 type matrix operations. This will remove any constraint on the number of layers needed to approximate the inhomogeneous profiles. This 4×4 matrix technique to solve the cylindrical multilayered problem was first introduced by Yeh and Lindgren in 1977 [14]. The analysis given here will follow this work. This matrix approach to solve the planar multilayered problem has also been used by Yeh et al. [15].

5.4.1 Formulation of the Problem

The geometry of the problem is shown in Fig. 5.7. We assume that the expressions for the field components of all modes are multiplied by the factor $\exp(-jn\theta - j\beta z + j\omega t)$, which will be suppressed throughout. The modal discussion given in Sect. 5.1 is still valid. In other words, there are TE_{0m},TM_{0m}, HE_{nm}, and EH_{nm} modes that can be supported by this layered system.

Dividing the inhomogeneous circular cylinder guide into $l+1$ regions, as shown in Fig. 5.7, we may write the expressions for the tangential fields in these regions as follows:

In region 1

$$\begin{bmatrix} E_z^{(1)} \\ \eta H_z^{(1)} \\ E_\theta^{(1)} \\ \eta H_\theta^{(1)} \end{bmatrix} = \begin{bmatrix} c_1(r) & 0 & 0 & 0 \\ 0 & d_1(r) & 0 & 0 \\ e_1(r) & f_1(r) & 0 & 0 \\ g_1(r) & h_1(r) & 0 & 0 \end{bmatrix} \begin{bmatrix} C_1 \\ D_1 \\ 0 \\ 0 \end{bmatrix} . \tag{5.82}$$

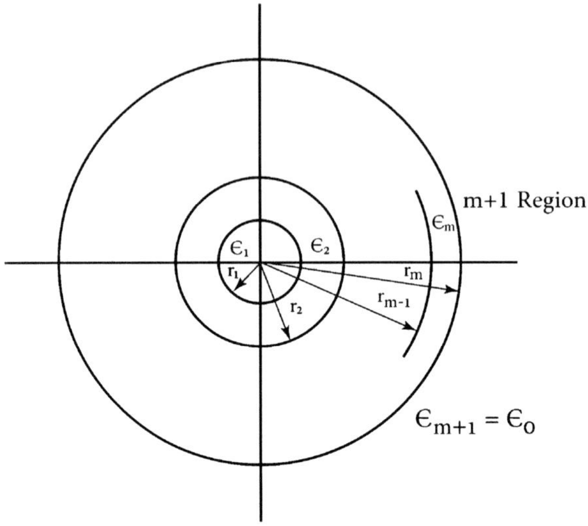

Figure 5.7. Geometry of the inhomogeneous circular cylinder

In region l $(l > 1)$

$$\begin{bmatrix} E_z^{(l)} \\ \eta H_z^{(l)} \\ E_\theta^{(l)} \\ \eta H_\theta^{(l)} \end{bmatrix} = \begin{bmatrix} c_l(r) & 0 & c_l'(r) & 0 \\ 0 & d_l(r) & 0 & d_l'(r) \\ e_l(r) & f_l(r) & e_l'(r) & f_l'(r) \\ g_l(r) & h_l(r) & g_l'(r) & h_l'(r) \end{bmatrix} \begin{bmatrix} C_l \\ D_l \\ C_l' \\ D_l' \end{bmatrix}. \tag{5.83}$$

In region $l+1$, (the outermost region),

$$\begin{bmatrix} E_z^{(l+1)} \\ \eta H_z^{(l+1)} \\ E_\theta^{(l+1)} \\ \eta H_\theta^{(l+1)} \end{bmatrix} = \begin{bmatrix} 0 & 0 & s(r) & 0 \\ 0 & 0 & 0 & \tau(r) \\ 0 & 0 & u(r) & v(r) \\ 0 & 0 & w(r) & \chi(r) \end{bmatrix} \begin{bmatrix} 0 \\ 0 \\ G \\ F \end{bmatrix}, \tag{5.84}$$

where

$$c_l^{(r)} = J_n(p_l r) \qquad c_l'^{(r)} = Y_n(p_l r),$$

$$d_l^{(r)} = \eta J_n(p_l r) \qquad d_l'^{(r)} = \eta Y_n(p_l r),$$

$$e_l^{(r)} = \frac{\beta n}{p_l^2 r} J_n(p_l r) \quad e_l'^{(r)} = \frac{\beta n}{p_l^2 r} Y_n(p_l r),$$

$$
\begin{aligned}
f_l^{(r)} &= \frac{j\omega\mu_0}{p_l} J_n'(p_l r) & f_l'^{(r)} &= \frac{j\omega\mu_0}{p_l} Y_n'(p_l r), \\
g_l^{(r)} &= -\frac{j\omega\epsilon_l\eta}{p_l} J_n'(p_l r) & g_l'^{(r)} &= -\frac{j\omega\epsilon_l\eta}{p_l} Y_n'(p_l r), \\
h_l^{(r)} &= \frac{\eta\beta n}{p_l^2 r} J_n(p_l r) & h_l'^{(r)} &= \frac{\eta\beta n}{p_l^2 r} Y_n(p_l r), \\
s(r) &= K_n(qr) & \tau(r) &= \eta K_n(qr), \\
u(r) &= -\frac{\beta n}{q^2 r} K_n(qr) & v(r) &= -\frac{j\omega\mu_0}{q} K_n'(qr), \\
w(r) &= \frac{j\omega\epsilon_{l+1}\eta}{q} K_n'(qr) & \chi(r) &= -\frac{\beta n\eta}{q^2 r} K_n(qr), \\
q^2 &= \beta^2 - \omega^2\mu_0\epsilon_{l+1} & p_l^2 &= \omega^2\mu_0\epsilon_l - \beta^2, \\
\eta &= (\mu_0/\epsilon_0)^{1/2},
\end{aligned}
\tag{5.85}
$$

and $C_1, D_1, \ldots, C_l, D_l, C_l', D_l', \ldots, G$, and F are arbitrary constants. It has been assumed that within each region the permittivity is a constant. The functions J_n, Y_n, and K_n are, respectively, the Bessel functions of the first kind, of the second kind, and the modified Bessel function of the second kind.

Matching the tangential electric and magnetic fields at the boundary surfaces, that is, $r = r_1, r_2, r_3, \ldots r_l$, (see Fig. 5.6) gives

$$
\begin{bmatrix} c_1(r_1) & 0 & 0 & 0 \\ 0 & 0 & d_1(r_1) & 0 \\ e_1(r_1) & 0 & f_1(r_1) & 0 \\ g_1(r_1) & 0 & h_1(r_1) & 0 \end{bmatrix}
\begin{bmatrix} C_1 \\ 0 \\ D_1 \\ 0 \end{bmatrix}
= \begin{bmatrix} c_2(r_1) & c_2'(r_1) & 0 & 0 \\ 0 & 0 & d_2(r_1) & d_2'(r_1) \\ e_2(r_1) & e_2'(r_1) & f_2(r_1) & f_2'(r_1) \\ g_2(r_1) & g_2'(r_1) & h_2(r_1) & h_2'(r_1) \end{bmatrix}
\begin{bmatrix} C_2 \\ C_2' \\ D_2 \\ D_2' \end{bmatrix},
$$

$$\begin{bmatrix} c_2(r_2) & c_2'(r_2) & 0 & 0 \\ 0 & 0 & d_2(r_2) & d_2'(r_2) \\ e_2(r_2) & e_2'(r_2) & f_2(r_2) & f_2'(r_2) \\ g_2(r_2) & g_2'(r_2) & h_2(r_2) & h_2'(r_2) \end{bmatrix} \begin{bmatrix} C_2 \\ C_2' \\ D_2 \\ D_2' \end{bmatrix}$$

$$= \begin{bmatrix} c_3(r_2) & c_3'(r_2) & 0 & 0 \\ 0 & 0 & d_3(r_2) & d_3'(r_2) \\ e_3(r_2) & e_3'(r_2) & f_3(r_2) & f_3'(r_2) \\ g_3(r_2) & g_3'(r_2) & h_3(r_2) & h_3'(r_2) \end{bmatrix} \begin{bmatrix} C_3 \\ C_3' \\ D_3 \\ D_3' \end{bmatrix},$$

$$\cdots\cdots\cdots\cdots\cdots\cdots\cdots\cdots\cdots\cdots\cdots$$

$$\cdots\cdots\cdots\cdots\cdots\cdots\cdots\cdots\cdots\cdots\cdots$$

$$\begin{bmatrix} c_{l-1}(r_{l-1}) & c_{l-1}'(r_{l-1}) & 0 & 0 \\ 0 & 0 & d_{l-1}(r_{l-1}) & d_{l-1}'(r_{l-1}) \\ e_{l-1}(r_{l-1}) & e_{l-1}'(r_{l-1}) & f_{l-1}(r_{l-1}) & f_{l-1}'(r_{l-1}) \\ g_{l-1}(r_{l-1}) & g_{l-1}'(r_{l-1}) & h_{l-1}(r_{l-1}) & h_{l-1}'(r_{l-1}) \end{bmatrix} \begin{bmatrix} C_{l-1} \\ C_{l-1}' \\ D_{l-1} \\ D_{l-1}' \end{bmatrix}$$

$$= \begin{bmatrix} c_l(r_{l-1}) & c_l'(r_{l-1}) & 0 & 0 \\ 0 & 0 & d_l(r_{l-1}) & d_l'(r_{l-1}) \\ e_l(r_{l-1}) & e_l'(r_{l-1}) & f_l(r_{l-1}) & f_l'(r_{l-1}) \\ g_l(r_{l-1}) & g_l'(r_{l-1}) & h_l(r_{l-1}) & h_l'(r_{l-1}) \end{bmatrix} \begin{bmatrix} C_l \\ C_l' \\ D_l \\ D_l' \end{bmatrix},$$

$$\begin{bmatrix} c_l(r_l) & c_l'(r_l) & 0 & 0 \\ 0 & 0 & d_l(r_l) & d_l'(r_l) \\ e_l(r_l) & e_l'(r_l) & f_l(r_l) & f_l'(r_l) \\ g_l(r_l) & g_l'(r_l) & h_l(r_l) & h_l'(r_l) \end{bmatrix} \begin{bmatrix} C_l \\ C_l' \\ D_l \\ D_l' \end{bmatrix} = \begin{bmatrix} s(r_l) & 0 & 0 & 0 \\ 0 & 0 & \tau(r_l) & 0 \\ u(r_l) & 0 & v(r_l) & 0 \\ w(r_l) & 0 & \chi(r_l) & 0 \end{bmatrix} \begin{bmatrix} G \\ 0 \\ F \\ 0 \end{bmatrix}. \tag{5.86}$$

It is clear that if we set the determinant of the above set of simultaneous equations to zero, we may obtain the dispersion relation for the inhomogeneous dielectric cylinder problem. However, the size of the resultant determinant is directly proportional to the number of layers that is used. To circumvent the problem, we follow the method developed by Yeh and Lindgren [14]. Realizing the fact that the array $\begin{bmatrix} C_2 \\ C_2' \\ D_2 \\ D_2' \end{bmatrix}$ can be expressed in terms of $\begin{bmatrix} C_3 \\ C_3' \\ D_3 \\ D_3' \end{bmatrix}$, etc., we can develop the following chain equation:

$$M_1(r_1)\begin{bmatrix} C_1 \\ 0 \\ D_1 \\ 0 \end{bmatrix} = M_2(r_1)M_2^{-1}(r_2)\cdots M_l(r_{l-1})M_l^{-1}(r_l)M_{l+1}(r_l)\begin{bmatrix} G \\ 0 \\ F \\ 0 \end{bmatrix}, \tag{5.87}$$

where

$$M_1(r_1) = \begin{bmatrix} c_1(r_1) & 0 & 0 & 0 \\ 0 & 0 & d_1(r_1) & 0 \\ e_1(r_1) & 0 & f_1(r_1) & 0 \\ g_1(r_1) & 0 & h_1(r_1) & 0 \end{bmatrix},$$

$$M_l(r_{l-1}) = \begin{bmatrix} c_l(r_{l-1}) & c_l'(r_{l-1})0 & 0 & 0 \\ 0 & 0 & d_l(r_{l-1}) & d_l'(r_{l-1}) \\ e_l(r_{l-1}) & e_l'(r_{l-1}) & f_l(r_{l-1}) & f_l'(r_{l-1}) \\ g_l(r_{l-1}) & g_l'(r_{l-1}) & h_l(r_{l-1}) & h_l'(r_{l-1}) \end{bmatrix},$$

$$M_l^{-1}(r_l) = \begin{bmatrix} c_l(r_l) & c_l'(r_l) & 0 & 0 \\ 0 & 0 & d_l(r_l) & d_l'(r_l) \\ e_l(r_l) & e_l'(r_l) & f_l(r_l) & f_l'(r_l) \\ g_l(r_l) & g_l'(r_l) & h_l(r_l) & h_l'(r_l) \end{bmatrix}^{-1},$$

$$M_{l+1}(r_l) = \begin{bmatrix} s(r_l) & 0 & 0 & 0 \\ 0 & 0 & \tau(r_l) & 0 \\ u(r_l) & 0 & v(r_l) & 0 \\ w(r_l) & 0 & \chi(r_l) & 0 \end{bmatrix}.$$

Hence, we are dealing only with matrices of size 4×4. This rearrangement eliminates the need to compute large matrices. The size of our present matrix is independent of the number of layers that we use. Hence, we may use as many layers as we wish to achieve the desired accuracy. Rewriting (5.86) gives

$$\begin{bmatrix} c_1(r_1) & 0 & -M_{11} & -M_{13} \\ 0 & d_1(r_1) & M_{21} & M_{23} \\ e_1(r_1) & f_1(r_1) & -M_{31} & -M_{33} \\ g_1(r_1) & h_1(r_1) & -M_{41} & -M_{43} \end{bmatrix}\begin{bmatrix} C_1 \\ D_1 \\ G \\ F \end{bmatrix} = 0, \tag{5.88}$$

where M_{ij} are elements of the 4×4 matrix M, defined as follows:

$$M = M_2(r_1)M_2^{-1}(r_2)M_3(r_2)M_3^{-1}(r_3)\cdots M_l(r_{l-1})M_l^{-1}(r_l)M_{l+1}(r_l). \tag{5.89}$$

Setting the determinant of (5.88) to zero, one obtains the dispersion relation from which the propagation constants of various modes along an inhomogeneous dielectric guide may be found. In other words, the propagation constants of various modes correspond to the roots of the dispersion relation. Given all known constants, such as the frequency, the size of the dielectric cylinder, the dielectric constant, thicknesses of the layers, etc., the roots can be found by the well-known Newton's method.

The Poynting flux in the lth region of a dielectric cylinder can be found from the following expression:

$$S_z(l) = \frac{1}{2}\,\mathrm{Re}\left[E_r^{(l)} H_\theta^{(l)*} - E_\theta^{(l)} H_r^{(l)*}\right]. \tag{5.90}$$

Here, $E_r^{(l)}$ and $H_r^{(l)}$ can be found by substituting $E_z^{(l)}$ and $H_z^{(l)}$ from (5.82)–(5.84) in (5.1) and (5.3), and $E_\theta^{(l)}$ and $H_\theta^{(l)}$ have been given in (5.82)–(5.84).

It is noted that the modal structures of the radially stratified dielectric cylinder remains the same as those for a solid dielectric cylinder, that is, circularly symmetric TE_{0m} and TM_{0m} as well as the asymmetric HE_{nm} and EH_{nm} modes may be supported by the radially inhomogeneous cylinder. The dominant mode with no cutoff frequency is still the HE_{11} mode.

5.4.2 Selected Examples

To demonstrate the viability of this technique to treat the cylinder with radially inhomogeneous dielectric problem, we have carried out computations for the following refractive-index profiles: (a) step-index core, (b) parabolic index core, and (c) doughnut-index core (see Fig. 5.8).

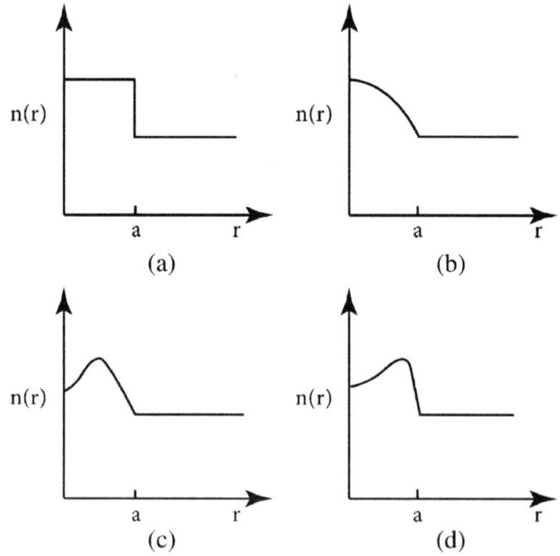

Figure 5.8. Refractive index profile of several fibers of interest

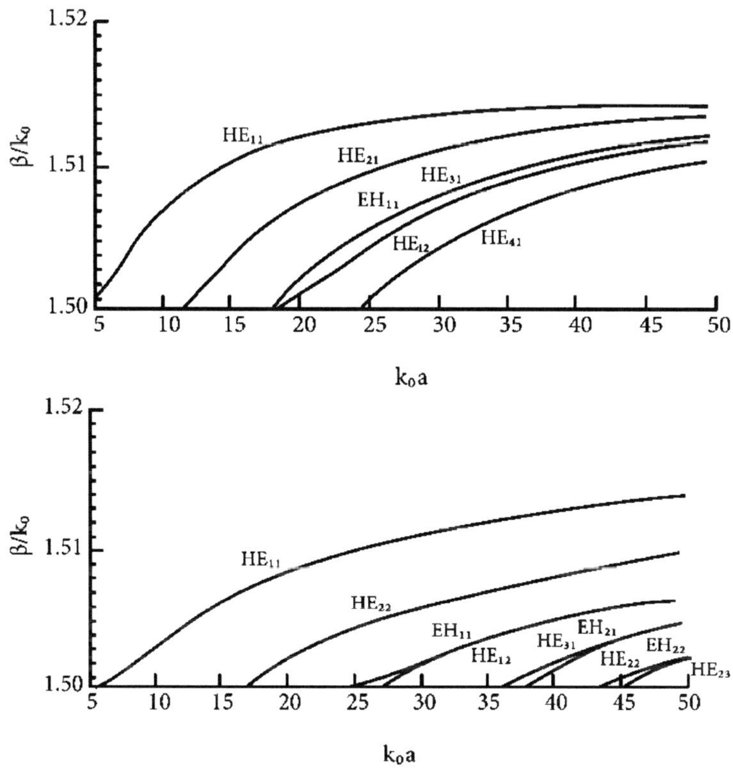

Figure 5.9. Dispersion curves for (*top*) homogeneous core fiber with $n_1 = 1.515$ and $n_2 = 1.500$; (*bottom*) parabolic core fiber (five-layer) with $n_{\mathrm{MAX}} = 1.515$ and $n_2 = 1.500$. The radius of the core is a [14]

Optical fibers were chosen for this demonstration. To achieve the desired accuracy (better than 1%), five layers of stratification must be used for the dominant HE_{11} mode. The accuracy could be improved by the use of an unequal distribution of stratification, that is, use large increments where the profile is flat and small increments where the profile is steep. Even better accuracy could be achieved with ten layers of stratification. For this case, the results were relatively insensitive to a minor shift in a layer's boundary location. Figure 5.9 gives the normalized propagation constant β/k_0 vs. the normalized frequency k_0a curves for several lower order modes of a homogeneous fiber and of an inhomogeneous fiber with parabolic index variation. From Fig. 5.9, one notices that the excursion of β/k_0 from the cutoff value to far above the cutoff value occurs over a much larger range of k_0a when the profile is parabolic rather than homogeneous. This agrees with the minimal

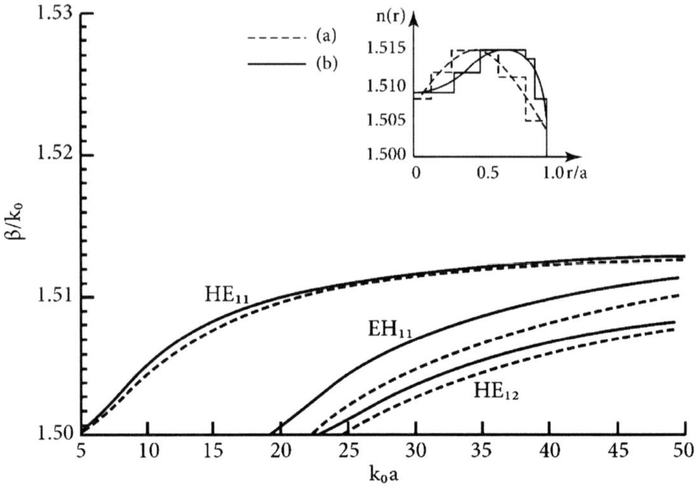

Figure 5.10. Comparison of dispersion curves for two fibers with different doughnut index profiles [14]

dispersion predictions provided by geometrical optics. The other interesting feature of the parabolic profile is that a number of modes have merged or possess nearly degenerate dispersion curves.

Figure 5.10 presents dispersion diagrams for fiber with a doughnut-index profile. We notice from the curves that the dispersion is intermediate between the homogeneous core and the parabolic core. Profile (a) is less dispersive than (b), which follows since the average value of the core index of (a) is less than that of (b).

Finally, the HE_{11} mode dispersion characteristics of the three profiles (step, parabolic, and doughnut) are compared in Fig. 5.11. This shows that for a given radius, a guide is less dispersive as the profile departs from a step profile.

In terms of dispersion, the refractive index profile has a pronounced effect on the guide dispersion characteristic. Of the profiles studied, the parabolic profile demonstrates the least dispersion. The factors that minimize pulse distortion in a parabolic profile guide can be deduced from generated dispersion curves. When we compare the parabolic guide with the step profile guide, the important differences are that for a parabolic guide (1) each mode is much less dispersive, (2) many modes have degenerate dispersion curves, and (3) for any core size a and index difference Δn, the total number of propagating modes has been reduced, and these modes are mostly degenerate.

Let us now consider the Poynting flux distribution (S_z) for several profiles that we treated above. The results are presented primarily at two operating conditions, near cutoff and far above cutoff, since the flux confinement is intermediate for

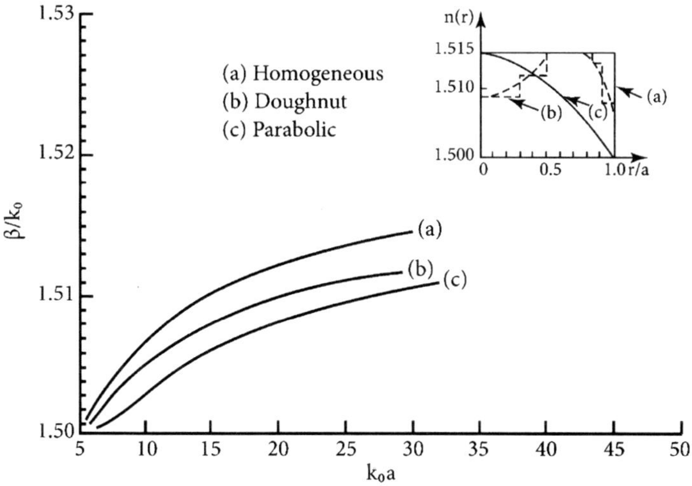

Figure 5.11. Comparison of HE_{11} mode dispersion curves for homogeneous, doughnut, and parabolic refractive-index profiles [14]

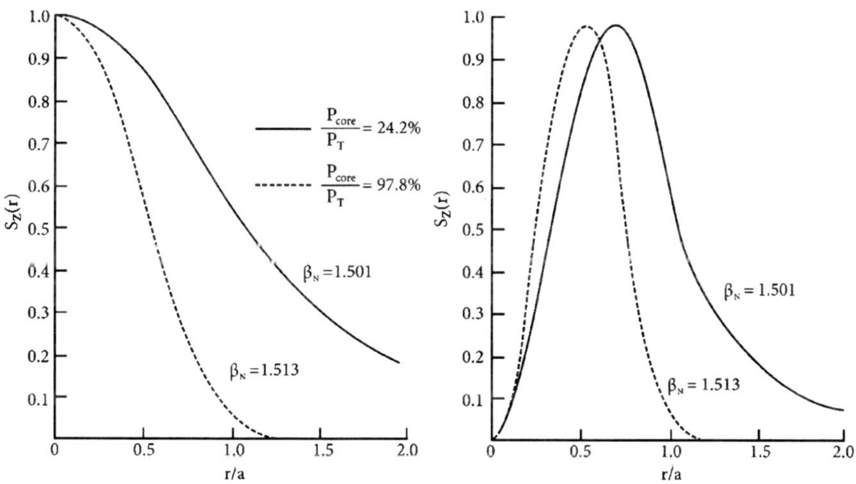

Figure 5.12. (*Left*) Poynting flux characteristics of HE_{11} mode. (*Right*) Poynting flux characteristics of HE_{21} mode. Fiber profile is homogeneous, $n_1 = 1.515$ and $n_2 = 1.500$ [14]

other operating conditions. Figure 5.12 shows the Poynting flux (S_z) for a fiber with homogeneous index profile and Fig. 5.13 shows S_z for a fiber with doughnut index profile. The interesting feature of the inhomogeneous core guides is that the Poynting flux concentrates around the region of highest refractive index as the frequency is increased. In the case of the doughnut profiles at far above cutoff, the

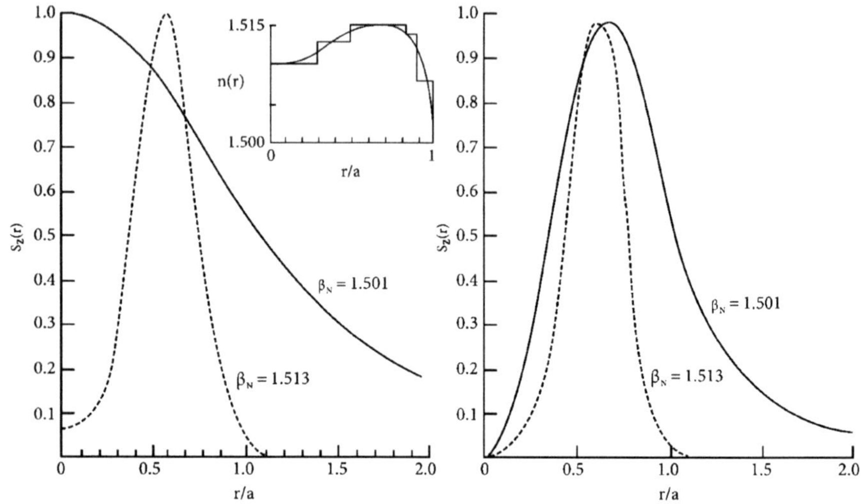

Figure 5.13. (*Left*) Poynting flux characteristics of HE_{11} mode. (*Right*) Poynting flux characteristics of HE_{21} mode. Fiber profile is doughnut (five-layer), $n_1 = 1.515$ and $n_2 = 1.500$ [14]

middle core region takes on the behavior of a slab guide, where the peak refractive index serves as the core and the layers inside and outside appear as the equivalent of a slab with upper and lower cladding regions.

In terms of field confinement, the parabolic profile guide has a desirable property: When operating at single mode the parabolic profile guide with a core radius 1.66 times larger than a step profile guide has the same spot size as that of a step profile guide. This comparison was made on the basis that both guides are operating at the same frequency and wave number ($\beta/k_0 = 1.507$, where $1.500 < \beta/k_0 < 1.515$). If we connect the two guides with mismatched core sizes, we may expect efficient transfer of energy at the junction since the spot sizes, frequency, and wave numbers all match. And finally, a comparison of the guide dispersions when operated under the above conditions indicates that the parabolic guide is still less dispersive than the step guide.

The doughnut profile guide was also studied in terms of dispersion and field confinement. The dispersion characteristics are intermediate to a step and parabolic profile guide, and the modes are not fully degenerate although the dispersion curves do show movement toward degeneracy. Hence, we expect that multimode pulse distortion would also be intermediate to the other two types of guides. The doughnut guide has interesting focusing properties that confirm the geometrical optics model of doughnut guidance. Even the fundamental HE_{11} mode at high frequencies focuses onto the high index region as if the high index region were a slab waveguide.

Thus, the doughnut guide has interesting possibilities as a transition from fiber to slab guide and an n-port power divider. For instance, we could join fiber of any profile to doughnut profile guide. Then, allow the doughnut guide to become highly multimoded by an increase in diameter. When sufficiently multimoded, the doughnut profile guided high-index region can be divided into azimuthal segments, each segment corresponding to a slab waveguide. The azimuthal segments can correspond to n-ports of a power divider with equal or unequal division.

And finally, it is interesting to note some general properties of fiber power density, S. First, to evaluate $S_z(r)$, one must use accurate values for ω-β solutions; otherwise, large errors in S_z occur at the layer boundaries. Second, near cutoff, the percentage of power carried in the core vs. that in the cladding does not approach zero for all modes: for instance, the HE_{13} mode carries 69% of the power in the core even near cutoff. Third, far above cutoff, almost all the power is carried in the core, which confirms the geometrical optics model of the fiber. And fourth, for a given mode and fiber profile, there is a relationship between γ, where

$$\gamma = \frac{\beta/k_0 - n_2}{n_1 - n_2}; \quad n_2 < \beta/k_0 < n_1, \tag{5.91}$$

and the percentage of power carried in the core region that is independent of Δn. For example, the HE_{11} mode operated at $\gamma = 0.0666$ has 24.2% of the power in the core whether Δn is 1% or 2%.

5.4.3 Hollow Cylindrical Dielectric Waveguide

The geometry of a hollow dielectric waveguide is shown in Fig. 5.14.

A hollow core of permittivity ϵ_0 is surrounded by a concentric layer of dielectric material having dielectric constant ϵ_1. The region exterior to the layer is free-space. The analysis for the modal structures can be carried out in a similar manner as in Sect. 5.4.1, with $l = 2$, $\epsilon_1 = \epsilon_0$, $\epsilon_2 = \epsilon_1$, and $\epsilon_3 = \epsilon_0$. The axial fields in regions 1, 2, and 3 are

Region 1: $(r \leq r_1)$

$$E_z^{(1)} = C_1 J_n(p_1 r),$$

$$H_z^{(1)} = D_1 J_n(p_1 r),$$

$$p_1^2 = \omega^2 \mu_0 \epsilon_0 - \beta^2.$$

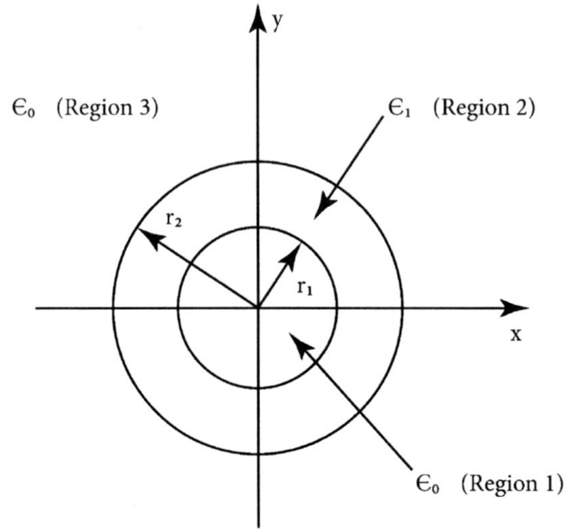

Figure 5.14. The hollow dielectric waveguide

Region 2: $(r_1 \leq r \leq r_2)$

$$E_z^{(2)} = C_2 J_n(p_2 r) + C_2' Y_n(p_2 r),$$

$$H_z^{(2)} = D_2 J_n(p_2 r) + D_2' Y_n(p_2 r),$$

$$p_2^2 = \omega^2 \mu_0 \epsilon_2 - \beta^2.$$

Region 3: $(r \geq r_2)$

$$E_z^{(3)} = G K_n(qr),$$

$$H_z^{(3)} = F K_n(qr),$$

$$q^2 = \beta^2 - \omega^2 \mu_0 \epsilon_0,$$

where C_1, D_1, C_2, C_2', D_2, D_2', G, and F are arbitrary constants. It is important to note that $\beta^2 > \omega^2 \mu_0 \epsilon_0$. So $p_1^2 = -\left|\beta^2 - \omega^2 \mu_0 \epsilon_0\right|$. Then, $p_1 = j\left|\beta^2 - \omega^2 \mu_0 \epsilon_0\right|^{1/2}$. The function $J_n(p_1 r)$ becomes $J_n(j\,|p_1 r|) = e^{jn\pi/2} I_n(|p_1 r|)$. Here, $I_n(|p_1 r|)$ is the modified Bessel function of the first kind. All transverse fields can be derived from (5.1)–(5.4). The dispersion relation can be found from (5.88) by setting $l = 2$, $\epsilon_1 = \epsilon_0, \epsilon_2 = \epsilon_1$, and $\epsilon_3 = \epsilon_0$. The matrix M becomes

$$M = M_2(r_1) M_2^{-1}(r_2) M_3(r_2). \tag{5.92}$$

In a similar manner as described in Sect. 5.4.1, one may obtain the propagation characteristics of various modes on a hollow dielectric tube structure.

It is of historical interest to note that the earliest theoretical analysis was carried out by Zachoval in 1933 [16]. He considered the propagation of a circularly symmetric TM wave along a lossless circular dielectric tube. Two years later, Liska [17] verified Zachoval's work experimentally. A more complete treatment on the theory of dielectric tube waveguides was given by Astraham in 1949 [18], in which both symmetric and asymmetric propagating waves were considered. He also substantiated his theoretical results by experimental data. Independently, Unger in 1959 [19] reported his investigation on the same subject and showed that a dielectric tube with a thin wall could support the dominant mode with very little loss. But the radial field extent was very large. Promising applications of dielectric tube waveguides may be found in the field of millimeter wave cavity resonator and beam coupling structure. Hollow core dielectric waveguide made with multilayered dielectric sheaths has also been studied by Ibanescu et al. [20] as a way to confine light in the hollow core region. Their analysis was carried out using the 4×4 matrix method [14, 15].

5.5 Experimental Determination of Propagation Characteristics of Circular Dielectric Waveguides

At microwave frequencies, the most sensitive way to measure the propagation characteristics of a given mode on a dielectric waveguide is to use a dielectric waveguide cavity resonator [6, 21]. This cavity consists of a straight length of dielectric waveguide supporting the desired mode, with parallel shorting plates at both ends. At a resonant frequency of such a cavity, the guide wavelength, λ_g, is obtained from the cavity spacing, and the attenuation constant, α, can be obtained from the measured Q. The cavity method also provides an accurate determination of the dielectric properties of the waveguide material.

5.5.1 Ultrahigh Q Dielectric Rod Resonant Cavity

The geometry of the dielectric rod resonator, including a schematic of the measurement system, is shown in Fig. 5.16.

The dielectric rod resonant cavity consists of a dielectric waveguide of length d terminated at its ends by sufficiently large, flat, and highly reflecting plates that are perpendicular to the axis of the guide. For a circular step-index dielectric rod, the HE_{11} mode is the dominant mode for this dielectric waveguide. Electromagnetic energy is coupled into and out of the resonator through small coupling holes at the centers of the reflecting plates. For best results, the holes are made small enough such that coupling losses are smaller than the ohmic wall losses. At resonance, the length of the cavity d must be $n\lambda_g/2$ (n an integer), where λ_g is the guide wavelength of the particular mode under consideration. By measuring the resonant frequencies of the cavity, one may obtain the guide wavelength of that mode in the dielectric waveguide. The propagation constant, β, of that mode is related to λ_g and v_p, the phase velocity, as follows:

$$\beta = \frac{2\pi}{\lambda_g} = \frac{\omega}{v_p}. \tag{5.93}$$

The Q of a resonator is indicative of the energy storage capability of a structure relative to the associated energy dissipation from various loss mechanisms, such as those due to the imperfection of the dielectric material and the finite conductivity of the end plates. The common definition of Q is applicable to the dielectric rod resonator and is given by [21]

$$Q = \frac{\omega \overline{W}}{\overline{P}}, \tag{5.94}$$

where ω is the angular frequency, $\overline{W}$ is the total time average stored energy, and $\overline{P}$ is the time average power loss.

For the case under study, with carefully machined dielectric rods and proper cavity alignment, the time average power dissipation $\overline{P}$ consists of two parts, the power loss due to the dielectric rod and that due to the metal walls, namely,

$$\overline{P} = \overline{P}_{\text{dielectric}} + \overline{P}_{\text{wall}}. \tag{5.95}$$

The power dissipation due to the dielectric rod is given by

$$\overline{P}_{\text{dielectric}} = \frac{1}{2}\sigma_{\text{d}} \int_0^{\text{d}} \int_{A_{\text{d}}} (\mathbf{E}_1 \cdot \mathbf{E}_1^*)\, dA\, dz, \tag{5.96}$$

where $\mathbf{E}_1$ is the electric field within the dielectric rod, σ_d is the conductivity of the dielectric, A_d is the cross-sectional area of the dielectric rod, and the asterisk denotes the complex conjugate. The ohmic losses at both walls are given by

$$\overline{P}_{\text{wall}} = 2\left(\frac{R_s}{2}\right) \int_{A_w} (\mathbf{H}_t \cdot \mathbf{H}_t^*)\, dA, \tag{5.97}$$

where $R_s = \sqrt{\omega\mu/2\sigma_r}$, the wall resistivity; σ_r is the conductivity of the reflector metal, and H_t is the tangential component of the magnetic field along the metal wall. Here, A_w is the area of the conducting wall, outside of which the electromagnetic fields are negligible. There is also a loss due to the coupling hole, but, as in this experiment, the coupling can be negligibly small, such that the primary wall losses are the ohmic losses. The time average energy stored is given by

$$\overline{W} = 2\overline{W}_{\mathrm{m}} = \mu_0 \int_V (\mathbf{H} \cdot \mathbf{H}^*)\, dV, \tag{5.98}$$

where V is the volume of the cavity, $\overline{W}_{\mathrm{m}}$ is the time average magnetic field energy, and $\mathbf{H}$ is the total field. Equations (5.96) and (5.97) can be rearranged to obtain

$$\frac{1}{Q} = \frac{\overline{P}}{\omega\overline{W}} = \frac{\overline{P}_{\mathrm{dielectric}}}{\omega\overline{W}} + \frac{\overline{P}_{\mathrm{wall}}}{\omega\overline{W}} = \frac{1}{Q_{\mathrm{d}}} + \frac{1}{Q_{\mathrm{w}}}. \tag{5.99}$$

The term Q_{d} is the Q factor if the end plates were perfectly conducting and Q_{w} is the Q factor if the dielectric was lossless. $\overline{P}_{\mathrm{dielectric}}$ and $\overline{W}$ are proportional to the length of the cavity and so from (5.99) we have

$$Q_{\mathrm{d}} = \frac{\omega\overline{W}}{\overline{P}_{\mathrm{dielectric}}} = \frac{1}{2\frac{\epsilon_1}{\epsilon_0}\tan\delta_{\mathrm{d}}}\frac{C_{\mathrm{t}}}{C_{\mathrm{d}}}, \tag{5.100}$$

$$Q_{\mathrm{w}} = \frac{\omega\overline{W}}{\overline{P}_{\mathrm{wall}}} = \frac{d}{2\delta_r}\frac{C_{\mathrm{t}}}{C_{\mathrm{w}}}, \tag{5.101}$$

where tan δ_{d} $(= \sigma/\omega\mu)$ is the loss tangent of the dielectric rod and $\delta_{\mathrm{r}}(= 2R_{\mathrm{s}}/\omega\mu)$ is the skin depth of the metallic end plates. The ratios $C_{\mathrm{t}}/C_{\mathrm{d}}$ and $C_{\mathrm{t}}/C_{\mathrm{w}}$ are dimensionless quantities involving integrals of the fields in the transverse directions.

It is noted that Q_{d} is independent of the length of the cavity, whereas Q_{w} is proportional to the length. For a long cavity, $Q_{\mathrm{w}} \gg Q_{\mathrm{d}}$ and $Q \approx Q_{\mathrm{d}}$. By measuring the Q of the cavity with $Q_{\mathrm{w}} \gg Q_{\mathrm{d}}$, we can obtain the attenuation constant α of the given mode.

We note that using the dielectric waveguide technique, with modern instrumentation, a $Q > 30,000$ can readily be measured. This value of Q corresponds to a dielectric loss tangent of the order of 10^{-5}. In the measurement setup, shown in Fig. 5.15, a swept signal is transmitted through the waveguide cavity via very small coupling holes and detected by a spectrum analyzer. The reflecting plates are sufficiently large enough ($>15\lambda_0$) that the fields beyond the plates are insignificant. The output is a series of narrow transmitted resonances at $f_1, f_2, \ldots, f_m$, with half-power bandwidths $\Delta f_1, \Delta f_2, \cdots, \Delta f_m$, respectively (see Fig. 5.16).

At each resonant frequency, the guide wavelength is given by

$$\lambda_{\mathrm{gm}} = \frac{2d}{n_m} \tag{5.102}$$

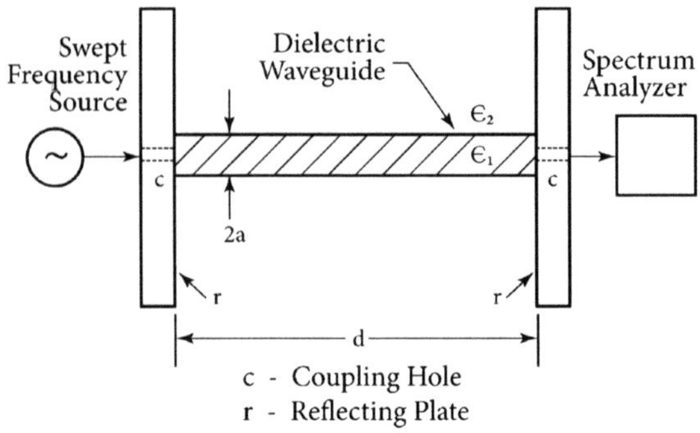

Figure 5.15. The schematic of the dielectric cavity resonator, including the measurement setup

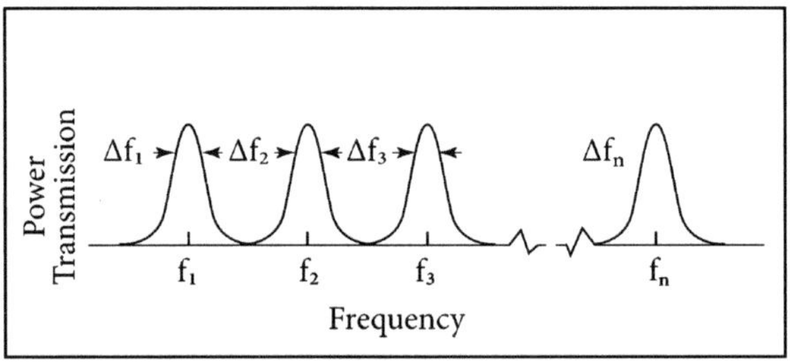

Figure 5.16. Power output of a swept input signal through a dielectric waveguide in a parallel-plate resonator

and Q by

$$Q_m = \frac{f_m}{\Delta f_m}, \tag{5.103}$$

where d is the length of the cavity and n_m is the number of guide half-wavelengths in the cavity for the mth resonant frequency. Since n_m is an integer, the measured accuracy of the guide wavelength is commensurate with the measured accuracy of the cavity length. The number n_m need only be determined to the nearest integer, and this is readily done by measuring the approximate guide wavelength with a probe, or by changing the cavity spacing and noting the shift in the resonant frequencies.

From previous discussion,

$$\frac{1}{Q_m} = \frac{1}{Q_{\rm d}} + \frac{1}{Q_{\rm w}}, \tag{5.104}$$

where Q_m is the measured Q of the mth resonant frequency; recalling that Q_d is independent of the cavity length, whereas Q_w is proportional to the cavity length. For the different dielectric waveguides used in this study, the calculated Q_w ranged from $18,000d$ to $21,000d$, where d is the length in centimeter. The cavity length was long enough such that

$$Q_\mathrm{w} \gg Q_\mathrm{d}. \tag{5.105}$$

The general relation between α and Q for a short-circuited low-loss waveguide has been derived in Sect. 3.7. It is

$$\alpha = 8.686 \frac{v_\mathrm{p}}{v_\mathrm{g}} \frac{\beta}{2Q} \quad (\mathrm{dB\,m^{-1}}), \tag{5.106}$$

where $v_\mathrm{p} = \omega/\beta$, and $v_\mathrm{g} = \partial\omega/\partial\beta$. For a dielectric rod waveguide [22],

$$\alpha = 4.343\,\omega\sqrt{\mu\epsilon_0}\tan\delta_\mathrm{d}\,(\epsilon_r\,R) \qquad (\mathrm{dB\,m^{-1}}), \tag{5.107}$$

where

$$R = \left| \frac{\displaystyle\int_{A_\mathrm{d}} (\mathbf{E}_1 \cdot \mathbf{E}_1^*)\, dA}{\displaystyle\sqrt{\frac{\mu_0}{\epsilon_0}}\,\mathrm{Re}\int_A \mathbf{e}_z \cdot (\mathbf{E}_\mathrm{t} \times \mathbf{H}_\mathrm{t}^*)\, dA} \right|. \tag{5.108}$$

It is known that $\epsilon_r = \epsilon_1/\epsilon_0$, A_d is the cross-sectional area of the core region of the dielectric waveguide, A is the total cross-sectional area of the cavity, $\mathbf{E}_1$ is the electric field within the dielectric rod, $\mathbf{e}_z$ is the unit vector along the direction of propagation, and $\mathbf{E}_\mathrm{t}$ and $\mathbf{H}_\mathrm{t}$ are the total transverse fields. The quantity R is a frequency dependent geometrical factor, which can be computed. The loss tangent can be obtained from (5.107) and (5.108),

$$\tan\delta_\mathrm{d} = \frac{\dfrac{v_\mathrm{p}}{v_\mathrm{g}}\dfrac{\beta}{2Q}}{\omega\sqrt{\mu_0\epsilon_0}\,\epsilon_\mathrm{r}\,R}. \tag{5.109}$$

From the measurements, Q, β, v_p, and v_g are determined. The group velocity can be derived by measuring $\beta's$ at adjacent resonant frequencies in a long cavity, and $v_\mathrm{g} \approx \Delta\omega/\Delta\beta$. a circular dielectric waveguide, one can calculate R for different values of ϵ_r. This is shown in Fig. 5.17. Hence, from the measurements of a dielectric rod in a high Q cavity, in addition to the propagation characteristics of the rod, accurate measurements of the electrical properties $(\epsilon_\mathrm{r}, \tan\delta_\mathrm{d})$ of the dielectric material can be obtained.

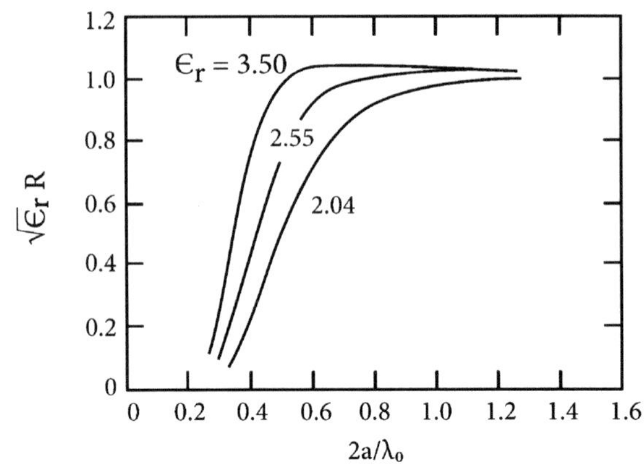

Figure 5.17. Plots of the attenuation factor R in a circular dielectric cylunder waveguide of radius a for different relative permittivities

5.5.2 Measured Results

Circular dielectric rod waveguides were made of Teflon, rexolite, polystyrene, polyethylene, and polypropylene. The diameters ranged from 0.4 to 0.63 cm and the lengths from 15.2 to 20.3 cm. These waveguides were placed in a parallel plate resonator. A swept frequency signal at Ka-band (26.5–40 GHz) was coupled into and out of the resonator through small coupling holes and detected by a spectrum analyzer. With this coupling, only the HE_{11} dominant mode was excited. This was verified by mapping the fields outside the dielectric waveguide with an electric probe.

At each resonance the Q is measured. The results are shown in Fig. 5.18. Because n_m is known to be an integer, it can be readily determined by measuring the guide wavelength approximately with a probe. Once n_m is known, the guide wavelengths at the various resonant frequencies are accurately determined, and the ω-β diagram can be generated, and α can be determined from (5.106). Here the dielectric waveguide had a circular cross-section and the following procedure was utilized. Once the guide wavelength and the waveguide dimensions were known, ϵ_r was determined from the dispersion relation (5.36). Using the ϵ_r value for Teflon and a rod diameter of 0.635 cm, we can calculate v_p and v_g from (5.36). The comparison between the calculated and measured values of v_p and v_g is shown in the right plot of Fig. 5.18.

The measured group velocity was obtained from the slope of the ω-β diagram. The attenuation coefficient was calculated from (108) and $\tan\delta_d$ was obtained from (5.109).

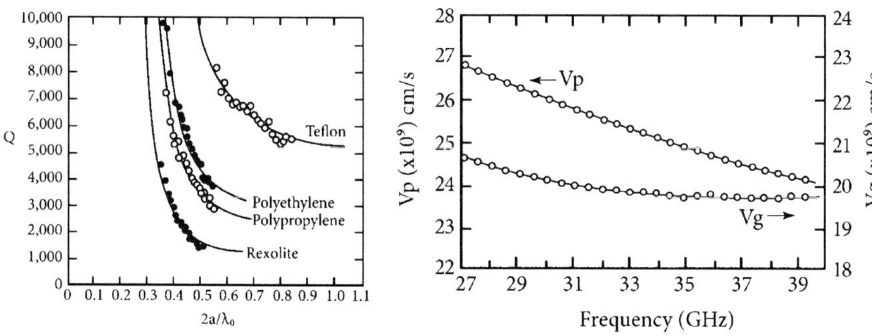

Figure 5.18. (*Left*) Measured Q's of the different circular dielectric waveguides. The solid line is the theoretical Q_d curve using the permittivities in Table 5.2. (*Right*) Comparison of measured and calculated group and phase velocities for a Teflon rod waveguide of diameter 0.635 cm. The solid lines are calculated and the measurements are indicated by circles [21]

Table 5.2 Measured relative permittivities and loss tangents, K_a band

Material	Estimates with standard error	
	ϵ_r	$10^3 \tan\delta$
Teflon	2.0422 ± 0.0006	0.217 ± 0.006
Polypropylene	2.261 ± 0.001	0.50 ± 0.03
Polyethylene	2.302 ± 0.003	0.38 ± 0.02
Polystyrene	2.542 ± 0.001	0.87 ± 0.07
Rexolite	2.548 ± 0.001	0.89 ± 0.07

The measured relative permittivities and loss tangents at different resonant frequencies for various materials are shown in Fig. 5.19. The average values with the corresponding standard deviations are given in Table 5.2 [23, 24].

The corresponding attenuation coefficients for the rod waveguides are shown in Fig. 5.20. In Fig. 5.20 are also shown plots of the half power bandwidths at the different resonances for two lengths of the 0.635 cm diameter Teflon waveguide. The plot indicates that the wall losses are much smaller than the dielectric losses. If the wall losses were significant, the Q's of the shorter length waveguide would have been noticeably lower at the lower frequencies and the derived loss tangents in Fig. 5.19 would have been noticeably higher. As a further check, the insertion losses of the resonator system were measured near 27, 33, and 39 GHz. The measured insertion losses were -71, -63, and -51 dB, respectively, at these three frequencies.

It is clear that for low-loss performance in circular dielectric waveguides, we should use small diameter rods made with small relative permittivity loss tangent.

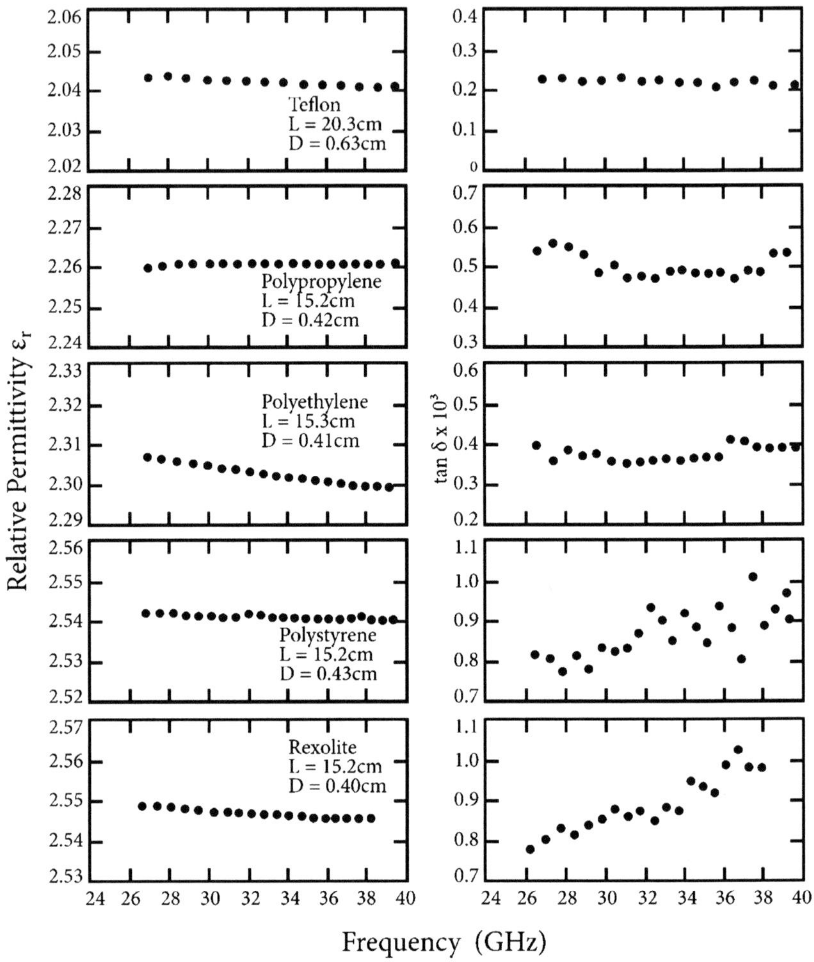

Figure 5.19. Derived values of ϵ_r and $\tan\delta$ from the measurements for different dielectric materials [21]

At Ka-band, the attenuation in a dielectric rod waveguide for small $2a/\lambda$ can be less than that of a conventional metallic rectangular waveguide. However, the guidance, due to bend losses and dimensional imperfections, is not as good. Because the surface resistivity of metals is proportional to the square root of frequency, the performance of dielectric waveguides relative to metallic waveguides improve as the frequency increases. This is shown in Fig. 5.21. The attenuation coefficient of different silver rectangular waveguides and of circular Teflon waveguides at the indicated frequencies are plotted in the figure. The assumption is that for the Teflon rod $2a/\lambda_0 = 0.4$ at the indicated frequencies.

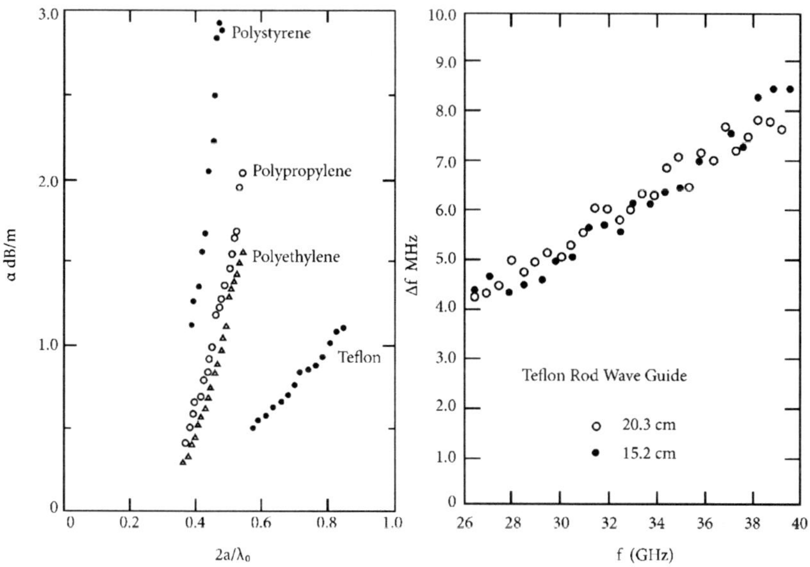

Figure 5.20. (*Left*) Measured attenuation coefficients for the different waveguides. Polystyrene and Rexolite have similar attenuation characteristics. (*Right*) Plot of half-power transmission bandwidths for two different lengths (6 in. and 8 in.) for circular Teflon waveguides [21]

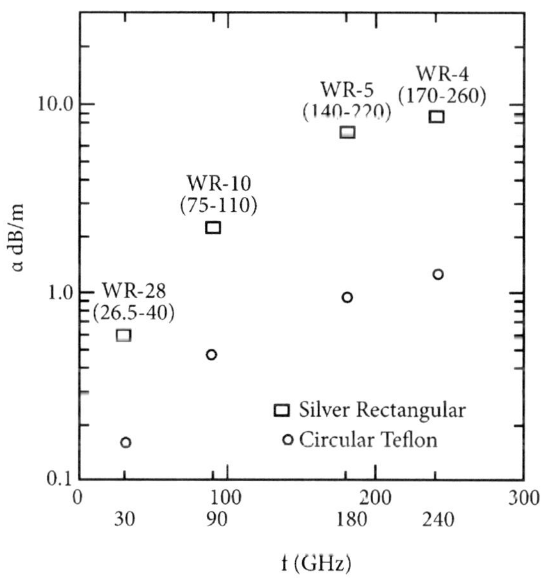

Figure 5.21. Comparison of attenuation coefficients of silver rectangular and Teflon circular waveguides at the indicated frequencies. The waveguide frequency range of the designated metal waveguides are shown in parentheses. For the dielectric waveguide, it is assumed that $2a/\lambda_0 = 0.4$ at the indicated frequencies [21]

5.6 Summary and Conclusions

The most important family of dielectric waveguides is that of the circular dielectric waveguide. Not only are these structures of importance in communication networks, but also the existence of analytical solutions for these structures provides thorough understanding of guiding characteristics of waves on them, which in turn promotes further inventions and developments to enhance the capabilities of modern communication networks. Here we shall list some of the salient features of our analyses on the circular dielectric waveguide given in this chapter.

1. All modes are hybrid, containing all components of the circular cylindrical fields, except for the circularly symmetric TE or TM modes.

2. The dominant mode is the asymmetric HE_{11} mode, which has a zero cutoff frequency and a bandwidth of $2.405\, c/\left[a(n_1^2 - n_2^2)\right]$ rad s^{-1}. Here, c is the speed of light in vacuum, a is the radius of the dielectric core, n_1 is the index of refraction of the core, and n_2 is the index of refraction of the cladding. The bandwidth is defined as the frequency band in which only the dominant HE_{11} can exist alone.

3. Since the ω-β curve for this HE_{11} mode is not a straight line, the signal carried by this mode will be broadened due to waveguide mode dispersion. Furthermore, the frequency dependent nature of the waveguide material will contribute to additional broadening. We note, however, that the dispersion behavior can be optimized through proper manipulation of the core index profile and choice of material.

4. For low values of normalized frequency, $V\left[= (\omega a/c)(n_1^2 - n_2^2)^{1/2}\right]$, that is, $V < 1$, significant amount of guided power for the dominant HE_{11} mode resides outside the core region, implying that the guided fields decay slowly away from the core surface and that the mode also experiences low attenuation since only a small amount of total guided power is carried inside the lossy core region. For high values of normalized frequency operating in the single mode region, that is, $2 < V < 2.4$, (the strongly guided case), more than 80% of the guided power is propagating within the core region, resulting in very rapidly decaying fields in the cladding region and an attenuation constant approaching that of a plane wave propagating in an infinite medium consisting of the core material.

5. For the strongly guided case, (a) the guided power of the HE_{11} mode may be transmitted through a curved dielectric waveguide with negligible radiation loss, provided that the curvature is larger than a few free-space wavelengths, (b) the transverse field pattern inside the core approaches that of a linearly polarized plane

wave with uniform intensity distribution, and (c) the guide wavelength also approaches that of a plane wave in a medium with the core dielectric constant.

6. The characteristic impedance of the HE_{11} mode, using the usual definition of the ratio of the transverse electric field to the transverse magnetic field, is not a constant. Knowing this limitation is important, because the treatment of waveguides as circuits is rather widespread.

7. The dispersion behavior of a given mode may be tailored through proper manipulation of the core index profile. The analysis may be carried out in a straightforward manner using the multilayered approach given in this chapter [14].

8. There is a tendency for modal fields to migrate towards the higher index region when the contrast ratio of higher to lower index regions is small or moderate (i.e., <2) and these fields will migrate towards the interface region when the contrast ratio is larger than 3. In other words, for a dielectric rod supporting only the dominant mode, the power density peaks inside the core region of the rod when the ratio of core index/cladding index is less than 2, and peaks at the boundary of the core and cladding when the ratio is greater than 3.

References

1. D. Hondros and P. Debye, "Electromagnetische wellen in dielektrischen drahtes," Ann. der Physik **32**, 465 (1910)
2. J. R. Carson, S. P. Mead, and S. A. Schelkunoff, "Hyperfrequency waveguides – Mathematical theory," Bell Syst. Tech. J. **15**, 310 (1936)
3. Mallach, "Dielektrische richstrahler," Bericht des V. I. F. S (1943)
4. G. F. Wegener, "Ausbreitungsgeschwindigkeit wellenwiderstand und dampfung elektromagnetischer wellen an dielektrischen Zylindern," Dissertation, Air Material Command Microfilm ZWB/FB/RE/2018, R8117F831 (1946)
5. W. M. Elsasser, "Attenuation in a dielectric circular rod," J. Appl. Phys. **20**, 1193 (1949)
6. C. H. Chandler, "An investigation of dielectric rod as waveguides," J. Appl. Phys. **20**, 1188 (1949)
7. C. Yeh, "A relation between α and Q," Proc. IRE **50**, 2143 (1962)
8. D. D. King, "Dielectric image line," J. Appl. Phys. **23**, 699 (1952)
9. D. D. King and A. P. Schlesinger, "Losses in dielectric image lines," IRE Trans. Microw. Theory Tech. **MTT-5**, 31 (1957)

10. E. Snitzer, "Cylindrical dielectric waveguide modes," J. Opt. Soc. Am. **51**, 491 (1961); E. Snitzer and H. Osterberg, "Observed dielectric waveguide modes in the visible spectrum," J.Opt. Soc. Am. **51**, 499 (1961)
11. G. Keiser, "Optical Fiber Communications," McGraw-Hill, New York (1991)
12. G. Goubau, "Surface waves and their application to transmission lines," J. Appl. Phys. **21**, 119 (1950)
13. A. Sommerfeld, "An oscillating dipole above a finitely conducting plane," Ann. der Physik **28**, 665 (1909)
14. C. Yeh and G. Lindgren, "Computing the propagation characteristics of radially stratified fibers – An efficient method," Appl. Opt. **16**, 483 (1977)
15. P. Yeh, "Optical Waves in Layered Media," Wiley, New York (1988); P. Yeh, A. Yariv, and C. S. Hong, "Electromagnetic propagation in periodic stratified media – General theory, " J. Opt. Soc. Am. **67**, 423 (1977)
16. L. Zachoval, "Electromagnetic wave propagation along a cylindrical dielectric tube," Rozpravy ces Akademie II, **47**, 34 (1933)
17. J. Liska, "Experiments on the propagation of electromagnetic wave along a dielectric tube," Casopis pro Pestovani Matematiky a Fysiky **63**, 69 (1934)
18. M. M. Astraham, "Dielectric tube waveguides," Dissertation, Northwestern University, Illinois (1949)
19. H. G. Unger, "Dielectric tube as waveguides," Arch. elek Ubertragang **8**, 241 (1954)
20. M. Ibanescu, S. G. Johnson, M. Soljaĉiĉ, J. D. Joannopoulos, Y. Fink, O. Weisberg, T. D. Engeness, S. A. Jacobs, and M. Skorobogatiy, "Analysis of mode structure in hollow dielectric waveguide fibers," Phys. Rev. E **67**, 046608 (2003)
21. F. Shimabukuro and C. Yeh, "Attenuation measurement of very low loss dielectric waveguides by the cavity resonator method applicable in the millimeter/submillimeter wavelength range," IEEE Trans. Microw. Theory Tech. **MTT-36**, 1160 (1988)
22. F. E. Borgnis and C. H. Papas, "Handbuch der Physik," **16**, Springer, Berlin Heidelberg New York (1958)
23. M. N. Afsar, "Dielectric measurement of millimeter wave materials," IEEE Trans. Microw. Theory Tech. **MTT-32**, 1598 (1984)
24. F. I. Shimabukuro, S. Lazar, M. R. Chernik, and H. B. Dyson, "A quasi-optical method for measuring the complex permittivity of materials," IEEE Trans. Microw. Theory Tech. **MTT-32**, 659 (1984)

6

ELLIPTICAL DIELECTRIC WAVEGUIDES

The guiding properties of a circular dielectric waveguide have been considered in detail in the preceding chapter. It is of interest, however, to understand how the propagation characteristics of the guided wave are affected when the circular cylinder is deformed and when the circular symmetry no longer exists. No analytic solution exists for the problem of wave propagation along a dielectric waveguide of arbitrary cross-sectional shape. Nevertheless, a deformed circular cylinder can, in general, be approximated by an elliptical cylinder. Depending upon the eccentricity of the elliptical cylinder, it can take the form of a circular cylinder or the form of a flat ribbon. The exact solution, although very involved, does exist for the elliptical dielectric waveguide problem. One notes that this solution is the only available exact solution for a noncircular dielectric waveguide of finite cross-section. Detailed theoretical as well as experimental results for the dominant modes and analytical results for higher order modes on an elliptical dielectric waveguides were given by Yeh [1], who provided the most thorough (analytical and theoretical) treatment for the propagation of the dominant modes on an elliptical dielectric waveguide. Lynbimov et al. [2] and Piefke [3] also derived the characteristic equations for various modes on an elliptical dielectric waveguide.

We shall first provide the fundamental theory of wave propagation along an elliptical dielectric rod. The distinct differences between the analytic solutions for a circular dielectric waveguide and for an elliptical dielectric waveguide are discussed in detail. The method developed by Yeh [1] to assure that the solutions of the wave equation will satisfy all the boundary conditions on the surface of the elliptical dielectric rod will be followed. The dispersion equations for all guided modes are obtained. These equations can be used to obtain the propagation characteristics of these modes. It is shown analytically that there exist two nondegenerate modes that possess no cutoff-frequency. They are called the dominant modes. We consider in detail the propagation characteristics of these modes. It is shown that all the principal modes on an elliptical dielectric rod degenerate smoothly to the well known modes on the circular dielectric guide, as the eccentricity of the elliptical rod approaches zero. In addition to providing the numerical results for the propagation characteristics of the two dominant modes, their field configurations, attenuation properties, as well as power distributions will be analyzed [4]. Selected results for higher order modes are also given.

6.1 Formulation of the Problem

The elliptical dielectric waveguide consists of a core of elliptical cross-section with dielectric constant ϵ_1 surrounded by a cladding of lower dielectric constant ϵ_0. Both regions are assumed to be lossless with the free-space permeability μ_0. It is further assumed that the exciting source is so far away that, in the region of interest, the surface wave dominates the radiated wave from the source. At a given frequency, a finite number of surface wave modes may be guided along this elliptical structure.

To analyze this problem, the elliptical cylinder coordinates (ξ, η, z), shown in Fig. 6.1, are introduced. In terms of the rectangular coordinates (x', y', z'), the elliptical cylinder coordinates are defined by the following relations:

$$\begin{aligned} x' &= q \cosh \xi \cos \eta, \\ y' &= q \sinh \xi \sin \eta, \\ z' &= z, \\ (0 \leq \ & \xi < \infty, \ \ 0 \leq \eta \leq 2\pi), \end{aligned} \tag{6.1}$$

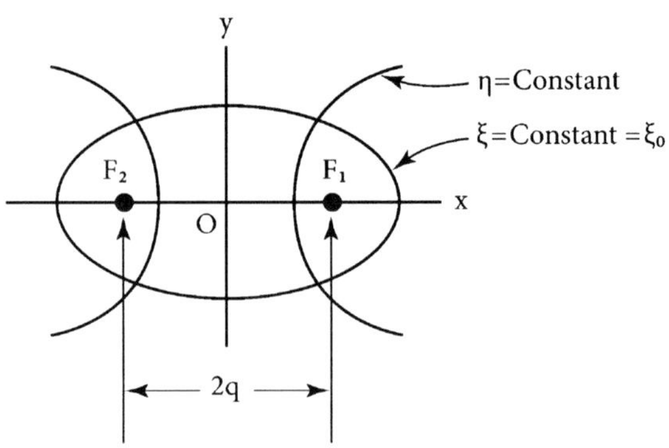

Figure 6.1. Cross section of elliptical dielectric waveguide. F_1 and F_2 are the foci of the ellipse. The distance between the foci is the focal distance $2q$

where q is the semifocal length of the ellipse. One of the confocal elliptical cylinders with $\xi = \xi_0$ is assumed to coincide with the boundary of the cylindrical core and the z-axis coincides with its longitudinal axis.

The harmonic time dependence of $e^{j\omega t}$ for all field components is assumed. We shall confine our treatment to waves propagating along the positive z-axis. In complex representation, these assumptions result in a multiplication of all wave functions by $e^{j\omega t - j\beta z}$. The propagation constant β is to be determined from the boundary conditions.

The solutions for the wave equation in elliptical cylindrical coordinates have been given in Sect. 2.5.3.3. Accordingly, the six field components in elliptical coordinates (ξ, η, z) are $\{E_\xi, E_\eta, E_z, H_\xi, H_\eta, H_z\}$. The wave equations satisfied by the axial fields (E_z, H_z) are

$$\left[\frac{1}{q^2(\cosh^2\xi - \cos^2\eta)}\left(\frac{\partial^2}{\partial\xi^2} + \frac{\partial^2}{\partial\eta^2}\right) + p^2\right]\begin{bmatrix} E_z \\ H_z \end{bmatrix} = 0, \tag{6.2}$$

with $p^2 = k^2 - \beta^2$ and $k^2 = \omega^2\mu\epsilon$. All transverse field components can be derived from the axial fields (E_z, H_z) as follows:

$$E_\xi = \frac{1}{p^2 d}\left(-j\beta\frac{\partial E_z}{\partial\xi} - j\omega\mu\frac{\partial H_z}{\partial\eta}\right), \tag{6.3}$$

$$E_\eta = \frac{1}{p^2 d}\left(-j\beta\frac{\partial E_z}{\partial\eta} + j\omega\mu\frac{\partial H_z}{\partial\xi}\right), \tag{6.4}$$

$$H_\xi = \frac{1}{p^2 d}\left(j\omega\epsilon\frac{\partial E_z}{\partial\eta} - j\beta\frac{\partial H_z}{\partial\xi}\right), \tag{6.5}$$

$$H_\eta = -\frac{1}{p^2 d}\left(j\omega\epsilon\frac{\partial E_z}{\partial\xi} + j\beta\frac{\partial H_z}{\partial\eta}\right), \tag{6.6}$$

with $d = q(\cosh^2\xi - \cos^2\eta)^{1/2}$.

Let us now review the solutions to (6.2). Consider the following partial differential equation:

$$\frac{\partial^2\Lambda}{\partial\xi^2} + \frac{\partial^2\Lambda}{\partial\eta^2} + \left[q^2\left(k^2 - \beta^2\right)\left(\sinh^2\xi + \sin^2\eta\right)\right]\Lambda = 0, \tag{6.7}$$

in which Λ may be H_z or E_z. To obtain the solutions of (6.7) one sets

$$\Lambda(\xi, \eta) = R(\xi)\Theta(\eta) \tag{6.8}$$

and substitutes (6.8) into (6.7). Applying the usual separation of variable procedure, one may separate (6.7) into the following two ordinary differential equations:

$$\frac{\partial^2\Theta(\eta)}{\partial\eta^2} + \left(c - 2\gamma^2\cos 2\eta\right)\Theta(\eta) = 0 \tag{6.9}$$

and

$$\frac{\partial^2 R(\xi)}{\partial\xi^2} - \left(c - 2\gamma^2\cosh 2\xi\right)R(\xi) = 0, \tag{6.10}$$

where c is the separation constant and $\gamma^2 = p^2q^2/4$. Equation (6.9) is the Mathieu differential equation. Equation (6.10), which follows from (6.9) by the transformation $\eta = \pm j\xi$, is the modified Mathieu differential equation [5].

For physically admissible single-valued electromagnetic fields, $\Lambda(\xi,\eta)$ must be a periodic function of η, of period π or 2π, and the separation constant c, in this case a function of γ^2, takes on an infinite set of characteristic values for every γ^2. When γ^2 is real the characteristic values are all real; since we are considering solutions in a lossless medium, only real values of c and γ^2 are of interest. Corresponding to $\gamma^2 = 0$ there are two independent periodic solutions, namely $\sin n\eta$ and $\cos n\eta$ with the separation constant $c = n^2$, where n is an integer. It can be shown that when γ^2 differs from zero, a characteristic value c determines one and only one periodic solution, which is either even or odd in η. The characteristic values c, giving rise to even and odd solutions, are denoted here by $a_n(\gamma^2)$ and $b_n(\gamma^2)$, respectively. The subscript n identifies those sets of characteristic values that approach n^2 as γ^2 approaches zero. It is known from the Sturmian theory of second order linear differential equations that solutions associated with $a_n(\gamma^2)$ and $b_n(\gamma^2)$ have n zeros in the interval $0 \le \eta \le \pi$ [6].

For arbitrary positive real values of γ^2, the periodic solutions of Mathieu equation (6.9) are

$$\Theta(\eta) = \left\{\begin{array}{c} ce_n(\eta,\gamma^2) \\ \\ se_n(\eta,\gamma^2) \end{array}\right\} \qquad \begin{array}{c} (a_n(\gamma^2)) \\ \\ (b_n(\gamma^2)) \end{array}, \tag{6.11}$$

and the corresponding solutions for the modified Mathieu equation (6.10) are

$$R(\xi) = \left\{\begin{array}{c} a_1 Ce_n(\xi,\gamma^2) + a_2 Fey_n(\xi,\gamma^2) \\ \\ b_1 Se_n(\xi,\gamma^2) + b_2 Gey_n(\xi,\gamma^2) \end{array}\right\} \qquad \begin{array}{c} (a_n(\gamma^2)) \\ \\ (b_n(\gamma^2)) \end{array}. \tag{6.12}$$

For arbitrary negative real values of γ^2, the periodic solutions of Mathieu equation (6.9) are

$$\Theta(\eta) = \left\{ \begin{array}{ll} ce_n^*(\eta, |\gamma|^2) & \begin{array}{ll} (a_n(|\gamma^2|)) & \text{with even } n \\ (b_n(|\gamma^2|)) & \text{with odd } n \end{array} \\ se_n^*(\eta, |\gamma|^2) & \begin{array}{ll} (b_n(|\gamma^2|)) & \text{with even } n \\ (a_n(|\gamma^2|)) & \text{with odd } n \end{array} \end{array} \right\}, \tag{6.13}$$

and the corresponding solutions for the modified Mathieu equation (6.10) are

$$R(\xi) = \left\{ \begin{array}{ll} c_1 Ce_n^*(\xi, |\gamma|^2) + c_2 Fek_n(\xi, |\gamma|^2) & \begin{array}{ll} (a_n(|\gamma^2|)) & \text{with even } n \\ (b_n(|\gamma^2|)) & \text{with odd } n \end{array} \\ d_1 Se_n^*(\xi, |\gamma|^2) + d_2 Gek_n(\xi, |\gamma|^2) & \begin{array}{ll} (b_n(|\gamma^2|)) & \text{with even } n \\ (a_n(|\gamma^2|)) & \text{with odd } n \end{array} \end{array} \right\}, \tag{6.14}$$

where $a_n(\gamma^2)$ and $b_n(\gamma^2)$ are the characteristic values for positive real values of γ^2, $a_n(|\gamma|^2)$ and $b_n(|\gamma|^2)$ are the characteristic values for negative real values of γ^2, and n is the order of the function [5]. The coefficients a_1, a_2, b_1, b_2, c_1, c_2, d_1, and d_2 are arbitrary constants.

The proper choice of the above solutions to represent the electromagnetic field of an elliptical dielectric cylinder depends upon the boundary conditions. For region 1, which is the space inside the dielectric rod, all fields must be finite. For region 0, which is the space outside the dielectric cylinder, in order that energy flows only along the axis of the cylinder, all field components must approach zero as the radial argument approaches infinity. So, for region 1 $(\xi_0 \geq \xi \geq 0)$, we must discard the functions $Fey_n(\xi, \gamma^2)$ and $Gey_n(\xi, \gamma^2)$, since they are infinite at the origin, that is, at $\xi = 0$. For region 0 $(\infty \geq \xi \geq \xi_0)$, the functions $Ce_n^*(\xi, |\gamma|^2)$ and $Se_n^*(\xi, |\gamma|^2)$ must also be discarded since they become infinite at $\xi = \infty$. Therefore, the appropriate solutions of the wave equation (6.2) are as follows:

$$\begin{aligned} H_z^{(1)}(\xi, \eta) &= \sum_{n=0}^{\infty} A_n Ce_n(\xi, \gamma_1^2) ce_n(\eta, \gamma_1^2) \\ &+ \sum_{n=1}^{\infty} A_n' Se_n(\xi, \gamma_1^2) se_n(\eta, \gamma_1^2) \qquad (\xi_0 \geq \xi \geq 0), \end{aligned} \tag{6.15}$$

$$H_z^{(0)}(\xi,\eta) = \sum_{n=0}^{\infty} L_n Fek_n(\xi, |\gamma_0|^2) ce_n^*(\eta, |\gamma_0^2|)$$
$$+ \sum_{n=1}^{\infty} L_n' Gek_n(\xi, |\gamma_0|^2) se_n^*(\eta, |\gamma_0^2|) \qquad (\infty \geq \xi \geq \xi_0), \quad (6.16)$$

$$E_z^{(1)}(\xi,\eta) = \sum_{n=0}^{\infty} B_n' Ce_n(\xi, \gamma_1^2) ce_n(\eta, \gamma_1^2)$$
$$+ \sum_{n=1}^{\infty} B_n Se_n(\xi, \gamma_1^2) se_n(\eta, \gamma_1^2) \qquad (\xi_0 \geq \xi \geq 0), \quad (6.17)$$

$$E_z^{(0)}(\xi,\eta) = \sum_{n=0}^{\infty} P_n' Fek_n(\xi, |\gamma_0|^2) ce_n^*(\eta, |\gamma_0^2|)$$
$$+ \sum_{n=1}^{\infty} P_n Gek_n(\xi, |\gamma_0|^2) se_n^*(\eta, |\gamma_0^2|) \qquad (\infty > \xi \geq \xi_0). \quad (6.18)$$

A_n, A_n', B_n, B_n', L_n, L_n', P_n, and P_n' are coefficients that are related by the boundary conditions and are functions of n, ω, γ_1^2, $|\gamma_0^2|$ and the nature of the exciting sources, but independent of the coordinates. γ_1^2 and $-\gamma_0^2$ are, respectively, $(k_1^2 - \beta^2)\, q^2/4$ and $(k_0^2 - \beta^2)\, q^2/4$, with $k_1^2 = \omega^2\mu\epsilon_1$ and $k_0^2 = \omega^2\mu\epsilon_0$. ϵ_1 is the dielectric constant of the cylinder and ϵ_0 is the dielectric constant of the surrounding medium. $\xi = \xi_0$ is the surface of the dielectric cylinder. All transverse field components for both regions can be derived from (6.3)–(6.6), using (6.15)–(6.18).

6.2 Boundary Conditions

The task of solving an electromagnetic wave boundary value problem is to find the finite and single-valued solutions that satisfy the source-free Maxwell equations and the boundary conditions. The boundary conditions are that the tangential components of the electric and magnetic fields must be continuous through any surface. If the region of interest is infinite, then the radiation condition must also be satisfied. The above conditions are necessary and sufficient. In the present problem, the continuity conditions in the elliptical cylindrical coordinates are

$$E_z^{(1)} = E_z^{(0)}, \quad (6.19)$$

$$H_z^{(1)} = H_z^{(0)}, \quad (6.20)$$

$$E_\eta^{(1)} = E_\eta^{(0)}, \tag{6.21}$$

$$H_\eta^{(1)} = H_\eta^{(0)}, \tag{6.22}$$

for $\xi = \xi_0$, $2\pi \geq \eta \geq 0$, and $\infty > z > -\infty$.

It is known that when Carson et al. [7] first solved the problem of a surface wave on a circular dielectric cylinder in 1936, they had to use a combination of TE and TM modal fields to satisfy the boundary conditions on the cylinder, resulting in the discovery of HE_{nm} modes when modes are not circularly symmetric. Here we encounter another obstacle, that is, the Mathieu angular functions used to represent the angular functions for the fields in elliptical cylindrical coordinates are not only functions of η, the angular coordinate, but also of the constitutive parameters (ϵ, μ) of the medium where they apply. This is unlike the case of the circular dielectric cylinder whose angular wave functions are represented by $e^{jn\theta}$, which are only functions of θ, the angular coordinate. The advantage of the circular case can be seen very clearly from the following example. Supposing that we wish to match the E_z component of the field on a circular dielectric cylinder at $r = a$, from (5.27), replacing q_2 by q_0 we have

$$\sum_{n=-\infty}^{\infty} C_n K_n\left(q_0 a\right) e^{jn\theta} = \sum_{n=-\infty}^{\infty} A_n J_n\left(p_1 a\right) e^{jn\theta}, \tag{6.23}$$

with $p_1^2 = k_1^2 - \beta^2$ and $q_0^2 = \beta^2 - k_0^2$. Multiplying both sides of (6.23) by $e^{-jl\theta}$ and integrating with respect to θ from 0 to 2π we have, due to the orthogonality of the angular exponential functions,

$$C_n K_n\left(q_0 a\right) = A_n J_n\left(p_1 a\right). \tag{6.24}$$

Equation (6.24) shows that (6.23) can be satisfied for each n separately, due to the fact that the angular exponential function is independent of the constitutive parameters of the medium.

Let us now consider the same situation for the elliptical dielectric cylinder. Substituting (6.17) and (6.18) in (6.19) and setting $\xi = \xi_0$, we find

$$\sum_{n=0}^{\infty} B_n' Ce_n(\xi_0, \gamma_1^2) ce_n(\eta, \gamma_1^2) + \sum_{n=1}^{\infty} B_n Se_n(\xi_0, \gamma_1^2) se_n(\eta, \gamma_1^2)$$
$$= \sum_{r=0}^{\infty} P_r' Fek_r(\xi_0, |\gamma_0^2|) ce_r^*(\eta, |\gamma_0^2|) + \sum_{r=1}^{\infty} P_r Gek_r(\xi_0, |\gamma_0^2|) se_r^*(\eta, |\gamma_0^2|). \tag{6.25}$$

Because of the orthogonality property of the even and odd angular Mathieu functions, (6.25) may be written as two separate equations, one corresponding to the even type modes and the other to the odd type modes. These equations are

$$\sum_{n=0}^{\infty} B'_n Ce_n(\xi_0, \gamma_1^2) ce_n(\eta, \gamma_1^2) = \sum_{r=0}^{\infty} P'_r Fek_r(\xi_0, |\gamma_0^2|) ce_r^*(\eta, |\gamma_0^2|) \tag{6.26}$$

and

$$\sum_{n=1}^{\infty} B_n Se_n(\xi_0, \gamma_1^2) se_n(\eta, \gamma_1^2) = \sum_{r=1}^{\infty} P_r Gek_r(\xi_0, |\gamma_0^2|) se_r^*(\eta, |\gamma_0^2|). \tag{6.27}$$

For the present demonstration let us use (6.27). Multiplying both sides of (6.27) by $se_l(\eta, \gamma_1^2)$ and integrating with respect to η from 0 to 2π and using the orthogonality relations of the Mathieu functions, we obtain

$$B_l Se_l(\xi_0, \gamma_1^2) = \frac{\sum_{r=1}^{\infty} P_r Gek_r(\xi_0, |\gamma_0^2|) \int_0^{2\pi} se_r^*(\eta, |\gamma_0^2|) se_l(\eta, \gamma_1^2) d\eta}{\int_0^{2\pi} se_l^2(\eta, \gamma_1^2) d\eta}$$

$$(l = 1, 3, 5 \ldots). \tag{6.28}$$

It becomes immediately apparent that the infinite summation in (6.28) must remain, implying that each coefficient B_l is coupled to an infinite number of coefficients P_r. Similar procedures may be applied to the remaining boundary conditions, (6.20)–(6.22). Unlike the circular case, (6.27) can no longer be satisfied for each n. Therefore, to satisfy the boundary conditions for the elliptical dielectric cylinder an infinite number of arbitrary constants must be coupled to each other, resulting in an infinite set of linear algebraic equations from which the dispersion relation can be derived. This approach of using the orthogonality property of Mathieu functions in eliminating the angular dependence in the boundary conditions such as in (6.27) was first used by Yeh [1]. It provides an exact means of solving the elliptical dielectric waveguide problem.

The above added complexity means that not only must all field components (E_ξ, E_η, E_z, H_ξ, H_η, H_z) be present for a guided mode on an elliptical dielectric cylinder, but also each field component of each guided mode must be represented by an infinite order of angular and radial Mathieu functions.

Let us now describe another approach in eliminating the angular dependence in the boundary conditions such as in (6.27). Lynbimov et al. [2] made use of the orthogonality property of trigonometric functions, as follows. Substituting expressions of Mathieu functions in terms of trigonometric functions,

$$se_n(\eta, \gamma_1^2) = \sum_{m=1}^{\infty} {}_nA_m(\gamma_1^2) \sin m\eta, \tag{6.29}$$

$$se_n^* \left(\eta, \left|\gamma_0^2\right|\right) = \sum_{m=1}^{\infty} {}_nA_m^*(\left|\gamma_0^2\right|) \sin m\eta, \tag{6.30}$$

in (6.27) yields

$$\sum_{n=0}^{\infty} B_n Se_n(\xi_0, \gamma_1^2) \sum_{m=1}^{\infty} {}_nA_m(\gamma_1^2) \sin m\eta$$
$$= \sum_{n=0}^{\infty} P_n Gek_n(\xi_0, \left|\gamma_0^2\right|) \sum_{m=1}^{\infty} {}_nA_m^*(\left|\gamma_0^2\right|) \sin m\eta, \tag{6.31}$$

where ${}_nA_m$ and ${}_nA_m^*$ are known constant expansion coefficients. Multiplying both sides of (6.31) by sin $l\eta$ and using the orthogonality relations of the trigonometric functions, we obtain

$$\sum_{n=0}^{\infty} B_n Se_n(\xi_0, \gamma_1^2) \sum_{m=1}^{\infty} {}_nA_m(\gamma_1^2)$$
$$= \sum_{n=0}^{\infty} P_n Gek_n(\xi_0, \left|\gamma_0^2\right|) \sum_{m=1}^{\infty} {}_nA_m^*(\left|\gamma_0^2\right|). \tag{6.32}$$

The angular dependence of (6.31) is eliminated. Comparing (6.32) with (6.28) shows that we must deal with larger matrices and double summations in (6.32) rather than the single summation in (6.28). Although we recognize that Yeh's approach and that of Lynbimov et al. must yield the same results, the ability to use only a single summation rather than the double summation makes Yeh's approach a more desirable one. Our treatment will follow Yeh's approach.

6.3 Mode Classifications

For a circular dielectric waveguide it is well known that the pure TE or TM waves can exist only if the fields are independent of the angular coordinates. These circularly symmetric waves are designated by H_{0m} for the pure TE waves and E_{0m} for the pure TM waves. The subscript 0 signifies the angular variations and m signifies the mth root of the characteristic equation. The coexistence of E and H waves is required to satisfy the boundary conditions if the field is a function of the angular coordinate. These asymmetric waves are then designated by HE_{nm} if the cross-sectional field pattern resembles that of a H wave and by EH_{nm} if the cross-sectional field pattern resembles that of an E wave at frequencies far from cutoff. The subscripts n and m denote, respectively, the number of cyclic variations with θ and the mth root of the characteristic equation. These hybrid asymmetric modes discussed above are doubly degenerate since an equally valid solution results if $\sin n\theta$ is replaced by $\cos n\theta$, and $\cos n\theta$ by $-\sin n\theta$.

Let us now consider the fields for the elliptical dielectric waveguide. To satisfy the boundary conditions at the boundary surface $\xi = \xi_0$, both E_z and H_z must be present. Hence, all modes on an elliptical dielectric guide are hybrid. Physically speaking, the presence of E_z in a predominantly H wave or vice versa assumes the return path for electric or magnetic lines of force. (The circularly symmetric TE or TM waves on a circular dielectric rod are exceptions since the electric and magnetic lines of force of these circularly symmetric waves have already formed closed loops.) Because of the asymmetry of the elliptical cylinder, it is possible to have two orientations for the field configurations. Thus, a hybrid wave on an elliptical dielectric rod will be designated by a pre-subscript e or o, indicating an even or an odd wave. The axial magnetic and electric field of an even wave are represented by even and odd Mathieu functions, respectively. The hybrid waves are designated by ${}_{\mathrm{e,o}}\mathrm{HE}_{n'm'}$ if the cross-sectional field pattern resembles that of an H wave, and by ${}_{\mathrm{e,o}}\mathrm{EH}_{n'm'}$ if the cross-sectional field pattern resembles that of an E wave at frequencies far from cutoff. (A more exact definition based on the relative strength of the H_z and E_z components has been given by Snitzer [7].) The subscripts n' and m' denote, respectively, the number of cyclic variations around the elliptical cross-section and the m'th root of the characteristic equation. It is noted that ${}_{\mathrm{e,o}}\mathrm{HE}_{n'm'}$ or ${}_{\mathrm{e,o}}\mathrm{EH}_{n'm'}$ modes on an elliptical rod degenerate, respectively, to HE_{nm} or EH_{nm} modes on a circular rod as the eccentricity of the ellipse becomes zero and n' becomes n. The indices n' and m' attached to the elliptical mode designation always mean n' cyclic variations around the cross-section of guide structure and the m'th root of the characteristic equation.

It is important to note the distinction between the subscript n' in the mode designation for the ${}_{\mathrm{e,o}}\mathrm{HE}_{n'm'}$ of an elliptical dielectric waveguide and that in the mode designation for the HE_{nm} mode of a circular dielectric waveguide. We note that the cyclical variation around the circular cross-section of each mode on a circular dielectric guide can be represented by a single trigonometric function with the same cyclical variation, that is, the n for HE_{nm} mode is the same n in the $\sin n\theta$ or $\cos n\theta$ term associated with that mode. On the other hand, in general, the cyclical variation around the elliptical cross-section of each mode on an elliptical dielectric rod *cannot* be represented by a single angular Mathieu function with the same cyclical variation, that is, the n' for the ${}_{\mathrm{e,o}}\mathrm{HE}_{n'm'}$ mode is *not the same* as the n in the $se_n(\eta)$ or $ce_n(\eta)$ term associated with that mode, except for the case when the ellipse degenerates into a circle; it requires, in general, a complete infinite series of $se_n(\eta)$ or $ce_n(\eta)$ to represent the cyclical variation of each ${}_{\mathrm{e,o}}\mathrm{HE}_{n'm'}$ mode. Nevertheless, one may still designate, or define, the subscript n' in the ${}_{\mathrm{e,o}}\mathrm{HE}_{n'm'}$ as the n' cyclical variation of the mode around the cross-section of the guide and n' does not necessarily equal to n, the index number for the angular Mathieu function. As long as this definition is understood, ambiguity can be avoided.

There is a limitation to this nomenclature, ${}_{\mathrm{e,o}}\mathrm{HE}_{n'm'}$, in classifying higher order modes for flatter elliptical dielectric cylinders. This is because lower indices $(n'm')$ do not necessarily correspond to lower order modes in the sense of lower cutoff frequencies. A different way of classifying modes for flatter elliptical dielectric cylinders is to base the numbering, or the order of modes, purely on the value for the cutoff frequencies, that is, the lower the cutoff frequency for a given elliptical cylinder, the lower is the mode order. However, this type of mode classification may also imply that mode order (or mode numbers) may be different for elliptical cylinders with different ellipticities.

6.4 The Dispersion Relations

${}_{e}HE_{n'm'}$ Wave

According to the definition given in the previous section, the most general expressions for the axial magnetic and electric fields of an ${}_{\mathrm{e}}\mathrm{HE}_{n'm'}$ wave are the following:

For region 1 $(0 \le \xi \le \xi_0)$

$$H_z^{(1)} = \sum_{m=0}^{\infty} A_m Ce_m(\xi) ce_m(\eta)\,,$$

$$E_z^{(1)} = \sum_{m=1}^{\infty} B_m Se_m(\xi) se_m(\eta)\,, \tag{6.33}$$

and for region 0 ($\xi_0 \leq \xi < \infty$)

$$H_z^{(0)} = \sum_{r=0}^{\infty} L_r Fek_r(\xi) ce_r^*(\eta),$$

$$E_z^{(0)} = \sum_{m=1}^{\infty} P_r Gek_r(\xi) se_r^*(\eta), \tag{6.34}$$

where A_n, B_n, L_r, and P_r are arbitrary constants and the abbreviations

$$\begin{aligned}
Ce_l(\xi) &= Ce_l(\xi, \gamma_1^2) & ce_l(\eta) &= ce_l(\eta, \gamma_1^2),\\
Se_l(\xi) &= Se_l(\xi, \gamma_1^2) & se_l(\eta) &= se_l(\eta, \gamma_1^2),\\
Fek_l(\xi) &= Fek_l(\xi, |\gamma_0^2|) & ce_l^*(\eta) &= ce_l(\eta, |\gamma_0^2|),\\
Gek_l(\xi) &= Gek_l(\xi, |\gamma_0^2|) & se_l^*(\eta) &= se_l(\eta, |\gamma_0^2|)
\end{aligned} \tag{6.35}$$

have been used. Also, $\gamma_1^2 = (k_1^2 - \beta^2)q^2/4$, $\gamma_0^2 = (\beta^2 - k_0^2)q^2/4$, $k_1^2 = \omega^2\mu_0\epsilon_1$, and $k_0^2 = \omega^2\mu_0\epsilon_0$. The transverse components of the field can be derived from (6.3)–(6.6). They are the following:

$$E_{\xi 1} = \frac{1}{p_1^2 d}\sum_m \left[-j\beta B_m Se_m'(\xi) se_m(\eta) - j\omega\mu A_m Ce_m(\xi) ce_m'(\eta)\right], \tag{6.36a}$$

$$E_{\eta 1} = \frac{1}{p_1^2 d}\sum_m \left[-j\beta B_m Se_m(\xi) se_m'(\eta) + j\omega\mu A_m Ce_m'(\xi) ce_m(\eta)\right], \tag{6.36b}$$

$$H_{\xi 1} = \frac{1}{p_1^2 d}\sum_m \left[j\omega\epsilon_1 B_m Se_m(\xi) se_m'(\eta) - j\beta A_m Ce_m'(\xi) ce_m(\eta)\right], \tag{6.36c}$$

$$H_{\eta 1} = -\frac{1}{p_1^2 d}\sum_m \left[j\omega\epsilon_1 B_m Se_m'(\xi) se_m(\eta) + j\beta A_m Ce_m(\xi) ce_m'(\eta)\right], \tag{6.36d}$$

$$E_{\xi 0} = -\frac{1}{q_0^2 d}\sum_r \left[-j\beta P_r Gek_r'(\xi) se_r^*(\eta) - j\omega\mu L_r Fek_r(\xi) ce_r^{*\prime}(\eta)\right], \tag{6.36e}$$

$$E_{\eta 0} = -\frac{1}{q_0^2 d}\sum_r \left[-j\beta P_r Gek_r(\xi) se_r^{*\prime}(\eta) + j\omega\mu L_r Fek_r'(\xi) ce_r^*(\eta)\right], \tag{6.36f}$$

$$H_{\xi 0} = -\frac{1}{q_0^2 d}\sum_r \left[j\omega\epsilon_0 P_r Gek_r(\xi) se_r^{*\prime}(\eta) - j\beta L_r Fek_r'(\xi) ce_r^*(\eta)\right], \tag{6.36g}$$

$$H_{\eta 0} = \frac{1}{q_0^2 d} \sum_r \left[j\omega\epsilon_0 P_r Gek_r'(\xi) se_r^*(\eta) + j\beta L_r Fek_r(\xi) ce_r^{*\prime}(\eta) \right], \tag{6.36h}$$

with

$$\begin{aligned} p_1^2 &= k_1^2 - \beta^2 \qquad k_1^2 = \omega^2 \mu \epsilon_1, \\ q_0^2 &= \beta^2 - k_0^2 \qquad k_0^2 = \omega^2 \mu \epsilon_0. \end{aligned} \tag{6.37}$$

Equating the tangential electric and magnetic fields at the boundary surface $\xi = \xi_0$ gives

$$\sum_{m=0}^{\infty} A_m Ce_m(\xi_0) ce_m(\eta) = \sum_{r=0}^{\infty} L_r Fek_r(\xi_0) ce_r^*(\eta), \tag{6.38}$$

$$\sum_{m=1}^{\infty} B_m Se_m(\xi_0) se_m(\eta) = \sum_{r=1}^{\infty} P_r Gek_r(\xi_0) se_r^*(\eta), \tag{6.39}$$

$$\sum_{m=1}^{\infty} \left\{ A_m \left(1 + \frac{\gamma_1^2}{\gamma_0^2}\right) Ce_m(\xi_0) ce_m'(\eta) + \frac{\omega\epsilon_1}{\beta} B_m Se_m'(\xi_0) se_m(\eta) \right\}$$

$$= -\frac{\gamma_1^2}{\gamma_0^2} \sum_{r=1}^{\infty} \frac{\omega\epsilon_0}{\beta} P_r Gek_r'(\xi_0) se_r^*(\eta), \tag{6.40}$$

$$\sum_{m=1}^{\infty} \left\{ A_m \frac{\omega\mu}{\beta} Ce_m'(\xi_0) ce_m(\eta) - B_m \left(1 + \frac{\gamma_1^2}{\gamma_0^2}\right) Se_m(\xi_0) se_m'(\eta) \right\}$$

$$= -\frac{\gamma_1^2}{\gamma_0^2} \sum_{r=0}^{\infty} \frac{\omega\mu}{\beta} L_r Fek_r'(\xi_0) se_r^*(\eta). \tag{6.41}$$

The prime denotes the derivatives with respect to ξ_0 or η, as the case may be. It is noted that in contrast with the circular dielectric cylinder case the angular functions in the elliptical case are functions not only of η, the angular coordinates, but also of the electrical properties of the medium in which they apply. Consequently, the summation signs and the angular Mathieu functions in the above equations may not be omitted. To eliminate the η dependence, we shall make use of the orthogonality properties of angular Mathieu functions as discussed in Sect. 6.2. Multiplying both sides of (6.38) and (6.40) by $ce_n(\eta)$ and both sides of (6.39) and (6.41) by $se_n(\eta)$, integrating with respect to η from 0 to 2π, and applying the orthogonality relations of the angular Mathieu functions leads to

$$A_n a_n = \sum_{r=0}^{\infty}{}' L_r l_r \alpha_{r,n}, \tag{6.42}$$

$$B_n b_n = \sum_{r=1}^{\infty}{}' P_r p_r \beta_{r,n}, \tag{6.43}$$

$$\frac{\omega\epsilon_1}{\beta} B_n b_n' + \left(1 + \frac{\gamma_1^2}{\gamma_0^2}\right) \sum_{r=1}^{\infty}{}' A_r a_r \chi_{r,n} = -\frac{\gamma_1^2}{\gamma_0^2}\frac{\omega\epsilon_0}{\beta} \sum_{r=1}^{\infty}{}' P_r p_r' \beta_{r,n}, \tag{6.44}$$

$$\frac{\omega\mu}{\beta} A_n a_n' - \left(1 + \frac{\gamma_1^2}{\gamma_0^2}\right) \sum_{r=0}^{\infty}{}' B_r b_r \nu_{r,n} = -\frac{\gamma_1^2}{\gamma_0^2}\frac{\omega\mu_0}{\beta} \sum_{r=1}^{\infty}{}' L_r l_r' \alpha_{r,n} \tag{6.45}$$

$$(n = 0, 2, 4, \ldots, \text{ or } n = 1, 3, 5, \ldots),$$

where the abbreviations

$$\begin{aligned}
a_n &= Ce_n(\xi_0) \quad & a_n' &= \frac{d}{d\xi_0}\left[Ce_n(\xi_0)\right], \\
b_n &= Se_n(\xi_0) \quad & b_n' &= \frac{d}{d\xi_0}\left[Se_n(\xi_0)\right], \\
l_r &= Fek_r(\xi_0) \quad & l_r' &= \frac{d}{d\xi_0}\left[Fek_r(\xi_0)\right], \\
p_r &= Gek_r(\xi_0) \quad & p_r' &= \frac{d}{d\xi_0}\left[Gek_r(\xi_0)\right]
\end{aligned} \tag{6.46}$$

have been used. This problem can also be solved by multiplying both sides of (6.38) and (6.41) by $ce_n^*(\eta)$ and both sides of (6.39) and (6.40) by $se_n^*(\eta)$, integrating with respect to η from 0 to 2π, and applying the orthogonality relations of the angular Mathieu functions. Similar results as those given here can be obtained. The prime over a_n, b_n, l_r, or p_r denotes the derivative with respect to ξ_0, while the prime over the summation sign indicates that odd or even integer values of r are to be taken according as n is odd or even. $\alpha_{r,n}$, $\beta_{r,n}$, $\chi_{r,n}$, and $\nu_{r,n}$ are given by the following:

$$\alpha_{r,n} = \int_0^{2\pi} ce_r^*(\eta) ce_n(\eta) d\eta \Big/ \int_0^{2\pi} ce_n^2(\eta) d\eta,$$

$$\beta_{r,n} = \int_0^{2\pi} se_r^*(\eta) se_n(\eta) d\eta \Big/ \int_0^{2\pi} se_n^2(\eta) d\eta,$$

$$\chi_{r,n} = \int_0^{2\pi} ce'_r(\eta) se_n(\eta) d\eta \Big/ \int_0^{2\pi} se_n^2(\eta) d\eta,$$

$$\nu_{r,n} = \int_0^{2\pi} sc'_r(\eta) cc_n(\eta) d\eta \Big/ \int_0^{2\pi} ce_n^2(\eta) d\eta. \tag{6.47}$$

Simplifying (6.42)–(6.45) and making the identifications

$$g_{m,n} = \left(1 + \frac{\gamma_1^2}{\gamma_0^2}\right) l_m \sum_{r=1}^{\infty}{}' \chi_{r,n} \alpha_{m,r},$$

$$s_{m,n} = -\left(1 + \frac{\gamma_1^2}{\gamma_0^2}\right) p_m \sum_{r=1}^{\infty}{}' \nu_{r,n} \beta_{m,r},$$

$$h_{m,n} = \frac{\omega\epsilon_1}{\beta} \frac{b'_n}{b_n} p_m \beta_{m,n} + \frac{\gamma_1^2}{\gamma_0^2} \frac{\omega\epsilon_0}{\beta} p'_m \beta_{m,n},$$

$$t_{m,n} = \frac{\omega\mu}{\beta} \frac{a'_n}{a_n} l_m \alpha_{m,n} + \frac{\gamma_1^2}{\gamma_0^2} \frac{\omega\mu}{\beta} l'_m \alpha_{m,n}, \tag{6.48}$$

one obtains

$$\sum_{m=0}^{\infty}{}' [L_m g_{m,n} + P_m h_{m,n}] = 0, \tag{6.49}$$

$$\sum_{m=0}^{\infty}{}' [L_m t_{m,n} + P_m s_{m,n}] = 0, \tag{6.50}$$

$$(n = 1, 3, 5, \ldots \text{ or } n = 0, 2, 4, \ldots).$$

Equations (6.49) and (6.50) are two sets of infinite homogeneous linear algebraic equations in L_m and P_m. For a nontrivial solution the determinant of these equations must vanish. The roots of this infinite determinant provide the values from which the propagation constant β can be determined. For example, the infinite determinant for the ${}_{\mathrm{e}}\mathrm{HE}_{n'm'}$ modes with odd n' and $n = 1, 3, 5, \ldots$ is

$$\begin{vmatrix} g_{1,1} & h_{1,1} & g_{3,1} & h_{3,1} & g_{5,1} & h_{5,1} & \cdot & \cdot \\ t_{1,1} & s_{1,1} & t_{3,1} & s_{3,1} & t_{5,1} & s_{5,1} & \cdot & \cdot \\ g_{1,3} & h_{1,3} & g_{3,3} & h_{3,3} & g_{5,3} & h_{5,3} & \cdot & \cdot \\ t_{1,3} & s_{1,3} & t_{3,3} & s_{3,3} & t_{5,3} & g_{5,3} & \cdot & \cdot \\ g_{1,5} & h_{1,5} & g_{3,5} & h_{3,5} & g_{5,5} & h_{5,5} & \cdot & \cdot \\ t_{1,5} & s_{1,5} & t_{3,5} & s_{3,5} & t_{5,5} & s_{5,5} & \cdot & \cdot \\ \cdot & \cdot & \cdot & \cdot & \cdot & \cdot & \cdot & \cdot \\ \cdot & \cdot & \cdot & \cdot & \cdot & \cdot & \cdot & \cdot \end{vmatrix} = 0. \tag{6.51a}$$

A similar infinite determinant for ${}_{\mathrm{e}}\mathrm{HE}_{n'm'}$ modes with even indices n' and $n = 0, 2, 4, \ldots$ may also be obtained. It is

$$\begin{vmatrix} g_{0,0} & h_{0,0} & g_{2,0} & h_{2,0} & g_{4,0} & h_{4,0} & \cdot & \cdot \\ t_{0,0} & s_{0,0} & t_{2,0} & s_{2,0} & t_{4,0} & s_{4,0} & \cdot & \cdot \\ g_{0,2} & h_{0,2} & g_{2,2} & h_{2,2} & g_{4,2} & h_{4,2} & \cdot & \cdot \\ t_{0,2} & s_{0,2} & t_{2,2} & s_{2,2} & t_{4,2} & g_{4,2} & \cdot & \cdot \\ g_{0,4} & h_{0,4} & g_{2,4} & h_{2,4} & g_{4,4} & h_{4,4} & \cdot & \cdot \\ t_{0,4} & s_{0,4} & t_{2,4} & s_{2,4} & t_{4,4} & s_{4,4} & \cdot & \cdot \\ \cdot & \cdot & \cdot & \cdot & \cdot & \cdot & \cdot & \cdot \\ \cdot & \cdot & \cdot & \cdot & \cdot & \cdot & \cdot & \cdot \end{vmatrix} = 0. \tag{6.51b}$$

Because of the extreme complexity of these infinite determinants, they can only be solved by the method of successive approximations [8]. It is found numerically that the first root of $n' = 1$, $m' = 1$ mode is governed principally by the expression

$$\begin{vmatrix} g_{1,1} & h_{1,1} \\ t_{1,1} & s_{1,1} \end{vmatrix} = 0$$

as long as the elliptical cross section is not too flat (i.e., $\xi_0 > 0.5$). For higher order modes, successive approximation should start from the expression

$$\begin{vmatrix} g_{n,n} & h_{n,n} \\ t_{n,n} & s_{n,n} \end{vmatrix} = 0.$$

It has been assumed that $n' = n$.

${}_{o}HE_{n'm'}$ *Wave*

The general expressions for the axial magnetic and electric fields of an ${}_{\mathrm{o}}\mathrm{HE}_{n'm'}$ wave are the following:

For region 1 $(0 \le \xi \le \xi_0)$

$$H_z^{(1)} = \sum_{m=1}^{\infty} C_m Se_m(\xi)\, se_m(\eta),$$

$$E_z^{(1)} = \sum_{m=0}^{\infty} D_m Ce_m(\xi)\, ce_m(\eta), \tag{6.52}$$

and for region 0 ($\xi_0 \leq \xi < \infty$)

$$H_z^{(0)} = \sum_{r=1}^{\infty} G_r Gek_r\,(\xi)\, se_r^*(\eta),$$

$$E_z^{(0)} = \sum_{r=0}^{\infty} F_r Fek_r\,(\xi)\, ce_r^*(\eta), \tag{6.53}$$

where C_m, D_m, G_r, and F_r are the arbitrary constants. The transverse fields are the following:

$$E_\xi^{(1)} = \frac{1}{p_1^2 d}\sum_m \left[-j\beta B_m Ce_m'(\xi)ce_m(\eta) - j\omega\mu A_m Se_m(\xi)se_m'(\eta)\right], \tag{6.54a}$$

$$E_\eta^{(1)} = \frac{1}{p_1^2 d}\sum_m \left[-j\beta B_m Ce_m(\xi)ce_m'(\eta) + j\omega\mu A_m Se_m'(\xi)se_m(\eta)\right], \tag{6.54b}$$

$$H_\xi^{(1)} = \frac{1}{p_1^2 d}\sum_m \left[j\omega\epsilon_1 B_m Ce_m(\xi)se_m'(\eta) - j\beta A_m Se_m'(\xi)se_m(\eta)\right], \tag{6.54c}$$

$$H_\eta^{(1)} = -\frac{1}{p_1^2 d}\sum_m \left[j\omega\epsilon_1 B_m Ce_m'(\xi)ce_m(\eta) + j\beta A_m Se_m(\xi)se_m'(\eta)\right], \tag{6.54d}$$

$$E_\xi^{(0)} = -\frac{1}{q_0^2 d}\sum_r \left[-j\beta P_r Fek_r'(\xi)se_r^*(\eta) - j\omega\mu L_r Gek_r(\xi)se_r^{*\prime}(\eta)\right], \tag{6.54e}$$

$$E_\eta^{(0)} = -\frac{1}{q_0^2 d}\sum_r \left[-j\beta P_r Fek_r(\xi)se_r^{*\prime}(\eta) + j\omega\mu L_r Gek_r'(\xi)se_r^*(\eta)\right], \tag{6.54f}$$

$$H_\xi^{(0)} = -\frac{1}{q_0^2 d}\sum_r \left[j\omega\epsilon_0 P_r Fek_r(\xi)se_r^{*\prime}(\eta) - j\beta L_r Fek_r'(\xi)se_r^*(\eta)\right], \tag{6.54g}$$

$$H_\eta^{(0)} = \frac{1}{q_0^2 s}\sum_r \left[j\omega\epsilon_0 P_r Fek_r'(\xi)ce_r^*(\eta) + j\beta L_r Gek_r(\xi)se_r^{*\prime}(\eta)\right], \tag{6.54h}$$

with $p_1^2 = k_1^2 - \beta^2$, $q_0^2 = \beta^2 - k_0^2$, $k_1^2 - \omega^2\mu\epsilon_1$, $k_0^2 - \omega^2\mu\epsilon_0$.

Upon matching the boundary conditions at $\xi = \xi_0$ and applying the similar mathematical operations as for the ${}_e\mathrm{HE}_{n'm'}$ mode, one can easily obtain the characteristic equation for the ${}_o\mathrm{HE}_{n'm'}$ wave. For example, the determinantal equation for the ${}_o\mathrm{HE}_{n'm'}$ modes with n' odd and $n = 1, 3, 5, \ldots$ is

$$\begin{vmatrix}
g^*_{1,1} & h^*_{1,1} & g^*_{3,1} & h^*_{3,1} & g^*_{5,1} & h^*_{5,1} & \cdot & \cdot \\
t^*_{1,1} & s^*_{1,1} & t^*_{3,1} & s^*_{3,1} & t^*_{5,1} & s^*_{5,1} & \cdot & \cdot \\
g^*_{1,3} & h^*_{1,3} & g^*_{3,3} & h^*_{3,3} & g^*_{5,3} & h^*_{5,3} & \cdot & \cdot \\
t^*_{1,3} & s^*_{1,3} & t^*_{3,3} & s^*_{3,3} & t^*_{5,3} & g^*_{5,3} & \cdot & \cdot \\
g^*_{1,5} & h^*_{1,5} & g^*_{3,5} & h^*_{3,5} & g^*_{5,5} & h^*_{5,5} & \cdot & \cdot \\
t^*_{1,5} & s^*_{1,5} & t^*_{3,5} & s^*_{3,5} & t^*_{5,5} & s^*_{5,5} & \cdot & \cdot \\
\cdot & \cdot & \cdot & \cdot & \cdot & \cdot & \cdot & \cdot \\
\cdot & \cdot & \cdot & \cdot & \cdot & \cdot & \cdot & \cdot
\end{vmatrix} = 0. \tag{6.55a}$$

A similar infinite determinant for the ${}_0\mathrm{HE}_{n'm'}$ modes with n' even and $n = 0, 2, 4, 6, \ldots$ may also be obtained. It is

$$\begin{vmatrix}
g^*_{0,0} & h^*_{0,0} & g^*_{2,0} & h^*_{2,0} & g^*_{4,0} & h^*_{4,0} & \cdot & \cdot \\
t^*_{0,0} & s^*_{0,0} & t^*_{2,0} & s^*_{2,0} & t^*_{4,0} & s^*_{4,0} & \cdot & \cdot \\
g^*_{0,2} & h^*_{0,2} & g^*_{2,2} & h^*_{2,2} & g^*_{4,2} & h^*_{4,2} & \cdot & \cdot \\
t^*_{0,2} & s^*_{0,2} & t^*_{2,2} & s^*_{2,2} & t^*_{4,2} & g^*_{4,2} & \cdot & \cdot \\
g^*_{0,4} & h^*_{0,4} & g^*_{2,4} & h^*_{2,4} & g^*_{4,4} & h^*_{4,4} & \cdot & \cdot \\
t^*_{0,4} & s^*_{0,4} & t^*_{2,4} & s^*_{2,4} & t^*_{4,4} & s^*_{4,4} & \cdot & \cdot \\
\cdot & \cdot & \cdot & \cdot & \cdot & \cdot & \cdot & \cdot \\
\cdot & \cdot & \cdot & \cdot & \cdot & \cdot & \cdot & \cdot
\end{vmatrix} = 0, \tag{6.55b}$$

where

$$\begin{aligned}
g^*_{m,n} &= \left(1 + \frac{\gamma_1^2}{\gamma_0^2}\right) p_m \sum_{r=1}^{\infty} \nu_{r,n} \beta_{m,r}, \\
s^*_{m,n} &= -\left(1 + \frac{\gamma_1^2}{\gamma_0^2}\right) l_m \sum_{r=1}^{\infty} \chi_{r,n} \alpha_{m,r}, \\
h^*_{m,n} &= \frac{\omega\epsilon_1}{\beta} \frac{a'_n}{a_n} l_m \alpha_{m,n} + \frac{\gamma_1^2}{\gamma_0^2} \frac{\omega\epsilon_0}{\beta} l'_m \alpha_{m,n}, \\
t^*_{m,n} &= \frac{\omega\mu}{\beta} \frac{b'_n}{b_n} p_m \beta_{m,n} + \frac{\gamma_1^2}{\gamma_0^2} \frac{\omega\mu}{\beta} p'_m \beta_{m,n}.
\end{aligned} \tag{6.56}$$

To simplify the notations, the following dimensionless quantities are introduced:

$$x_1^2 = q^2 \cosh^2 \xi_0 \left(k_1^2 - \beta^2\right) = 4 \cosh^2 \xi_0 \, \gamma_1^2, \tag{6.57}$$

$$y_1^2 = -q^2 \cosh^2 \xi_0 \left(k_0^2 - \beta^2\right) = 4 \cosh^2 \xi_0 \, \gamma_0^2. \tag{6.58}$$

Hence the infinite determinants are functions of x_1, y_1, ξ_0, and ϵ_1/ϵ_0 only.

6.4.1 Cutoff Frequencies of Modes

It is known that x_1 and y_1 are the roots of the dispersion relations. Combining (6.57) and (6.58) we arrive at the propagation constant

$$\beta = \frac{1}{q\cosh\xi_0}\left(q^2\cosh^2\xi_0\, k_1^2 - x_1^2\right)^{1/2} = \frac{1}{q\cosh\xi_0}\left(q^2\cosh^2\xi_0 k_0^2 + y_1^2\right)^{1/2}$$
$$= \frac{1}{q\cosh\xi_0}\left(\frac{x_1^2 + \frac{\epsilon_1}{\epsilon_0}y_1^2}{\frac{\epsilon_1}{\epsilon_0} - 1}\right)^{1/2}. \tag{6.59}$$

To have a guided wave, β^2, x_1^2, and y_1^2 must all be real and positive. The fact that x_1^2 and y_1^2 must all be real and positive offers a way to determine the upper and lower bounds of the propagation constant β. According to (6.57) and (6.58), $\beta^2 \leq k_1^2$ and $\beta^2 \geq k_0^2$. Thus, $k_0 \leq \beta \leq k_1$. One recalls that the positive and real values of γ_0^2 or y_1^2 indicate that the field intensities outside the dielectric rod decay with increasing distance from the surface of the guide. If y_1^2 is negative and real, the expressions for the field components will indicate the presence of an outgoing radial wave at a large distance from the surface of the dielectric rod, which can only come from an infinitely long (in the z-direction) line type source located at some finite ξ. Such sources have not been postulated in the assumptions. Thus y_1^2 must be real for all surface guided waves and consequently the lowest permissible value of y_1^2 is zero. The propagation constant and the frequency corresponding to this value are

$$\beta_{\text{cutoff}} = \frac{x_1}{q\cosh\xi_0\sqrt{\left(\frac{\epsilon_1}{\epsilon_0} - 1\right)}} \tag{6.60}$$

and

$$\omega_{\text{cutoff}} = \frac{x_1}{q\cosh\xi_0\sqrt{\left(\frac{\epsilon_1}{\epsilon_0} - 1\right)\mu\epsilon_0}}, \tag{6.61}$$

respectively. x_1 corresponds to the root of the characteristic equation with $y_1^2 = 0$. The frequency defined by (6.61) is called the cutoff frequency of the wave, since below such frequency the mode no longer exists on the dielectric guide. Physically it means that below this cutoff frequency the structure can no longer support such a wave and thereby ceases to be a binding medium.

The approximate expressions of the modified Mathieu functions for small x_1 and y_1 are quite complicated and involved [5]. The derivation and the explicit forms of these expressions are given in [9]. It can be shown that for small values of x_1 and y_1 to the first order approximation,

$$\begin{aligned} \alpha_{r,n} \sim \beta_{r,n} &\sim 1 \qquad \text{when } r = n \\ &\sim 0 \qquad \text{when } r \neq n \end{aligned} \tag{6.62a}$$

and

$$\begin{aligned} \nu_{m,n} \sim -\chi_{m,n} &\sim n \qquad \text{when } m = n \\ &\sim 0 \qquad \text{when } m \neq n. \end{aligned} \tag{6.62b}$$

The mode order n' also becomes n. Hence, the infinite determinant (6.51) for the ${}_{\rm e}{\rm HE}_{n'm'}$ modes with odd n'collapses and reduces approximately to

$$x_1^2 \approx \frac{\left[n\left(1+\frac{\epsilon_0}{\epsilon_1}\right)+\left(\frac{\epsilon_0}{\epsilon_1}\tanh\xi_0\,Q_1+\coth\xi_0\,Q_2\right)\right]4\left(n^2-1\right)}{\frac{e^{2\xi_0}}{\cosh^2\xi_0}\left[(n+1)+(n-1)\,e^{-4\xi_0}\right]} \tag{6.63}$$

when $n \geq 3$ (n is odd) and

$$x_1^2 \approx \frac{8\left[\left(1+\frac{\epsilon_0}{\epsilon 1}\right)+\left(\frac{\epsilon_0}{\epsilon_1}\tanh\xi_0+\coth\xi_0\right)\right]}{-\frac{\epsilon_0}{\epsilon_1}\frac{e^{2\xi_0}}{\cosh^2\xi_0}\ln\left(\frac{e^{\alpha}y_1\,e^{\xi_0}}{2\cosh\xi_0}\right)\left[3-2\,e^{-2\xi_0}\right]} \tag{6.64}$$

when $n = 1$, where Q_1 and Q_2 are finite constants and α is the Euler constant. The subscript n' is now equal to n.

Upon inspection of (6.63) we may immediately conclude that the right hand side of the equation is always positive and nonzero and is not necessarily small for all values of ξ_0 and ϵ_0/ϵ_1. Thus the imposed small x_1 argument is not valid and x_1 must be determined from the original characteristic equation (6.51) with $y_1 = 0$. The same conclusion may be reached from $n' \geq 2$ even modes.

From (6.64) it is noted that as y_1 approaches zero, $\ln\left(\frac{e^{\alpha}y_1 e^{\xi_0}}{2\cosh\xi_0}\right)$ approaches $-\infty$; thus the right hand side of this equation approaches $+0$. In other words, as y_1 approaches zero, x_1 also approaches zero and the imposed small x_1 approximation is valid. Therefore, the cutoff frequency of the ${}_{\rm e}{\rm HE}_{11}$ mode is zero [see (6.59)].

Similarly, for the ${}_{\mathrm{o}}\mathrm{HE}_{n'm'}$ mode, we have

$$x_1^2 \approx \frac{4(n^2-1)\left[n\left(1+\frac{\epsilon_0}{\epsilon_1}\right)+\left(\tanh\xi_0\,Q_1+\frac{\epsilon_0}{\epsilon_1}\coth\xi_0\,Q_2\right)\right]}{\frac{e^{2\xi_0}}{\cosh^2\xi_0}\left[(n+1)+(n-1)\,e^{-4\xi_0}\right]} \tag{6.65}$$

when $n \geq 3$ (n is odd) and

$$x_1^2 \approx \frac{8\left[\left(1+\frac{\epsilon_0}{\epsilon_1}\right)+\left(\tanh\xi_0+\frac{\epsilon_0}{\epsilon_1}\coth\xi_0\right)\right]}{-\frac{\epsilon_0}{\epsilon_1}\frac{e^{2\xi_0}}{\cosh^2\xi_0}\ln\left(\frac{e^{\alpha}y_1\,e^{\xi_0}}{2\cosh\xi_0}\right)\left[3-2\,e^{-2\xi_0}\right]} \tag{6.66}$$

when $n = 1$. The subscript n' becomes n.

We observe that for $n' \geq 3$ ${}_{\mathrm{o}}\mathrm{HE}_{n'm'}$ wave, the right hand side of (6.65) is always positive and nonzero; thus x_1 is also positive and nonzero. The same conclusion applies for the $n' \geq 2$ ${}_{\mathrm{o}}\mathrm{HE}_{n'm'}$ waves. For the $n' = 1$ ${}_{\mathrm{o}}\mathrm{HE}_{11}$ wave, according to (6.66), as y_1 approaches zero, x_1 must also approach zero. Therefore, no cutoff frequency for the ${}_{\mathrm{o}}\mathrm{HE}_{11}$ mode exists.

The above analysis shows that unlike the case for the circular dielectric rod, no simple cutoff equations can be found for the surface wave modes along the elliptical dielectric rod. Cutoff conditions must be computed from the rather complicated determinantal equations. But, if we impose the conditions of small x_1 as well as small y_1, an expression for x_1 as a function of y_1, ξ_0, n, and ϵ_0/ϵ_1 may be derived from the determinantal equations. We can show that the imposed small x_1 condition as $y_1 \rightarrow 0$ is only valid if $n' = 1$ for both ${}_{\mathrm{e}}\mathrm{HE}_{11}$ and ${}_{\mathrm{o}}\mathrm{HE}_{11}$ modes. Furthermore, as $y_1 \rightarrow 0$, x_1 also approaches zero for $n' = 1$ mode, although x_1 approaches zero slower for flatter rods for the ${}_{\mathrm{e}}\mathrm{HE}_{11}$ mode, x_1 approaches zero faster for flatter rods for the ${}_{\mathrm{o}}\mathrm{HE}_{11}$ mode. Hence, there are only two nondegenerate modes on an elliptical dielectric rod, namely, the ${}_{\mathrm{e}}\mathrm{HE}_{11}$ and ${}_{\mathrm{o}}\mathrm{HE}_{11}$ mode that possess zero cutoff frequency. Additional numerical computations also show that as the elliptical dielectric cylinder becomes flatter, the cutoff frequencies for higher order even ${}_{\mathrm{e}}\mathrm{HE}_{n'm'}$ modes are higher and those for higher order odd ${}_{\mathrm{o}}\mathrm{HE}_{n'm'}$ modes are lower.

6.4.2 Transition to Circular Cross-Section

As an ellipse degenerates to a circle, its semifocal length q tends to zero while ξ_0 approaches infinity so that the product $q\cosh\xi_0$ or $q\sinh\xi_0$ or $qe^{\xi_0}/2$ tends to a constant r_0, which is the radius of the degenerated circle. The degenerate forms of the Mathieu and modified Mathieu functions can be found from [5].

Using these degenerate expressions one obtains the following degenerate forms for the factors appearing in the characteristic equations:

$$a_n \sim b_n \sim Ce_n(\xi_0) \sim Se_n(\xi_0) \sim J_n(x_1),$$

$$a'_n \sim b'_n \sim Ce'_n(\xi_0) \sim Se'_n(\xi_0) \sim x_1 J'_n(x_1),$$

$$l_r \sim p_r \sim Fek_r(\xi_0) \sim Gek_r(\xi_0) \sim K_r(y_1),$$

$$l'_r \sim p'_r \sim Fek'_r(\xi_0) \sim Gek'_r(\xi_0) \sim y_1 K_r(y_1), \tag{6.67}$$

$$\alpha_{r,n} \sim \beta_{r,n} \sim \quad \begin{cases} 1 \text{ when } r = n \\ 0 \text{ when } r \neq n \end{cases}$$

$$\nu_{r,n} \sim -\chi_{r,n} \sim \quad \begin{cases} r \text{ when } r = n \\ 0 \text{ when } r \neq n, \end{cases}$$

where $x_1^2 = r_0^2\left(k_1^2 - \beta^2\right)$ and $y_1^2 = r_0^2\left(\beta^2 - k_0^2\right)$. All terms in the infinite determinants vanish except those inside the dashed boundary [see (6.51a) and (6.55a)]. We note that the degenerate forms of (6.51a) and (6.55a) are identical; hence, ${}_e\text{HE}_{11}$ wave and ${}_o\text{HE}_{11}$ wave are degenerate on a circular cylinder. The degenerated infinite determinant becomes

$$\prod_n (g_{nn}s_{nn} - h_{nn}t_{nn}) = 0 \tag{6.68}$$

or

$$(g_{nn}s_{nn} - h_{nn}t_{nn}) = 0, \tag{6.69}$$

with $n = 0, 1, 2, 3, \ldots$ representing all possible hybrid waves on a circular dielectric cylinder. Substituting the degenerate expressions for g_{nn}, s_{nn}, h_{nn}, and t_{nn} in (6.69) gives

$$\left[\frac{1}{x_1}\frac{J'_n(x_1)}{J_n(x_1)} + \frac{1}{y_1}\frac{K'_n(y_1)}{K_n(y_1)}\right]\left[\frac{1}{x_1}\frac{J'_n(x_1)}{J_n(x_1)} + \frac{\epsilon_0}{\epsilon_1}\frac{1}{y_1}\frac{K'_n(y_1)}{K_n(y_1)}\right]$$
$$-n^2\frac{\left(x_1^2 + y_1^2\right)\left(y_1^2 + x_1^2\dfrac{\epsilon_0}{\epsilon_1}\right)}{x_1^4 y_1^4} = 0, \tag{6.70}$$

which is exactly the characteristic equation for an HE_{nm} wave on a circular dielectric cylinder. When $n = 0$, (6.69) becomes

$$(g_{0,0}s_{0,0} - h_{0,0}t_{0,0}) = 0 \tag{6.71}$$

or

$$\left[\frac{1}{x_1}\frac{J_0'(x_1)}{J_0(x_1)} + \frac{1}{y_1}\frac{K_0'(y_1)}{K_0(y_1)}\right]\left[\frac{1}{x_1}\frac{J_0'(x_1)}{J_0(x_1)} + \frac{\epsilon_0}{\epsilon_1}\frac{1}{y_1}\frac{K_0'(y_1)}{K_0(y_1)}\right] = 0, \tag{6.72}$$

which is the characteristic equation for the TE_{0m} and TM_{0m} waves on a circular rod.

6.4.3 Approximate Characteristic Equations

Examination of the infinite determinants representing the characteristic equations for the ${}_{\mathrm{e,o}}\mathrm{HE}_{n'm'}$ modes shows that the coefficients responsible for coupling the elements of the determinants are $\alpha_{r,n}$, $\beta_{r,n}$, $\nu_{r,n}$, and $\chi_{r,n}$, which are the expansion coefficients for the infinite series expressing one periodic angular Mathieu function in one medium in terms of all the periodic angular Mathieu functions in a different medium. In other words,

$$ce_r^*(\eta) = \sum_n \alpha_{r,n} ce_n(\eta), \tag{6.73a}$$

$$se_r^*(\eta) = \sum_n \beta_{r,n} se_n(\eta), \tag{6.73b}$$

$$ce_r'(\eta) = \sum_n \chi_{r,n} se_n(\eta), \tag{6.73c}$$

$$se_r'(\eta) = \sum_n \nu_{r,n} ce_n(\eta). \tag{6.73d}$$

Another way of interpreting the mode structure on an elliptical dielectric waveguide is to state that the cyclical variation of a given mode around the cross-section of the waveguide cannot be represented by a single angular Mathieu function of a given order regardless whether the angular function is the one in the core region or in the cladding region. That cyclical variation must be represented by an infinite series of angular Mathieu functions in either region (core or cladding). On the other hand, it is realized that very significant simplification is obtained if one simply assumes that $ce_r^*(\eta) \approx ce_r(\eta)$, $se_r^*(\eta) \approx se_r(\eta)$, $ce_r'(\eta) \approx -se_r(\eta)$, and $se_r'(\eta) \approx ce_r(\eta)$

and that $n' = n$. Using this assumption, we have

$$\alpha_{r,n} \approx \beta_{r,n} = \begin{cases} 1 \text{ when } r = n \\ 0 \text{ when } r \neq n \end{cases}, \tag{6.74a}$$

$$\nu_{r,n} \sim -\chi_{r,n} \sim \begin{cases} n \text{ when } r = n \\ 0 \text{ when } r \neq n \end{cases}. \tag{6.74b}$$

The infinite determinants become

$$(g_{nn}s_{nn} - h_{nn}t_{nn}) = 0 \tag{6.75}$$

for the ${}_{\mathrm{e}}\mathrm{HE}_{n'm'}$ modes, and

$$(g^*_{nn}s^*_{nn} - h^*_{nn}t^*_{nn}) = 0 \tag{6.76}$$

for the ${}_{\mathrm{o}}\mathrm{HE}_{n'm'}$ modes. Or,

$$\left[\frac{1}{\gamma_1^2}\frac{Ce'_n(\xi_0)}{Ce_n(\xi_0)} + \frac{1}{\gamma_0^2}\frac{Fek'_n(\xi_0)}{Fek_n(\xi_0)}\right]\left[\frac{1}{\gamma_1^2}\frac{Se'_n(\xi_0)}{Se_n(\xi_0)} + \frac{\epsilon_0}{\epsilon_1}\frac{1}{\gamma_0^2}\frac{Gek'_n(\xi_0)}{Gek_n(\xi_0)}\right]$$

$$= n^2\frac{(\gamma_0^2 + \gamma_1^2)\left(\gamma_0^2 + \frac{\epsilon_0}{\epsilon_1}\gamma_1^2\right)}{\gamma_0^4\gamma_1^4} \tag{6.77}$$

for the ${}_{\mathrm{e}}\mathrm{HE}_{n'm'}$ modes, and

$$\left[\frac{1}{\gamma_1^2}\frac{Se'_n(\xi_0)}{Se_n(\xi_0)} + \frac{1}{\gamma_0^2}\frac{Gek'_n(\xi_0)}{Gek_n(\xi_0)}\right]\left[\frac{1}{\gamma_1^2}\frac{Ce'_n(\xi_0)}{Ce_n(\xi_0)} + \frac{\epsilon_0}{\epsilon_1}\frac{1}{\gamma_0^2}\frac{Fek'_n(\xi_0)}{Fek_n(\xi_0)}\right]$$

$$= n^2\frac{(\gamma_0^2 + \gamma_1^2)\left(\gamma_0^2 + \frac{\epsilon_0}{\epsilon_1}\gamma_1^2\right)}{\gamma_0^4\gamma_1^4} \tag{6.78}$$

for the ${}_{\mathrm{o}}\mathrm{HE}_{n'm'}$ modes. Note that the subscript n' is now n.

These greatly simplified characteristic equations can be used to find the propagation constants of various even and odd modes. To determine the accuracy of their results we have compared the exact results for the propagation constants of the dominant ${}_{\mathrm{e}}\mathrm{HE}_{11}$ and ${}_{\mathrm{o}}\mathrm{HE}_{11}$ modes, for an elliptical dielectric cylinder with relative dielectric constant 2.5 immersed in free-space, obtained according to the infinite determinants, with the approximate results calculated using the above simplified equations. Numerical computations show that the results are well within 10–15% of each other, even for an eccentricity ratio (major axis/minor axis) as high as 4. Of course, if the ellipse is close to a circular shape, the results are almost identical, as expected. The results are also much closer in the low normalized frequency range (normalized to the size of the guide).

6.4.4 Propagation Characteristics

The solutions of the characteristic equations will now be considered. It can be seen that the transcendental characteristic equations are of the form

$$f\left(\xi_0, \frac{\epsilon_0}{\epsilon_1}, y_1, x_1\right) = 0. \tag{6.79}$$

Knowing ξ_0, which determines the eccentricity of the elliptical cross-section, and ϵ_0/ϵ_1, which is the ratio of dielectric constant of the surrounding medium to that of the dielectric rod, (6.79) reduces to

$$g(y_1, x_1) = 0, \tag{6.80}$$

where ξ_0 and ϵ_0/ϵ_1 are given constants. y_1 and x_1 are related to the major axis of the rod, the frequency, the propagation constant, and the characteristics of the medium by the relations

$$y_1^2 = -q^2 \cosh^2 \xi_0 \left(k_0^2 - \beta^2\right),$$

$$x_1^2 = q^2 \cosh^2 \xi_0 \left(k_1^2 - \beta^2\right). \tag{6.81}$$

For propagating waves, y_1 and x_1 must both be positive and real. Roots of (6.51) and (6.55) can most readily be solved by the successive approximation method [8]. Furthermore, for these dominant modes y_1 can vary from 0 to $+\infty$, while x_1 varies from 0 to some finite positive constant. For example, for the dominant ${}_{\mathrm{e}}\mathrm{HE}_{11}$ mode, the value of the determinant was first obtained using a (2×2) determinant centered at columns (1,2) and row (1,2) (i.e., the first two columns and the first two rows of the determinant), then using a (4×4) determinant, etc., until a converged value is obtained. For the higher order ${}_{\mathrm{e}}\mathrm{HE}_{31}$ mode, the value of the determinant must first be obtained using a (2×2) determinant centered at columns (3,4) and row (3,4), then using a (4×4) determinant centered at columns (3,4) and row (3,4), etc., until a converged value is obtained. It was found (numerically) that the infinite determinants converge rather rapidly within the present region of interests (i.e., $0 \leq x_1 \leq 5$ and $0 \leq y_1 \leq 3$). This rate of convergence was obtained for all modes. An 8×8 determinant was the largest one used to obtain a two significant figure accuracy.

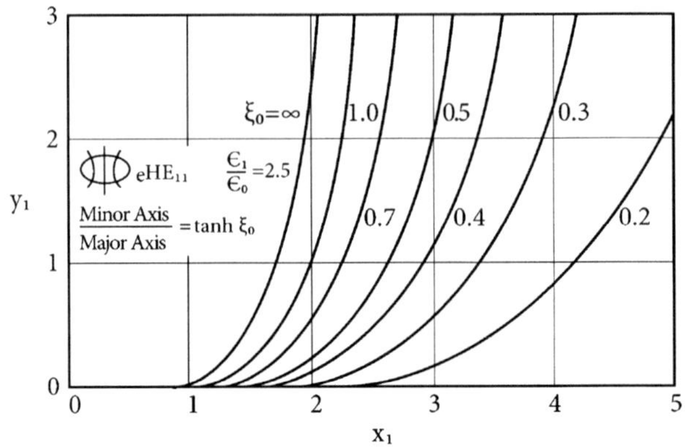

Figure 6.2. Roots of the characteristic equation for the ${}_e\mathrm{HE}_{11}$ wave [1]

6.4.4.1 The Even Dominant ${}_e\mathrm{HE}_{11}$ Mode

The roots of the determinantal equation for the ${}_e\mathrm{HE}_{11}$ mode are shown in Fig. 6.2 for the case $\epsilon_1/\epsilon_0 = 2.5$.

Figure 6.3 gives the relation between the normalized guide wavelength λ/λ_0 and the normalized major axis (NMA) $2q\cosh\xi_0/\lambda_0$ for various values of ξ_0 with $\epsilon_1/\epsilon_0 = 2.5$. $\lambda_0 = 2\pi/\left(\omega\sqrt{\mu\epsilon_0}\right)$ is the free-space wavelength. As expected, no cutoff frequency exists for this mode, and the cutoff frequency approaches zero slower as the elliptical cross-section becomes flatter. For small values of NMA, which corresponds to small values of y_1, the guide wavelength λ approaches the free-space wavelength λ_0, and the radial Mathieu functions describing the fields outside the dielectric rod decay very slowly; physically it means that the field strength of this wave falls off very slowly away from the rod and only a small part of the total energy is transported within the dielectric rod. For large values of NMA, which corresponds to large values of y_1, the guide wavelength approaches asymptotically to the characteristic wavelength of the material, $\lambda_M = \lambda_0/\sqrt{\epsilon_1/\epsilon_0}$, and the radial Mathieu functions describing the fields outside the rod disappear very quickly and almost all the energy is transported within the dielectric rod.

Figure 6.3 also shows curves of the normalized guide wavelength as functions of normalized cross-sectional area (NCSA) $(2q\cosh\xi_0/\lambda_0)^2\tanh\xi_0$ for various ξ_0 with $\epsilon_1/\epsilon_0 = 2.5$. For a fixed value of NCSA, λ/λ_0 is larger for flatter elliptical cross-section. This behavior suggests that the field intensity is more concentrated in a circular rod. It is also noted that the flatter is the cross-section, the smaller is the variation of λ/λ_0 as a function of NCSA.

The effect of the variation of relative dielectric constant ϵ_1/ϵ_0 on the propagation constant can be seen readily from Fig. 6.4, in which λ/λ_0 is plotted against

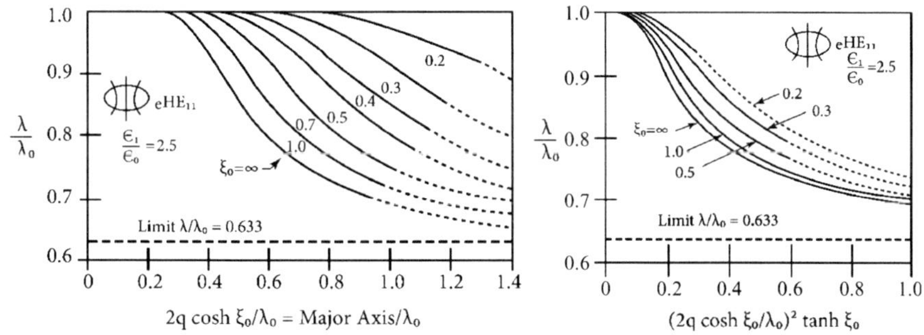

Figure 6.3. Normalized guide wavelength λ/λ_0 for the ${}_{e}HE_{11}$ wave as a function of (*left*) the major axis and (*right*) the normalized cross-sectional area [1]

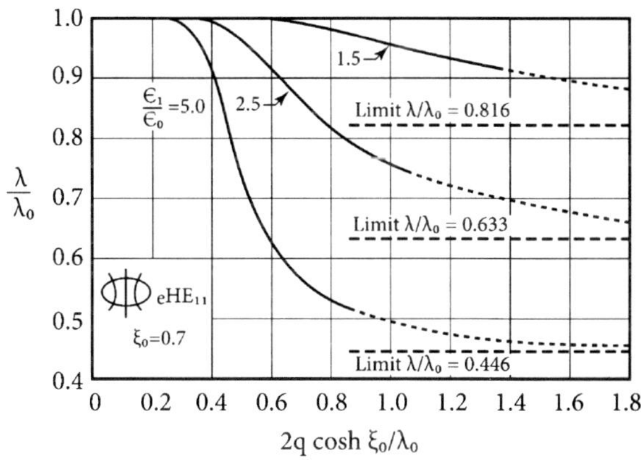

Figure 6.4. Normalized guide wavelength of the ${}_{e}HE_{11}$ wave as a function of normalized major axis for various values of ϵ_1/ϵ_0 [1]

NMA for various values of ϵ_1/ϵ_0 with $\xi_0 = 0.7$. As ϵ_1/ϵ_0 becomes larger, λ approaches the characteristic wavelength of the rod material $\lambda_M = \lambda_0/\sqrt{\epsilon_1/\epsilon_0}$ faster. This is the same behavior exhibited by the dominant wave propagating along a circular dielectric rod.

6.4.4.2 The Odd Dominant ${}_{o}HE_{11}$ Mode

The roots of the determinantal equation, the normalized guide wavelength λ/λ_0 as a function of NMA and NCSA for the ${}_{o}HE_{11}$ are shown, respectively, in Figs. 6.5 and 6.6.

The general characteristics of the guide wavelength as a function of NMA, NCSA, ξ_0, and ϵ_1/ϵ_0 are very similar to those indicated by the even dominant wave. However, unlike the case for the ${}_{e}HE_{11}$ wave, it seems that the elliptical rod is a better binding geometry than a circular rod, since for a fixed value or NCSA,

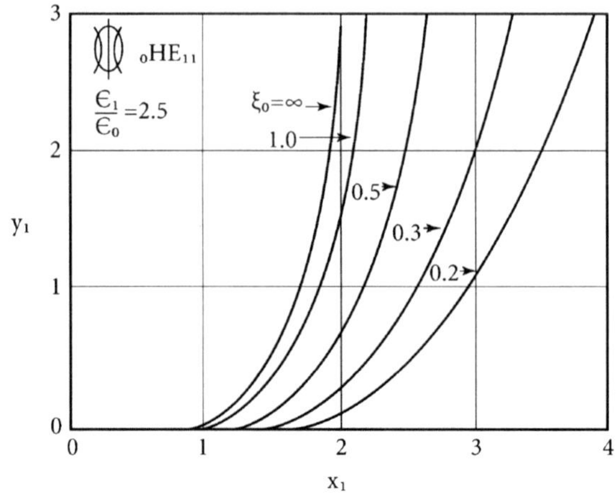

Figure 6.5. Roots of the characteristic equation for the $_0$HE$_{11}$ wave [1]

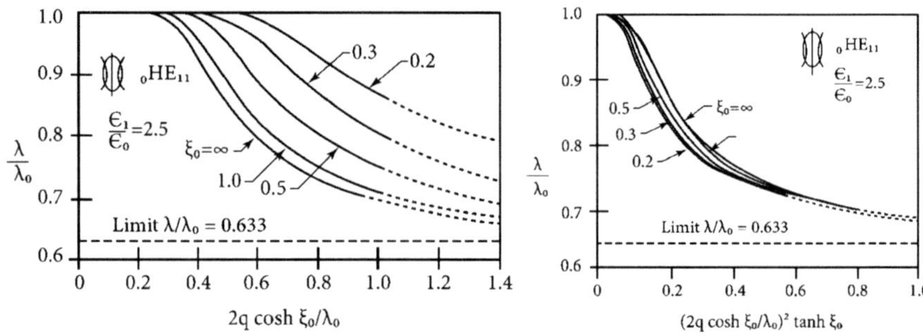

Figure 6.6. Normalized guide wavelength λ/λ_0 for the $_0$HE$_{11}$ wave as a function of (*left*) the normalized major axis and (*right*) the normalized cross-sectional area [1]

λ/λ_0 is smaller for flatter cross-section. The fact that the curves for various values of ξ_0 are quite close to each other suggests that the field intensity is quite uniform for this odd dominant mode in the dielectric rod.

It should be noted that the propagation characteristics of the $_e$HE$_{11}$ wave and the $_0$HE$_{11}$ wave pass smoothly to that of the circular HE$_{11}$ wave as $\xi_0 \to \infty$.

6.4.4.3 Higher Order $_{e,o}$HE$_{n'm'}$ Modes

Higher order modes on an elliptical dielectric waveguide and their cutoff frequencies have been studied by numerous researchers [10–13]. When the ellipse is nearly a circle, such as $b/a = 0.9$, where b is the minor axis and a is the major axis, it is necessary to use only a 4×4 determinant or (6.77) and (6.78) to calculate the roots of the dispersion relations. Results for the case of an $\epsilon_1/\epsilon_0 = 2.26$ dielectric cylinder with $b/a = 0.9$ are displayed in Fig. 6.7.

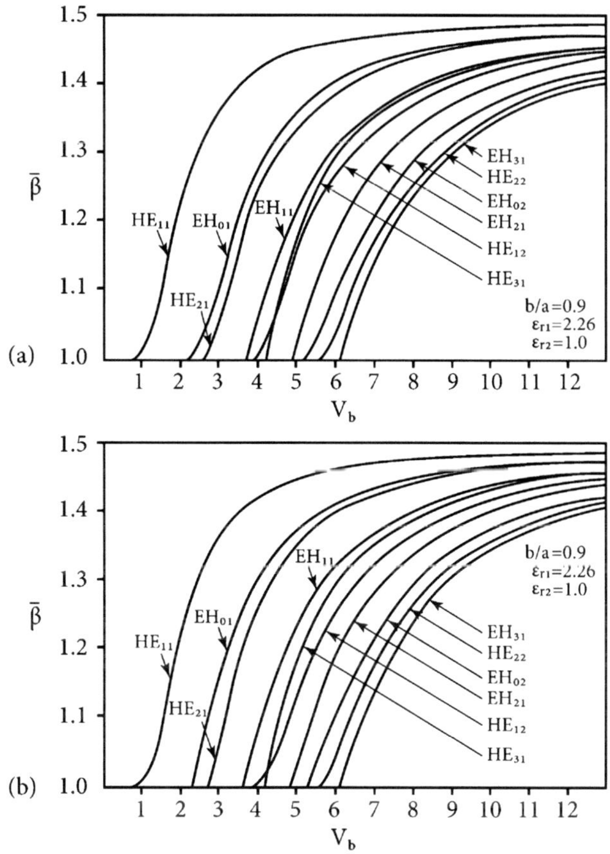

Figure 6.7. Normalized propagation constants on a dielectric rod ϵ_1=2.26 in free space with ellipticity $a/b = 0.9$ for (*top*) odd modes and (*bottom*) even modes. A 4×4 determinant using Mathieu functions was used [11]

Cutoff frequencies for several higher order modes have been obtained by Rengarajan [12] for core relative dielectric constant 2.1316 and cladding relative dielectric constant 1.7956. He used the determinants up to order 9. Results are shown in Fig. 6.8.

6.4.5 Field Configurations of the Dominant Modes

The field configuration for a specific mode can be calculated from the mode function for that mode. The field distribution is given by field lines in which the direction of the line at a point gives the direction of the field and the density of lines the magnitude. On the one hand, unlike the case for the circular dielectric rod, the elliptical dielectric rod cannot support the circularly symmetric modes. On the other hand, the dominant $_e\mathrm{HE}_{11}$ and the $_o\mathrm{HE}_{11}$ modes on the elliptical rod bear

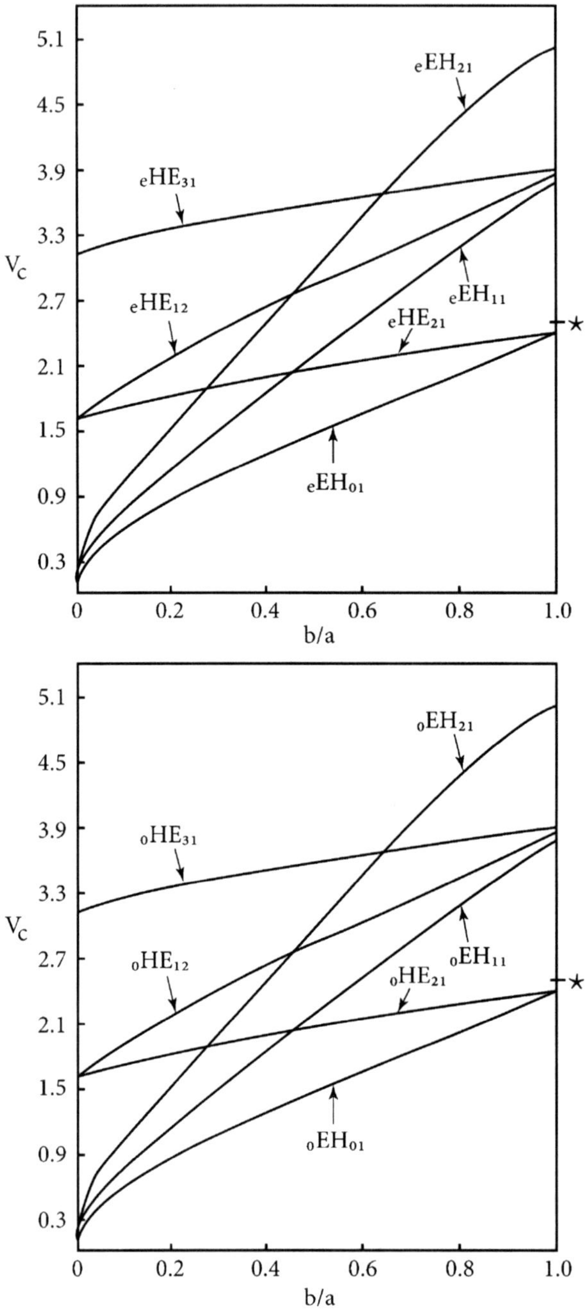

Figure 6.8. Higher mode cutoff for elliptical fibers $\Delta n = 0.1$. The asterisk denotes the exact cutoff for the circular HE_{21} mode [12]

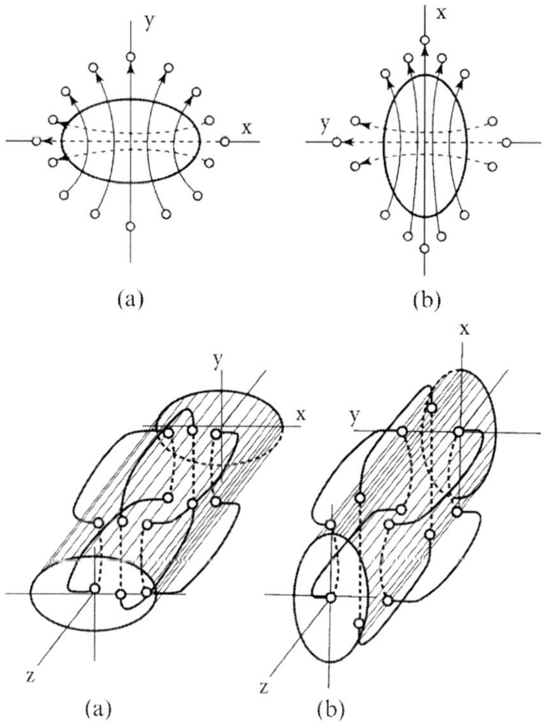

Figure 6.9. Cross-sectional field configuration of (**a**) the ${}_{e}HE_{11}$ wave and (**b**) the ${}_{o}HE_{11}$ wave. Solid lines represent the electric lines of force and the dashed lines represent the magnetic lines of force. For the bottom figure, only electric lines of force are shown

very close resemblance to the HE_{11} mode on the circular rod. It is also noted that the field distribution in an elliptical dielectric rod are very similar to those in an elliptical metallic waveguide. The significant difference is that in the former case the electric and magnetic lines of force must form closed loops outside the dielectric rod and in the latter case the electric field is normal to the metallic boundary surrounding the core, and the magnetic field is parallel to it.

Figure 6.9a,b shows the transverse cross-sectional field lines for the ${}_{e}HE_{11}$ and ${}_{o}HE_{11}$ waves. Perspective views of the electric line of force are also shown in the figure. The fact that the cross-sectional field configurations are similar to the corresponding dominant waves in the metal tube waveguides suggests a simple method of exciting these dominant dielectric waves.

6.4.6 Attenuation Calculation

It is of interest to learn if changing the cross-sectional shape of a dielectric waveguide can affect the attenuation characteristics of a guided mode. It has already been shown that the dispersion behavior of a dominant mode can be significantly altered by the ellipticity of the guide cross-section. Figures 6.4 and 6.8 show that for the same cross-sectional area, depending on the polarization of the dominant mode, the propagation characteristics can be either significantly altered or very slightly altered.

Although the computation for the attenuation constant was rather laborious, it was nevertheless carried out by Yeh [4]. The perturbation formula (3.20) was used:

$$\alpha = 8.686 \left(\frac{\pi}{\lambda_0}\right) \tan\delta_1 \, \epsilon_{r1} \, R \qquad (\mathrm{dB\,m}^{-1}), \tag{6.82}$$

$$R = \frac{\displaystyle\int_{A_1} \left(\mathbf{E}^{(1)} \cdot \mathbf{E}^{(1)*}\right) dA}{\sqrt{\dfrac{\mu_0}{\epsilon_0}} \left[\displaystyle\int_{A_1} \mathbf{e}_z \cdot \left(\mathbf{E}_\mathrm{t}^{(1)} \times \mathbf{H}_\mathrm{t}^{(1)*}\right) dA + \int_{A_0} \mathbf{e}_z \cdot \left(\mathbf{E}_\mathrm{t}^{(0)} \times \mathbf{H}_\mathrm{t}^{(0)*}\right) dA\right]}.$$

Here, the superscript 1 and 0 refer, respectively, to the core region and the cladding region of the guide. $\mathbf{e}_z$ is the unit vector in the direction of propagation, A_1 and A_0 are, respectively, the cross-sectional areas of the core and surrounding region, $\mathbf{E}$ and $\mathbf{H}$ are the electric and magnetic field vectors of a single guided mode under consideration, and ϵ_{1r} and $\tan\delta_1$ are the relative dielectric constant and loss tangent of the dielectric core region. For example, for a single ${}_{\mathrm{e,o}}\mathrm{HE}_{n'm'}$ mode, we have

$${}_{\mathrm{e,o}}\mathbf{E}^{(1),(0)}(\xi,\eta) = {}_{\mathrm{e,o}}E_\xi^{(1),(0)} \mathbf{e}_\xi + {}_{\mathrm{e,o}}E_\eta^{(1),(0)} \mathbf{e}_\eta + {}_{\mathrm{e,o}}E_z^{(1),(0)} \mathbf{e}_z, \tag{6.83a}$$

$${}_{\mathrm{e,o}}\mathbf{H}^{(1),(0)}(\xi,\eta) = {}_{\mathrm{e,o}}H_\xi^{(1),(0)} \mathbf{e}_\xi + {}_{\mathrm{e,o}}H_\eta^{(1),(0)} \mathbf{e}_\eta + {}_{\mathrm{e,o}}H_z^{(1),(0)} \mathbf{e}_z. \tag{6.83b}$$

As an example, the results of this lengthy calculation for the ${}_\mathrm{e}\mathrm{HE}_{11}$ with $\epsilon_{r1} = 2.085$ (Teflon) and $a/b = 1, 2, 3$ are displayed in Fig. 6.10, where the loss factor $\epsilon_{r1} R$ is plotted as a function of the normalized cross-sectional area $A(\epsilon_{r1} - 1)/\lambda_0^2$, where $A = \pi ab$ for an ellipse. It is seen that a mere flattening of a circular dielectric rod along the maximum intensity of the electric field lines for the ${}_\mathrm{e}\mathrm{HE}_{11}$ mode can improve the loss factor $\epsilon_{r1} R$ (hence α) by a factor of 2 or more.

6.5 Weakly Guiding Elliptical Dielectric Waveguides [13–15]

Because of the advent of optical fiber communication, the case of special interest is one where the refractive index difference between the core and its cladding of optical fibers is very small, that is, $(n_1/n_0 - 1) \ll 1$, where n_1 is the core index

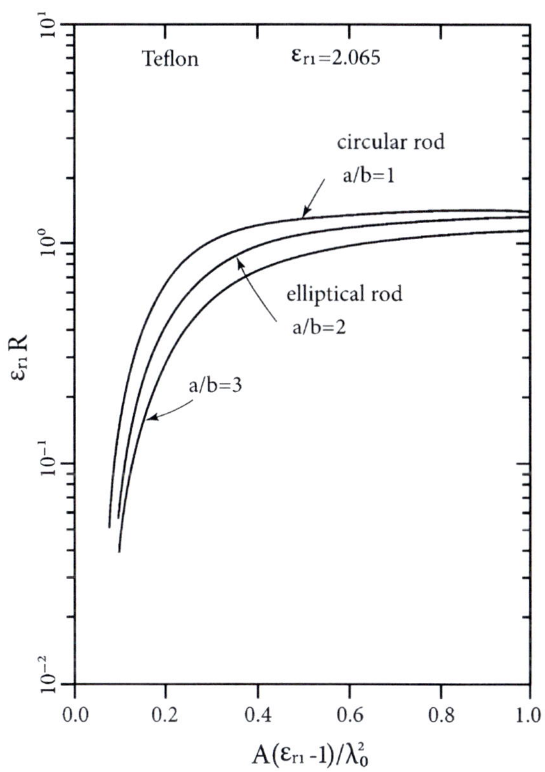

Figure 6.10. Configuration loss factor $\epsilon_1 R$ as a function of normalized area for an elliptical Teflon rod supporting the dominant ${}_e\mathrm{HE}_{11}$ mode. Here A is the cross-sectional area, λ_0 is the free-space wavelength, a is the semi-major axis of the elliptical rod, and b is the semi-minor axis. Note that the flatter rod yields a smaller configuration loss factor for the same cross-sectional area

of refraction and n_0 is the cladding index of refraction. Fibers possessing this type of index of refraction characteristics are called weakly guiding fibers. One notes, however, since most of the guided power is contained within the fiber core region, transmitted power is still very well guided by this "weakly guiding" fiber. In other words, very little power is leaked to the radiation wave.

The circular weakly guiding fiber case was treated by Snyder [16, 17] who introduced a smallness parameter θ_p defined as

$$\theta_\mathrm{p} = \left(1 - \frac{\beta^2}{k_1^2}\right)^{1/2}, \tag{6.84}$$

where β is the propagation constant, $k_1^2 = \omega^2 \mu_0 \epsilon_1$, and ϵ_1 is the dielectric constant of the core region.

Using the definitions for γ_1^2 and γ_0^2 as given in (6.57) and (6.58), we have

$$U^2 = q^2 \cosh^2 \xi_0 \left(k_1^2 - \beta^2\right) = 4\gamma_1^2 \cosh^2 \xi_0, \tag{6.85}$$

$$W^2 = -q^2 \cosh^2 \xi_0 \left(k_0^2 - \beta^2\right) = 4\gamma_0^2 \cosh^2 \xi_0, \tag{6.86}$$

$$\begin{aligned} V^2 = U^2 + W^2 &= q^2 \cosh^2 \xi_0 \, \omega^2 \mu \epsilon_1 \left(1 - \frac{\epsilon_0}{\epsilon_1}\right), \\ &= q^2 \cosh^2 \xi_0 \, \omega^2 \mu \epsilon_1 \, \delta, \end{aligned} \tag{6.87}$$

$$\theta_p = \sqrt{\delta} \frac{U}{V} = \left(1 - \frac{\beta^2}{k_1^2}\right)^{1/2}. \tag{6.88}$$

These parameters are similar to those defined by Snyder in his treatment of the circular fiber case. In fact, these parameters are identical to Snyder's when the elliptical cylinder degenerates to a circular one.

In a similar manner as that carried out by Snyder for the circular fiber case, we shall expand all functions in terms of the smallness parameter θ_{p} if $(n_1/n_0 - 1) \ll 1$. In other words, when $\theta_{\mathrm{p}} \ll 1$, considerable simplification results. Hence, we have

$$-\gamma_0^2 = \gamma_1^2 - \frac{\omega^2 \mu \epsilon_1 \, \delta}{4} = \gamma_1^2 + O\left(\theta_{\mathrm{p}}^2\right), \tag{6.89}$$

$$ce_r^*(\eta, \gamma_0^2) = ce_r\left(\eta, \gamma_1^2\right) + O\left(\theta_{\mathrm{p}}^2\right), \tag{6.90}$$

$$se_r^*(\eta, \gamma_0^2) = se_r\left(\eta, \gamma_1^2\right) + O\left(\theta_{\mathrm{p}}^2\right), \tag{6.91}$$

$$\alpha_{r,m} = \frac{\displaystyle\int_0^{2\pi} ce_r\left(\eta, \gamma_1^2\right) ce_n\left(\eta, \gamma_1^2\right) d\eta}{\displaystyle\int_0^{2\pi} ce_n^2\left(\eta, \gamma_1^2\right) d\eta} + O\left(\theta_{\mathrm{p}}^2\right) = C_{r,m}^{(1)} \Delta_{r,m} + O\left(\theta_{\mathrm{p}}^2\right), \tag{6.92}$$

$$\beta_{r,m} = \frac{\displaystyle\int_0^{2\pi} se_r\left(\eta, \gamma_1^2\right) se_n\left(\eta, \gamma_1^2\right) d\eta}{\displaystyle\int_0^{2\pi} se_n^2\left(\eta, \gamma_1^2\right) d\eta} + O\left(\theta_{\mathrm{p}}^2\right) = C_{r,m}^{(2)} \Delta_{r,m} + O\left(\theta_{\mathrm{p}}^2\right), \tag{6.93}$$

$$1 + \frac{\gamma_1^2}{\gamma_0^2} = O\left(\theta_{\mathrm{p}}^2\right), \tag{6.94}$$

where $C_{r,m}^{(1),(2)}$ are known constants and $\Delta_{r,m}$ is the Kronecker delta, which is zero when $r \neq n$ and is unity when $r = n$.

Substituting (6.89)–(6.94) in (6.48) and (6.56) gives

$$ {}_{\mathrm{e,o}}g_{m,n} = O\left(\theta_{\mathrm{p}}^2\right), \tag{6.95} $$

$$ {}_{\mathrm{e,o}}s_{m,n} = O\left(\theta_{\mathrm{p}}^2\right), \tag{6.96} $$

$$ {}_{\mathrm{e,o}}h_{m,n} = {}_{\mathrm{e,o}}D_{m,n}^{(1)}\Delta_{m,n} + O\left(\theta_{\mathrm{p}}^2\right), \tag{6.97} $$

$$ {}_{\mathrm{e,o}}t_{m,n} = {}_{\mathrm{e,o}}D_{m,n}^{(2)}\Delta_{m,n} + O\left(\theta_{\mathrm{p}}^2\right), \tag{6.98} $$

where ${}_{\mathrm{e,o}}D_{m,n}^{(1),(2)}$ can be obtained readily from (6.48) and (6.56).

Inserting (6.95)–(6.98) in (6.48) and (6.56), one may show that all off-diagonal terms are all of the order of $O\left(\theta_p^2\right)$. For example, from (6.48) and (6.56) one has

$$ \left| \begin{array}{cc|cc|cc} O\left(\theta_{\mathrm{p}}^2\right) & {}_{\mathrm{e,o}}h_{1,1} & O\left(\theta_{\mathrm{p}}^2\right) & O\left(\theta_{\mathrm{p}}^2\right) & \cdot & \cdot \\ {}_{\mathrm{e,o}}t_{1,1} & O\left(\theta_{\mathrm{p}}^2\right) & O\left(\theta_{\mathrm{p}}^2\right) & O\left(\theta_{\mathrm{p}}^2\right) & \cdot & \cdot \\ \hline O\left(\theta_{\mathrm{p}}^2\right) & O\left(\theta_{\mathrm{p}}^2\right) & O\left(\theta_{\mathrm{p}}^2\right) & {}_{\mathrm{e,o}}h_{3,3} & \cdot & \cdot \\ O\left(\theta_{\mathrm{p}}^2\right) & O\left(\theta_{\mathrm{p}}^2\right) & {}_{\mathrm{e,o}}t_{3,3} & O\left(\theta_{\mathrm{p}}^2\right) & \cdot & \cdot \\ \hline \cdot & \cdot & \cdot & \cdot & \cdot & \cdot \\ \cdot & \cdot & \cdot & \cdot & \cdot & \cdot \end{array} \right| = 0. \tag{6.99} $$

Ignoring terms of the order of $O\left(\theta_{\mathrm{p}}^2\right)$, the zero order dispersion relation for ${}_{\mathrm{e}}\mathrm{HE}_{nm}$ and ${}_{\mathrm{o}}\mathrm{HE}_{nm}$ modes are, respectively,

$$ \frac{1}{\gamma_1^2}\frac{Ce_n'(\xi_0,\gamma_1^2)}{Ce_n(\xi_0,\gamma_1^2)} + \frac{1}{\gamma_0^2}\frac{Fek_n'(\xi_0,|\gamma_0^2|)}{Fek_n(\xi_0,|\gamma_0^2|)} = 0 \tag{6.100} $$

and

$$ \frac{1}{\gamma_1^2}\frac{Se_n'(\xi_0,\gamma_1^2)}{Se_n(\xi_0,\gamma_1^2)} + \frac{1}{\gamma_0^2}\frac{Gek_n'(\xi_0,|\gamma_0^2|)}{Gek_n(\xi_0,|\gamma_0^2|)} = 0. \tag{6.101} $$

These are the simplified dispersion relations that we are seeking. These relations are valid if $(n_1/n_0 - 1) \ll 1$. It can be seen that (6.100) and (6.101) are much simpler expressions for the dispersion relations than the infinite determinants represented by (6.51) and (6.55).

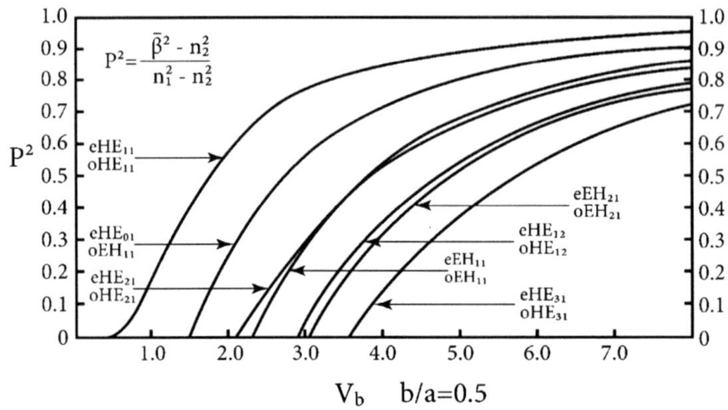

Figure 6.11. Propagation in elliptical guide with small dielectric difference. $n_1 \approx n_2$, $V_b = (2\pi/\lambda_0)\left(n_1^2 - n_2^2\right)^{1/2}$ [14]

To verify the accuracy of the approximate dispersion relations, numerical computations were carried out for several specific cases using first the exact dispersion relations and then the approximate ones [13]. For the cases considered in which $n_1/n_0 \lesssim 1$, the approximate dispersion relations yield results that are within 2% of the exact values, and hence indistinguishable from the exact results.

Additional work on weakly guiding elliptical core fiber has been carried out by Eyges et al. [14] using the extended boundary condition method, by Shaw et al. [15] using scalar wave solutions and characteristic numbers of Mathieu equation, by Saad [18] using a point matching technique discussed in Chap. 7, and by Yeh and Manshadi [19] using the beam propagation method discussed in Chap. 12. Representative results by Eyges et al. [14] are shown in Fig. 6.11.

6.6 Experimental Results

The cavity resonator method discussed in Chap. 5 can be used readily to measure the propagation constant and the attenuation constant of a given mode supported by a dielectric waveguide of arbitrary cross-section. The normalized propagation constants of a few modes along an elliptical dielectric rod with $\epsilon_{r1} = 2.3$, $\epsilon_{r2} = 1$, and $b/a = 0.2$ as a function of normalized frequency $V\left[= (2\pi b/\lambda_0)\left(\epsilon_{r2} - \epsilon_{r2}\right)^{1/2}\right]$ were measured by Dyott [10,20] using the resonator method. Here ϵ_{r1} is the relative dielectric constant of the rod, ϵ_{r2} is the relative dielectric constant of the surrounding medium, b is the semi-minor axis, and a is the semimajor axis of the elliptical cross-section. Results are displayed in Fig. 6.12.

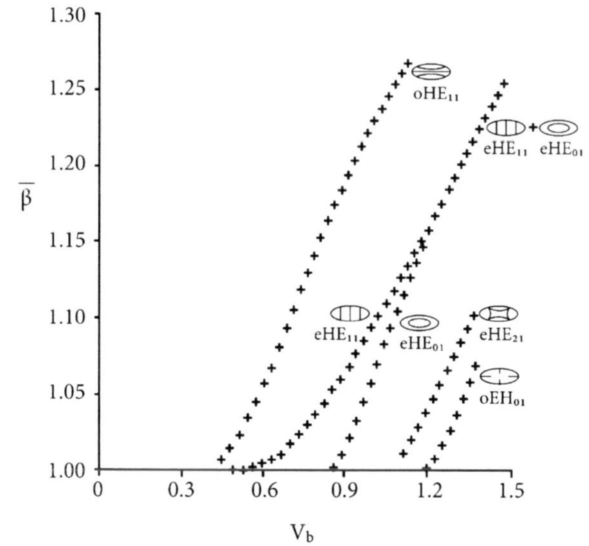

Figure 6.12. Normalized propagation constants of first few modes in elliptical dielectric waveguide. $\epsilon_1/\epsilon_0 = 2.3$; $\epsilon_2/\epsilon_0 = 1.0$; $b/a = 0.2$; $\overline{\beta} = \beta/k_0$; and $V_b = (2\pi/\lambda_0)\,(n_1^2 - n_2^2)$. Experimental points were determined from the microwave resonator [10]

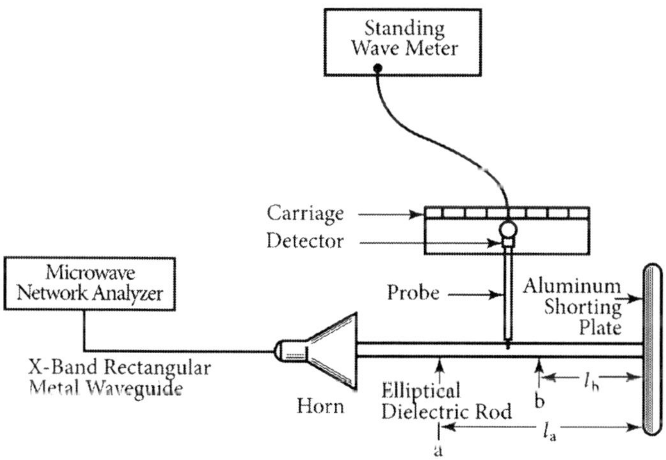

Figure 6.13. Standard standing wave measuring apparatus

A standard standing wave technique can also be used to perform the measurements (see Fig. 6.13) [21]. According to the setup, the guide wavelength and the attenuation constant can be measured. The formula relating the attenuation factor α with the standing wave ratio can be derived as follows: It is known that

$$A = 5\log_{10}\left(\frac{P_1}{P_3}\right) \quad \text{(dB)},$$

where P_1 and P_3 are, respectively, the input and reflected power of the guide; and

$$\frac{P_1}{P_3} = \left(\frac{r-1}{r+1}\right)^2,$$

where r is the standing wave ratio at the probe. Therefore, we have

$$\alpha = \frac{10}{l}\log_{10}\left[\left(\frac{r-1}{r+1}\right)^2\right] \quad (\text{dB m}^{-1}),$$

in which l is the length of the guide as indicated in Fig. 6.13. To take into account the loss due to the imperfections of the shorting plate, one notes that the attenuation measured at point a is $A_a = \alpha l_a + D$, and similarly the attenuation measured at

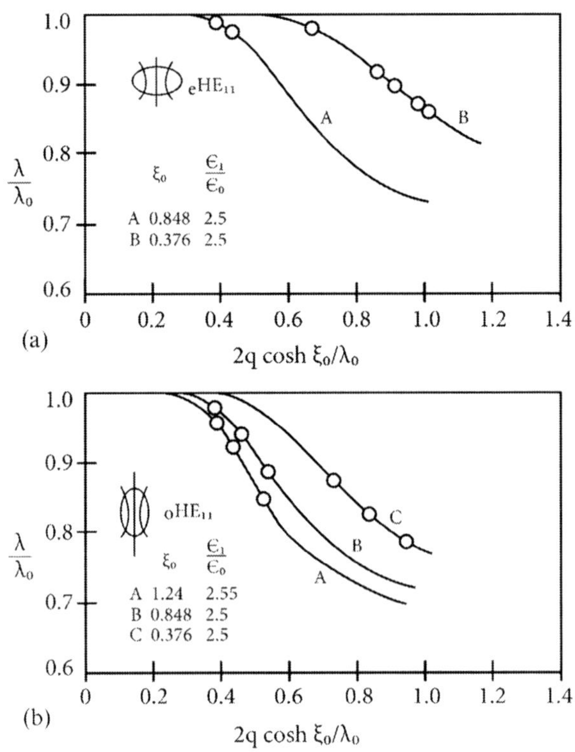

Figure 6.14. Normalized guide wavelength as a function of normalized major axis for the (*top*) $_{\mathrm{e}}\mathrm{HE}_{11}$ wave and (*bottom*) $_{\mathrm{o}}\mathrm{HE}_{11}$ wave [1]

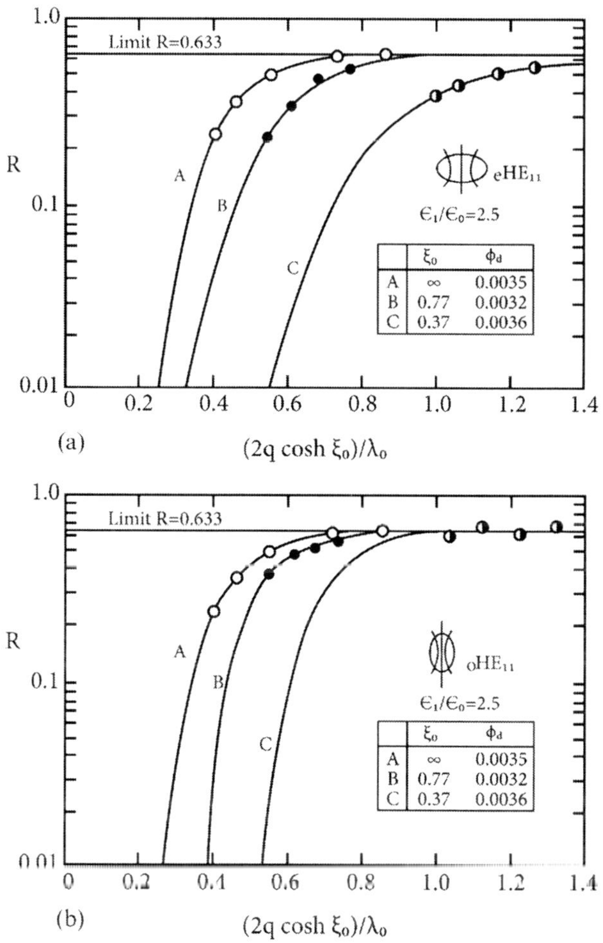

Figure 6.15. R as a function of $(2q\cosh\xi_0)/\lambda_0$ for the ${}_{e}HE_{11}$ wave. R as a function of $(2q\cosh\xi_0)/\lambda_0$ for the ${}_{o}HE_{11}$ wave [4]

point b is $A_b = \alpha l_b + D$, where D is the loss of the reflecting plate. Combining these two equations and eliminating D one gets

$$\alpha = \frac{(A_a - A_b)}{l_a - l_b}.$$

The distance between adjacent minima of the standing wave on the rod is $\lambda_g/2$, where λ_g is the guide wavelength of the dominant mode. Measured results are shown in Figs. 6.14 and 6.15. Excellent agreement with the calculated results is seen.

6.7 Comments

It is worth noting that the elliptical dielectric waveguide is the only noncircular dielectric waveguide that possesses an analytic solution. By changing the eccentricity of the elliptical cross-section, the waveguide can alter from a roundish circular waveguide to a flattish ribbon-like planar waveguide. Because of the asymmetry of the elliptical cross-section with respect to its major and minor axes, there can exist two polarizations for the dominant ${}_e\mathrm{HE}_{n'm'}$ and ${}_o\mathrm{HE}_{n'm'}$ modes with different propagation constants. This canonical solution provides the necessary reference for which solutions from all numerical methods or other approximate techniques must be compared.

Unlike the familiar case of a circular dielectric waveguide or even the case of an elliptical metallic waveguide, the angular mode number corresponding to the cyclical variation of the modal fields no longer possesses the one-to-one correspondence to the order of the analytic modal functions representing the modal fields. This characteristic makes it more difficult to classify the mode numbers. We must resort to observing the physical cyclical variation of the modal fields around the cross-section of the guide and then assign the modal numbers accordingly. The description of the modal fields is further complicated by the fact that each modal field must be represented by an infinite series of the analytic angular and radial Mathieu functions. No single order analytic special function representing a single modal field can be found for the elliptical dielectric waveguide. The complication is also manifested in the calculation and labeling of the cutoffs for the higher order modes. The same difficulty is carried over to weakly guiding elliptical fibers [10, 15].

References

1. C. Yeh, "Elliptical dielectric waveguides," J. Appl. Phys. **33**, 3235 (1962)
2. L. A. Lynbimov, G. I. Veselov, and N. A. Bei, "Dielectric waveguide with elliptical cross-section," Radio Eng. Electron. (USSR) **6**, 1668 (1961)
3. G. Piefke, "Grundlagen sur Berechnung der Ubertragung-seigenschafter elliptischer Wellenleiter," A. E. U. **18**, 4 (1964)
4. C. Yeh, "Attenuation in a dielectric elliptical cylinder," IEEE Trans. Antenn. Propag. **AP-11**, 177 (1963)
5. N. W. McLachlan, "Theory and Application of Mathieu Functions," University Press, Oxford (1951); J. Meixner and F. W. Schafke, "Mathieu-Funktionen und Spharoid-Funktionen, " Springer, Berlin Heidelberg New York(1954); National Bureau of Standards, "Tables Relating to Mathieu Functions," Columbia University Press, New York (1951); M. Abramowitz and I. A. Stegun, eds., "Handbook of Math Functions

with Formulas, Graphs, and Mathematical Tables," National Bureau of Standards, Washington DC, Applied Math. Series (1964)

6. E. L. Ince, "Ordinary Differential Equations," Longmans, London (1927)
7. E. Snitzer, "Cylindrical dielectric waveguide modes," J. Opt. Soc. Am. **51**, 491 (1961); J. R. Carson, S. P. Mead, and S. A. Schelkunoff, "Hyperfrequency waveguides – Mathematical theory," Bell Syst. Tech. J. **15**, 310 (1936)
8. L. Kantorovich and V. Krylov, "Approximate Methods of Higher Analysis," Interscience Publishers, New York (1958)
9. C. Yeh, "Electromagnetic surface-wave propagation along a dielectric cylinder of elliptic cross section", Technical Report No. 27, Antenna Laboratory, California Institute of Technology, Pasadena, CA (1962)
10. R. B. Dyott, "Elliptical Fiber Waveguide," Artech House, Boston (1995)
11. J. E. Lewis and G. Deshpande, "Modes on elliptical cross-section dielectric tube waveguides," Microw. Opt. Acoust. **3**, 147 (1979)
12. S. F. Rengarajan, "On higher order mode cutoff frequencies in elliptical step index fibers," IEEE Trans. Microw. Theory Tech. **MTT-37**, 1244 (1989)
13. C. Yeh, "Modes in weakly guiding elliptical optical fibers," Opt. Quantum Electron. **8**, 43 (1976)
14. L. Eyges, P. Gianino, and P. Wintersteiner, "Modes of dielectric waveguides of arbitrary cross sectional shape," J. Opt. Soc. Am. **69**, 1226 (1979)
15. J. K. Shaw, W. M. Henry, and W. R. Winfrey, "Weakly guiding analysis of elliptical core step index waveguides based on the characteristic numbers of Mathieu's equation," J. Lightwave Tech. **13**, 2359 (1995)
16. A. W. Snyder, "Asymptotic expressions for eigenfunctions and eigenvalues of a dielectric or optical waveguide," IEEE Trans. Microw. Theory Tech. **MTT-17**, 1130 (1969)
17. D. Gloge, "Weakly guiding fibers," Appl. Opt. **10**, 2252 (1971)
18. S. M. Saad, "On the higher order modes of elliptical fibers," IEEE Trans. Microw. Theory Tech. **MTT-33**, 1110 (1985)
19. C. Yeh and F. Manshadi, "On weakly guiding single-mode optical waveguides," J. Lightwave Tech. **3**, 199 (1985)
20. R. B. Dyott, "Cutoff of the first higher order modes in elliptical dielectric waveguides: An experimental approach," Electron. Lett. **26**, 1721 (1990)
21. E. L. Ginzton, "Microwave Measurements," McGraw-Hill, New York (1957)

7

APPROXIMATE METHODS

For dielectric waveguides, the only canonical shapes with closed surfaces in the transverse plane that possess exact analytic solutions to guided wave problems are the circular cylinder and the elliptical cylinder, as discussed in previous chapters. An exact analytic solution does not exist for the case of wave propagation along a dielectric waveguide of rectangular shape. The need for approximate techniques to solve the problem associated with rectangular or more general shaped dielectric structures is apparent. Two approximate techniques will be discussed in particular: (a) the Marcatili approach [1] and (b) the circular harmonic point matching technique [2]. Other notable approximate techniques by Schlosser and Unger [3], using rectangular harmonics, by Eyges, et al., using the extended boundary condition method [4], and by Shaw et al. [5], using a variational approach will not be discussed here due to their computational complexity. Other computational techniques are discussed in Chap. 15.

7.1 Marcatili's Approximate Method [1]

The a priori assumption of Marcatili's approach is that for a well guided mode in a dielectric waveguide, most of its guided power is contained within the core region of the guide. Very little power of that mode resides in the cladding region of the guide. So, if the boundary conditions are satisfied by most of the assumed fields that carry most of the guided power, then these assumed fields may be used to approximate the solution for the problem. For a rectangular waveguide shown in Fig. 7.1, very little guided power is contained in the corner regions of the guide, that is, the shaded regions. Marcatili formulated an approximate solution to this problem by ignoring the matching of fields along the edges of the shaded areas.

7.1.1 Approximate Solution for a Rectangular Dielectric Waveguide

By ignoring the matching of the fields along the shaded boundaries, shown in Fig. 7.1, we may obtain the expressions for the fields that must satisfy the Maxwell

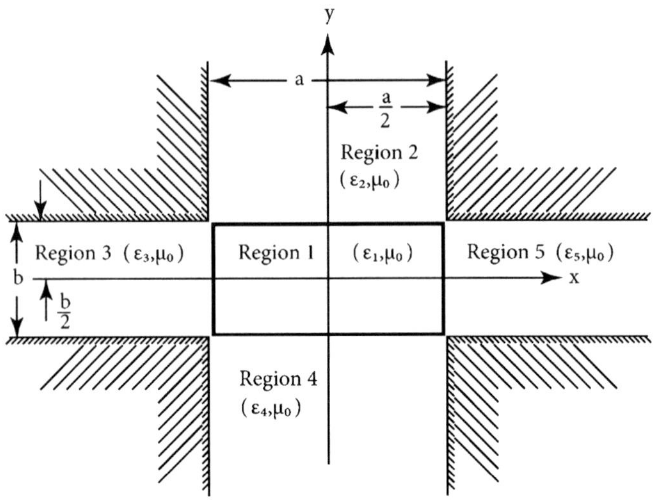

Figure 7.1. Cross-sectional rectangular geometry for Marcatili's approximate approach

equations and the boundary conditions along the sides of the core region, defined as region 1. From the knowledge we gained in Chap. 4, it appears that there can exist two independent families of modes on this rectangular dielectric waveguide: one family has most of its electric field polarized in the y-direction, designated as E^y_{nm} modes, and the other has most of its electric field in the x-direction, designated as E^x_{nm} modes. The subscripts n and m represent, respectively, the number of extrema that the field components for this mode have along the x and y directions. The components of the E^y_{nm} modes are $E_y,\ E_x,\ E_z,\ H_x$, and H_z, with $H_y = 0$. The components of the E^x_{nm} modes are $E_x,\ E_z,\ H_x,\ H_y$, and H_z, with $E_y = 0$.

It is perhaps instructive to review the concept of mode classification discussed earlier in Sect. 2.5. The usual practice in the study of wave propagation along a straight waveguide is to express all modal fields in terms of the longitudinal fields (E_z and H_z) and classify (or designate) the modes according to the presence or absence of the longitudinal field components. For example, we name modes with $E_z = 0$ as TE (transverse electric) modes, or modes with $H_z = 0$ as TM (transverse magnetic) modes, or modes with both E_z and H_z present as HE (hybrid) modes. These designations are independent of the coordinate systems used to describe the waveguides. It is recognized, however, that since the fields are linear, the superposition theorem applies. This means that combinations of any of the above mentioned modal fields are allowed. So new types of modes consisting of linear combinations of the established TE, TM, or HE modes may be constructed depending on the suitability of the situation. For example, for the present problem, it is recognized that the modal fields of interest are those whose electric fields are polarized mostly in the x or y direction in the transverse plane. Hence, one may define the modes according to the orientation of the dominant component of the electric field. Thus, the designation E^y_{nm} modes refers to the dominant electric field as the E_y field. It follows that all other modal fields should be expressed in terms of E_y with $H_y = 0$. The field components for the E^y_{nm} modes are E_x, E_y, E_z, H_x, and H_z, with $H_y = 0$. The companion independent set of fields for the E^x_{nm} modes are E_x, E_z, H_x, H_y, and H_z, with $E_y = 0$. The complete set of fields is the sum of the E^y_{nm} and E^x_{nm} modal fields. (Referring to the discussion on Debye potentials in Sect. 2.5.1, one notes that the E^y_{nm} and E^x_{nm} modes may be identified with the case $\mathbf{a} = \mathbf{e}_y$.) It is further recognized that all modal field lines (electric as well as magnetic) in and around a dielectric waveguide must form closed loops. So, if E_y is the dominant electric field for E^y_{nm} modes, then the other dominant fields must be E_z, H_x, and H_z, while E_x and H_y may be neglected. Similarly, if E_x is the dominant electric field for E^x_{nm} modes, then the other dominant fields must be E_z, H_y, and H_z, while E_y and H_x may be neglected.

7.1.1.1 The E^y_{nm} Modes

By definition, $H_y = 0$ for all E^y_{nm} modes. Substituting this into the time-harmonic Maxwell equations (2.95)–(2.100) in a simple medium (ϵ, μ) and with the assumption that the factor $e^{j\omega t - j\beta z}$ is attached to all field components and suppressed, we have, in the rectangular coordinates,

$$E_x = \frac{1}{j\omega\epsilon}\frac{\partial H_z}{\partial y}, \tag{7.1}$$

$$E_y = -\frac{1}{j\omega\epsilon}\frac{\partial H_z}{\partial x} - \frac{\beta}{\omega\epsilon}H_x, \tag{7.2}$$

$$E_z = -\frac{1}{j\omega\epsilon}\frac{\partial H_x}{\partial y}, \tag{7.3}$$

$$H_x = -\frac{1}{j\omega\mu}\left[\frac{\partial E_z}{\partial y} + j\beta E_y\right], \tag{7.4}$$

$$0 = -\frac{1}{j\omega\mu}\left[-\frac{\partial E_z}{\partial x} - j\beta E_x\right], \tag{7.5}$$

$$H_z = -\frac{1}{j\omega\mu}\left[-\frac{\partial E_y}{\partial x} - \frac{\partial E_x}{\partial y}\right], \tag{7.6}$$

with

$$\nabla^2 E_{x,y,z} + k^2 E_{x,y,z} = 0, \tag{7.7}$$

$$\nabla^2 H_{x,y,z} + k^2 H_{x,y,z} = 0, \tag{7.8}$$

$$k^2 = \omega^2 \mu \epsilon.$$

Let us express all field components in terms of E_y. Inserting (7.3) into (7.4) gives

$$H_x = -\frac{1}{k^2}\frac{\partial^2 H_x}{\partial y^2} - \frac{\beta}{\omega\mu}E_y = \frac{k_y^2}{k^2}H_x - \frac{\beta}{\omega\mu}E_y$$

or

$$H_x = -\frac{1}{k^2 - k_y^2}\frac{\beta k^2}{\omega\mu}E_y. \tag{7.9}$$

Substituting (7.9) in (7.3) yields

$$E_z = \frac{\beta}{j(k^2 - k_y^2)}\frac{\partial E_y}{\partial y}. \tag{7.10}$$

Inserting (7.1) into (7.6) gives

$$H_z = -\frac{1}{j\omega\mu}\frac{\partial E_y}{\partial x}\left(\frac{k^2}{k^2 - k_y^2}\right). \tag{7.11}$$

Substituting (7.11) in (7.1) yields

$$E_x = \frac{1}{k^2 - k_y^2}\frac{\partial^2 E_y}{\partial x \partial y}. \tag{7.12}$$

In the above derivation, we have assumed that the y dependence of all fields is $e^{\pm j k_y y}$, where k_y is the separation constant for the wave equation (7.7) or (7.8). Let us summarize the above results.

$$E_x = \frac{1}{s^2}\frac{\partial^2 E_y}{\partial x \partial y}, \tag{7.13}$$

$$E_z = \frac{\beta}{j s^2}\frac{\partial E_y}{\partial y}, \tag{7.14}$$

$$H_x = -\frac{\beta k^2}{\omega\mu s^2}E_y, \tag{7.15}$$

$$H_z = -\frac{1}{j\omega\mu}\frac{\partial E_y}{\partial x}\left(\frac{k^2}{k^2-k_y^2}\right), \tag{7.16}$$

$$H_y = 0, \tag{7.17}$$

with

$$s^2 = k^2 - k_y^2. \tag{7.18}$$

Before we proceed, let us evoke Marcatili's approximation [1]. By restricting the cases of interest to the condition $(n_1/n_{2,3,4,5} - 1) \ll 1$, only modes made of plane wavelets impinging at grazing angles on the surface of medium are guided. This means $k_{x,y} \ll k_z$. Thus, E_x is of O($k_x k_y$) while (E_z, H_z) are of O(k_z) and (E_y, H_x) are of O(1), where O means the order of. Therefore, we may assume $E_x \approx 0$ for (7.13).

The solution for E_y is found from the wave equation (7.7). The y-component of the fields for the E^y_{nm} modes for the five regions as shown in Fig. 7.1 are

$$E_y^{(1)} = A_1\cos(k_{x1}x+\alpha)\cos(k_{y1}y+\gamma), \tag{7.19a}$$

$$E_y^{(2)} = A_2\cos(k_{x2}x+\alpha)\,e^{-jk_{y2}y}, \tag{7.19b}$$

$$E_y^{(3)} = A_3\cos(k_{y3}y+\gamma)\,e^{jk_{x3}x}, \tag{7.19c}$$

$$E_y^{(4)} = A_4\cos(k_{x4}x+\alpha)\,e^{jk_{y4}y}, \tag{7.19d}$$

$$E_y^{(5)} = A_5\cos(k_{y5}y+\gamma)\,e^{-jk_{x5}x}, \tag{7.19e}$$

$$H_y^{(\nu)} = 0, \tag{7.20}$$

where

$$k_{x\nu}^2 + k_{y\nu}^2 + \beta^2 - \omega^2\mu\epsilon_\nu = k_\nu^2, \tag{7.21}$$

$$\nu = 1, 2, 3, 4, 5,$$

A_ν are arbitrary constants, and α and γ locate the field maxima and minima in region 1, the core region. All other components of the fields can be derived from (7.13)–(7.16).

Let us now match the tangential electric and magnetic fields at the boundaries: At $y = b/2$,

$$H_x^{(1)} = H_x^{(2)} \tag{7.22a}$$

or

$$-\frac{\beta k_1^2}{\omega\mu s_1^2} A_1 \cos(k_{x1}x + \alpha)\cos\left(k_{y1}\frac{b}{2} + \gamma\right) = \frac{\beta k_2^2}{\omega\mu s_2^2} A_2 \cos(k_{x2}x + \alpha)\, e^{-jk_{y2}b/2}, \tag{7.22b}$$

and

$$E_z^{(1)} = E_z^{(2)} \tag{7.23a}$$

or

$$\frac{\beta}{js_1^2}(-k_{y1})\, A_1 \cos(k_{x1}x + \alpha)\sin\left(k_{y1}\frac{b}{2} + \gamma\right) = \frac{\beta}{js_2^2}(-jk_{y2})\, A_2 \cos(k_{x2}x + \alpha)\, e^{-jk_{y2}b/2}, \tag{7.23b}$$

with

$$s_\nu^2 = k_\nu^2 - k_{y\nu}^2, \tag{7.24}$$

$$k_\nu^2 = \omega^2\mu\epsilon_\nu, \tag{7.25}$$

$$\nu = 1, 2.$$

To match the fields at all values of x, we must assume

$$k_{x1} = k_{x2} = k_x. \tag{7.26}$$

Substituting (7.21) in (7.24) yields

$$s_\nu^2 = k_\nu^2 - k_{y\nu}^2 = k_{x\nu}^2 + \beta^2.$$

Since $k_{x1} = k_{x2}$, then $s_1^2 = s_2^2$. Equations (7.22a) and (7.22b) become

$$A_1 \cos\left(k_{y1}\frac{b}{2} + \gamma\right) = \frac{k_2^2}{k_1^2} A_2\, e^{-jk_{y2}b/2}, \tag{7.27}$$

$$A_1 \sin\left(k_{y1}\frac{b}{2} + \gamma\right) = j\frac{k_{y2}}{k_{y1}} A_2\, e^{-jk_{y2}b/2}. \tag{7.28}$$

For a nontrivial solution, the determinant of the above equations must be set to zero, or

$$\tan\left(k_{y1}\frac{b}{2}+\gamma\right)=j\frac{\epsilon_1}{\epsilon_2}\frac{k_{y2}}{k_{y1}}. \tag{7.29}$$

In a similar manner, matching the boundary conditions at $y=-b/2$, one obtains

$$\tan\left(k_{y1}\frac{b}{2}-\gamma\right)=j\frac{\epsilon_1}{\epsilon_4}\frac{k_{y4}}{k_{y1}}, \tag{7.30}$$

where

$$k_{x1}=k_{x4}=k_x \tag{7.31}$$

has been assumed.

At $x - a/2$,

$$E_y^{(1)}=E_y^{(5)} \tag{7.32a}$$

or

$$A_1\cos\left(k_{x1}\frac{a}{2}+\alpha\right)=A_5\,e^{-jk_{x5}a/2}, \tag{7.32b}$$

and

$$H_z^{(1)}=H_z^{(5)} \tag{7.33a}$$

or

$$A_1k_{x1}\sin\left(k_{x1}\frac{a}{2}+\alpha\right)=jA_5k_{x5}\,e^{-jk_{x5}a/2}. \tag{7.33b}$$

Again we must assume

$$k_{y1}=k_{x5}=k_y \tag{7.34}$$

in order to match the fields at all values of y. Setting the determinant of (7.32b) and (7.33b) to zero yields the dispersion relation

$$\tan\left(k_x\frac{a}{2}+\alpha\right)=j\frac{k_{x5}}{k_x}. \tag{7.35}$$

In a similar manner, matching the boundary conditions at $x=-a/2$, one obtains

$$\tan\left(k_x\frac{a}{2}-\alpha\right)=j\frac{k_{x3}}{k_x}, \tag{7.36}$$

where

$$k_{y1}=k_{y3}=k_y \tag{7.37}$$

has been assumed.

To summarize, the dispersion relations are the following:

from (7.29) and (7.30),

$$\tan\left(k_{y1}\frac{b}{2} \pm \gamma\right) = j\frac{\epsilon_1}{\epsilon_{2,4}}\frac{k_{y2,y4}}{k_y} \tag{7.38}$$

and from (7.35) and (7.36),

$$\tan\left(k_x\frac{a}{2} \pm \alpha\right) = j\frac{k_{x5,x3}}{k_x}. \tag{7.39}$$

In these equations, the upper sign goes with the first subscript and the lower sign goes with the second subscript. From (7.21), (7.26), (7.31), (7.34), and (7.37), we have

$$k_{x\nu}^2 + k_{y\nu}^2 + \beta^2 = k_\nu^2, \qquad \nu = 1, 2, 3, 4, 5, \tag{7.40}$$

$$\begin{aligned} k_{y2}^2 &= k_2^2 - \beta^2 - k_{x2}^2 = k_2^2 - \beta^2 - k_x^2, \\ &= k_2^2 - \beta^2 - (k_1^2 - \beta^2 - k_{y1}^2) \\ &= k_2^2 - k_1^2 + k_y^2, \end{aligned} \tag{7.41}$$

$$\begin{aligned} k_{y4}^2 &= k_4^2 - \beta^2 - k_{x4}^2 = k_4^2 - \beta^2 - k_x^2, \\ &= k_4^2 - \beta^2 - (k_1^2 - \beta^2 - k_{y1}^2), \\ &= k_4^2 - k_1^2 + k_y^2, \end{aligned} \tag{7.42}$$

$$\begin{aligned} k_{x3}^2 &= k_3^2 - \beta^2 - k_{y3}^2 = k_3^2 - \beta^2 - k_y^2, \\ &= k_3^2 - \beta^2 - (k_1^2 - \beta^2 - k_{x1}^2), \\ &= k_3^2 - k_1^2 + k_x^2, \end{aligned} \tag{7.43}$$

$$\begin{aligned} k_{x5}^2 &= k_5^2 - \beta^2 - k_{y5}^2 = k_5^2 - \beta^2 - k_y^2, \\ &= k_5^2 - \beta^2 - (k_1^2 - \beta^2 - k_{x1}^2), \\ &= k_5^2 - k_1^2 + k_x^2. \end{aligned} \tag{7.44}$$

Rewriting (7.38) and (7.39), we have

$$\cot\left(k_y\frac{b}{2} \pm \gamma\right) = -j\frac{\epsilon_{2,4}}{\epsilon_1}\frac{k_y}{k_{y2,y4}} \tag{7.45a}$$

or

$$\tan\left(k_y\frac{b}{2} \pm \gamma - \frac{m\pi}{2}\right) = -\frac{\epsilon_{2,4}}{\epsilon_1}\frac{k_y}{|k_{y2,y4}|}, \tag{7.45b}$$

and

$$\cot\left(k_x\frac{a}{2} \pm \alpha\right) = -j\frac{k_x}{k_{x5,x3}} \tag{7.46a}$$

or

$$\tan\left(k_x \frac{a}{2} \pm \alpha - \frac{n\pi}{2}\right) = -\frac{k_x}{|k_{x5,x3}|}. \tag{7.46b}$$

Simplifying gives

$$k_y b = m\pi - \tan^{-1}\left(\frac{\epsilon_2}{\epsilon_1}\frac{k_y}{\sqrt{k_1^2 - k_2^2 - k_y^2}}\right) - \tan^{-1}\left(\frac{\epsilon_4}{\epsilon_1}\frac{k_y}{\sqrt{k_1^2 - k_4^2 - k_y^2}}\right), \tag{7.47}$$

$$k_x a = n\pi - \tan^{-1}\left(\frac{k_x}{\sqrt{k_1^2 - k_3^2 - k_x^2}}\right) - \tan^{-1}\left(\frac{k_x}{\sqrt{k_1^2 - k_5^2 - k_x^2}}\right), \tag{7.48}$$

where n and m are arbitrary integers characterizing the order of the propagating mode. The wave numbers $k_x a$ and $k_y b$, for given $k_1 a$, $k_1 b$, ϵ_2/ϵ_1, ϵ_3/ϵ_1, ϵ_4/ϵ_1, and ϵ_5/ϵ_1, can be found from the transcendental equations (7.47) and (7.48). The propagation constant β can be found from (7.40):

$$\beta a = \left[(k_1 a)^2 - (k_x a)^2 - \left((k_y b)\cdot\frac{a}{b}\right)^2\right]^{1/2}. \tag{7.49}$$

7.1.1.2 The E^x_{nm} Modes

The above analysis can be repeated for the E^x_{nm} modes by simply making the following substitutions:

$$\begin{aligned} E_{x,y,z} \quad &\text{by} \quad H_{x,y,z}, \\ \mu \quad &\text{by} \quad -\epsilon \end{aligned} \tag{7.50}$$

and vice versa. The dispersion relations for the E^x_{nm} modes are

$$k'_y b = m\pi - \tan^{-1}\left(\frac{k'_y}{\sqrt{k_1^2 - k_2^2 - k'^2_y}}\right) - \tan^{-1}\left(\frac{k'_y}{\sqrt{k_1^2 - k_4^2 - k'^2_y}}\right), \tag{7.51}$$

$$k'_x a = n\pi - \tan^{-1}\left(\frac{\epsilon_3}{\epsilon_1}\frac{k'_x}{\sqrt{k_1^2 - k_3^2 - k'^2_x}}\right) - \tan^{-1}\left(\frac{\epsilon_5}{\epsilon_1}\frac{k'_x}{\sqrt{k_1^2 - k_5^2 - k'^2_x}}\right), \tag{7.52}$$

$$\beta' a = \left[(k_1 a)^2 - (k'_x a)^2 - \left((k'_y b)\cdot\frac{a}{b}\right)^2\right]^{1/2}, \tag{7.53}$$

where the prime represents the E^x_{nm} modes.

7.1.2 Examples

The propagation characteristics for a rectangular dielectric structure can now be calculated according to the above approximate dispersion relations. The important assumption is that most of the modal power must be confined within the core region of the guide. Otherwise the approximate dispersion relations, obtained by ignoring the matching of the fields along the dashed boundaries, would provide inaccurate results. Limiting most of the modal power within the core region also means that the mode cutoff condition cannot be obtained from the approximate dispersion relations. Nevertheless, because of the simplicity of these approximate expressions and the fact that the dielectric media surrounding the four sides of the rectangular core may have different dielectric constants, these approximate dispersion relations derived by Marcatili [1] are very useful.

Let us now consider a few examples where $\epsilon_2 = \epsilon_3 = \epsilon_4 = \epsilon_5 < \epsilon_1$. Inspection of the field expression for E^y_{nm} and E^x_{nm} modes can provide an indication on the variations of these fields for several lower order modes. They are shown in Fig. 7.2a,b.

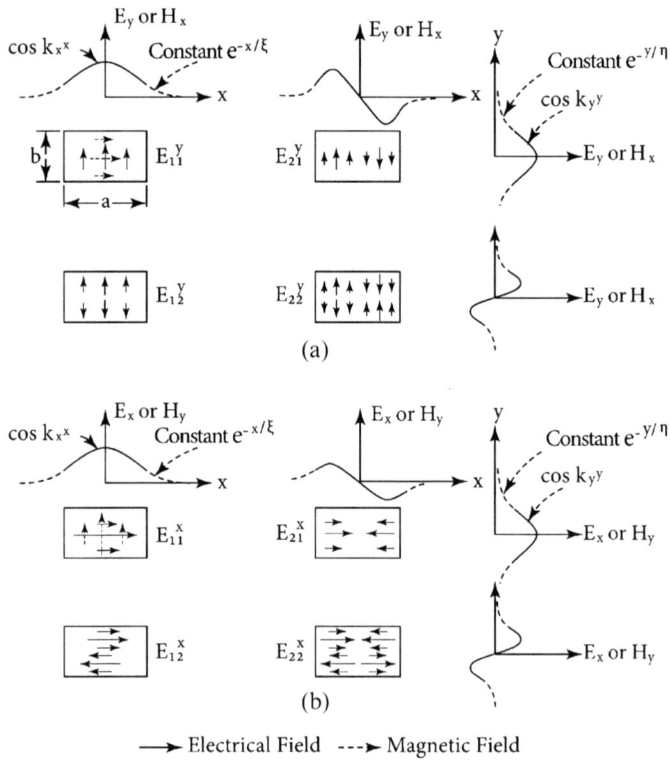

Figure 7.2. (**a**) Field configuration of E^y_{nm} modes. (**b**) Field configuration of E^x_{nm} modes [1]

Calculations based on the approximate dispersion relations using a normalized propagation constant, which is defined as

$$\Lambda = \frac{\beta^2 - k_4^2}{k_1^2 - k_4^2}, \tag{7.54}$$

where $k_1 = \omega\sqrt{\mu\epsilon_1}$ and $k_4 = \omega\sqrt{\mu\epsilon_4}$, ϵ_1 and ϵ_4 are, respectively, the dielectric constants of the core and surrounding medium, and the normalized frequency is given by

$$B = \frac{2b}{\lambda}\left(\frac{\epsilon_1}{\epsilon_0} - \frac{\epsilon_4}{\epsilon_0}\right)^{1/2}, \tag{7.55}$$

where b is the height of the guide and λ is the free-space wavelength, are carried out for a few lower order modes. Results are shown in Figs. 7.3 and 7.4. Plotted in these figures are also the results obtained by the circular harmonic expansion method, which will be discussed in the following section. Very good agreements are found when most of the modal power is confined in the core region, or for

$$\Lambda = \frac{\beta^2 - k_4^2}{k_1^2 - k_4^2} \leq 0.5. \tag{7.56}$$

As expected, a large discrepancy occurs near the cutoff region, since the approximate Marcatili approach is invalid in that region.

The versatility of Marcatili's approach to treat the rectangular structure is seen in an example where the rectangular waveguide is embedded in several different dielectric media such as the case of a channel waveguide. The results are shown in Fig. 7.5. Figure 7.6 shows a comparison between Marcatili's results for the E_{11}^x and E_{11}^y modes and those obtained according to the finite-element numerical method. Very good agreement was found except when Λ is less than 0.5, where the guided power can extend further away from the core region.

7.2 The Circular Harmonics Method

It is known from our discussion on the solutions of the wave equation in circular cylindrical coordinates that the circular harmonics series expansion solutions representing the electromagnetic fields inside and outside the core region of a dielectric waveguide as shown in Sect. 5.2 constitute a complete set of solutions that satisfy Maxwell's equations. The boundary conditions are satisfied by matching the inside and outside electric and magnetic fields that are tangential at appropriate points on the boundary. This approach will yield a set of linear equations from which the propagation characteristics for various modes may be obtained [2].

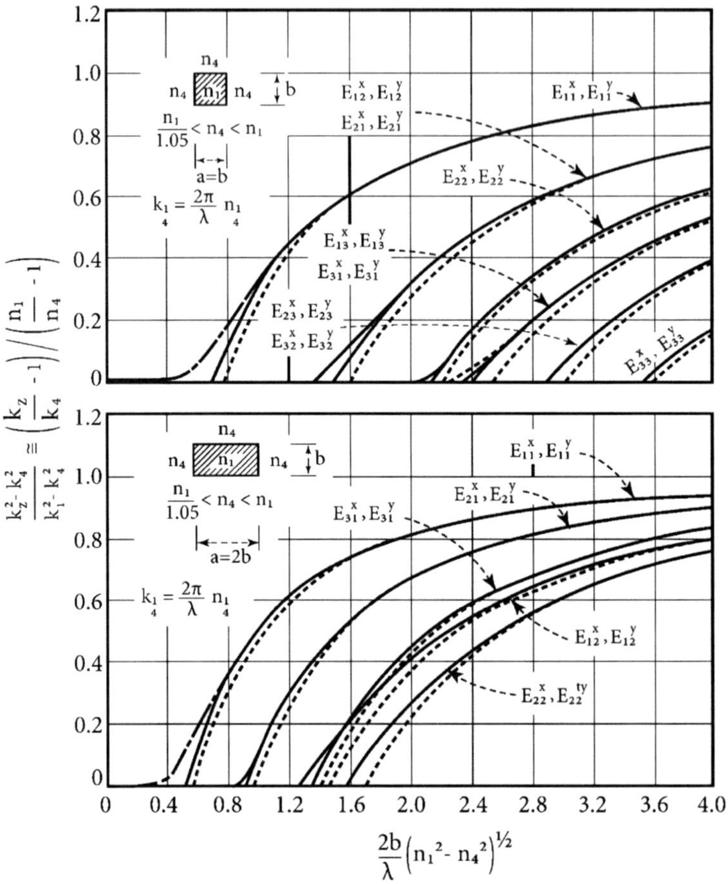

Figure 7.3. Propagation constant for different modes and guides. Solid lines represent Marcatili's transcendental equation solutions. Dashed lines represent Marcatili's closed form solutions. The dot-dashed lines represent Goell's computer results. $\epsilon_1/\epsilon_4 \approx 1.1$ [1]

Expressing the longitudinal electric and magnetic fields in terms of circular harmonics, we have

$$E_z^{(1)} = \sum_{n=0}^{\infty} a_n J_n(pr) \sin(n\theta + \varphi_n)\, e^{-j\beta z + j\omega t}, \tag{7.57}$$

$$H_z^{(1)} = \sum_{n=0}^{\infty} b_n J_n(pr) \sin(n\theta + \psi_n)\, e^{-j\beta z + j\omega t} \tag{7.58}$$

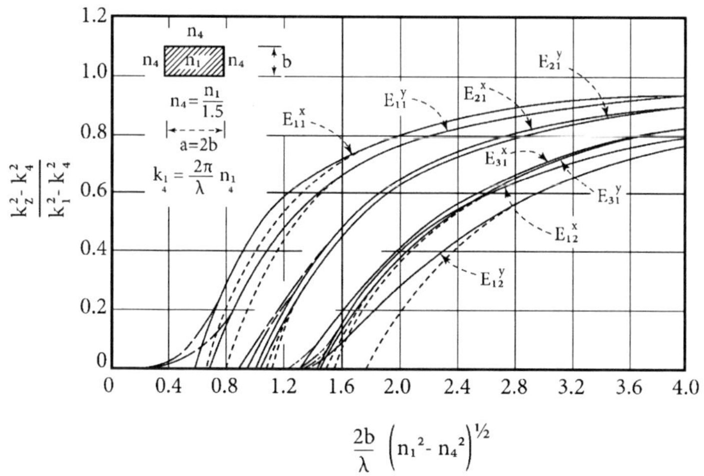

Figure 7.4. Propagation constant on different channel guides. Solid lines represent Marcatili's transcendental equation solutions. Dashed lines represent Marcatili's closed form solutions. The dot-dashed lines represent Goell's computer results. $\epsilon_1/\epsilon_4 \approx 2.25$ [1]

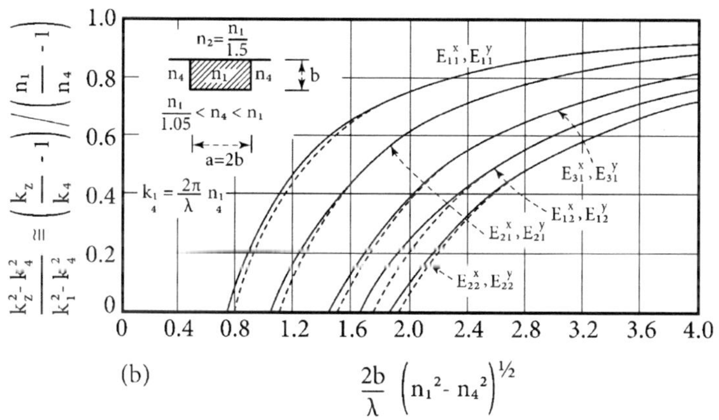

Figure 7.5. Propagation constant on different channel guides. Solid lines represent Marcatili's transcendental equation solutions and dashed lines represent Marcatili's closed form solutions [1]

inside the dielectric core, and

$$E_z^{(0)} = \sum_{n=0}^{\infty} c_n K_n(qr) \sin(n\theta + \varphi_n)\, e^{-j\beta z + j\omega t}, \tag{7.59}$$

$$H_z^{(1)} = \sum_{n=0}^{\infty} d_n K_n(qr) \sin(n\theta + \psi_n)\, e^{-j\beta z + j\omega t} \tag{7.60}$$

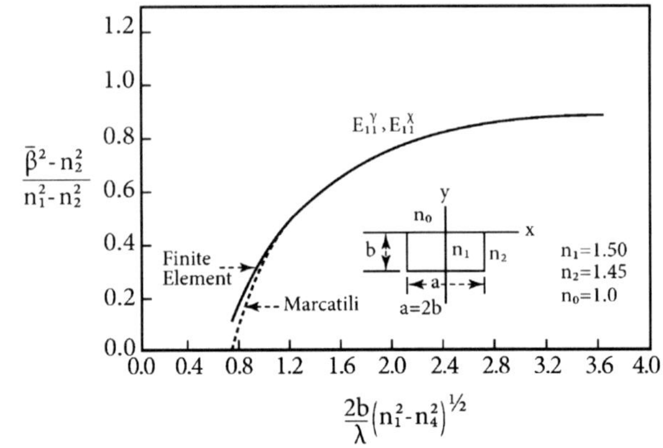

Figure 7.6. Comparison of the finite element results for the lowest order mode with Marcatili's approximate results for the channel waveguide. $\overline{\beta} = \beta/k_0$ [6]

outside the core, where

$$
\begin{aligned}
p &= \left(k_1^2 - \beta^2\right)^{1/2}, \\
q &= \left(\beta^2 - k_0^2\right)^{1/2}, \\
k_1 &= \omega\left(\mu\epsilon_1\right)^{1/2}, \\
k_0 &= \omega\left(\mu\epsilon_0\right)^{1/2},
\end{aligned}
\tag{7.61}
$$

and a_n, b_n, c_n, and d_n are arbitrary constants. J_n and K_n are the nth order Bessel functions and modified Bessel functions, respectively, and φ_n and ψ_n are arbitrary phase angles. The transverse fields can be found easily from Maxwell's equations. As a specific example, let us consider the case of a rectangular dielectric waveguide as shown in Fig. 7.7. It is noted that the component of the electric field tangent to the rectangular core is given by

$$
E_t = \pm\left(E_r \sin\theta + E_\theta \cos\theta\right) \begin{cases} -\theta_c < \theta < \theta_c \\ \pi - \theta_c < \theta < \pi + \theta_c \end{cases}
\tag{7.62}
$$

or

$$
E_t = \pm\left(-E_r \cos\theta + E_\theta \sin\theta\right) \begin{cases} \theta_c < \theta < \pi - \theta_c \\ \pi - \theta_c < \theta < -\theta_c \end{cases},
\tag{7.63}
$$

where θ_c is given in Fig. 7.7. E_r and E_θ are, respectively, the radial component and the θ component of the electric field. Similar expressions exist for the tangential magnetic field.

Matching the tangential electric and magnetic fields along the boundary of the rectangular core gives, in matrix form,

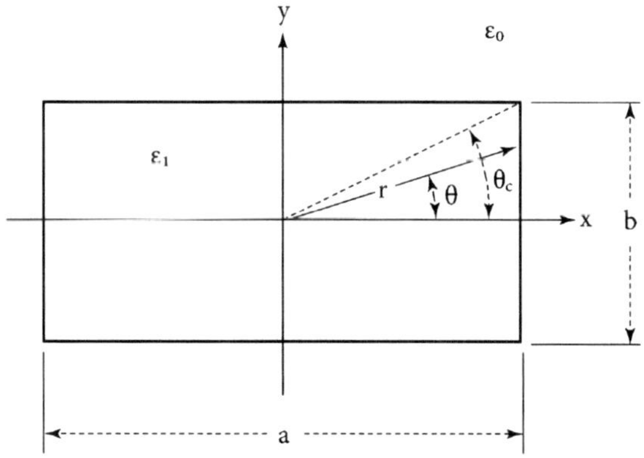

Figure 7.7. Dimensions and coordinate system

$$\begin{bmatrix} G_1 & 0 & -G_2 & 0 \\ 0 & G_3 & 0 & -G_4 \\ G_5 & G_6 & -G_3 & -G_8 \\ G_9 & G_{10} & -G_{11} & -G_{12} \end{bmatrix} \begin{bmatrix} A \\ B \\ C \\ D \end{bmatrix} = 0, \tag{7.64}$$

where $A,\ B,\ C,$ an D are N element column matrices of the $a_n,\ b_n,\ c_n,$ and d_n mode coefficients, respectively. The elements of the $m \times n$ matrices $G_1, G_2, \ldots, G_{12}$ are given by

$$\begin{aligned} g_{1mn} &= JS, & g_{7mn} &= \beta\,(Q'SR + QCT)\,, \\ g_{2mn} &= KS, & g_{8mn} &= -\omega\mu\,(QSR + Q'CT)\,, \\ g_{3mn} &= JC, & g_{9mn} &= \omega\epsilon_1\,(PCR - P'ST), \\ g_{4mn} &= KC, & g_{10mn} &= -\beta\,(P'CR - PST)\,, \\ g_{5mn} &= -\beta\,(P'SR + PCT), & g_{11mn} &= -\omega\epsilon_0\,(QCR - Q'ST), \\ g_{6mn} &= \omega\mu\,(PSR + P'CT), & g_{12mn} &= \beta\,(Q'CR - QST), \end{aligned} \tag{7.65}$$

where

$$\left.\begin{aligned} S &= \sin(n\theta_m + \varphi) \\ C &= \cos(n\theta_m + \varphi) \end{aligned}\right\} \text{ for } \varphi = 0 \text{ or } \varphi = \pi/2,$$

$$J = J_n(pr_m), \qquad K = K_n(qr_m),$$

$$P = \frac{nJ_n(pr_m)}{p^2 r_m}, \quad Q = \frac{nK_n(qr_m)}{q^2 r_m},$$

$$P' = \frac{J_n'(pr_m)}{p}, \quad Q' = \frac{K_n'(qr_m)}{q},$$

and

$$\left.\begin{aligned} R &= \sin\theta_m \\ T &= \cos\theta_m \\ r_m &= \frac{a}{2}\cos\theta_m \end{aligned}\right\} \theta < \theta_c\,, \qquad \left.\begin{aligned} R &= -\cos\theta_m \\ T &= \sin\theta_m \\ r_m &= \frac{b}{2}\sin\theta_m \end{aligned}\right\} \theta > \theta_c. \tag{7.66}$$

For $\theta = \theta_c$, $R = \cos(\theta_m + \pi/4)$, $T = \cos(\theta_m - \pi/4)$, and $r_m = \left(a^2 + b^2\right)^{1/2} \epsilon/4$. Any given mode must consist of either even or odd harmonics. For the odd harmonic cases, $\theta_m = (m - \frac{1}{2})\pi/2N;\ m = 1, 2, \ldots, N$, where N is the number of space harmonics. For the even harmonic cases, if the aspect ratio is unity, that is, $a/b = 1$, $\theta_m = (m - \frac{1}{2})\pi/2N;\ m = 1, 2, \ldots, N$, for the field components with even symmetry about the x-axis; and $\theta_m = (m - N - \frac{1}{2})\pi/2(N-1)$; $m = N+1, N+2, \ldots, (2N-1)$, for the field components with odd symmetry about the x-axis; if the aspect ratio is other than unity, θ_m were chosen according to the first formula, except that the first and last points for the odd z-component were omitted.

Modes are designated as E^y_{sq} if in the limit of short wavelengths their electric field is parallel to the y-axis and as E^x_{sq} if in the limit their electric field is parallel to the x-axis. The s and q subscripts are used to designate the number of maxima in the x and y directions, respectively. The dominant E^y_{11} and E^x_{11} modes correspond, respectively, to the $_{\mathrm{e}}\mathrm{HE}_{11}$ and $_{\mathrm{o}}\mathrm{HE}_{11}$ modes on an elliptical dielectric waveguide [7].

The dispersion relation of various modes is obtained by equating the determinant of (7.64) to zero. Numerical computation shows that for a small aspect ratio, only a few (up to 5) harmonics are needed to achieve the desired three-place accuracy, but for larger aspect ratios the convergence is not very good. According to Goell [2], the computed results are believed to be accurate to better than 2%. The normalized propagation constant Λ is plotted against the normalized frequency in Figs. 7.8 and 7.9 for various modes and various a/b ratios. The following normalized parameters are defined:

$$\text{Normalized Propagation Constant} \quad \Lambda = \frac{(\lambda_0/\lambda)^2 - 1}{\epsilon_1/\epsilon_0 - 1},$$

$$\text{Normalized Frequency} \quad F = \frac{2b}{\lambda_0}\left(\epsilon_1/\epsilon_0 - 1\right)^{1/2},$$

with $\beta = 2\pi/\lambda$ and $k_0 = 2\pi/\lambda_0$. Figure 7.8 shows that E^y_{sq} and E^x_{sq} modes are degenerate as $\epsilon_1 \to \epsilon_0$. The degeneracy disappears when $\epsilon_1/\epsilon_0 = 2.25$ with $a/b = 2$ as shown in Fig. 7.8. The effect of the change of relative dielectric constant on the propagation characteristics of the dominant modes can be seen from Fig. 7.9. It is noted that if one plots Λ as a function of the normalized cross-sectional area

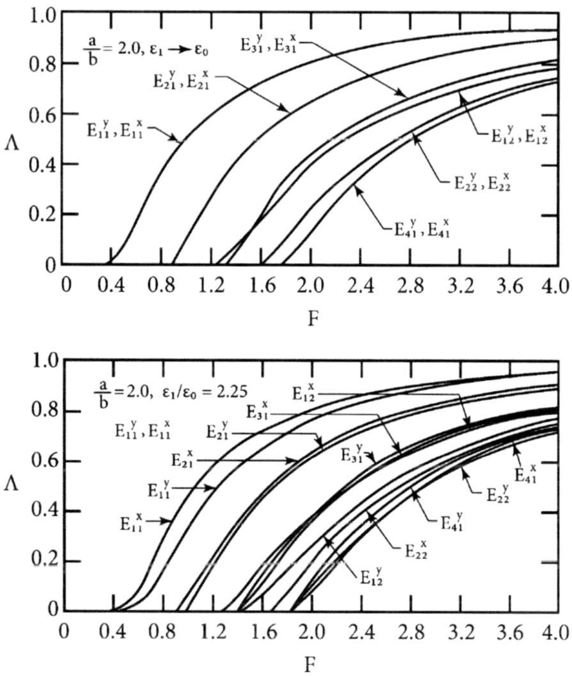

Figure 7.8. Normalized propagation constant Λ as a function of normalized frequency F for various modes on a rectangular dielectric waveguide with $a/b = 2$ and $\epsilon_1/\epsilon_0 \approx 1$ (*upper plot*) and $\epsilon_1/\epsilon_0 = 2.25$ (*lower plot*) [2]

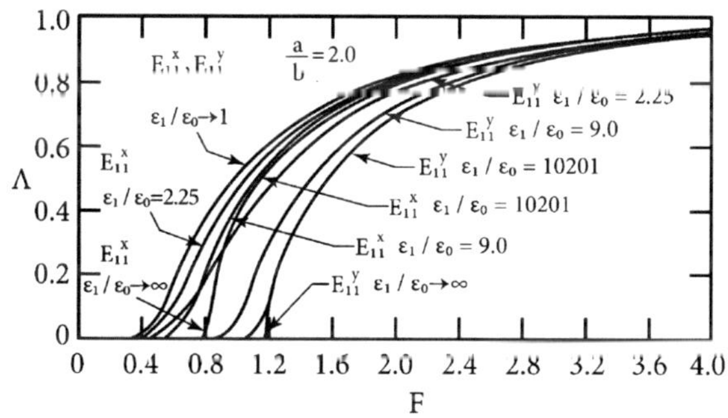

Figure 7.9. Normalized propagation constant Λ as a function of normalized frequency F for a rectangular dielectric waveguide with aspect ratio $a/b = 2$ for various values of ϵ_1/ϵ_0 [2]

$(F'^2 a/b)$ for a fixed ϵ_1/ϵ_0, the variation between the propagation curves of a lower order mode for various a/b ratios is quite small.

A sketch of the cross-sectional field configurations for various modes are given in Fig. 7.10.

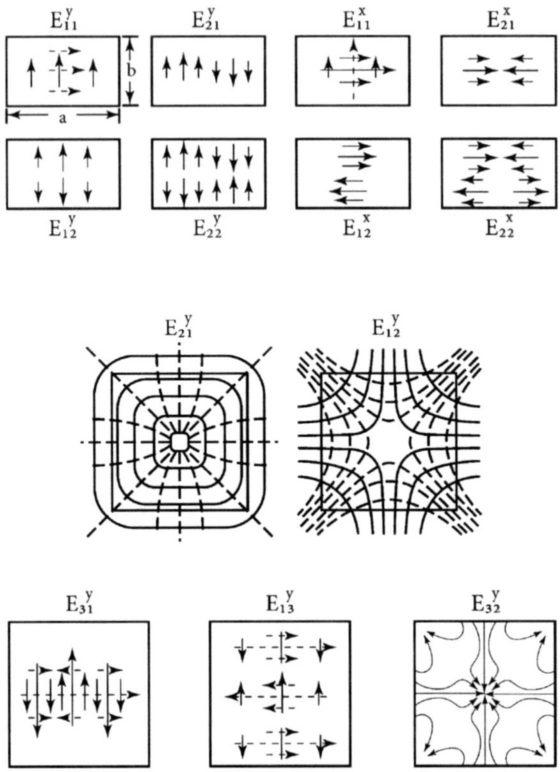

Figure 7.10. A sketch of the cross-sectional field configurations for several lower order modes. Solid lines represent electric field lines and dashed lines represent magnetic field lines [2]

It is of interest to compare the results obtained by Goell (the circular harmonic method), Marcatili (the planar approximate method), and Yeh et al. (the finite element method). The results are displayed in Fig. 7.11 for the dominant E^y_{11} (or ${}_eHE_{11}$) mode. Much closer agreement was found between Goell's and Yeh's results, especially for $\Lambda < 0.5$, as expected. The limitation of Goell's approach is that the circular harmonic expansion converges best when the cross-sectional shape is close to that of a circle. Poor convergence of the expansion results when the aspect ratio of the cross-sectional shape becomes larger than 2 or 3.

7.3 Experimental Measurements

The most thorough experimental investigation of wave propagation on a rectangular dielectric rod was carried out by Schlesinger and King [8]. They made measurements in the microwave frequency range on seventeen samples of rectangular polystyrene rods ($\epsilon_1/\epsilon_0 = 2.56$) using the image line configuration and the

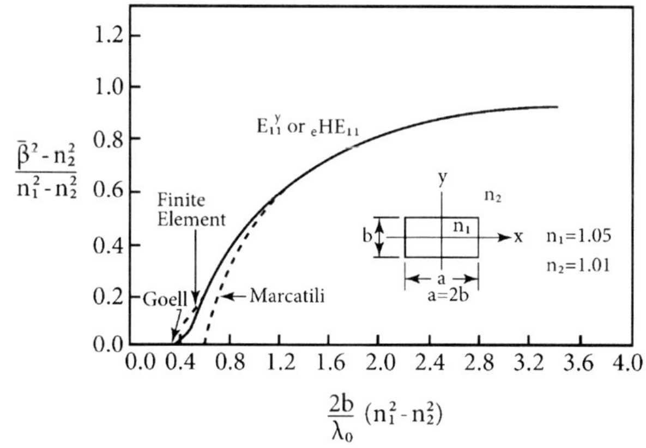

Figure 7.11. Comparison of the finite element results for the $_eHE_{11}$ mode with Goell's results and with Marcatili's approximate results for the rectangular fiber guide [6]

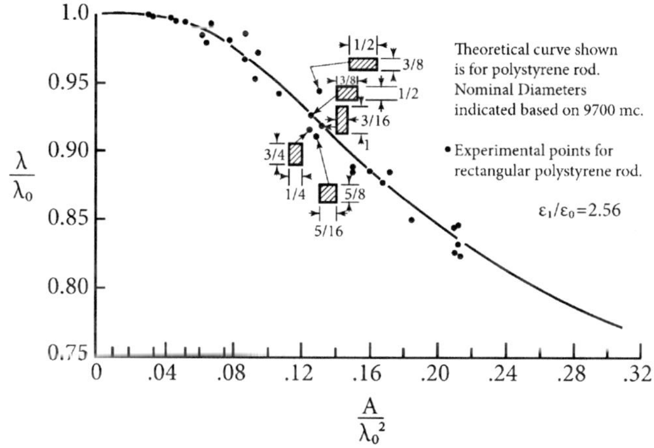

Figure 7.12. Experimental results for the HE_{11} dominant mode propagation for various rectangular polystyrene rods. A is the cross-sectional area [8]

flat plate resonator. Results were obtained for dominant modes. Figure 7.12 shows the dependence of λ/λ_0 on the normalized cross-sectional area (NCSA= ab/λ_0^2) of the rectangular dielectric rod for various samples, where a and b are shown in the figure. It can be seen that if the ratio of major axis to minor axis is not too large, the measured points lie within 3% of a curve drawn for a circular dielectric rod with an equivalent cross-sectional area. It is further noted that given a certain area with moderate ratio of major axis to minor axis, the orientation (i.e., whether the $_eHE_{11}$ or $_oHE_{11}$ mode was excited) has only a slight effect within the approximate 3%. This result is also confirmed by the analysis given in Chap. 6 on elliptical dielectric

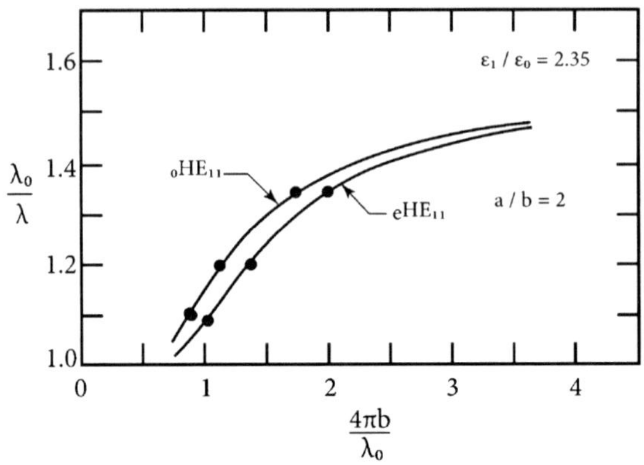

Figure 7.13. Comparison of measured results (*points*) by Schlosser and Unger [3] and the analytical results (*curves*) by Yeh [7]

waveguides and in Sect. 7.1 on rectangular dielectric waveguides. Additional experimental data for the ${}_{e}HE_{11}$ and ${}_{0}HE_{11}$ modes were given by Schlosser and Unger [3] using rectangular polyethylene rods ($\epsilon_1/\epsilon_0 = 2.35$ and major axis/minor axis = 2.0). Results are plotted in Fig. 7.13. Solid curves are calculated according to the elliptical dielectric waveguide theory given earlier and the points represent experimental results. It can be seen that very close agreements are obtained.

References

1. E. A. J. Marcatili, "Dielectric rectangular waveguide and directional couplers for integrated optics," Bell Syst. Tech. J. **48**, 2071 (1969)
2. J. E. Goell, "A circular-harmonic computer analysis of rectangular dielectric waveguides," Bell Syst. Tech. J. **48**, 2133 (1969)
3. W. Schlosser and H. G. Unger, "Advances in Microwaves," Academic Press, New York (1966)
4. L. Eyges, P. Gianino, and P. Wintersteiner, "Modes of dielectric waveguides of arbitrary cross sectional shapes," J. Opt. Soc. Am. **69**, 1226 (1979)
5. C. B. Shaw, B. T. French, and C. Warner III, "Further research on optical transmission lines," Sci. Rep. No. 2, Autonetics Report No. C7-929/501, Air Force Contract AF449(638)-1504 AD625 501
6. C. Yeh, K. Ha, S. B. Dong, and W. P. Brown, "Single-mode optical waveguides," Appl. Opt. **18**, 1490 (1979)
7. C. Yeh, "Elliptical dielectric waveguides." J. Appl. Phys. **33**, 3235 (1962)
8. S. P. Schlesinger and D. D. King, "Dielectric image lines," IRE Trans. Microw. Theory Tech. **MTT-6**, 291 (1958)

8

INHOMOGENEOUS DIELECTRIC WAVEGUIDES

When the permittivity and permeability of a medium are functions of space, that is, $\epsilon = \epsilon(\mathbf{r})$ and $\mu = \mu(\mathbf{r})$, the time harmonic, source-free vector wave equations governing the field vectors in this medium become

$$\nabla \times \nabla \times \mathbf{E} - \frac{\nabla\mu(\mathbf{r})}{\mu(\mathbf{r})} \times \nabla \times \mathbf{E} - \omega^2\epsilon(\mathbf{r})\mu(\mathbf{r})\mathbf{E} = \mathbf{0}, \tag{8.1}$$

$$\nabla \times \nabla \times \mathbf{H} - \frac{\nabla\epsilon(\mathbf{r})}{\epsilon(\mathbf{r})} \times \nabla \times \mathbf{H} - \omega^2\epsilon(\mathbf{r})\mu(\mathbf{r})\mathbf{H} = \mathbf{0}, \tag{8.2}$$

and the divergence equations become

$$\nabla \cdot [\epsilon(\mathbf{r})\mathbf{E}] = 0, \tag{8.3}$$

$$\nabla \cdot [\mu(\mathbf{r})\mathbf{H}] = 0. \tag{8.4}$$

This medium is called an inhomogeneous medium. Only under certain special spatial variations of $\epsilon(\mathbf{r})$ and $\mu(\mathbf{r})$ can the vector wave equation be solved by the separation of variables method. In this chapter we shall discuss some of these special cases. Specific examples will be given when analytic solutions can be displayed.

8.1 Debye Potentials for Inhomogeneous Medium [1]

In spite of the complexity of the vector wave equations for an inhomogeneous medium, the electromagnetic fields in this medium are still linear, that is, superposition applies. Thus, we may choose to generate a complete set of electromagnetic fields using arbitrary Debye potentials. For example, let us introduce the Debye potentials as follows:

$$\mathbf{E}^{(\mathrm{I})}(\mathbf{r}) = \nabla \times [\mathbf{a}\Psi(\mathbf{r})], \tag{8.5}$$

$$\mathbf{H}^{(\mathrm{II})}(\mathbf{r}) = \nabla \times [\mathbf{a}\Phi(\mathbf{r})], \tag{8.6}$$

where $\mathbf{a}$ is a unit vector or a position vector $\mathbf{r}$ and $\Phi(\mathbf{r})$ and $\Psi(\mathbf{r})$ are the Debye potentials. The superscripts (I) or (II) represent the two types of linearly independent fields. The combination of these two types of fields will yield the complete set of electromagnetic fields. All components of $\mathbf{E}$ and $\mathbf{H}$ fields can be obtained from the Maxwell equations using (8.5) and (8.6):

$$\mathbf{H}^{(\mathrm{I})}(\mathbf{r}) = \frac{j}{\omega\mu(r)}\nabla \times \nabla \times [\mathbf{a}\,\Psi(\mathbf{r})], \tag{8.7}$$

$$\mathbf{E}^{(\mathrm{II})}(\mathbf{r}) = -\frac{j}{\omega\epsilon(r)}\nabla \times \nabla \times [\mathbf{a}\,\Phi(\mathbf{r})]. \tag{8.8}$$

By properly choosing $\mathbf{a}$, we can reduce the vector wave equations to scalar wave equations even for certain inhomogeneous medium. In the following we show the specific cases for which this reduction is possible.

8.1.1 Rectangular Coordinates (x, y, z)

Let us consider the special situation where

$$\epsilon(\mathbf{r}) = \epsilon(\xi), \tag{8.9}$$

$$\mu(\mathbf{r}) = \mu(\xi). \tag{8.10}$$

Here, ξ may be x, y, or z. To satisfy the divergence conditions, we choose $\mathbf{a} = \mathbf{e}_\xi$ in (8.5) and (8.6), that is,

$$\begin{aligned}\nabla \cdot \left[\epsilon(\xi)\mathbf{E}^{(\mathrm{I})}(\mathbf{r})\right] &= \mathbf{E}^{(\mathrm{I})}(\mathbf{r}) \cdot \nabla\epsilon(\xi) + \epsilon(\xi)\nabla \cdot \mathbf{E}^{(I)}(\mathbf{r}), \\ &= \{\nabla \times [\mathbf{e}_\xi\,\Psi(\mathbf{r})]\} \cdot \nabla\epsilon(\xi) + \epsilon(\xi)\nabla \cdot \{\nabla \times [\mathbf{e}_\xi\,\Psi(\mathbf{r})]\}, \\ &= 0. \end{aligned}\tag{8.11}$$

This is because the first term shows that the vectors in the bracketed term are transverse to $\mathbf{e}_\xi$ and $\nabla\epsilon(\xi)$ is a vector in $\mathbf{e}_\xi$. Thus the dot product term is zero. Since $\nabla \cdot \nabla \times \mathbf{A} = 0$, the second term is identically zero. A similar argument can be used to show that

$$\nabla \cdot \left[\mu(\xi)\mathbf{H}^{(\mathrm{II})}(\mathbf{r})\right] = 0. \tag{8.12}$$

We shall now derive the wave equation satisfied by the Debye potentials. Substituting (8.5) in (8.6) yields

$$\nabla \times \nabla \times \nabla \times [\mathbf{e}_\xi\,\Psi(\mathbf{r})] - \frac{\nabla\mu(\xi)}{\mu(\xi)} \times \nabla \times \nabla \times [\mathbf{e}_\xi\,\Psi(\mathbf{r})] \\ - \omega^2\epsilon(\xi)\mu(\xi)\nabla \times [\mathbf{e}_\xi\,\Psi(\mathbf{r})] = 0. \tag{8.13}$$

Carrying out the vector operations in Cartesian coordinates and setting each component of the unit vectors to zero, we can show that

$$\nabla^2 \Psi(\mathbf{r}) - \frac{\partial \mu(\xi)}{\partial \xi} \frac{1}{\mu(\xi)} \frac{\partial \Psi(\mathbf{r})}{\partial \xi} + \omega^2 \epsilon(\xi) \mu(\xi) \Psi(\mathbf{r}) = 0. \tag{8.14}$$

Substituting (8.6) in (8.2) yields

$$\nabla \times \nabla \times \nabla \times [\mathbf{e}_\xi \, \Phi(\mathbf{r})] - \frac{\nabla \epsilon(\xi)}{\epsilon(\xi)} \times \nabla \times \nabla \times [\mathbf{e}_\xi \, \Phi(\mathbf{r})] \\ - \omega^2 \epsilon(\xi) \mu(\xi) \nabla \times [\mathbf{e}_\xi \, \Phi(\mathbf{r})] = 0. \tag{8.15}$$

Simplifying gives

$$\nabla^2 \Phi(\mathbf{r}) - \frac{\partial \epsilon(\xi)}{\partial \xi} \frac{1}{\epsilon(\xi)} \frac{\partial \Phi(\mathbf{r})}{\partial \xi} + \omega^2 \epsilon(\xi) \mu(\xi) \Phi(\mathbf{r}) = 0. \tag{8.16}$$

Equations (8.14) and (8.16) are scalar wave equations. Note that these equations are valid only for Cartesian coordinates. They can be solved by the separation of variables.

8.1.2 Spherical Coordinates (r, θ, φ)

Let us consider the special situation where

$$\epsilon(\mathbf{r}) = \epsilon(r), \tag{8.17}$$

$$\mu(\mathbf{r}) = \mu(r). \tag{8.18}$$

We note that the divergence conditions may be satisfied if we choose $\mathbf{a} = \mathbf{e}_r$ in (8.5)–(8.8), where $\mathbf{e}_r$ is the radial unit vector in spherical coordinates. Substituting (8.5) in (8.1) yields

$$\nabla \times \nabla \times \nabla \times [\mathbf{e}_r \, \Psi(\mathbf{r})] - \frac{\nabla \mu(r)}{\mu(r)} \times \nabla \times \nabla \times [\mathbf{e}_r \, \Psi(\mathbf{r})] \\ - \omega^2 \epsilon(r) \mu(r) \nabla \times [\mathbf{e}_r \, \Psi(\mathbf{r})] = 0. \tag{8.19}$$

Carrying out the vector operations in spherical coordinates yields

$$\frac{\partial^2 \Psi(\mathbf{r})}{\partial r^2} + \frac{1}{r^2 \sin\theta} \frac{\partial}{\partial \theta} \left(\sin\theta \, \frac{\partial \Psi(\mathbf{r})}{\partial \theta} \right) + \frac{1}{r^2 \sin^2\theta} \frac{\partial^2 \Psi(\mathbf{r})}{\partial \phi^2} \\ - \frac{\partial \mu(r)}{\partial r} \frac{1}{\mu(r)} \frac{1}{r} \frac{\partial \Psi(\mathbf{r})}{\partial r} + \omega^2 \epsilon(r) \mu(\mathbf{r}) \Psi(\mathbf{r}) = 0. \tag{8.20}$$

Similarly, we get

$$\frac{\partial^2\Phi(\mathbf{r})}{\partial r^2}+\frac{1}{r^2\sin\theta}\frac{\partial}{\partial\theta}\left(\sin\theta\,\frac{\partial\Phi(\mathbf{r})}{\partial\theta}\right)+\frac{1}{r^2\sin^2\theta}\frac{\partial^2\Phi(\mathbf{r})}{\partial\phi^2}$$
$$-\frac{\partial\epsilon(r)}{\partial r}\frac{1}{\epsilon(r)}\frac{1}{r}\frac{\partial\Phi(\mathbf{r})}{\partial r}+\omega^2\epsilon(r)\mu(\mathbf{r})\Phi(\mathbf{r})=0. \quad (8.21)$$

Equations (8.20) and (8.21) can be solved by the separation of variables [2–4].

8.1.3 Circular Cylindrical Coordinates (ρ, θ, z)

Let us consider the following two situations:

(a) $\epsilon(\mathbf{r})=\epsilon(z),\ \ \mu(\mathbf{r})=\mu(z)$

Following the same procedures as done for the Cartesian coordinates, substituting $\mathbf{a}=\mathbf{e}_z$ in (8.5)–(8.8), we obtain

$$\mathbf{E}^{(\mathrm{I})}(\rho,\theta,z)=\nabla\times\left[\mathbf{e}_z\Psi(\rho,\theta,z)\right], \quad (8.22)$$

$$\mathbf{H}^{(\mathrm{I})}(\rho,\theta,z)=\frac{j}{\omega\mu(z)}\nabla\times\nabla\times\left[\mathbf{e}_z\Psi(\rho,\theta,z)\right], \quad (8.23)$$

$$\mathbf{H}^{(\mathrm{II})}(\rho,\theta,z)=\nabla\times\left[\mathbf{e}_z\Phi(\rho,\theta,z)\right], \quad (8.24)$$

$$\mathbf{E}^{(\mathrm{II})}(\rho,\theta,z)=-\frac{j}{\omega\epsilon(z)}\nabla\times\nabla\times\left[\mathbf{e}_z\Phi(\rho,\theta,z)\right], \quad (8.25)$$

$$\frac{\partial^2\Psi(\rho,\theta,z)}{\partial r^2}-\frac{\partial\mu(z)}{\partial z}\frac{1}{\mu(z)}\frac{\partial\Psi(\rho,\theta,z)}{\partial z}+\omega^2\epsilon(z)\mu(z)\Psi(\rho,\theta,z)=0, \quad (8.26)$$

$$\frac{\partial^2\Phi(\rho,\theta,z)}{\partial r^2}-\frac{\partial\epsilon(z)}{\partial z}\frac{1}{\epsilon(z)}\frac{\partial\Phi(\rho,\theta,z)}{\partial z}+\omega^2\epsilon(z)\mu(z)\Phi(\rho,\theta,z)=0. \quad (8.27)$$

We note that Ψ and Φ may be recognized as the regular Hertz potentials. Of course, they must satisfy a more complicated set of scalar wave equations, (8.26) and (8.27) for inhomogeneous media. These equations can be solved by the separation of variables method.

(b) $\epsilon(\mathbf{r})=\epsilon(\rho),\ \ \mu(\mathbf{r})=\mu(\rho)$

Let us consider the circularly symmetric case with $\partial/\partial\theta=0$. Substituting $\mathbf{a}=\mathbf{e}_\rho$ in (8.5)–(8.8) yields

$$\mathbf{E}^{(\mathrm{I})}(\rho, z) = \nabla \times [\mathbf{e}_\rho \Psi(\rho, z)], \tag{8.28}$$

$$\mathbf{H}^{(\mathrm{I})}(\rho, z) = \frac{j}{\omega\mu(\rho)} \nabla \times \nabla \times [\mathbf{e}_\rho \Psi(\rho, z)], \tag{8.29}$$

$$\mathbf{H}^{(\mathrm{II})}(\rho, z) = \nabla \times [\mathbf{e}_\rho \Phi(\rho, z)], \tag{8.30}$$

$$\mathbf{E}^{(\mathrm{II})}(\rho, \theta, z) = -\frac{j}{\omega\epsilon(\rho)} \nabla \times \nabla \times [\mathbf{e}_\rho \Phi(\rho, z)], \tag{8.31}$$

$$\frac{\partial}{\partial\rho}\frac{1}{\rho}\frac{\partial}{\partial\rho}[\rho\Psi(\rho,z)] - \frac{1}{\mu(\rho)}\frac{\partial\mu(\rho)}{\partial\rho}\frac{1}{\rho}\frac{\partial}{\partial\rho}[\rho\Psi(\rho,z)], \\ + \frac{\partial^2\Psi(\rho,z)}{\partial z^2} + \omega^2\epsilon\rho)\mu(\rho)\Psi(\rho,z) = 0 \tag{8.32}$$

$$\frac{\partial}{\partial\rho}\frac{1}{\rho}\frac{\partial}{\partial\rho}[\rho\Phi(\rho,z)] - \frac{1}{\epsilon(\rho)}\frac{\partial\epsilon(\rho)}{\partial\rho}\frac{1}{\rho}\frac{\partial}{\partial\rho}[\rho\Phi(\rho,z)] \\ + \frac{\partial^2\Phi(\rho,z)}{\partial z^2} + \omega^2\epsilon(\rho)\mu(\rho)\Phi(\rho,z) = 0. \tag{8.33}$$

Again these equations may be solved by the separation of variables method. Note that this approach is not applicable for noncircularly symmetric waves where $\partial/\partial\theta \neq 0$.

8.2 Applications

In this section we shall provide a few illustrative examples on the propagation of guided waves on structures with inhomogeneous dielectrics [5, 6].

One of the significant virtues of the separation of variables method is its ability to reduce partial differential equations to total differential equations, whose solutions may be expressed in terms of known analytic special functions [7]. One specific example is the modified form of the Bessel equation [7].

$$x^2\frac{d^2 f}{dx^2} + (1-2\alpha)\,x\,\frac{df}{dx} + \left[\beta^2\gamma^2 x^{2\gamma} + (\alpha^2 - p^2\gamma^2)\right] f = 0 \\ p = n = \text{integer}, \tag{8.34}$$

which has its solutions

$$f = Ax^\alpha J_p(\beta x^\gamma) + Bx^\alpha J_{-p}(\beta x^\gamma) \qquad p \neq \text{integer}, \tag{8.35a}$$

$$f = Ax^\alpha J_n(\beta x^\gamma) + Bx^\alpha Y_n(\beta x^\gamma) \qquad p = n = \text{integer}, \tag{8.35b}$$

where A and B are arbitrary constants and α, β, γ, and p are all known given constants.

8.2.1 Structures with Transverse Inhomogeneity

8.2.1.1 Wave Propagation along a Dielectric Slab with $\epsilon(x)$ and μ_0 Immersed in Free-Space

This structure is shown in Fig. 8.1. The inhomogeneous dielectric slab has a thickness of $2d$ with a permittivity of $\epsilon(x)$ and permeability of μ_0. It is immersed in a medium with (ϵ_0, μ_0). We wish to obtain the propagation characteristics of surface waves supported by this structure propagating in the z-direction. For this problem, $\partial/\partial y = 0$. From Sect. 8.1.1 we learn that there are two independent sets of waves: (I) and (II). For the inhomogeneous region a $(-d \le x \le d)$, identifying ξ as x, we have

$$\mathbf{E}^{(\mathrm{I})\mathrm{a}}(x,z) = \nabla \times \left[\mathbf{e}_x \Psi^{\mathrm{a}}(x,z)\right], \tag{8.36}$$

$$\mathbf{H}^{(\mathrm{I})\mathrm{a}}(x,z) = \frac{j}{\omega\mu_0} \nabla \times \nabla \times \left[\mathbf{e}_x \Psi^{\mathrm{a}}(x,z)\right], \tag{8.37}$$

$$\nabla^2 \Psi^{\mathrm{a}}(x,z) + \omega^2 \epsilon(x)\mu_0 \Psi^{\mathrm{a}}(x,z) = 0, \tag{8.38}$$

and

$$\mathbf{H}^{(\mathrm{II})\mathrm{a}}(x,z) = \nabla \times \left[\mathbf{e}_x \Phi^{\mathrm{a}}(x,z)\right], \tag{8.39}$$

$$\mathbf{E}^{(\mathrm{II})\mathrm{a}}(x,z) = -\frac{j}{\omega\epsilon(x)} \nabla \times \nabla \times \left[\mathbf{e}_x \Phi^{\mathrm{a}}(x,z)\right], \tag{8.40}$$

$$\nabla^2 \Phi^{\mathrm{a}}(x,z) - \frac{\partial \epsilon(x)}{\partial x} \frac{1}{\epsilon(x)} \frac{\partial \Phi^{\mathrm{a}}(x,z)}{\partial x} + \omega^2 \epsilon(x)\mu_0 \Phi^{\mathrm{a}}(x,z) = 0, \tag{8.41}$$

whose solutions are

$$\Psi^{\mathrm{a}}(x,z) = \left[A_1 f^{(1)}(x) + A_2 f^{(2)}(x)\right] e^{-j\beta^{(\mathrm{I})} z}, \tag{8.42}$$

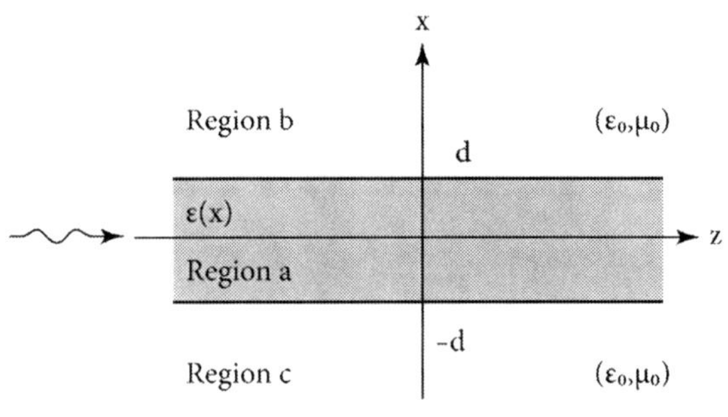

Figure 8.1. Dielectric slab with permittivity variation in the transverse direction

$$\Phi^{\mathrm{a}}(x,z) = \left[B_1 g^{(1)}(x) + B_2 g^{(2)}(x)\right] e^{-j\beta^{(\mathrm{II})}z}. \tag{8.43}$$

Here A_1, A_2, B_1, and B_2 are arbitrary constants and $\beta^{(\mathrm{I})}$ and $\beta^{(\mathrm{II})}$ are respectively, the propagation constants of type (I) and type (II) waves. The functions $f^{(1),(2)}$ and $g^{(1),(2)}$ satisfy, respectively, the following differential equations

$$\frac{d^2 f^{(1),(2)}(x)}{dx^2} + \left[\omega^2\epsilon(x)\mu_0 - \beta^{(I)2}\right] f^{(1),(2)}(x) = 0 \tag{8.44}$$

and

$$\frac{d^2 g^{(1),(2)}(x)}{dx^2} - \frac{d\epsilon(x)}{dx}\frac{1}{\epsilon(x)}\frac{dg^{(1),(2)}(x)}{dx} + \left[\omega^2\epsilon(x)\mu_0 - \beta^{(\mathrm{II})2}\right] g^{(1),(2)}(x) = 0. \tag{8.45}$$

Knowing $\epsilon(x)$ we can readily obtain the appropriate solutions of the above total second-order differential equations. For some special cases, we may identify these solutions as known special functions [7]. From (8.36) and (8.37), and using (8.42), the tangential field components for the type (I) wave are

$$E_y^{(\mathrm{I})\mathrm{a}}(x,z) = -j\beta^{(\mathrm{I})}\left[A_1 f^{(1)}(x) + A_2 f^{(2)}(x)\right] e^{-j\beta^{(\mathrm{I})}z}, \tag{8.46}$$

$$H_z^{(\mathrm{I})\mathrm{a}}(x,z) = \frac{1}{\omega\mu_0}\beta^{(\mathrm{I})}\left[A_1 f^{(1)\prime}(x) + A_2 f^{(2)\prime}(x)\right] e^{-j\beta^{(\mathrm{I})}z}, \tag{8.47}$$

where the symbol prime signifies the derivative with respect to its argument. From (8.39) and (8.40), and using (8.43), the tangential field components for the type (II) wave are

$$H_y^{(\mathrm{II})\mathrm{a}}(x,z) = -j\beta^{(\mathrm{II})}\left[B_1 g^{(1)}(x) + B_2 g^{(2)}(x)\right] e^{-j\beta^{(\mathrm{II})}z}, \tag{8.48}$$

$$E_z^{(\mathrm{II})\mathrm{a}}(x,z) = -\frac{1}{\omega\epsilon(x)}\beta^{(\mathrm{II})}\left[B_1 g^{(1)\prime}(x) + B_2 g^{(2)\prime}(x)\right] e^{-j\beta^{(\mathrm{II})}z}, \tag{8.49}$$

where A_1, A_2, B_1, and B_2 are arbitrary constants. In region b ($x > d$), the required tangential field components are

$$E_y^{(\mathrm{I})\mathrm{b}}(x,z) = -j\beta^{(\mathrm{I})}C_1\, e^{-q^{(\mathrm{I})}x} e^{-j\beta^{(\mathrm{I})}z}, \tag{8.50}$$

$$H_z^{(\mathrm{I})\mathrm{b}}(x,z) = -\frac{1}{\omega\mu_0}\beta^{(\mathrm{I})}q^{(\mathrm{I})}C_1\, e^{-q^{(\mathrm{I})}x} e^{-j\beta^{(\mathrm{I})}z}, \tag{8.51}$$

$$H_y^{(\mathrm{II})\mathrm{b}}(x,z) = -j\beta^{(\mathrm{II})} D_1\, e^{-q^{(\mathrm{II})}x} e^{-j\beta^{(\mathrm{II})}z}, \tag{8.52}$$

$$E_z^{(\mathrm{II})\mathrm{b}}(x,z) = \frac{1}{\omega\epsilon_0}\beta^{(\mathrm{II})} q^{(\mathrm{II})} D_1\, e^{-q^{(\mathrm{II})}x} e^{-j\beta^{(\mathrm{II})}z}, \tag{8.53}$$

where C_1 and D_1 are arbitrary constants, $q^{(\mathrm{I})2} = \beta^{(\mathrm{I})2} - k_0^2$, and $q^{(\mathrm{II})2} = \beta^{(\mathrm{II})2} - k_0^2$. In region c ($x < -d$), the required tangential field components are

$$E_y^{(\mathrm{I})\mathrm{c}}(x,z) = -j\beta^{(\mathrm{I})} E_1\, e^{q^{(\mathrm{I})}x} e^{-j\beta^{(\mathrm{I})}z}, \tag{8.54}$$

$$H_z^{(\mathrm{I})\mathrm{c}}(x,z) = \frac{1}{\omega\mu_0}\beta^{(\mathrm{I})} q^{(\mathrm{I})} E_1\, e^{q^{(\mathrm{I})}x} e^{-j\beta^{(\mathrm{I})}z}, \tag{8.55}$$

$$H_y^{(\mathrm{II})\mathrm{c}}(x,z) = -j\beta^{(\mathrm{II})} F_1\, e^{q^{(\mathrm{II})}x} e^{-j\beta^{(\mathrm{II})}z}, \tag{8.56}$$

$$E_z^{(\mathrm{II})\mathrm{c}}(x,z) = -\frac{1}{\omega\epsilon_0}\beta^{(\mathrm{II})} q^{(\mathrm{II})} F_1\, e^{q^{(\mathrm{II})}x} e^{-j\beta^{(\mathrm{II})}z}, \tag{8.57}$$

where C_1 and D_1 are arbitrary constants.

Matching the appropriate tangential fields at $x = \pm d$ yields

$$\begin{bmatrix} f^{(1)}(d) & f^{(2)}(d) & -e^{-q^{(\mathrm{I})}d} & 0 \\ f^{(1)\prime}(d) & f^{(2)\prime}(d) & q^{(\mathrm{I})}\, e^{-q^{(\mathrm{I})}d} & 0 \\ f^{(1)}(-d) & f^{(2)}(-d) & 0 & e^{-q^{(\mathrm{I})}d} \\ f^{(1)\prime}(-d) & f^{(2)\prime}(-d) & 0 & -q^{(\mathrm{I})}\, e^{-q^{(\mathrm{I})}d} \end{bmatrix} \begin{bmatrix} A_1 \\ A_2 \\ C_1 \\ E_1 \end{bmatrix} = 0, \tag{8.58}$$

$$\begin{bmatrix} g^{(1)}(d) & g^{(2)}(d) & -e^{-q^{(\mathrm{II})}d} & 0 \\ \dfrac{\epsilon_0 g^{(1)\prime}(d)}{\epsilon(d)} & \dfrac{\epsilon_0 g^{(2)\prime}(d)}{\epsilon(d)} & q^{(\mathrm{II})}\, e^{-q^{(\mathrm{II})}d} & 0 \\ g^{(1)}(-d) & g^{(2)}(-d) & 0 & e^{-q^{(\mathrm{II})}d} \\ \dfrac{\epsilon_0 g^{(1)\prime}(-d)}{\epsilon(-d)} & \dfrac{\epsilon_0 g^{(2)\prime}(-d)}{\epsilon(-d)} & 0 & -q^{(\mathrm{II})}\, e^{-q^{(\mathrm{II})}d} \end{bmatrix} \begin{bmatrix} B_1 \\ B_2 \\ D_1 \\ F_1 \end{bmatrix} = 0. \tag{8.59}$$

Setting the determinants of these simultaneous linear equations to zero gives the dispersion relations for type (I) and type (II) waves:

$$\left[f^{(1)\prime}(d) + q^{(\mathrm{I})} f^{(1)}(d)\right] \left[f^{(2)\prime}(-d) - q^{(\mathrm{I})} f^{(2)}(-d)\right]$$
$$= \left[f^{(2)\prime}(d) + q^{(\mathrm{I})} f^{(2)}(d)\right] \left[f^{(1)\prime}(-d) - q^{(\mathrm{I})} f^{(1)}(-d)\right], \quad (8.60)$$

$$\left[\frac{\epsilon_0}{\epsilon(d)} g^{(1)\prime}(d) + q^{(\mathrm{II})} g^{(1)}(d)\right] \left[\frac{\epsilon_0}{\epsilon(-d)} g^{(2)\prime}(-d) - q^{(\mathrm{II})} g^{(2)}(-d)\right]$$
$$= \left[\frac{\epsilon_0}{\epsilon(d)} g^{(2)\prime}(d) + q^{(\mathrm{II})} g^{(2)}(d)\right] \left[\frac{\epsilon_0}{\epsilon(-d)} g^{(1)\prime}(-d) - q^{(\mathrm{II})} g^{(1)}(-d)\right]. \quad (8.61)$$

The formal solutions to this inhomogeneous slab problem are obtained. The complete guided fields are the linear combination of the type (I) and type (II) fields. We can easily show that if the inhomogeneous dielectric variation is a constant, the above results degenerate to the results given in Sect. 4.2 for the homogeneous slab case. In that case, $f^{(1)} \to \sin q^{(\mathrm{I})} x$, $g^{(1)} \to \sin q^{(\mathrm{II})} x$, $f^{(2)} \to \cos q^{(\mathrm{I})} x$, and $g^{(2)} \to \cos q^{(\mathrm{II})} x$.

8.2.1.2 Waves in Metallic Rectangular Waveguide Filled with Transversely Inhomogeneous Dielectrics

This problem was treated earlier without the use of Debye potentials given here [8–12]. To match the boundary conditions, a linear combination of two Hertz potentials must be used. We recall that the Hertz potential is a special case of the Debye potentials where the Hertz potentials are derived using $\mathbf{e}_\xi = \mathbf{e}_z$ and z is the direction of propagation of the guided waves. Our earlier discussion on Debye potentials in inhomogeneous media shows that by choosing $\mathbf{e}_\xi$ to be in the same direction as the permittivity and permeability variation, the divergence equations for the inhomogeneous medium are automatically satisfied. This approach will generate a new independent set of wave types, which can be used to satisfy the boundary conditions without resorting to the need for the reconstruction of fields using linear combination of the Hertz fields [6–10]. In the following, we show our simple approach to the problem.

The geometry of the problem is shown in Fig. 8.2. According to Sect. 8.1.1, (8.5)–(8.8), (8.14), and (8.16), the complete set of fields for the $(\epsilon(x), \mu_0)$ medium in rectangular coordinates is given as follows:

$$\mathbf{E}^{(\mathrm{I})}(x, y, z) = \nabla \times [\mathbf{e}_x \Psi(x, y, z)], \quad (8.62)$$

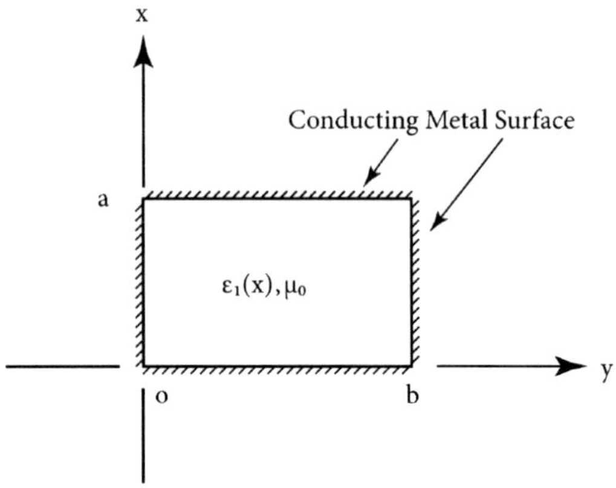

Figure 8.2. Rectangular metallic waveguide filled with a inhomogeneous dielectric. a and b are the waveguide dimensions

$$\mathbf{H}^{(\mathrm{I})}(x,y,z) = \frac{j}{\omega\mu_0}\nabla\times\nabla\times\left[\mathbf{e}_x\,\Psi(x,y,z)\right], \tag{8.63}$$

$$\mathbf{H}^{(\mathrm{II})}(x,y,z) = \nabla\times\left[\mathbf{e}_x\,\Phi(x,y,z)\right], \tag{8.64}$$

$$\mathbf{E}^{(\mathrm{II})}(x,y,z) = -\frac{j}{\omega\epsilon(x)}\nabla\times\nabla\times\left[\mathbf{e}_x\,\Phi(x,y,z)\right], \tag{8.65}$$

$$\Psi(x,y,z) = f^{(1),(2)}(x)\begin{bmatrix}\sin p^{(\mathrm{I})}y\\ \cos p^{(\mathrm{I})}y\end{bmatrix} e^{\pm j\beta^{(\mathrm{I})}z}, \tag{8.66}$$

$$\Phi(x,y,z) = g^{(1),(2)}(x)\begin{bmatrix}\sin p^{(\mathrm{II})}y\\ \cos p^{(\mathrm{II})}y\end{bmatrix} e^{\pm j\beta^{(\mathrm{II})}z}, \tag{8.67}$$

with

$$\frac{d^2 f^{(1),(2)}(x)}{dx^2} + \left[\omega^2\epsilon(x)\mu_0 - p^{(\mathrm{I})2} - \beta^{(\mathrm{I})2}\right] f^{(1),(2)}(x) = 0, \tag{8.68}$$

$$\frac{d^2 g^{(1),(2)}(x)}{dx^2} - \frac{d\epsilon(x)}{dx}\frac{1}{\epsilon(x)}\frac{dg^{(1),(2)}(x)}{dx} + \left[\omega^2\epsilon(x)\mu_0 - p^{(\mathrm{II})2} - \beta^{(\mathrm{II})2}\right] g^{(1),(2)}(x) = 0. \tag{8.69}$$

The boundary conditions on the conducting walls require that the tangential electric fields be zero.

$$E_x(\text{at } y = 0, b) = 0, \tag{8.70}$$

$$E_y(\text{at } x = 0, a) = 0, \tag{8.71}$$

$$E_z(\text{at } x = 0, a, \ y = 0, b) = 0. \tag{8.72}$$

For type (I) wave, we have

$$\Psi(x, y, z) = \left[A_1 f^{(1)}(x) + A_2 f^{(2)}(x)\right] \cos\left(\frac{m\pi y}{b}\right) e^{-j\beta^{(\mathrm{I})} z}, \tag{8.73}$$

$$E_y(x, y, z) = (-j\beta^{(\mathrm{I})}) \left[A_1 f^{(1)}(x) + A_2 f^{(2)}(x)\right] \cos\left(\frac{m\pi y}{b}\right) e^{-j\beta^{(\mathrm{I})} z}, \tag{8.74}$$

$$E_z(x, y, z) = \left[A_1 f^{(1)}(x) + A_2 f^{(2)}(x)\right] \left(\frac{m\pi}{b}\right) \sin\left(\frac{m\pi y}{b}\right) e^{-j\beta^{(\mathrm{I})} z}, \tag{8.75}$$

with $p^{(\mathrm{I})} = m\pi/b$, $m = 1, 2, \ldots$. To satisfy the boundary conditions at $x = 0, a$, we must have

$$A_1 f^{(1)}(0) + A_2 f^{(2)}(0) = 0, \tag{8.76}$$

$$A_1 f^{(1)}(a) + A_2 f^{(2)}(a) = 0, \tag{8.77}$$

or

$$f^{(1)}(0) f^{(2)}(a) - f^{(1)}(a) f^{(2)}(0) = 0, \tag{8.78}$$

$$A_1 = -A_2 \frac{f^{(2)}(0)}{f^{(1)}(0)}. \tag{8.79}$$

Equation (8.78) gives the eigenvalues for the eigenfunctions $f^{(1),(2)}(x)$. Note that all necessary boundary conditions are satisfied by the above Set (I) fields in a straightforward manner.

For type (II) wave, we have

$$\Phi(x, y, z) = \left[B_1 g^{(1)}(x) + B_2 g^{(2)}(x)\right] \sin\left(\frac{m\pi y}{b}\right) e^{-j\beta^{(\mathrm{II})} z}, \tag{8.80}$$

$$E_x(x, y, z) = \frac{j}{\omega\epsilon(x)} \left[(m\pi/b)^2 + \beta^{(\mathrm{II})2}\right] \times \left[B_1 g^{(1)}(x) + B_2 g^{(2)}(x)\right] \sin\left(\frac{m\pi y}{b}\right) e^{-j\beta^{(\mathrm{II})} z}, \tag{8.81}$$

$$E_y(x, y, z) = -\frac{\beta^{(\mathrm{II})}}{\omega\epsilon(x)} \frac{m\pi}{b} \left[B_1 g^{(1)\prime}(x) + B_2 g^{(2)\prime}(x)\right] \cos\left(\frac{m\pi y}{b}\right) e^{-j\beta^{(\mathrm{II})}z}, \tag{8.82}$$

$$E_z(x, y, z) = -\frac{\beta^{(\mathrm{II})}}{\omega\epsilon(x)} \left[B_1 g^{(1)\prime}(x) + B_2 g^{(2)\prime}(x)\right] \left(\frac{m\pi}{b}\right) \sin\left(\frac{m\pi y}{b}\right) e^{-j\beta^{(\mathrm{II})}z}, \tag{8.83}$$

with $p^{(\mathrm{II})} = m\pi/b$, $m = 1, 2, \ldots$. To satisfy the boundary conditions at $x = 0, a$, we must have

$$B_1 g^{(1)\prime}(0) + B_2 g^{(2)\prime}(0) = 0, \tag{8.84}$$

$$B_1 g^{(1)\prime}(a) + B_2 g^{(2)\prime}(a) = 0, \tag{8.85}$$

or

$$g^{(1)\prime}(0) g^{(2)\prime}(a) - g^{(1)\prime}(a) g^{(2)\prime}(0) = 0, \tag{8.86}$$

$$B_1 = -B_2 \frac{g^{(2)\prime}(0)}{g^{(1)\prime}(0)}. \tag{8.87}$$

Equation (8.86) gives the eigenvalues for the eigenfunction $g^{(1),(2)}(x)$. Again, all necessary boundary conditions are satisfied by the above Set (II) fields in a straightforward manner.

The complete fields can be obtained from the two independent sets of fields, (I) and (II). This approach shows that the two independent sets of fields that can satisfy the boundary conditions in this inhomogeneous medium with $\epsilon(x)$ are Set (I) with $E_x = 0$ and Set (II) with $H_x = 0$. For this inhomogeneous $\epsilon(x)$ medium, the usual set of Hertz fields with $E_z = 0$ or $H_z = 0$ will not be able to satisfy the boundary conditions by itself – a combination of the two Hertz sets must be used.

8.2.1.3 Circularly Symmetric Waves Along a Cylindrical Radially Inhomogeneous Dielectric Cylinder

The geometry of the problem is shown in Fig. 8.3.

According to Sect. 8.1.3(b), the circularly symmetric fields in the radially inhomogeneous medium (region a) are

$$\mathbf{E}^{(\mathrm{I})\mathrm{a}}(\rho, z) = \nabla \times \left[\mathbf{e}_\rho \Psi^{\mathrm{a}}(\rho, z)\right], \tag{8.88}$$

$$\mathbf{H}^{(\mathrm{I})\mathrm{a}}(\rho, z) = \frac{j}{\omega\mu_0} \nabla \times \nabla \times \left[\mathbf{e}_\rho \Psi^{\mathrm{a}}(\rho, z)\right], \tag{8.89}$$

$$\mathbf{H}^{(\mathrm{II})\mathrm{a}}(\rho, z) = \nabla \times \left[\mathbf{e}_\rho \Phi^{\mathrm{a}}(\rho, z)\right], \tag{8.90}$$

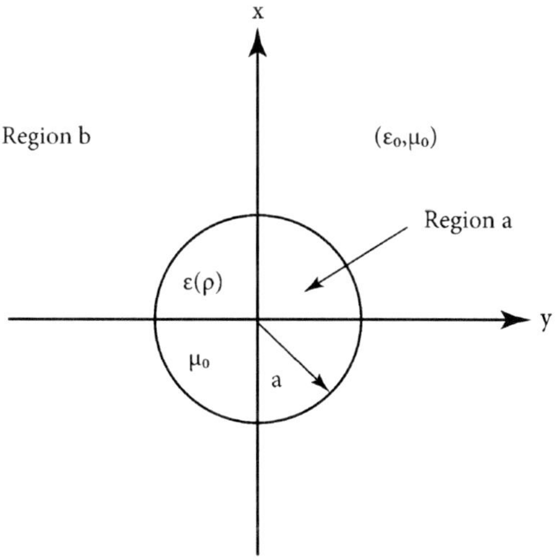

Figure 8.3. Radially inhomogeneous dielectric cylinder

$$\mathbf{E}^{(\mathrm{II})\mathrm{a}}(\rho, z) = -\frac{j}{\omega\epsilon(\rho)}\nabla \times \nabla \times \left[\mathbf{e}_\rho \Phi^{\mathrm{a}}(\rho, z)\right], \tag{8.91}$$

$$\Psi^{\mathrm{a}}(\rho, \theta, z) = A_1 f^{(1)}(\rho)\, e^{-j\beta^{(\mathrm{I})}z}, \tag{8.92}$$

$$\Phi^{\mathrm{a}}(\rho, \theta, z) = B_1 g^{(1)}(\rho)\, e^{-j\beta^{(\mathrm{II})}z}, \tag{8.93}$$

with

$$\frac{d}{d\rho}\frac{1}{\rho}\frac{d}{d\rho}\left[\rho f^{(1)}(\rho)\right] + \left[\omega^2\epsilon(\rho)\mu_0 - \beta^{(\mathrm{I})2}\right] f^{(1)}(\rho) = 0, \tag{8.94}$$

$$\frac{d}{d\rho}\frac{1}{\rho}\frac{d}{d\rho}\left[\rho g^{(1)}(\rho)\right] - \frac{1}{\epsilon(\rho)}\frac{d\epsilon(\rho)}{d\rho}\frac{1}{\rho}\frac{d}{d\rho}\left[\rho g^{(1)}(\rho)\right] + \left[\omega^2\epsilon(\rho)\mu_0 - \beta^{(\mathrm{II})2}\right] g^{(1)}(\rho) = 0, \tag{8.95}$$

where A_1 and B_1 are arbitrary constants. From (8.88)–(8.91), we find the fields tangential to the boundary at $\rho = a$:

For region a ($\rho \leq a$),

$$E_\theta^{(\mathrm{I})\mathrm{a}}(\rho, z) = A_1\left(-j\beta^{(\mathrm{I})}\right) f^{(1)}(\rho)\, e^{-j\beta^{(\mathrm{I})}z}, \tag{8.96}$$

$$H_z^{(\mathrm{I})\mathrm{a}}(\rho, z) = A_1 \frac{\beta^{(\mathrm{I})}}{\omega\mu_0\rho}\frac{\partial}{\partial\rho}\left[\rho f^{(1)}(\rho)\right] e^{-j\beta^{(\mathrm{I})z}}, \tag{8.97}$$

$$H_\theta^{(\mathrm{II})\mathrm{a}}(\rho,z) = B_1 \left(-j\beta^{(\mathrm{II})}\right) g^{(1)}(\rho)\, e^{-j\beta^{(\mathrm{II})}z}, \tag{8.98}$$

$$E_z^{(\mathrm{II})\mathrm{a}}(\rho,z) = B_1 \left(-\frac{\beta^{(\mathrm{II})}}{\omega\epsilon(\rho)}\right) \frac{1}{\rho}\frac{\partial}{\partial\rho}\left[\rho g^{(1)}(\rho)\right] e^{-j\beta^{(\mathrm{II})}z}. \tag{8.99}$$

For region b ($\rho \geq a$),

$$E_\theta^{(\mathrm{I})\mathrm{a}}(\rho,z) = C_1 \left(-j\beta^{(\mathrm{I})}\right) K_0(q^{(\mathrm{I})}\rho)\, e^{-j\beta^{(\mathrm{I})}z}, \tag{8.100}$$

$$H_z^{(\mathrm{I})\mathrm{a}}(\rho,z) = C_1 \frac{\beta^{(\mathrm{I})}}{\omega\mu_0\rho}\frac{\partial}{\partial\rho}\left[\rho K_0(q^{(\mathrm{I})}\rho)\right] e^{-j\beta^{(\mathrm{I})}z}, \tag{8.101}$$

$$H_\theta^{(\mathrm{II})\mathrm{b}}(\rho,z) = D_1 \left(-j\beta^{(\mathrm{II})}\right) K_0(q^{(\mathrm{II})}\rho)\, e^{-j\beta^{(\mathrm{II})}z}, \tag{8.102}$$

$$E_z^{(\mathrm{II})\mathrm{b}}(\rho,z) = D_1 \left(-\frac{\beta^{(\mathrm{II})}}{\omega\epsilon_0}\right) \frac{1}{\rho}\frac{\partial}{\partial\rho}\left[\rho K_0(q^{(\mathrm{II})}\rho)\right] e^{-j\beta^{(\mathrm{II})}z}, \tag{8.103}$$

where A_1, B_1, C_1, and D_1 are arbitrary constants, $K_0(u)$ is the modified Bessel function of argument u and order zero, $q^{(\mathrm{I})2} = \beta^{(\mathrm{I})2} - \omega^2\mu_0\epsilon_0$, and $q^{(\mathrm{II})2} = \beta^{(\mathrm{II})2} - \omega^2\mu_0\epsilon_0$.

Matching the tangential fields at $\rho = a$ gives

$$A_1 f^{(1)}(a) = C_1 K_0(q^{(\mathrm{I})}a), \tag{8.104}$$

$$A_1 \frac{\partial}{\partial\rho}\left[\rho f^{(1)}(\rho)\right]|_{\rho=a} = C_1 \frac{\partial}{\partial\rho}\left[\rho K_0(q^{(\mathrm{I})}\rho)\right]|_{\rho=a}\,, \tag{8.105}$$

$$B_1 g^{(1)}(a) = D_1 K_0(q^{(\mathrm{II})}a), \tag{8.106}$$

$$B_1 \frac{\epsilon_0}{\epsilon(a)}\frac{\partial}{\partial\rho}\left[\rho g^{(1)}(\rho)\right]|_{\rho=a} = D_1 \frac{\partial}{\partial\rho}\left[\rho K_0(q^{(\mathrm{II})}\rho)\right]|_{\rho=a}\,. \tag{8.107}$$

Setting the determinant of (8.104)–(8.105) and the determinant of (8.106)–(8.107) to zero yields the dispersion relations for the type (I) and type (II) waves:

type (I)

$$f^{(1)}(a)\frac{\partial}{\partial\rho}\left[\rho K_0(q^{(\mathrm{I})}\rho)\right]|_{\rho=a} = K_0(q^{(\mathrm{I})}a)\frac{\partial}{\partial\rho}\left[\rho f^{(1)}(\rho)\right]|_{\rho=a}\,, \tag{8.108}$$

type (II)

$$g^{(1)}(a)\frac{\partial}{\partial\rho}\left[\rho K_0(q^{(\mathrm{II})}\rho)\right]|_{\rho=a} = \frac{\epsilon_0}{\epsilon(a)}K_0(q^{(\mathrm{II})}a)\frac{\partial}{\partial\rho}\left[\rho g^{(1)}(\rho)\right]|_{\rho=a}\,. \tag{8.109}$$

8.2.2 Structures with Longitudinal Inhomogeneity

We now consider the special case where $\epsilon = \epsilon(z)$, $\mu = \mu_0$. Here, z is the longitudinal coordinate. According to Sect. 8.1, two independent type of fields exist:

Type (I) (TE Wave)

$$\mathbf{E}^{(\mathrm{I})}(u_1, u_2, z) = \nabla \times \left[\mathbf{e}_z \Psi(u_1, u_2, z)\right], \tag{8.110}$$

$$\mathbf{H}^{(\mathrm{I})}(u_1, u_2, z) = \frac{j}{\omega\mu_0} \nabla \times \nabla \times \left[\mathbf{e}_z \Psi(u_1, u_2, z)\right], \tag{8.111}$$

$$\left(\nabla_{\mathrm{t}}^2 + \frac{\partial^2}{\partial z^2}\right) \Psi(u_1, u_2, z) + \omega^2 \mu_0 \epsilon(z) \Psi(u_1, u_2, z) = 0. \tag{8.112}$$

Type (II) (TM Wave)

$$\mathbf{H}^{(\mathrm{II})}(u_1, u_2, z) = \nabla \times \left[\mathbf{e}_z \Phi(u_1, u_2, z)\right], \tag{8.113}$$

$$\mathbf{E}^{(\mathrm{II})}(u_1, u_2, z) = -\frac{j}{\omega\epsilon(z)} \nabla \times \nabla \times \left[\mathbf{e}_z \Phi(u_1, u_2, z)\right], \tag{8.114}$$

$$\left(\nabla_{\mathrm{t}}^2 + \frac{\partial^2}{\partial z^2}\right) \Phi(u_1, u_2, z) - \frac{1}{\epsilon(z)} \frac{\partial \epsilon(z)}{\partial z} \frac{\partial \Phi(u_1, u_2, z)}{\partial z} + \omega^2 \mu_0 \epsilon(z) \Psi(u_1, u_2, z) = 0. \tag{8.115}$$

Here ∇_t is the transverse del operator in the curvilinear coordinates and (u_1, u_2) are the transverse coordinates. The separation of variables method can be used to find $\Psi(u_1, u_2, z)$ and $\Phi(u_1, u_2, z)$.

(a) In rectangular coordinates (x, y, z)

$$\Psi(u_1, u_2, z) = \left\{ \begin{array}{c} \sin sx \\ \\ \cos sx \end{array} \right\} \left\{ \begin{array}{c} \sin wy \\ \\ \cos wy \end{array} \right\} U^{(1),(2)}(z), \tag{8.116}$$

$$\Phi(u_1, u_2, z) = \left\{ \begin{array}{c} \sin px \\ \\ \cos px \end{array} \right\} \left\{ \begin{array}{c} \sin qy \\ \\ \cos qy \end{array} \right\} V^{(1),(2)}(z). \tag{8.117}$$

(b) In circular cylindrical coordinates (ρ, θ, z)

$$\Psi(u_1, u_2, z) = \left\{ \begin{array}{c} J_n\,(p\rho) \\ \\ Y_n\,(p\rho) \end{array} \right\} e^{\pm jn\theta}\, U^{(1),(2)}(z), \tag{8.118}$$

$$\Phi(u_1, u_2, z) = \left\{ \begin{array}{c} J_n\,(q\rho) \\ \\ Y_n\,(q\rho) \end{array} \right\} e^{\pm jn\theta}\, V^{(1),(2)}(z). \tag{8.119}$$

(c) In elliptical cylindrical coordinates (ξ, η, z)

$$\Psi(u_1, u_2, z) = \left\{ \begin{array}{c} Ce_n\left(\xi, \gamma_a^2\right),\; Fey_n\left(\xi, \gamma_a^2\right) \\ \\ Se_n\left(\xi, \gamma_a^2\right),\; Gey_n\left(\xi, \gamma_a^2\right) \end{array} \right\} \left\{ \begin{array}{c} ce_n\left(\eta, \gamma_a^2\right) \\ \\ se_n\left(\eta, \gamma_a^2\right) \end{array} \right\} U^{(1),(2)}(z), \tag{8.120}$$

$$\Phi(u_1, u_2, z) = \left\{ \begin{array}{c} Ce_n\left(\xi, \gamma_b^2\right),\; Fey_n\left(\xi, \gamma_b^2\right) \\ \\ Se_n\left(\xi, \gamma_b^2\right),\; Gey_n\left(\xi, \gamma_b^2\right) \end{array} \right\} \left\{ \begin{array}{c} ce_n\left(\eta, \gamma_b^2\right) \\ \\ se_n\left(\eta, \gamma_b^2\right) \end{array} \right\} V^{(1),(2)}(z). \tag{8.121}$$

The functions $U(z)$ and $V(z)$ satisfy the following equations:

$$\frac{d^2 U^{(1),(2)}(z)}{dz^2} + \left[\omega^2 \mu_0 \epsilon(z) - \kappa_1^2\right] U^{(1),(2)}(z) = 0 \tag{8.122}$$

and

$$\frac{d^2 V^{(1),(2)}(z)}{dz^2} - \frac{d\epsilon(z)}{dz}\frac{1}{\epsilon(z)}\frac{\partial V^{(1),(2)}(z)}{\partial z} + \left[\omega^2 \mu_0 \epsilon(z) - \kappa_2^2\right] V^{(1),(2)}(z) = 0, \tag{8.123}$$

where κ_1^2 and κ_2^2 are separation constants. For example, in rectangular coordinates, $\kappa_1^2 = s^2 + w^2$, and $\kappa_2^2 = p^2 + q^2$ [see (8.116) and (8.117)].

8.2.2.1 Longitudinal Periodic Medium

We shall provide an in-depth treatment of a specific example [13, 14] on wave propagation in a longitudinally inhomogeneous medium whose dielectric constant variation is given by the following:

$$\epsilon(z) = A\epsilon_0 \left[1 - \delta\, \cos\left(\frac{2\pi z}{d}\right)\right], \tag{8.124}$$

$$\mu(z) = \mu_0, \tag{8.125}$$

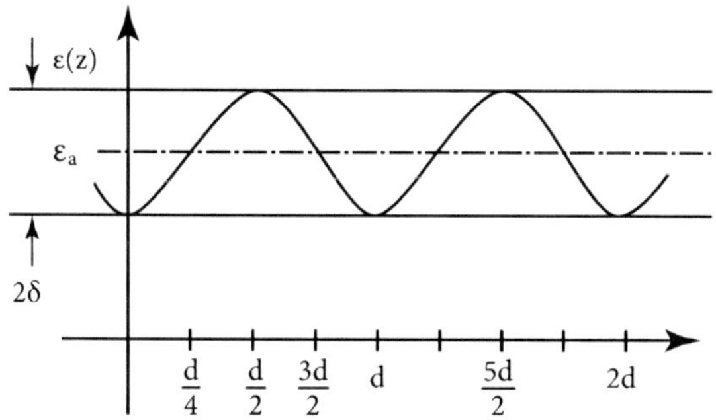

Figure 8.4. Variation of permittivity as a function of longitudinal distance. $\epsilon_a = A\epsilon_0$

where d is the periodicity of the dielectric variation, A and δ are known positive constants, $\epsilon_a = A\epsilon_0$, and $0 \leq \delta \leq 1$ (see Fig. 8.4).

Introducing the dimensionless variable, $\xi = \pi z/d$, $U^{(1),(2)}(\xi)$, and $V^{(1),(2)}(\xi)$ satisfy, respectively, the following differential equations:

$$\left\{\frac{d^2}{d\xi^2} + \left[\left(\frac{k_0 d}{\pi}\right)^2 \frac{\epsilon(\xi)}{\epsilon_0} - \left(\frac{s}{k_0}\right)^2 - \left(\frac{w}{k_0}\right)^2\right]\right\} U^{(1),(2)}(\xi) = 0 \tag{8.126}$$

and

$$\left\{\frac{d^2}{d\xi^2} - \frac{d\epsilon(\xi)}{d\xi}\frac{1}{\epsilon(\xi)}\frac{d}{d\xi} + \left[\left(\frac{k_0 d}{\pi}\right)^2 \frac{\epsilon(\xi)}{\epsilon_0} - \left(\frac{p}{k_0}\right)^2 \left(\frac{q}{k_0}\right)^2\right]\right\} V^{(1),(2)}(\xi) = 0. \tag{8.127}$$

Rewriting (8.127), one has

$$\left[\frac{d^2}{d\xi^2} + g(\xi)\right] W^{(1),(2)}(\xi) = 0, \tag{8.128}$$

where

$$W^{(1),(2)}(\xi) = \frac{1}{\sqrt{\epsilon(\xi)}} V^{(1),(2)}(\xi) \tag{8.129}$$

and

$$g(\xi) = \frac{1}{2\epsilon(\xi)}\frac{d^2\epsilon(\xi)}{d\xi^2} - \frac{3}{4}\frac{1}{\epsilon^2(\xi)}\left(\frac{d\epsilon(\xi)}{d\xi}\right)^2 + \left(\frac{k_0 d}{\pi}\right)^2 \frac{\epsilon(\xi)}{\epsilon_0} - \left(\frac{p}{k_0}\right)^2 - \left(\frac{q}{k_0}\right)^2. \tag{8.130}$$

It is seen that (8.126) or (8.128) is in the form

$$\left[\frac{d^2}{d\xi^2} + h(\xi)\right] F(\xi) = 0. \tag{8.131}$$

For type (I) waves

$$F(\xi) = U^{(1),(2)}(\xi) \tag{8.132}$$

and

$$h(\xi) = \left(\frac{k_0 d}{\pi}\right)^2 \frac{\epsilon(\xi)}{\epsilon_0} - \left(\frac{p}{k_0}\right)^2 - \left(\frac{q}{k_0}\right)^2, \tag{8.133}$$

and for type (II) waves

$$F(\xi) = W^{(1),(2)}(\xi) = \frac{1}{\sqrt{\epsilon(\xi)}} V^{(1),(2)}(\xi) \tag{8.134}$$

and $h(\xi) = g(\xi)$. Substituting (8.124) into (8.133) and (8.130) yields

$$h(\xi) = \left(\frac{k_0 d}{\pi}\right)^2 A\left(1 - \delta \cos 2\xi\right) - \left(\frac{s}{k_0}\right)^2 - \left(\frac{w}{k_0}\right)^2, \tag{8.135}$$

$$g(\xi) = \frac{2\delta \cos 2\xi}{1 - \delta \cos 2\xi} - \frac{3\delta^3 \sin^2 2\xi}{(1 - \delta \cos 2\xi)^2} + \left(\frac{k_0 d}{\pi}\right)^2 \left\{A - A\delta \cos 2\xi - \left[\left(\frac{p}{k_0}\right)^2 + \left(\frac{q}{k_0}\right)^2\right]\right\}. \tag{8.136}$$

It is noted that $h(\xi)$ and $g(\xi)$ are periodic functions, which may be represented by a Fourier series. For this specific dielectric variation (8.124), (8.126) is the Mathieu equation while (8.128) is the Hill equation [15].

As a specific example, let us treat the type(II) wave case [the type(I) case can be treated in a similar manner]. We note that $g(\xi)$, (8.136), is an even function, and it may be expanded in a Fourier cosine series.

$$g(\xi) = \theta_0 + 2\sum_{n=1}^{\infty} \theta_n \cos 2n\xi, \tag{8.137}$$

in which

$$\theta_0 = \left(\frac{k_0 d}{\pi}\right)^2 \left[A - \left(\frac{p}{k_0}\right)^2 - \left(\frac{q}{k_0}\right)^2\right] - \left[\frac{1}{\sqrt{1 - \delta^2}} - 1\right], \tag{8.138a}$$

$$\theta_1 = -\frac{\delta}{2}\left(\frac{k_0 d}{\pi}\right)^2 A + \frac{4b^3 - 2b}{b^2 - 1}, \tag{8.138b}$$

$$\theta_n = \frac{(3n+1)\, b^{n+2} - (3n-1)\, b^n}{b^2 - 1} \qquad (n \geq 2), \tag{8.138c}$$

$$b = \frac{1}{\delta} - \frac{1}{\delta}\sqrt{1 - \delta^2}. \tag{8.138d}$$

The above series converges absolutely for $0 \leq \delta \leq 1$. Substituting (8.137) in (8.128), one obtains

$$\left[\frac{d^2}{d\xi^2} + \theta_0 + 2\sum_{n=1}^{\infty} \theta_n \cos 2n\xi\right] W^{(1),(2)}(\xi) = 0, \tag{8.139}$$

which is the general form of the Hill equation [15]. It is known that two types of solutions for the Hill equation exist: one called the stable type and the other called the unstable type. To have propagating waves in the z-direction, only the stable type is allowed.

8.2.2.2 Solutions to the Hill Equation

With the help of Floquet's Theorem [16] concerning wave propagation in periodic media, the solutions of the Hill equation can be expressed in the following form:

$$W^{(1),(2)}(\xi) = e^{\pm 2j\beta\xi} \sum_{n=-\infty}^{\infty} C_n(\beta)\, e^{\pm 2jn\xi}, \tag{8.140}$$

where β and $C_n(\beta)$ are yet unknown coefficients. After substituting (8.140) in (8.139) and simplifying, one obtains the following recursion relations:

$$-\left(\beta + 2n\right)^2 C_n + \sum_{m=-\infty}^{\infty} \theta_m C_{nm} = 0, \tag{8.141}$$

$$(n = -\infty, \ldots, -2, -1, 0, 2, \ldots, \infty),$$

where $\theta_{-m} = \theta_m$. The above is a set of an infinite number of homogeneous linear algebraic equations in C_n. For a nontrivial solution to exist the determinant of these equations must vanish. This equation is called the characteristic equation of the Hill equation. Using an ingenious method described in Morse and Feshbach [17] it is possible to simplify this characteristic equation to give

$$\sin^2(\pi\beta/2) = \Delta(0)\sin^2\left(\pi\sqrt{\theta_0}/2\right), \tag{8.142}$$

where $\Delta(0)$ is the determinant of the matrix $[M]$ whose elements are

$$M_{mm} = 1.$$
$$M_{mn} = \frac{-\theta_{m-n}}{4m^2 - \theta_0}, \qquad m \neq n.$$

The characteristic number β can be obtained from (8.142). Real values of β yield stable solutions to the Hill equation, while complex values of β produce unstable solutions. Physically speaking, the stable solutions correspond to modulated propagating waves, and the unstable solutions correspond to damped or growing waves. (The fields for the growing waves do not satisfy the radiation condition at infinity, hence they must be discarded.)

Numerical computation has been carried out for (8.142). Results of the computation are given in terms of a "stability diagram," which is customary in the study of Hill type equations. Figure 8.5 shows the "stability diagram" for the case $\delta = 0.4$. The unshaded areas in these figures are the "stable regions" wherein β is purely real; the shaded areas are the "unstable regions" wherein β is complex. It is noted that the value of β in the unshaded regions is bounded by

$$m \leq |\beta| \leq m + 1 \qquad (m = 0, 1, 2, 3, \ldots), \tag{8.143}$$

so that the value of m may be used to label the appropriate regions in Fig. 8.5.

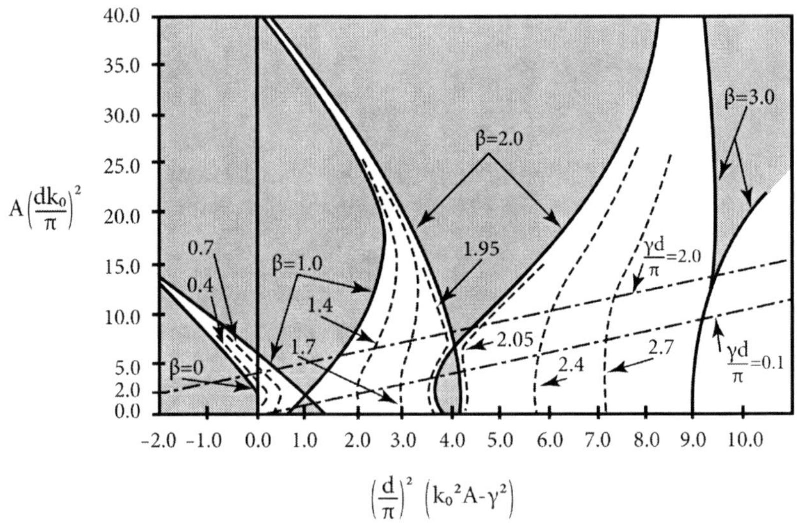

Figure 8.5. Stability chart for the Hill equation with $\delta = 0.4$. Unstable regions are shaded. Family of straight lines represent (8.148) for various values of $\gamma d/\pi$ [13]

It is interesting to note the difference between the stability diagram for the Mathieu equation and those given in Fig. 8.5. Unlike the Mathieu case, curves separating the stable and unstable regions do not necessarily meet at the abscissa. As a matter of fact, in some instances they cross over each other (as can be seen in these figures) near the point $\theta_0 = 4.0$. As δ increases, the overlapped region becomes larger.

8.2.2.3 Propagation Characteristics of Type (II) (TM) Waves in Periodic Structures

(a) Infinite region filled with sinusoidally stratified dielectric

The transverse magnetic field components of a type(II) (TM) wave in an infinite medium filled with sinusoidally stratified dielectrics can be obtained from (8.113) and (8.114):

$$H_x^{(\mathrm{II})} = \sum_{n=-\infty}^{\infty} jqC_n\, e^{jpx} e^{jqy} e^{j(\beta+2n)\pi z/d} \left(1 - \delta \cos \frac{2\pi z}{d}\right)^{\frac{1}{2}}, \quad (8.144)$$

$$H_y^{(\mathrm{II})} = \sum_{n=-\infty}^{\infty} -jpC_n\, e^{jpx} e^{jqy} e^{j(\beta+2n)\pi z/d} \left(1 - \delta \cos \frac{2\pi z}{d}\right)^{\frac{1}{2}}, \quad (8.145)$$

where the coefficients C_n can be determined from (8.141) in terms of C_0. C_0 is obtained from a normalization condition. All electric field components may be found from Maxwell equations.

Unlike the case of a type (II) wave propagating in an infinite homogeneous medium in which β is simply related to p and q by the following:

$$\beta^2 = k^2 - \gamma^2,$$

with $\gamma^2 = p^2 + q^2$ and $k^2 = \omega^2\mu\epsilon$, where μ and ϵ are the permeability and permittivity of the homogeneous medium, the propagation constant β for the inhomogeneous case is related to p and q through the stability diagrams given by Fig. 8.5. Real values of β as a function of real values of γ for a fixed value of A, δ, and $k_0 d$ are shown in Fig. 8.6. It is recalled that complex values of β indicate the presence of damped waves (i.e., nonpropagating waves). p and q are taken to be real. The unshaded regions in these figures indicate the regions in which β is real (i.e., regions in which propagating waves may exist). One notes from these figures that for very small values of $k_0 d$, say $k_0 d < 0.2$, as long as $\gamma^2 < k_0^2 A$, β is always real. However, as $k_0 d$ increases, there exists regions in which β is complex even though $\gamma^2 < k_0^2 A$.

The presence of these stop band and pass band regions is characteristic of wave propagation in periodic structures [16].

(b) Conducting waveguide filled longitudinally with sinusoidally stratified dielectric

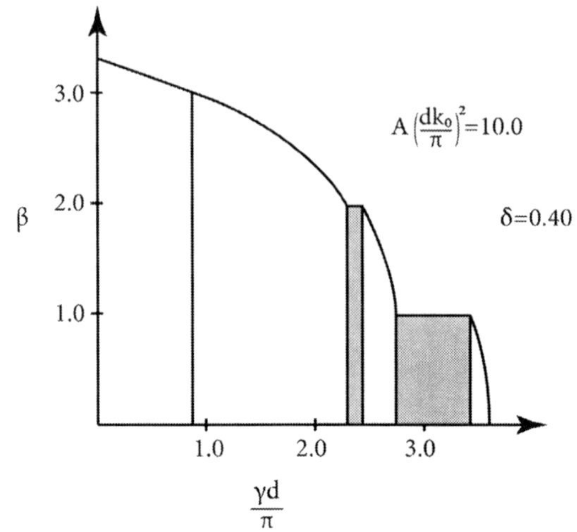

Figure 8.6. The propagation constant β as a function of $\gamma d/\pi$. Unstable regions are shaded [13]

It is assumed that a rectangular waveguide of dimension h_1 is filled completely with an inhomogeneous dielectric medium, which varies sinusoidally in the longitudinal direction. The general expressions for the transverse magnetic field components of a Type (II) wave are

$$H_x^{(\mathrm{II})} = \sum_{m=1}^{\infty}\sum_{r=1}^{\infty}\sum_{n=-\infty}^{\infty}\left[C_n^{m,r} \sin\frac{\pi m x}{h_1} \cos\frac{\pi r y}{h_2}\, e^{j(\beta^{n,r}+2n)\pi z/d} \left(\frac{r\pi}{h_2}\right)\left(1-\delta\cos\frac{2\pi z}{d}\right)^{\frac{1}{2}}\right], \quad (8.146)$$

$$H_y^{(\mathrm{II})} = \sum_{m=1}^{\infty}\sum_{r=1}^{\infty}\sum_{n=-\infty}^{\infty}\left[C_n^{m,r} \cos\frac{\pi m x}{h_1} \sin\frac{\pi r y}{h_2}\, e^{j(\beta^{n,r}+2n)\pi z/d} \left(-\frac{m\pi}{h_1}\right)\left(1-\delta\cos\frac{2\pi z}{d}\right)^{\frac{1}{2}}\right], \quad (8.147)$$

where $C_0^{m,r}$ are arbitrary constants and $C_n^{m,r}$ can be determined from (8.141) in terms of $C_0^{m,r}$. Expressions for the electric field components can be easily derived from the Maxwell equations.

To obtain the dispersion curves for this case, one combines (8.138a) and (8.138b)

$$\theta_0 + \frac{2\theta_1}{\delta} = \left(\frac{\pi d}{\pi}\right)^2 - \left[\frac{1}{\sqrt{1-\delta^2}} - 1\right] + \frac{4b^3 - 2b}{b^2 - 1}, \tag{8.148}$$

$$\gamma^2 = \left(\frac{m\pi}{h_1}\right)^2 + \left(\frac{r\pi}{h_2}\right)^2. \tag{8.149}$$

Expression (8.146) is the equation of a family of straight lines as shown in Fig. 8.5. For given values of $\gamma d/\pi$ and δ one can then obtain the ω-β diagram from these figures. The ω-β diagrams are given for $\gamma d/\pi = 0.1,\ 2.0$ and $\delta = 0.4$, in Fig. 8.7. The pass-band and stop-band characteristics can clearly be seen. It is interesting to note that the first pass-band starts at a frequency which is lower than the cutoff frequency for an identical waveguide filled with homogeneous dielectric material, which has a dielectric constant equal to the average value of the inhomogeneous dielectric.

The ω-β diagram given by Fig. 8.7 is also applicable for circular waveguides except γ^2 is given by $(\Gamma_{m,r}/\rho_0)^2$, where ρ_0 is the radius of the circular waveguide and $\Gamma_{m,r}$ are roots of the equation

$$J_m(\Gamma_{m,r}) = 0. \tag{8.150}$$

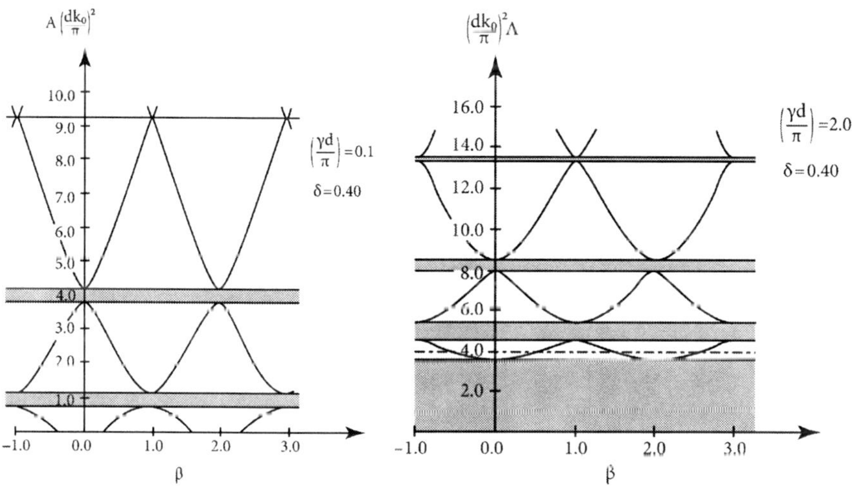

Figure 8.7. Frequency as a function of β with $\delta = 0.4$. Stop bands are shaded. The dot-dash line represents the cutoff frequency of a waveguide filled with a homogeneous dielectric medium with $\epsilon = \epsilon_a$ [13]

References

1. C. Yeh and Z. A. Kaprielian, "On inhomogeneously filled waveguides," USCEC Report, Elec. Eng. Dept., University of Southern California, Los Angeles (1963), p. 84
2. C. Yeh, "Dyadic Green's function for a radially inhomogeneous spherical medium," Phys. Rev. **131**, 2350 (1963)
3. P. J. Wyatt, "Scattering of electromagnetic plane waves from inhomogeneous spherically symmetric objects," Phys. Rev. **127**, 1837 (1962); L. I. Schiff, "Scattering of waves and particles by inhomogeneous regions," J. Opt. Soc. Am. **52**, 140 (1962)
4. C. T. Tai, "The electromagnetic theory of the spherical luneberg lens," Appl. Sci. Rec. Sec. **B 7**, 113 (1958)
5. A. K. Ghatak and M. S. Sodha, "The Inhomogeneous Optical Waveguide," Plenum Press, New York (1977)
6. T. Tamir, "Guided-wave Optoelectronics," Springer, Berlin Heidelberg New York (1988)
7. E. T. Whittaker and G. N. Watson, "A Course on Modern Analysis," Cambridge University Press, Oxford (1927); H. Wayland, "Differential Equations Applied in Science and Engineering," D. Van Nostrand, New York (1957)
8. R. E. Collin, "Field Theory of Guided Waves," McGraw-Hill, New York (1960)
9. I. L. Pincherle, "Electromagnetic waves in metal tubes filled longitudinally with two dielectrics," Phys. Rev. **66**, 118 (1944)
10. L. G. Chambers, "Propagation in waveguides filled longitudinally with two or more dielectrics," Brit. J. Appl. Phys. **4**, 39 (1953); L. G. Chambers, "An approximate method for the calculation of propagation constants for inhomogeneously filled waveguides," Quant. J. Mech. Appl. Math. **7**, 299 (1954)
11. R. M. Fano, L. J. Chu, and R. B. Adler, "Electromagnetic Fields, Energy, and Forces," Wiley, New York (1960)
12. R. B. Adler, "Waves in inhomogeneous cylindrical structures," Proc. IRE **40**, 339 (1952)
13. C. Yeh, K. F. Casey, and Z. A. Kaprielian, "Transverse magnetic wave propagation in sinusoidally stratified dielectric media," IEEE Trans. Microw. Theory Tech. **MTT-13**, 297 (1965)
14. T. Tamir, H.C. Wang, and A. A. Oliner, "Wave propagation in sinusoidally stratified dielectric media," IEEE Trans. Microwave Theory Tech. **MTT-12**, 323 (1965)
15. N. W. McLachlan, "Theory and Application of Mathieu Functions," University Press, Oxford, England (1951); J. Meixner and F. W. Schäfke, "Mathieusche funktionen spharoid funktionen," Springer, Berlin Heidelberg New York, Germany (1954)
16. L. Brillouin, "Wave Propagation in Periodic Structures," Dover, New York (1953)
17. P. M. Morse and H. Feshbach, "Methods of Mathematical Physics," McGraw-Hill, New York (1953)

9

OPTICAL FIBERS

One of the most important applications of dielectric waveguides is optical fibers. After the rigorous analysis of Carson et al. in 1936 [1] showing that a circular dielectric waveguide can support a hybrid dominant mode with no cutoff frequency, it languished as a practical electromagnetic wave guiding structure for almost 35 years. Occasionally, dielectric waveguides acted as novel flexible microwave guides or as material for classroom demonstrations, showing that waveguides could be made with nonmetallic material. Prior to the late 1960s, the most notable uses for dielectric waveguides as optical fibers were in flexible medical imaging endoscopes for a short distance or even in decorations for homes [2]. The whole field of optical communication links through optical fibers was awakened by the successful development of low-loss fibers (with losses less than 20 dB km^{-1}) in 1970 [3]. More recently, advances in the design of dispersion-flattened, dispersion-shifted, or dispersion-modified fibers [4] involving the use of multiple cladding layers and a tailoring of the refractive-index profile have enabled the successful operation of optical communication links in single-wavelength format or in wavelength division multiplexed (WDM) format in the 1.55 μm wavelength range with fiber loss of less than 0.2 dB km^{-1} [5]. Here we shall expand the electromagnetic wave analysis given in previous chapters to describe the propagation characteristics of fibers [6–10].

9.1 Weakly Guiding Optical Fibers

Although light waves can be guided in dielectric waveguides with arbitrary index contrast between the core region with index n_1 and the surrounding region with index n_2, provided that $n_1 > n_2$, most optical fibers for communication use possess index contrast of only a few per cent. The term "weakly guiding fiber" has been used to describe this type of fiber. It should be pointed out that "weakly guiding" does not mean that the optical power is guided very poorly or weakly so that the

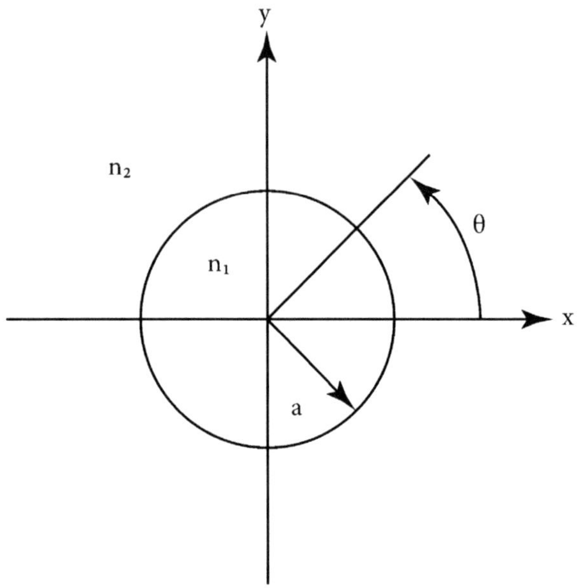

Figure 9.1. The geometry of a circular fiber with core index n_1 and cladding index n_2

guided power can easily radiate (escape) from the fiber with small disturbances or small imperfections, or that most of the guided optical power resides outside the core region of the fiber. It only means that the index contrast is very small. In other words, $\Delta = (n_1 - n_2)/n_1 \ll 1$. Here, for good guidance, most of the guided power is contained within the core region of the weakly guiding fiber.

Snyder [11] and Gloge [12] first investigated this situation with $\Delta \ll 1$ for a circular step-index fiber (see Fig. 9.1). They found that the dispersion relation (5.40) can be simplified and that a number of lower order modes become degenerate modes with similar propagation constants. These degenerate modes may be grouped together to form a group of nearly linearly polarized modes with only four dominant field components, that is, either (E_x, H_y, E_z, H_z) or (E_y, H_x, E_z, H_z). Note that field lines for these linearly polarized modes must still form closed loops. Table 9.1 shows the composition of the lower-order linearly polarized modes [10]. This composition is self-evident if one compares Figs. 9.2 and 9.3. There, the plots of the propagation constants as a function of V for a few lowest-order modes based on the exact dispersion relation with the traditional mode designation are displayed in Fig. 9.2. The same plots based on the linearly polarized modes are shown in Fig. 9.3.

According to Gloge [12], the four dominant field components for the linearly polarized modes (LP$_{nm}$ modes) are the following:

Table 9.1 Composition of the lower-order linearly polarized modes [10]

LP-mode designation	Traditional-mode designation and number of modes	Number of degenerate modes
LP_{01}	$HE_{11} \times 2$	2
LP_{11}	TE_{01}, TM_{01}, $HE_{21} \times 2$	4
LP_{21}	$EH_{11} \times 2$, $HE_{31} \times 2$	4
LP_{02}	$HE_{12} \times 2$	2
LP_{31}	$EH_{21} \times 2$, $HE_{41} \times 2$	4
LP_{12}	TE_{02}, TM_{02}, $HE_{22} \times 2$	4
LP_{41}	$EH_{31} \times 2$, $HE_{51} \times 2$	4
LP_{22}	$EH_{12} \times 2$, $HE_{32} \times 2$	4
LP_{03}	$HE_{13} \times 2$	2
LP_{51}	$EH_{41} \times 2$, $HE_{61} \times 2$	4

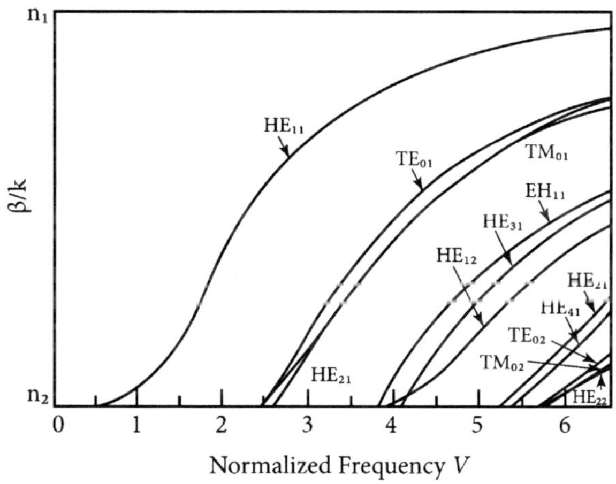

Figure 9.2. Plots of the propagation constant (in terms of β/k) as a function of V for a few of the lowest order modes. $V = (2\pi a/\lambda)(n_1^2 - n_2^2)^{1/2}$ [10]

For $r \leq a$

$$E_y^{(1)} = A_n \left[\frac{J_n(p_1 r)}{J_n(p_1 a)}\right] \cos n\theta \, e^{j\omega t - j\beta z}, \tag{9.1a}$$

$$H_x^{(1)} = A_n \left(\frac{Z_0}{n_1}\right) E_y^{(1)}, \tag{9.1b}$$

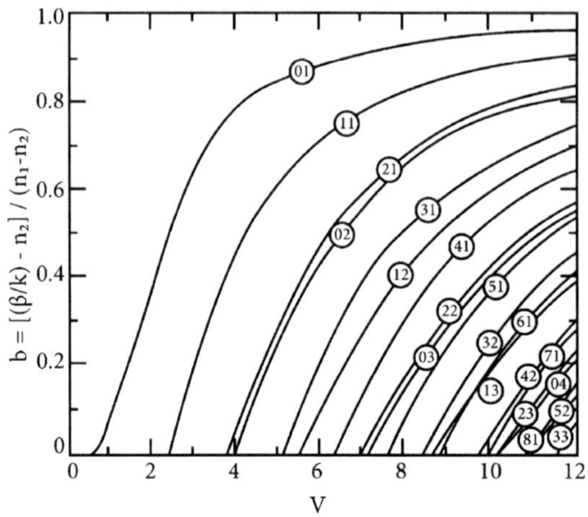

Figure 9.3. Plots of the propagation constant b as a function for various LP$_{jm}$ modes [12]

$$E_z^{(1)} = \frac{jZ_0}{k_0}\frac{1}{n_1^2}\frac{\partial H_x^{(1)}}{\partial y}, \tag{9.1c}$$

$$H_z^{(1)} = \frac{j}{k_0 Z_0}\frac{\partial E_y^{(1)}}{\partial x}, \tag{9.1d}$$

and for $r \geq a$

$$E_y^{(2)} = A_n \left[\frac{K_n(q_2 r)}{K_n(q_2 a)}\right] \cos n\theta \, e^{j\omega t - j\beta z}, \tag{9.2a}$$

$$H_x^{(2)} = A_n \left(\frac{Z_0}{n_2}\right) E_y^{(2)}, \tag{9.2b}$$

$$E_z^{(2)} = \frac{jZ_0}{k_0}\frac{1}{n_2^2}\frac{\partial H_x^{(2)}}{\partial y}, \tag{9.2c}$$

$$H_z^{(2)} = \frac{j}{k_0 Z_0}\frac{\partial E_y^{(2)}}{\partial x}, \tag{9.2d}$$

with

$$Z_0 = \sqrt{\mu_0/\epsilon_0} \quad k_0 = \omega\sqrt{\mu_0\epsilon_0} \quad n_1 = \sqrt{\epsilon_1/\epsilon_0} \quad n_2 = \sqrt{\epsilon_2/\epsilon_0}$$

$$p_1^2 = k_1^2 - \beta^2 \quad q_2^2 = \beta^2 - k_2^2$$

$$k_1^2 = \omega^2\mu_0\epsilon_1 \quad k_2^2 = \omega^2\mu_0\epsilon_2.$$

Matching the tangential electric and magnetic field components at $r = a$ in cylindrical coordinates yields the dispersion relation for the LP_{nm} modes,

$$\frac{p_1 J_{n-1}(p_1 a)}{J_n(p_1 a)} = -\frac{q_2 K_{n-1}(q_2 a)}{K_n(q_2 a)}. \tag{9.3}$$

From this relation, one can show that the dominant mode is the LP_{01} mode (corresponding to the traditional HE_{11} mode) with no cutoff frequency. All other higher order modes possess cutoff frequencies.

From (9.1a), the transverse electric field for the dominant LP_{01} (or HE_{11}) mode is

$$E_y^{(1)}(LP_{01}) = A_0 \left[\frac{J_0(p_1 r)}{J_0(p_1 a)}\right] e^{j\omega t - j\beta z}, \tag{9.4}$$

$$E_y^{(2)}(LP_{01}) = A_0 \left[\frac{K_0(q_2 r)}{K_0(q_2 a)}\right] e^{j\omega t - j\beta z}. \tag{9.5}$$

It is of interest to investigate the power flow distribution in the core and cladding regions of a step-index weakly guiding optical fiber as a function of V. Using the simplified approach given above, we have the power in the core region, P_{core}, and the power in the cladding region, P_{cladding}, as follows

$$\begin{aligned} P_{\text{core}} &= \frac{1}{2}\int_0^a \int_0^{2\pi} E_y^{(1)} H_x^{(1)*} r\, dr\, d\theta, \\ &= \frac{A_n^2}{2}\frac{Z_0}{n_1}\int_0^a \int_0^{2\pi} \left[\frac{J_n(p_1 r)}{J_n(p_1 a)}\right]^2 \cos^2 n\theta\, r\, dr\, d\theta, \end{aligned} \tag{9.6}$$

$$\begin{aligned} P_{\text{cladding}} &= \frac{1}{2}\int_a^\infty \int_0^{2\pi} E_y^{(2)} H_x^{(2)*} r\, dr\, d\theta, \\ &= \frac{A_n^2}{2}\frac{Z_0}{n_2}\int_a^\infty \int_0^{2\pi} \left[\frac{K_n(q_2 r)}{K_n(q_2 a)}\right]^2 \cos^2 n\theta\, r\, dr\, d\theta. \end{aligned} \tag{9.7}$$

The total power is

$$P = P_{\text{core}} + P_{\text{cladding}}. \tag{9.8}$$

The calculated results are shown in Fig. 9.4. It is seen that for the range $2 < V < 2.4$, (the single mode region) more than 75% of the modal power for the dominant LP_{01} or HE_{11} mode resides within the core region. This means that within this V range this mode is very well guided by the fiber. (Note that the cutoff for the next higher order mode is at $V = 2.405$).

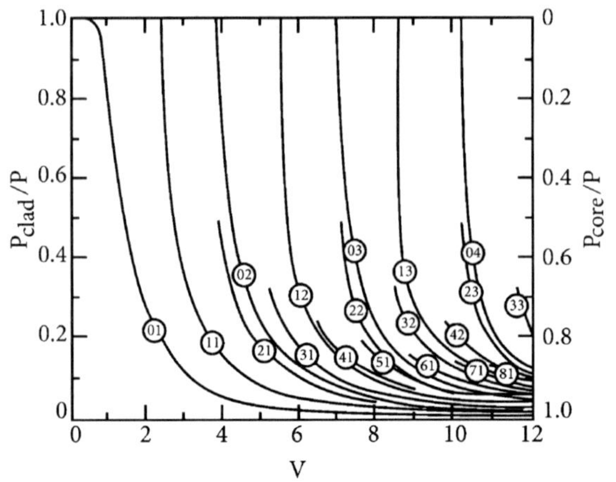

Figure 9.4. Fractional power flow in the cladding of a step-index optical fiber as a function of V. When $\nu \neq 1$, the curve numbers νm designate the $\mathrm{HE}_{\nu+1,m}$ modes and $\mathrm{HE}_{\nu-1,m}$ modes. For $\nu = 1$, the curve numbers νm give the HE_{2m}, TE_{0m}, and TM_{0m} modes [12]

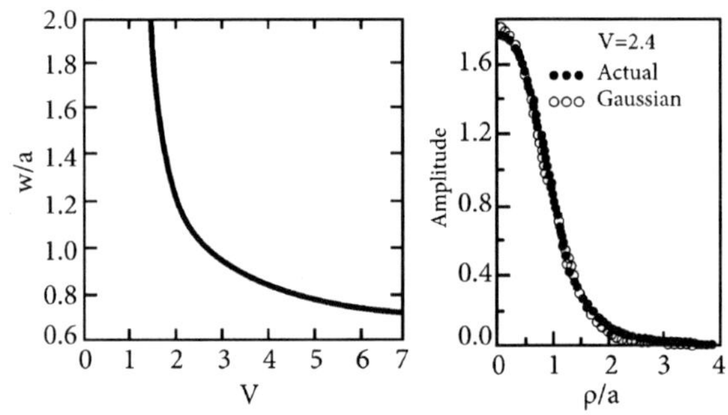

Figure 9.5. Variation of mode-width parameter, w, with V obtained by fitting the fundamental fiber mode to a Gaussian distribution. Traces on the right show the quality of fit for $V = 2.4$ [13]

Let us now compare the dominant mode transverse electric field distribution given by (9.4) and (9.5) with a Gaussian distribution given by

$$E_y = A\, e^{-r^2/w^2}\, e^{j\omega t - j\beta z}, \tag{9.9}$$

where

$$\frac{w}{a} = 0.65 + 1.619\, V^{-3/2} + 2.879\, V^{-6} \qquad (2 < V < 2.4). \tag{9.10}$$

Figure 9.5 provides the comparison [13]. Very close agreement can be observed. Therefore, the use of a Gaussian distribution for the dominant transverse electric field variation is justified for $2 < V < 2.4$.

The case for a weakly guiding ($n_1/n_2 \simeq 1$) and linearly polarized modes on an elliptical dielectric fiber has been treated by Shaw et al. [14]. Since Shaw et al.'s work, as well as Gloge's work [12] and Yeh et al.'s work [15], is based on solutions to the scalar wave equation, the conditions listed in Sect. 2.8 must be satisfied. On the other hand, the weakly guiding cases considered by Yeh [16] and Eyges [17] for modes on an elliptical dielectric fiber are not based on the scalar wave equation. Therefore, no scalar restrictions must be placed on their solutions.

Having made the above observations, it is important to note that for a wide operational range of interest for an optical fiber, the scalar wave representation of the modal fields is very adequate. Its simplicity enables us to maneuver the governing scalar equation to gain much insight in the propagation behavior of the guided wave, as shown in the following sections.

9.2 Dispersion

For a single mode fiber, the contributions to mode dispersion come from two sources: material dispersion and waveguide dispersion. Strictly speaking, one must deal with the dispersion effects simultaneously; that is, the frequency dependent index of refraction is inserted into the dispersion relation for a given mode and then the ω-β diagram is obtained for that mode. The dispersion coefficient D (3.40) is obtained by taking the second derivative of β with respect to ω from the ω-β diagram. Because of the slow variation of the index of refraction as well as the dispersion coefficient with respect to ω, it is possible to deal with material dispersion and waveguide dispersion independently.

9.2.1 Material Dispersion [18]

Material dispersion is caused by the variation of the index of refraction as a function of frequency [18]. To calculate this material dispersion, let us consider the propagation of a plane wave in an infinite dielectric medium. The propagation constant, β_{M}, of that plane wave is

$$\beta_{\mathrm{M}} = \frac{\omega}{c} n(\omega), \tag{9.11}$$

where the subscript M represents the material, ω is the frequency, and $n(\omega)$ is the frequency-dependent index of refraction of the dielectric medium. A typical variation of refractive index n with wavelength for fused silica is shown in Fig. 9.6. The first derivative of β_{M} is

$$\beta_{1\mathrm{M}} = \frac{d\beta_{\mathrm{M}}}{d\omega} = \frac{n}{c} + \frac{\omega}{c}\frac{dn}{d\omega} \tag{9.12}$$

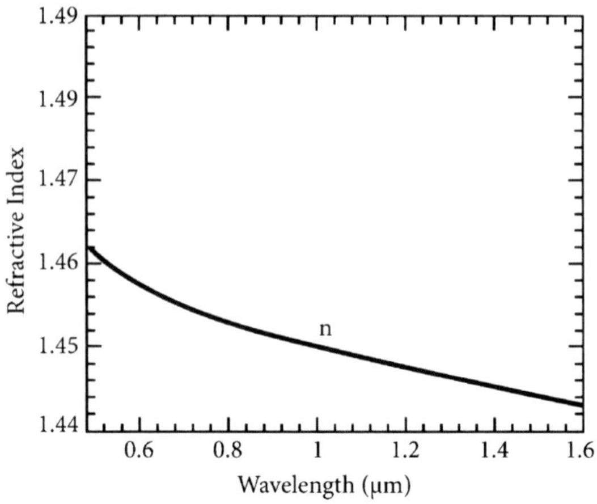

Figure 9.6. Variation of refractive index n with wavelength for fused silica [5]

and the second derivative is

$$\beta_{2\mathrm{M}} = \frac{d^2\beta_{\mathrm{M}}}{d\omega^2} = \frac{2}{c}\frac{dn}{d\omega} + \frac{\omega}{c}\frac{d^2n}{d\omega^2}. \tag{9.13}$$

The dispersion coefficient due to material dispersion, D_{M}, is defined as follows

$$\begin{aligned} D_{\mathrm{M}} &= -\frac{2\pi c}{\lambda^2}\beta_{2\mathrm{M}} = -\frac{\omega^2}{2\pi c}\beta_{2\mathrm{M}}, \\ &= -\frac{\omega^2}{\pi c^2}\left(2\frac{dn}{d\omega} + \omega\frac{d^2n}{d\omega^2}\right), \\ &= -\frac{2\pi}{\lambda^2}\left(2\frac{dn}{d\omega} + \omega\frac{d^2n}{d\omega^2}\right). \end{aligned} \tag{9.14}$$

9.2.2 Waveguide Dispersion [4]

Waveguide dispersion is caused by the curvature of the ω-β curve representing the propagation constant of a given propagating mode in a medium without material dispersion. The first and second derivatives of β_{W} with respect to ω are the following:

$$\beta_{1\mathrm{W}} = \frac{d\beta_{\mathrm{W}}}{d\omega}, \tag{9.15}$$

$$\beta_{2\mathrm{W}} = \frac{d^2\beta_{\mathrm{W}}}{d\omega^2}, \tag{9.16}$$

where the subscript W represents waveguide. These derivatives may be obtained numerically from the above mentioned ω-β diagram for the mode of interest. The dispersion coefficient due to waveguide dispersion, D_W, is defined as follows:

$$D_W = -\frac{2\pi}{\lambda^2}\beta_{2W} = -\frac{\omega^2}{2\pi c}\beta_{2W}. \tag{9.17}$$

9.2.3 Total Dispersion

The total dispersion D represents the sum of dispersions caused by the material dispersion and by the waveguide dispersion for that mode:

$$D = D_M + D_W. \tag{9.18}$$

Figure 9.7 shows the dispersion curves for a typical conventional step-index single mode fiber. The contributions of D_M and D_W to the total dispersion D can be seen clearly. We may identify two distinctive dispersion regimes: (a) the normal-dispersion regime where $\beta_2 > 0$ or $D < 0$ and (b) the anomalous-dispersion regime where $\beta_2 < 0$ or $D > 0$. There exists also a region where $\beta_2 \simeq 0$, called the zero-dispersion region. We note that waveguide dispersion is only important near the zero-dispersion wavelength.

The intrinsic material dispersion cannot be easily altered [18]. The workable remedy to minimize mode dispersion is to adjust the waveguide dispersion through

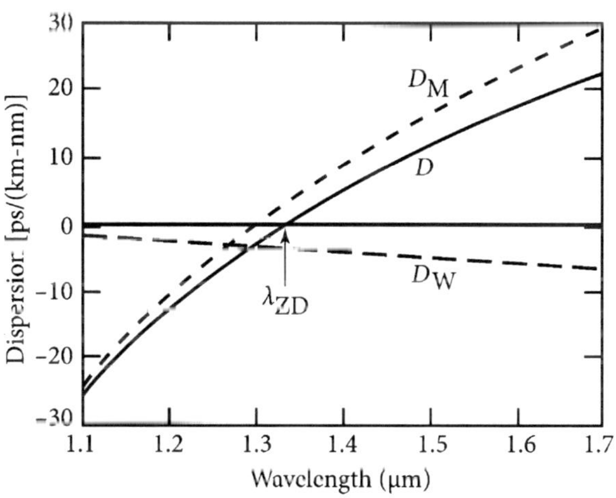

Figure 9.7. Total dispersion D and relative contributions of material dispersion D_M and waveguide dispersion D_W for a conventional single-mode fiber. The zero-dispersion wavelength shifts to a higher value because of the waveguide contribution [5]

index profile design [4]. Because of its ultra-low-loss property, silica has been chosen as the preferred material for single mode optical fiber. For a single-mode optical fiber, its attenuation is at a minimum near 1,550 nm while the dispersion for a simple step-index fiber is smallest at 1,300 nm. By adjusting the index profile, one may shift the wavelength at which the dispersion is a minimum or null to the wavelength where the attenuation is at a minimum [4]. Furthermore, when using WDM (wavelength division multiplexed) format, it is desirable to operate the single-mode fiber in its dispersion flattened region. Hence the need for dispersion-flattened fibers is apparent. Figure 9.8 shows a few representative index profiles that can provide the desired dispersion characteristics discussed earlier [10]. A sketch of the resultant

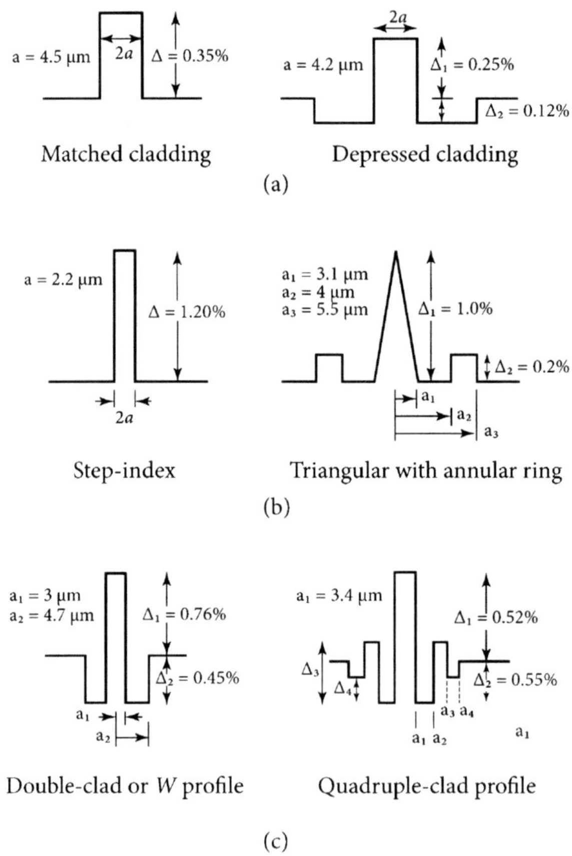

Figure 9.8. Representative index profiles for (**a**) 1,300-nm-optimized, (**b**) dispersion-shifted, and (**c**) dispersion-flattened single-mode fibers. Typical mode diameter is 9.5 μm [10]

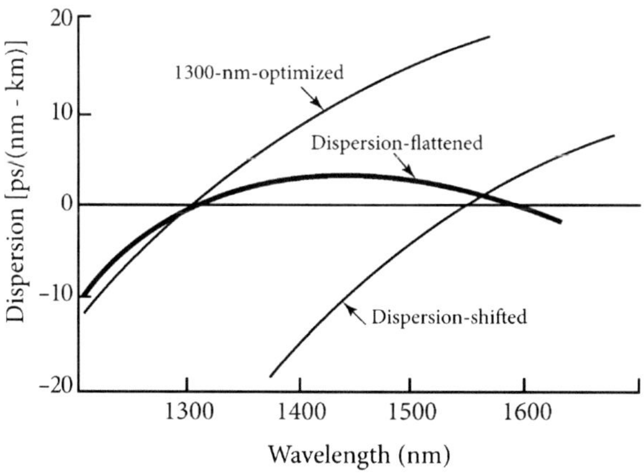

Figure 9.9. Typical dispersion characteristics for (**a**) 1,300-nm-optimized, (**b**) dispersion-shifted, and (**c**) dispersion-flattened single-mode fibers [10]

Table 9.2 Characteristics of several commercial fibers

Fiber type and trade name	A_{eff} (μm^2)	λ_{ZD} (nm)	D (C band) (ps km^{-1} nm^{-1})	Slope S (ps km^{-2} nm^{-2})
Corning SMF-28	80	1,302–1,322	16 to 19	0.090
Lucent AllWave	80	1,300–1,322	17 to 20	0.088
Alcatel ColorLock	80	1,300–1,320	16 to 19	0.090
Corning Vascade	101	1,300–1,310	18 to 20	0.060
Lucent TrueWave-RS	50	1,470–1,490	2.6 to 6	0.050
Corning LEAF	72	1,490–1,500	2 to 6	0.060
Lucent TrueWave-XL	72	1,570–1,580	−1.4 to −4.6	0.112
Alcatel TeraLight	65	1,440–1,450	5.5 to 10	0.058

dispersion behavior is shown in Fig. 9.9 [10]. The distinctive dispersion behavior of these types of fibers is readily seen. Dispersion nulls are found for all these fibers. The flatness of the dispersion-flattened fiber for a rather wide wavelength can be observed.

The dispersion characteristics of these fibers with special radially varying index profiles can be efficiently calculated using the analytic technique discussed in detail in Sect. 5.3.

Shown in Table 9.2 are several commercially available fibers with various desired characteristics. Note that

$$\beta_2 = -D\frac{\lambda^2}{2\pi c}, \tag{9.19}$$

$$\beta_3 = \left[S - \left(\frac{4\pi c}{\lambda^3}\right)\beta_2\right]\left(\frac{\lambda^2}{2\pi c}\right)^2, \tag{9.20}$$

where D and S have been given in Table 9.2. These dispersion values, β_2 and β_3, can be used to calculate the pulse broadening behavior of a data stream carried in this fiber. Detailed discussion of this kind of calculation is given in Sect. 9.4.

9.3 Attenuation

Another effect that would degrade the pulse shape of the data stream is attenuation. When a propagating single mode is attenuated due to the presence of fiber loss, the amplitudes of carrier envelopes are diminished, causing a decrease of signal-to-noise ratio. According to our earlier discussion, in Sect. 3.5 on the attenuation constant of a single mode, all field components are multiplied by the factor $e^{-\alpha z}$, where α is the attenuation constant of that mode. This attenuation constant can be calculated according to the perturbation technique as shown in Sect. 3.5. This attenuation constant is directly proportional to $\tan\delta$, the loss tangent of the material. For the optical fiber case, the material is silica. Since most of the guide power is within the core region of the optical fiber and the transverse field of the dominant single mode is mostly linearly polarized, the attenuation constant becomes (5.64). This is equivalent to the case of a plane wave propagating in an infinite medium with loss tangent $\tan\delta$ and index n, that is, the attenuation is

$$\alpha = 4.343\frac{2\pi}{\lambda}n \tan\delta \qquad \left(\text{dB m}^{-1}\right), \tag{9.21}$$

where λ is the free-space wavelength, n is the index of refraction of the core medium, and $\tan\delta$ is the loss tangent of the core region. Typical attenuation ranges for production-run single-mode fibers are shown in Fig. 9.10.

It is seen that fiber with attenuation as low as $0.2\,\text{dB m}^{-1}$ is obtainable. The loss and dispersion of Lucent's AllWave fiber is shown in Fig. 9.11.

9.4 The Propagation Equation

According to the analysis given in Chap. 5 for a step-index circular dielectric fiber, the dominant mode is a hybrid HE_{11} mode, or the LP_{01} mode, that possesses no cutoff frequency. For single-mode operation, the normalized frequency, V, must be less than 2.405. Here,

$$V = k_0 a(n_1^2 - n_2^2)^{1/2}, \tag{9.22}$$

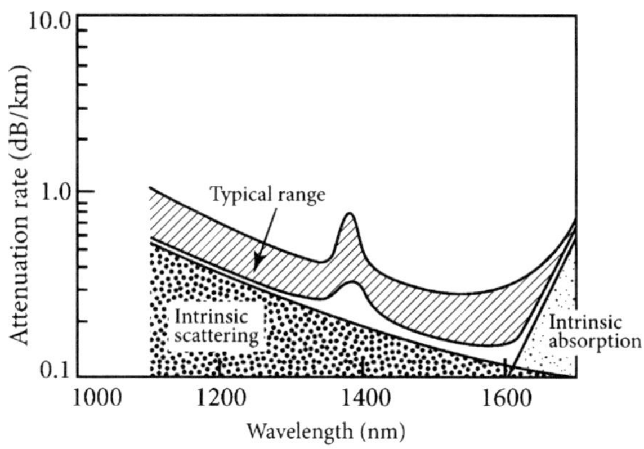

Figure 9.10. Typical spectral attenuation range for production-run single-mode fibers [19]

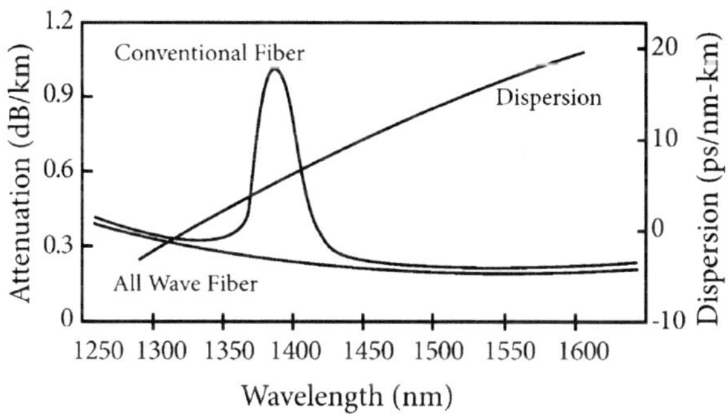

Figure 9.11. Loss and dispersion of the AllWave fiber. Loss of a conventional fiber is also shown for comparison [5]

where $k_0 = \omega\sqrt{\mu_0\epsilon_0}$, the free-space wave-number, a is the radius of the fiber core, n_1 and n_2 are, respectively, the indices of refraction of the core and cladding region. Detailed expressions for all field components of the HE_{11} modes have been given in (5.7)–(5.22). In general, the transverse electric field in the frequency domain under the weakly guiding approximation [12] can be written as follows

$$\mathbf{E}_t(r,\theta,\omega) = A(0,\omega)\mathbf{F}_\mathrm{t}(r,\theta)\, e^{-j\beta z} e^{j\omega t}, \tag{9.23}$$

where $\mathbf{F}_\mathrm{t}(r,\theta)$ is the transverse field distribution for the HE_{11} mode, which is a slowly varying function of ω; β is the propagation constant; and A is the initial amplitude of the field at $z = 0$. When referring to the LP_{01} mode,

$$\mathbf{F}_\mathrm{t}(r,\theta) = \mathbf{e}_y \exp(-r^2/w^2),$$

with

$$w/a = 0.65 + 1.619V^{-3/2} + 2.879V^{-6}.$$

The time-domain expression for the transverse electric field is

$$\mathbf{E}_{\mathrm{t}}^{(t)}(r,\theta,t) = \mathbf{F}_{\mathrm{t}}(r,\theta)\frac{1}{2\pi}\int_{-\infty}^{\infty} A^{(\omega)}(z,\omega)\, e^{j\omega t}\, d\omega, \tag{9.24}$$

with

$$A^{(\omega)}(z,\omega) = A^{(\omega)}(0,\omega)\, e^{-j\beta z}. \tag{9.25}$$

For the case of interest where the pulse spectral width,

$$\Delta\omega = \omega - \omega_0, \tag{9.26}$$

of the input signal is much smaller than ω_0, the carrier frequency $\mathbf{F}_t(r,\theta)$ may be considered to be independent of ω. From (9.24) the time dependent field amplitude is

$$A^{(t)}(z,t) = \frac{1}{2\pi}\int_{-\infty}^{\infty} A^{(\omega)}(z,\omega)\, e^{j\omega t} d\omega = \frac{1}{2\pi}\int_{-\infty}^{\infty} A^{(\omega)}(0,\omega)\, e^{-j\beta z} e^{j\omega t}\, d\omega. \tag{9.27}$$

Using the Taylor series expansion for the propagation constant β with the condition that $\Delta\omega\ (= \omega - \omega_0) \ll \omega_0$, we have

$$\beta(\omega) \approx \beta_0 + \beta_1(\Delta\omega) + \frac{\beta_2(\Delta\omega)^2}{2!} + \frac{\beta_3(\Delta\omega)^3}{3!} + \cdots, \tag{9.28}$$

with

$$\beta_m = \left.\frac{d^m\beta}{d\omega^m}\right|_{\omega=\omega_0} \qquad m = 1, 2, 3, \ldots. \tag{9.29}$$

Here,

$$\beta_1 = \frac{1}{v_{\mathrm{g}}} \tag{9.30}$$

where v_{g} is the group velocity of the mode; β_2 is related to the dispersion parameter, D, as follows:

$$D = -\frac{2\pi c}{\lambda^2}\beta_2 \qquad (\text{ps km}^{-1}\,\text{nm}^{-1}), \tag{9.31}$$

where λ is the free-space wavelength and c is the speed of light in a vacuum; β_3 is related to the differential-dispersion parameter, S, as follows:

$$S = \frac{dD}{d\lambda} = \left(\frac{2\pi c}{\lambda^2}\right)^2 \beta_3 + \frac{4\pi c}{\lambda^3}\beta_2. \tag{9.32}$$

As discussed in Sect. 3.6, β_2 is also recognized as the pulse broadening factor caused by the material (D_M) and the waveguide (D_W) dispersion of the fiber on the propagating mode. Thus the presence of β_2 and β_3 will alter the pulse shape of the mode as it propagates along the fiber.

Substituting (9.28) in (9.27) yields

$$A^{(t)}(z,t) = \frac{1}{2\pi}\int_{-\infty}^{\infty} A^{(\omega)}(0,\omega)\, \exp\Big[-j\beta_0 z - j\beta_1(\Delta\omega)z - j\beta_2\frac{(\Delta\omega)^2}{2!}z - j\beta_3\frac{(\Delta\omega)^3}{3!}z - \cdots + j\omega t\Big]\, d\omega. \tag{9.33}$$

Changing the variable from ω to $\Delta\omega$, with $\Delta\omega = \omega - \omega_0$, gives

$$\begin{aligned} A^{(t)}(z,t) &= \frac{1}{2\pi}\int_{-\infty}^{\infty} P(0,\Delta\omega)\, \exp\left[-j\beta_0 z - j\beta_1(\Delta\omega)z - j\beta_2\frac{(\Delta\omega)^2}{2!}z \right. \\ &\quad \left. -j\beta_3\frac{(\Delta\omega)^3}{3!}z - \cdots + j(\Delta\omega)t + j\omega_0 t\right] d(\Delta\omega), \\ &= e^{-j\beta_0 z}\, e^{j\omega_0 t}\left\{\frac{1}{2\pi}\int_{-\infty}^{\infty} P(0,\Delta\omega)\, \exp\left[-j\beta_1(\Delta\omega)z - j\beta_2\frac{(\Delta\omega)^2}{2!}z \right.\right. \\ &\quad \left.\left. -j\beta_3\frac{(\Delta\omega)^3}{3!}z - \cdots + j(\Delta\omega)t\right] d(\Delta\omega)\right\}. \end{aligned} \tag{9.34}$$

It has been assumed that $P(0,\Delta\omega) = A^{(\omega)}(0,\omega)$. Let

$$\begin{aligned} P^{(t)}(z,t) &= \frac{1}{2\pi}\int_{-\infty}^{\infty} P(0,\Delta\omega)\, \exp\left[-j\beta_1(\Delta\omega)z - j\beta_2\frac{(\Delta\omega)^2}{2!}z \right. \\ &\quad \left. -j\beta_3\frac{(\Delta\omega)^3}{3!}z\right] e^{j(\Delta\omega)t}\, d(\Delta\omega), \end{aligned} \tag{9.35}$$

then

$$A^{(t)}(z,t) = P^{(t)}(z,t)\, e^{-j(\beta_0 z - \omega_0 t)}. \tag{9.36}$$

The function $P^{(t)}(z,t)$ is identified as a slowly varying amplitude of the pulse envelope. It is the variation of the information-carrying pulse envelope as it propagates down the fiber that is of interest.

Let us now derive an equation, called the propagation equation, for this envelope function, $P^{(t)}(z,t)$. Taking the partial derivative of (9.35) with respect to z yields

$$\frac{\partial P^{(t)}}{\partial z} = \frac{1}{2\pi}\int_{-\infty}^{\infty} P(0,\Delta\omega)\left[-j\beta_1(\Delta\omega) - j\beta_2\frac{(\Delta\omega)^2}{2!}\right.$$

$$\left.-j\beta_3\frac{(\Delta\omega)^3}{3!}\right] e^{-j\chi z} e^{j(\Delta\omega)t}\, d(\Delta\omega), \tag{9.37}$$

where

$$\chi = \beta_1(\Delta\omega) + \beta_2\frac{(\Delta\omega)^2}{2!} + \beta_3\frac{(\Delta\omega)^3}{3!} + \cdots. \tag{9.38}$$

Realizing that

$$\frac{\partial^m P^{(t)}}{\partial t^m} = \frac{1}{2\pi}\int_{-\infty}^{\infty} P(0,\Delta\omega)(j\Delta\omega)^m\, e^{-j\chi z} e^{j(\Delta\omega)t}\, d(\Delta\omega) \tag{9.39}$$

$$(m = 1, 2, 3, \ldots),$$

we may write (9.37) as follows

$$\frac{\partial P^{(t)}}{\partial z} + \beta_1\frac{\partial P^{(t)}}{\partial t} - \frac{j\beta_2}{2}\frac{\partial^2 P^{(t)}}{\partial t^2} - \frac{\beta_3}{6}\frac{\partial^3 P^{(t)}}{\partial t^3} = 0. \tag{9.40}$$

This is the basic propagation equation that we have been seeking. Solution of this equation with the initial condition at $z = 0$, that is, $P^{(t)}(0,t)$, will provide the information on the evolution of this pulse inside a single-mode fiber. It is of interest to note that if $\beta_2 = \beta_3 = 0$, that is, the dispersionless case, (9.40) becomes

$$\frac{\partial P^{(t)}}{\partial z} + \beta_1\frac{\partial P^{(t)}}{\partial t} = 0, \tag{9.41}$$

which yields the solution for the undisturbed pulse propagating in the fiber with a propagation constant β and a group velocity $v_g = 1/\beta_1$. In other words,

$$P^{(t)}(z,t) = P^{(t)}(0, t - \beta_1 z). \tag{9.42}$$

Using the new coordinates,

$$t_p = t - \beta_1 z, \tag{9.43a}$$

$$z_p = z, \tag{9.43b}$$

we may transform (9.40) to a reference frame moving with the pulse:

$$\frac{\partial P^{(t)}}{\partial z_p} - \frac{j\beta_2}{2}\frac{\partial^2 P^{(t)}}{\partial t_p^2} - \frac{\beta_3}{6}\frac{\partial^3 P^{(t)}}{\partial t_p^3} = 0. \tag{9.44}$$

Without ambiguity, we may write (9.44) as

$$\frac{\partial P^{(t)}}{\partial z} - \frac{j\beta_2}{2}\frac{\partial^2 P^{(t)}}{\partial t^2} - \frac{\beta_3}{6}\frac{\partial^3 P^{(t)}}{\partial t^3} = 0 \tag{9.45}$$

as long as it is understood that all subsequent discussion will refer to cases where the reference frame moves with a velocity v_{g} along the fiber axis.

Assuming that the higher order dispersion β_3 is zero, (9.45) becomes

$$\frac{\partial P^{(t)}}{\partial z} = \frac{j\beta_2}{2}\frac{\partial^2 P^{(t)}}{\partial t^2}. \tag{9.46}$$

This equation is readily solved by using the Fourier-transform method. Taking the Fourier-transform of (9.46) with respect to t yields

$$\frac{\partial P}{\partial z} = -\frac{j\beta_2}{2}\omega^2 P, \tag{9.47}$$

where

$$P^{(t)}(z,t) = \frac{1}{2\pi}\int_{-\infty}^{\infty} P(z,\omega)\, e^{j\omega t}\, d\omega. \tag{9.48}$$

The solution of (9.47) is

$$P(z,\omega) = P(0,\omega)\,\exp(-\frac{1}{2}j\beta_2\omega^2 z), \tag{9.49}$$

where $P(0,\omega)$ is the inverse transform of the incident pulse $P^{(t)}(0,t)$ at $z = 0$:

$$P(0,\omega) = \int_{-\infty}^{\infty} P^{(t)}(0,t)\, e^{-j\omega t}\, dt. \tag{9.50}$$

So, given $P^{(t)}(0,t)$ at $z=0$, one may obtain $P^{(t)}(z,t)$ as follows:

$$P^{(t)}(z,t) = \frac{1}{2\pi}\int_{-\infty}^{\infty} P(0,\omega)\,\exp\left(-\frac{1}{2}j\beta_2\omega^2 z + j\omega t\right)\,d\omega,$$

$$= \frac{1}{2\pi}\int_{-\infty}^{\infty}\left[\int_{-\infty}^{\infty} P^{(t)}(0,t)\,e^{-j\omega t}\,dt\right]\exp\left(-\frac{1}{2}j\beta_2\omega^2 z + j\omega t\right)\,d\omega. \tag{9.51}$$

9.5 Selected Solutions to the Propagation Equation

Let us now consider the solutions for several input waveforms of interest.

(a) Gaussian Pulse

The input field is a Gaussian pulse.

$$P^{(t)}(0,t) = \exp(-t^2/2\tau_0^2), \tag{9.52}$$

where τ_0 is the pulse width between the two half-power points of the field. Another way to measure the pulse width is called the full width at half maximum (FWHM), which is defined as follows:

$$\tau_{\text{FWHM}} = 2(\ln 2)^{1/2}\tau_0 = 1.665\tau_0. \tag{9.53}$$

Substituting (9.52) in (9.51) and carrying out the integration yields

$$P^{(t)}(z,t) = \frac{\tau_0}{(\tau_0^2 + j\beta_2 z)^{1/2}}\exp\left[-\frac{t^2}{2(\tau_0^2 + j\beta_2 z)}\right]. \tag{9.54}$$

Now, the pulse width as a function of z is

$$\tau_1(z) = \tau_0\left[1 + (z/L_{\text{D}})^2\right]^{1/2}, \tag{9.55}$$

with $L_{\text{D}} = \tau_0^2/\left|\beta_2\right|$. This shows that regardless of the sign of β_2, a pulse is broadened as it propagates down the fiber. The rate of broadening depends on L_{D} and the shorter the pulse width, the faster it broadens. It is noted that (9.54) and (9.55) are identical to (3.37) and (3.39), respectively.

We also note from (9.54) that although the input pulse is an unchirped pulse, the output pulse becomes chirped. (Chirp means a pulse with nonzero phase modulation.)

(b) Chirped Gaussian Pulse

The input field is a chirped Gaussian pulse

$$P^{(t)}(0,t) = \exp\left\{-\left[(1-jC)/2\right]t^2/\tau_0^2\right\}, \tag{9.56}$$

where C is a chirp parameter. Substituting (9.56) in (9.51) and carrying out the integration, we have

$$P^{(t)}(z,t) = \left\{\frac{\tau_0}{\left[\tau_0^2 + j\beta_2 z(1-jC)\right]^{1/2}}\right\}\exp\left\{-\frac{(1-jC)t^2}{2\left[\tau_0^2 + j\beta_2 z(1-jC)\right]}\right\}, \tag{9.57}$$

with a pulse width of

$$\frac{\tau_1}{\tau_0} = \left[\left(1 + \frac{C\beta_2 z}{\tau_0^2}\right)^2 + \left(\frac{\beta_2 z}{\tau_0^2}\right)^2\right]. \tag{9.58}$$

It is noted that the Gaussian pulse shape is still maintained even with the presence of chirp. Figure 9.12 is a plot of τ_1/τ_0 vs. z/L_D for various values of C. Here, $L_D = \tau_0^2/|\beta_2|$ is called the dispersion length, where τ_0 represents the half-width at $1/e$ intensity point for the initial pulse. The width of an unchirped Gaussian pulse is increased by a factor of $\sqrt{2}$ at $z = L_D$. It shows that a narrowing of the

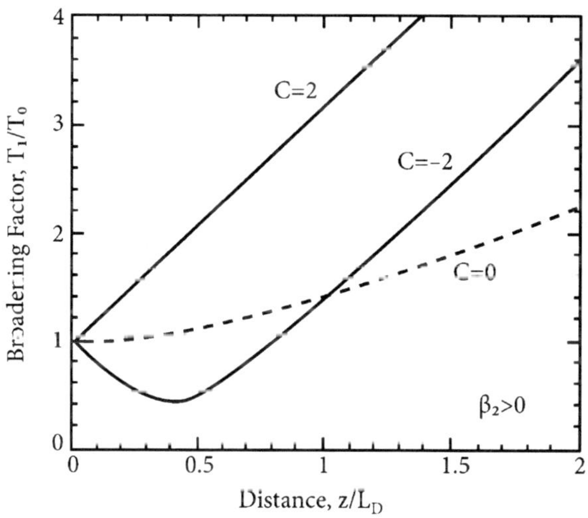

Figure 9.12. Variation of broadening factor with propagated distance for a chirped Gaussian input pulse. Dashed curve corresponds to the case of an unchirped Gaussian pulse. For $\beta_2 < 0$, the same curves are obtained if the sign of the chirp parameter C is reversed [5]

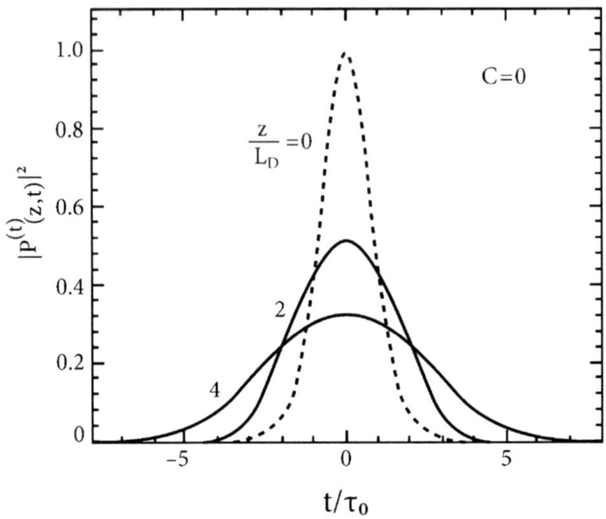

Figure 9.13. Pulse shapes at $z = 2L_\mathrm{D}$ and $z = 4L_\mathrm{D}$ of a pulse whose shape at $z = 0$ (*dashed curve*) is described by a "sech" profile [20]

pulse occurs initially when the chirp is negative. Then the effect of the dispersion-induced chirp overcomes that due to the initial negative chirp, leading to the eventual pulse broadening.

(c) Hyperbolic-Secant Pulse

The input is a hyperbolic-secant pulse

$$P^{(t)}(0,t) = \mathrm{sech}\left(\frac{t}{\tau_0}\right)\exp\left[\frac{jCt^2}{2\tau_0^2}\right], \tag{9.59}$$

where C is a chirp parameter. This pulse shape is of special interest because it occurs naturally in the context of optical solitons [20]. It also resembles the pulse emitted from mode-locked lasers. Substituting (9.59) in (9.51), we can obtain $P^{(t)}(z,t)$ numerically. Some results are shown in Fig. 9.13. The pulse broadening features for this "sech" pulse are very similar to those displayed by the "Gaussian" pulse discussed earlier.

9.6 Wavelength Division Multiplexed Beams (WDM)

So far we have been discussing the transmission of data pulses along a single wavelength beam in a single mode fiber. Because of the large bandwidth of a typical single mode fiber, multiple beams of different wavelengths with separate data pulses may also propagate simultaneously in this fiber, thereby increasing enormously its data carrying capacity (or the bandwidth). The complexity comes from multiplexing and demultiplexing only.

According to the theory on mode propagation on fibers with linear index medium discussed earlier, the superposition theorem applies: each beam may be treated separately and there is no interaction among the copropagating beams. It is of interest to note that, in general, pulses on different wavelength beams will propagate at different group velocities. Hence the pulses on these different wavelength beams will walk-off from one another. The walk-off length is defined as follows:

$$L_{\mathrm{w}} = \frac{\tau_0}{|d_{12}|},$$

$$d_{12} = \beta_1(\lambda_1) - \beta_1(\lambda_2) = \frac{1}{v_{\mathrm{g}}(\lambda_1)} - \frac{1}{v_{\mathrm{g}}(\lambda_2)},$$

where β_1 is the propagation constant, v_{g} is the group velocity of the pulse, and τ_0 is the pulse width. This walk-off length means the distance for which the pulse on one beam with λ_1 will totally overtake the pulse on the other beam with λ_2.

Because of the small size of optical fibers (typical core diameter of a few tens of microns), high light intensities in the fiber core can frequently occur. At high light intensities, the nonlinear property of the index of refraction for silica can no longer be considered negligible [21]. This means that the copropagating WDM beams may produce undesirable interactions with each other through the nonlinear index medium, unless the intensities of the beams can be kept below the nonlinear index threshold, or if the pulses on each beam can be kept away from those on any other beams. Other considerations are insertion loss, channel width, and cross talk. Insertion loss refers to the coupling losses and any intrinsic losses within the multiplexing elements. Channel-width must be chosen to insure that there is no interchannel interference due to source instability. Cross talk refers to the amount of signal coupling from one channel to another caused by imperfect multiplexing. It is known that photodetectors are usually sensitive over a broad range of wavelengths. To achieve good channel separations in demultiplexing, very good optical filters are required. These are important WDM system problems that must be addressed. Typical setup of a unidirectional WDM system and that of a bidirectional WDM system are shown in Fig. 9.14.

From the guided wave standpoint, it suffices to recognize the capability of single-mode optical fibers to support the guidance of WDM beams. Additional propagation problems including the nonlinear index effect will be addressed in Chap. 10.

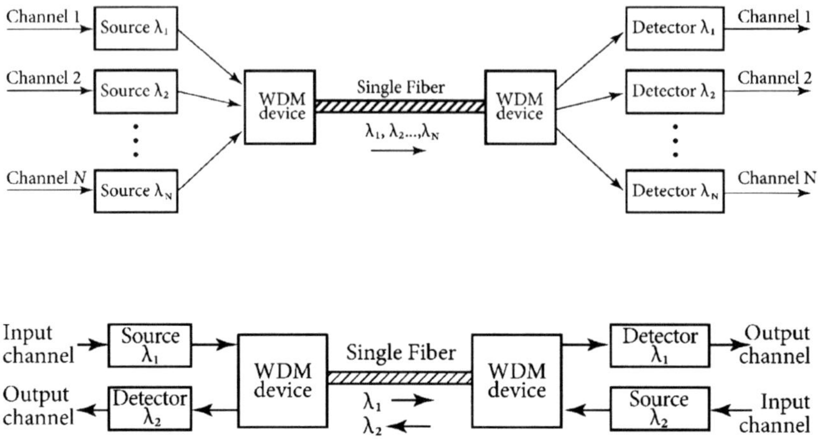

Figure 9.14. The top figure shows a unidirectional WDM system that combines N independent input signals for transmission over a single fiber. The bottom figure shows a schematic representation of a bidirectional WDM system in which two or more wavelengths are transmitted simultaneously in opposite directions over the same fiber

9.6.1 Bit-Parallel WDM Single-Fiber Link

Unlike the usual wavelength division multiplexed (WDM) format where input parallel pulses are first converted into a series of single pulses which are then launched on different wavelength beams into a single mode-fiber, the bit-parallel (BP) WDM format [22] requires no parallel to serial conversion of the input signal. Parallel pulses are launched simultaneously on different wavelength beams. Time alignment of the pulses for a given signal byte is very important.

There exists a competing non-WDM approach to transmit parallel bits – the fiber optic ribbon approach – where parallel bits are sent through corresponding parallel fibers in a ribbon format. However, it is very difficult to maintain time alignment of the parallel pulses due to practical difficulty in manufacturing identical uniform fibers. Thus, the single-fiber WDM format of transmitting parallel bits rather than a fiber ribbon format may be the media of choice.

As an example, the detailed design of a long distance (32 km) all-optical bit-parallel WDM single-fiber link with 12 bit-parallel channels having 1 Gb s^{-1} capacity is demonstrated. The speed–distance product for this link is 32 Gb s^{-1} km^{-1}, while the maximum speed–distance product for fiber ribbon is less than 100 Mb s^{-1} km^{-1}.

9.6.2 Elements of a 12-Bit Parallel WDM System [22]

Let us consider the design of the BP-WDM system. Because of the relatively broad pulse widths ($\sim$1 ns) and low power levels of the data pulses, nonlinear interaction

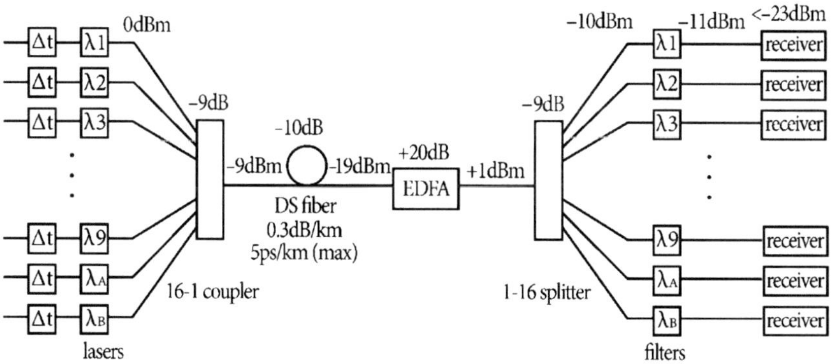

Figure 9.15. Block diagram for an all-optical 12 channel bit-parallel WDM single-fiber system [22]

of copropagating pulses can be considered to be negligible. It is expected that 12 separate beams will be used. Anticipating the use of erbium amplifiers, beam separation among these 12 beams must be limited by the useful bandwidth of the erbium amplifier, which ranges from 1,535 to 1,560 nm. Hence, separation between neighboring beams must be less than $15/12 = 2.08$ or 2 nm. A block diagram of the link is shown in Fig. 9.15.

9.6.2.1 The Transmitter

The transmitter of the system consists of 12 discreet distributed-feedback laser diodes and a 16-to-1 fiber coupler. Each laser element is selected to fall within the erbium gain bandwidth at a preselected $\Delta\lambda$ from its neighbors. To minimize system cost, the lasers are directly modulated with NRZ data at a rate up to 1 Gb s^{-1} each, for an aggregate of 1 Gb s^{-1}. The timing of the bits in any word are aligned at the input to the fiber link by adjusting the phase of the laser drive signal for each bit, using conventional electrical delay components. The optical power coupled into the fiber arms at the input to the 16-to-1 coupler is about 0 dB m (i.e., 1 mw).

9.6.2.2 The Single-Mode Fiber

Corning DS fiber is chosen to be the single-mode fiber for this system because of its desirable dispersion characteristics [23]. The dispersion characteristics of this fiber is shown in Fig. 9.16. It is seen that for the wavelength range of interest (1,535–1,560 nm), the dispersion coefficient $|\beta_2|$ is around 2 ps^2 km^{-1}. The difference of group velocities as a function of the wavelength of the beams have been measured and are displayed in Fig. 9.17. It is seen that the maximum difference in group velocity over the wavelength of interest is 5 ps km^{-1}. An erbium doped fiber amplifier (EDFA) is used to boost the power at the receiver.

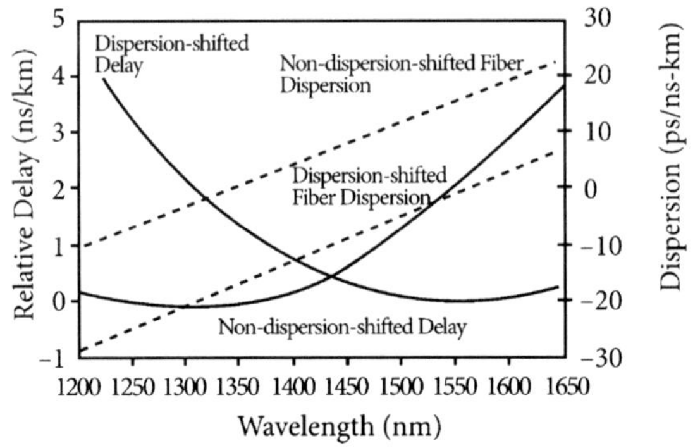

Figure 9.16. Dispersion characteristics of Corning DS fiber [22]

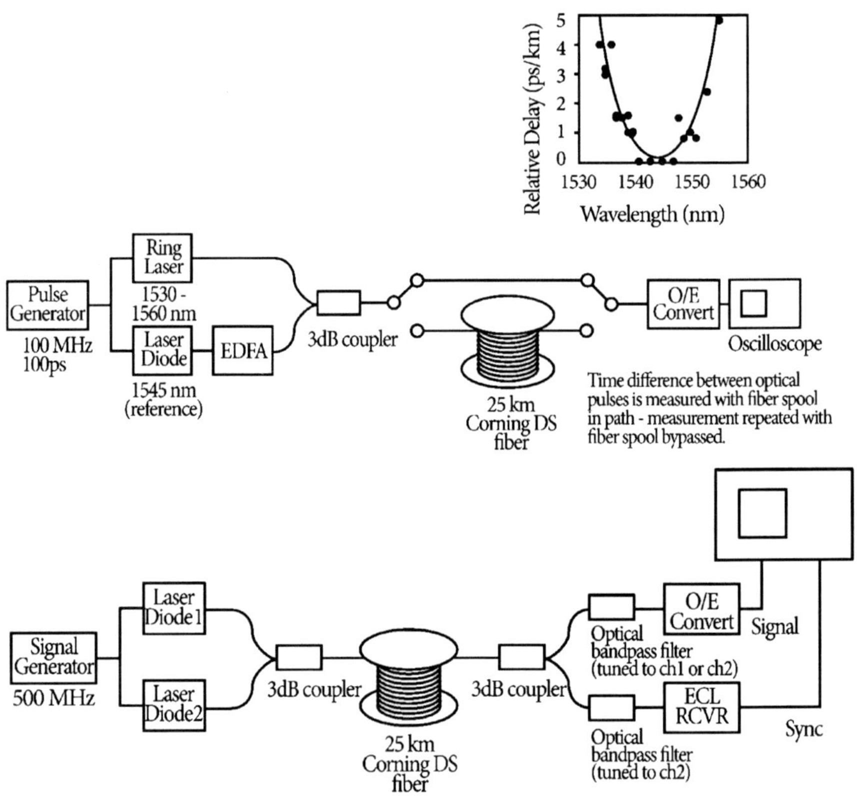

Figure 9.17. The top figure shows the measured group velocity differences for different wavelength beams. The sources are a tunable ring laser and a DFB laser diode at 1,545 nm. The bottom picture shows the experimental setup of bit-parallel nanosecond pulses propagating on two beams at 1,530 and 1,545 nm in a 25.2 km long Corning DS fiber [22]

9.6.2.3 The Receiver

The receiver of the system consists of a 1-to-16 fiber splitter, 12 optical bandpass filters, and 12 fiber-optic receivers.

9.6.3 Design Considerations

9.6.3.1 Wavelength Spacing Considerations

At 1 Gb s^{-1}, each bit path must have a minimum bandwidth of 2 GHz to reproduce the data. In estimating the spread of the optical spectrum of each laser element, a 4 GHz bandwidth will be assumed. The spread of each element's $\Delta\lambda$ is then 0.032 nm for a 4 GHz bandwidth, which is well within the 2 nm beam separation between neighboring beams. It should be noted that any spectral broadening of the pulse due to chirp or other factors will be much less than the 2 nm beam separation that has been used for this system. Furthermore, the 2 nm beam separation also lessens the demand on the optical bandpass filters used to separate the WDM beams at the receiver end.

9.6.3.2 Skew and Walk-off Considerations

At 1 Gb s^{-1}, the bit period is approximately 1 ns. For the worst case, the setup and hold time for standard ECL logic is 350 ps. This means that there is a leeway of $(1,000 - 350)/2 = 325$ ps, in which the pulses may drift away from each other. If one limits the skew or walk-off to half or 325 ps, then the maximum length of the fiber is $160/5 = 32$ km.

9.6.3.3 Loss Considerations

For a maximum length of 32 km, it is clear that an EDFA will be needed to increase the power at the receiver. As indicated in Fig. 9.15, a gain of 20 dB via the EDFA will provide a gain margin of more than 12 dB at the receiver.

9.6.4 Experimental Demonstration of a Two Wavelength BP-WDM System

The experimental setup is shown in Fig. 9.17. Two beams from two laser diodes, whose wavelengths are 1,530 and 1,545 nm, are modulated by nanosecond size pulses. These beams whose spectral shapes are displayed in Fig. 9.18 are coupled simultaneously into a Corning DS fiber. A picture of the pulses on these two beams before they were launched into the fiber is shown in Fig. 9.19 (top). It is seen that these nanosecond size pulses were well aligned at the entrance of the fiber link.

The spool of Corning DS fiber used for this experimental link was 25.2 km long. The output is displayed in Fig. 9.19 (bottom). One can readily measure the shift of the walk-off between these pulses – it was 200 ps or 6 ps km^{-1}. This result is consistent with our previous measurement displayed in Fig. 9.17. There, the walk-off was measured between a tunable ring laser and a 1,545 nm laser diode.

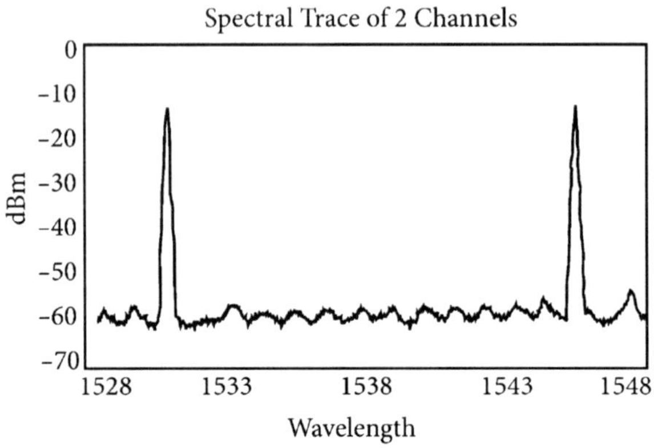

Figure 9.18. Spectral shapes for the two copropagating beams [22]

It is noted that the experimentally measured walk-off of 200 ps for this two wavelength BP-WDM demonstration is well within the allowable setup and hold time for the standard ECL logic, which is 350 ps or for a bit period of 1 ns.

This example shows that one can design a 32-km BP-WDM single-fiber data link with 1 Gb s^{-1} capacity. This is an all-optical link with byte-wide optical path, which can bypass any electrical bottleneck.

It was also shown through an actual experiment that nanosecond size pulses on two BP-WDM beams at 1,530 and 1,545 nm can be successfully transmitted through a 25.2 km long Corning DS fiber with an acceptable walk-off, which is well within the allowable setup and hold time of standard ECL logic circuits. As can be seen from Fig. 9.19, the maximum walk-off between any beams within the wavelength range of 1,530 and 1,545 nm is 200 ps. This result implies that 30 bit-parallel beams spaced 1 nm apart from 1,530 and 1,560 nm, each carrying a 1 GB s^{-1} signal, can be sent through a 25.2 km Corning DS fiber at an information rate of 30 Gb s^{-1}. This means that the speed–distance product for this link is about 94 Gb $(s\,km)^{-1}$.

9.7 Concluding Remarks

The use of optical fibers as broadband long-distance communication links has been well established. As internet traffic expands, demand for more and more bandwidth will only increase. In this chapter, we have provided the most basic fundamentals of wave-propagation on linear fibers. To develop higher bandwidth and longer distance transmission links, one may make use of the nonlinear properties of fibers. How this is done will be discussed in the next chapter.

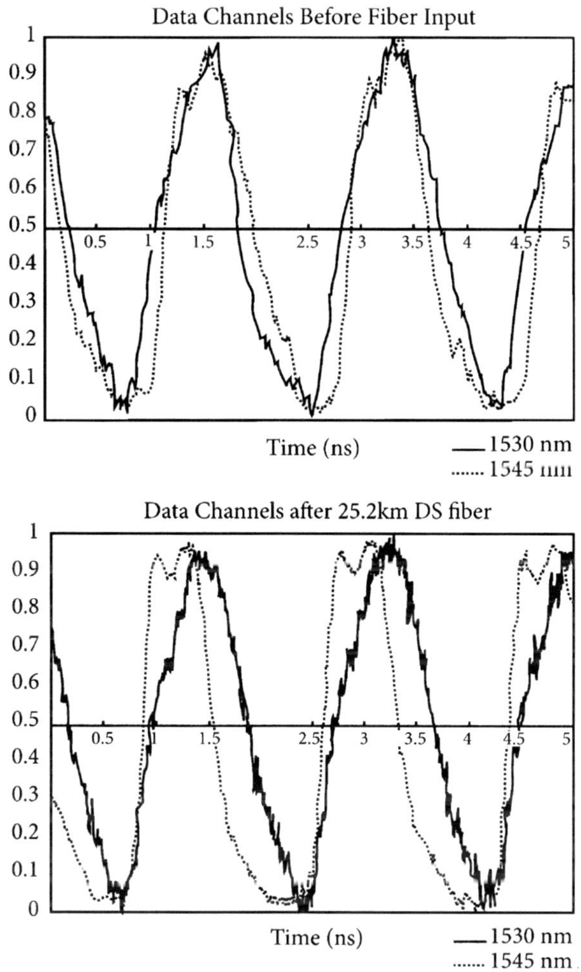

Figure 9.19. The upper figure shows a picture of the nanosecond pulses on the two copropagating beams before entering the fiber. These pulses are very well aligned at the entrance to the fiber link. The lower figure shows a picture of the nanosecond pulses on the two copropagating beams after passing through 25.2 km of the fiber. The alignment of the two pulses is shifted 200 ps at the output of the fiber link. This shift represents the walk-off among these different wavelength beams [22]

References

1. J. R. Carson, S. P. Mead, and S. A. Schelkunoff, "Hyperfrequency waveguides – Mathematical theory," Bell Syst. Tech. J. **15**, 310 (1936)

2. N. S. Kapany, "Fiber Optics," Academic Press, New York (1967)

3. F. P. Kapron, D. B. Keck, and R. D. Maurer, "Radiation losses in glass optical waveguides," Appl. Phys. Lett. **17**, 423 (1970); K. C. Kao and G. A. Hockam, "Dielectric fiber surface waveguides for optical frequencies," Proc. IEE **113**, 1151 (1966)

4. L. G. Cohen, C. Liu, and W. G. French, "Effect of temperature of transmission in lightguides," Elec. Lett. **15**, 334 (1979); C. T. Chang, "Minimum dispersion in a single-mode step-index optical fiber," Elec. Lett. **15**, 765 (1979); C. T. Chang, Appl. Opt. **18**, 2516 (1979); L. G. Cohen, W. L. Mammell, and S. Lumish, "Tailoring the shapes of dispersion spectra in single-mode fibers," Opt. Lett. **7**, 183 (1982); S. J. Jang, L. G. Cohen, W. L. Mammell, and M. A. Shaifi, "Experimental verification of ultra-wide bandwidth spectra in double-clad single mode fiber," Bell Syst. Tech. J. **61**, 385 (1982); V. A. Bhagavatula, M. S. Spotz, W. F. Love, and D. B. Keck, "Segmented-core single-mode fibres with low loss and low dispersion," Elec. Lett. **19**, 317 (1983)

5. G. P. Agrawal, "Fiber Optic Communication Systems," 3rd edn., Wiley, New York (2002)

6. E. Snitzer, "Cylindrical dielectric waveguide modes." J. Opt. Soc. Am. **51**, 491 (1961); E. Snitzer and H. Osterberg, "Observed dielectric waveguide modes in the visible spectrum," J. Opt. Soc. Am. **51**, 499 (1961)

7. D. Gloge, ed., "Optical Fiber Technology," IEEE Press, New York (1976)

8. D. Marcuse, "Light Transmission Optics," Van Nostrand, New York (1972); D. Marcuse, "Theory of Dielectric Optical Waveguide," Academic Press, New York (1974)

9. A. W. Snyder and J. D. Love, "Optical Waveguide Theory," Chapman and Hall, New York (1984)

10. G. Keiser, "Optical Fiber Communications," 2nd edn., McGraw-Hill, New York (1991)

11. A. W. Snyder, "Asymptotic expressions for eigenfunctions and eigenvalues of a dielectric or optical waveguide," IEEE Trans. Microw. Theory Tech. **MTT-17**, 1130 (1969)

12. D. Gloge, "Weakly guiding fibers," Appl. Opt, **10**, 2252 (1971)

13. D. Marcuse, "Gaussian approximation of the fundamental modes of graded-index fibers," J. Opt. Soc. Am. **68**, 103 (1978)

14. J. K. Shaw, W. M. Henry, and W. R. Winfrey, "Weakly guiding analysis of elliptical core step index waveguides based on the characteristic numbers of Mathieu's equations," J. Lightwave Tech. **13**, 2359 (1995)

15. C. Yeh and F. Manshadi, "On weakly guiding single-mode optical waveguides," J. Lightwave Tech. **3**, 199 (1985)

16. C. Yeh, "Modes in weakly guiding elliptical optical fibers," Opt. Quantum Electron. **8**, 43 (1976)

17. L. Eyges, P. Gianino, and P. Wintersteiner, "Modes of dielectric waveguides of arbitrary cross sectional shape," J. Opt. Soc. Am. **69**, 1226 (1979)

18. I. H. Malitson, J. Opt. Soc., "Interspecimen comparison of the refractive index of fused silica," Am. **55**, 1205 (1965); M. DiDomenico, "Material dispersion in optical fiber waveguides," Appl. Opt. **11**, 652 (1972)

19. D. B. Keck, "Fundamentals of optical waveguide fibers," IEEE Comm. Magazine **23**, 17 (1985)

20. G. P. Agrawal, "Nonlinear Fiber Optics," 3rd edn., Elsevier, Singapore (2004)

21. N. Bloembergen, "Nonlinear Optics," Benjamin, Reading, MA (1977); Y. R. Shen, "Principles of Nonlinear Optics," Wiley, New York (1984); R. W. Boyd, " Nonlinear Optics," Academic Press, San Diego (1992)

22. L. Bergman, J. Morookian, and C. Yeh, "An all-optical long distance multi-gbytes/s bit-parallel WDM single-fiber link," J. Lightwave Tech. **16**, 1577 (1998)

23. Corning Inc., "Single-Mode Dispersion," MM26, Opto-Electronics Group, Corning, New York (1996)

10

SOLITONS AND WDM SOLITONS

The discovery in 1973 that an optical soliton [1] on a single wavelength beam can exist in fiber is one of the most significant events since the perfection of low-loss optical fiber communication. This means that, in principle, data pulses may be transmitted in a fiber without degradation forever. This soliton discovery sets the ultimate goal for optical fiber communication on a single wavelength beam.

Another significant event is the development of wavelength division multiplexed (WDM) transmission in a single mode fiber [2]. This means that multiple beams of different wavelengths, each carrying its own data load, can propagate simultaneously in a single mode fiber. This WDM technique provides dramatic increase in the bandwidth of a fiber. However, because of the presence of complex nonlinear interaction between co-propagating pulses on different wavelength beams, it is no longer certain that WDM solitons can exist.

The existence of solitons is a blissful event in nature. It is a marvel that the delicate balance between the dispersion effect and the nonlinear effect can allow a specially shaped optical pulse to propagate in the fiber without degradation. This is called a temporal soliton [1]. It is an equal marvel that the delicate balance between the diffraction effect and the nonlinear effect can also allow a specially shaped pulse to propagate in a planar waveguide or array waveguides without degradation. This is called a spatial soliton [3]. They occur only on a single wavelength beam.

When beams with different wavelengths co-propagate in a single mode fiber, such as in the WDM case [2], interaction of pulses on different beams via the nonlinear cross phase modulation (CPM) effect (the Kerr effect) is usually instrumental in destroying the integrity of solitons on these multiplexed beams [4]. Other applications of the CPM effect in fiber will also be described here.

Starting from the fundamentals on the polarization of a nonlinear material medium and the Maxwell equations, we shall derive the propagation equation for the slowly varying amplitude for the transverse electric field of the dominant mode in a nonlinear fiber [4]. The propagation equation will be solved for several phenomena of interest: (a) the group velocity dispersion (GVD) effect [5]; (b) the nonlinear phenomenon of self-phase modulation (SPM) [6]; (c) the nonlinear phenomenon of cross-phase modulation (CPM or XPM) [7]; (d) the appearance of temporal optical solitons [1]; and (e) the existence of wavelength division multiplexed (WDM) solitons [8]. Several special applications on pulse shepherding [9], pulse compression [10], pulse generation [11], and bit parallel propagation [12] will also be discussed.

10.1 Nonlinear Refractive Index

According to the Maxwell constitutive relations, (2.34) and (2.35)

$$\mathbf{D} = \epsilon(E)\mathbf{E}, \tag{10.1}$$
$$\mathbf{B} = \mu(H)\mathbf{H}, \tag{10.2}$$

where the dielectric constant $\epsilon(E)$ and magnetic permeability $\mu(H)$ are functions of the field strengths for nonlinear material. For nonmagnetic dielectric material, optical nonlinearity is related to anharmonic motion of bound electrons under the influence of an applied field. This nonlinear material properties can be described by expanding the polarization $\mathbf{P}$ induced by electric dipoles in a power series in the field [13]:

$$\mathbf{P} = \epsilon_0 \left(\chi^{(1)} \cdot \mathbf{E} + \chi^{(2)} : \mathbf{EE} + \chi^{(3)} \vdots \mathbf{EEE} + \cdots \right), \tag{10.3}$$

where ϵ_0 is the free space dielectric constant and $\chi^{(i)}$ $(i = 1, 2, 3, \ldots)$ is the ith order susceptibility. The first term represents the dominant linear susceptibility, $\chi^{(1)}$, which is responsible for the linear part of the refractive index n and the attenuation coefficient α. The second term represents the second-order susceptibility, $\chi^{(2)}$, which is responsible for the second harmonic generation, sum frequency generation, etc. The third term represents the third-order susceptibility, $\chi^{(3)}$, which is responsible for the nonlinear part of the refractive index for silica fiber, four-wave mixing, third-order generation, etc. Since silica (SiO_2) is a symmetric molecule, $\chi^{(2)}$ is generally zero for optical fibers. Thus, the lowest order nonlinear effects in optical fibers are due to the third-order susceptibility, $\chi^{(3)}$. Assuming that, in general, phase matching is not present, the dominant nonlinear effect originates from the intensity dependence of the refractive index. In other words, for optical fibers, the index of refraction is

$$n(\omega, |E|^2) = n(\omega) + n_2 |E|^2, \tag{10.4}$$

where $n(\omega)$ is the linear index of refraction, n_2 is the nonlinear index coefficient, and $|E|^2$ is the optical intensity.

Referring to the definition

$$\mathbf{D} = \epsilon_0 \mathbf{E} + \mathbf{P}, \tag{10.5}$$

one has

$$\mathbf{D} = \epsilon \mathbf{E} = \epsilon_0 \mathbf{E} + \epsilon_0 \left(\chi^{(1)} \cdot \mathbf{E} + \chi^{(2)} : \mathbf{EE} + \chi^{(3)} \vdots \mathbf{EEE} + \cdots \right) \tag{10.6}$$

or, in the frequency domain,

$$\frac{\epsilon^{(\omega)}}{\epsilon_0} = \left(1 + \chi^{(1)(\omega)} \right) + \epsilon_{\mathrm{NL}}^{(\omega)}, \tag{10.7}$$

where $\chi^{(1)(\omega)}$ is the Fourier transform of $\chi^{(1)}$ and

$$\epsilon_{\mathrm{L}}^{(\omega)} = \left(1 + \chi^{(1)(\omega)} \right) \tag{10.8}$$

represents the linear part of the relative dielectric constant and $\epsilon_{\mathrm{NL}}^{(\omega)}$ represents the nonlinear part of the relative dielectric constant. By definition, $\epsilon_{\mathrm{L}}^{(\omega)}$ is related to the linear part of the index of refraction $n_{\mathrm{L}}^{(\omega)}$ and the absorption coefficient $\alpha^{(\omega)}$ as follows:

$$\epsilon_{\mathrm{L}}^{(\omega)} = \left[n_{\mathrm{L}}^{(\omega)} - j \frac{\alpha^{(\omega)} c}{2\omega} \right]^2, \tag{10.9}$$

or

$$n_{\mathrm{L}}^{(\omega)} = \left[1 + \mathrm{Re} \left(\chi^{(1)(\omega)} \right) \right]^{1/2}, \tag{10.10}$$

$$\alpha^{(\omega)} = \left[\frac{\omega}{c n_{\mathrm{L}}^{(\omega)}} \mathrm{Im} \left(\chi^{(1)(\omega)} \right) \right], \tag{10.11}$$

where the assumption $\mathrm{Im}\left(\chi^{(1)(\omega)}\right) \ll \left[1 + \mathrm{Re}\left(\chi^{(1)(\omega)}\right)\right]$ has been used. From (10.6) and (10.7), assuming that (a) the contribution of molecular vibrations due to $\chi^{(3)}$ can be neglected (this is true for pulse widths >1 ps for silica fibers), (b) the third-harmonic oscillation term in (10.6) is ignored (this is true when phase matching is negligible in optical fibers), and (c) $\chi^{(2)}$ is zero due to the symmetry of the SiO_2 molecule, we may obtain [13]

$$\epsilon_{\text{NL}}^{(\omega)} = \frac{3}{4}\chi^{(3)(\omega)} |E|^2, \tag{10.12}$$

where $|E|^2$ is the slowly varying electric field intensity for the envelope of the propagating pulse. It has also been assumed that the propagating field is linearly polarized in the transverse direction.

The relation between the relative dielectric constant and n is

$$\epsilon_{\text{L}}^{(\omega)} + \epsilon_{\text{NL}}^{(\omega)} = \left(n_{\text{L}}^{(\omega)} + n_{\text{NL}}^{(\omega)}\right)^2, \tag{10.13}$$

where $n_{\text{L}}^{(\omega)}$ and $n_{\text{NL}}^{(\omega)}$ represent, respectively, the linear and the nonlinear part of the index of refraction. Simplifying (10.13) and using the fact that $\epsilon_{\text{L}}^{(\omega)} \gg \epsilon_{\text{NL}}^{(\omega)}$ and $n_{\text{L}}^{(\omega)} \gg n_{\text{NL}}^{(\omega)}$, one has

$$n_{\text{L}}^{(\omega)} = \left[\epsilon_{\text{L}}^{(\omega)}\right]^{1/2}, \tag{10.14}$$

$$n_{\text{NL}}^{(\omega)} = \frac{1}{2}\frac{\epsilon_{\text{NL}}^{(\omega)}}{n_{\text{L}}^{(\omega)}} = \frac{3}{8}\frac{\chi^{(3)(\omega)} |E|^2}{n_{\text{L}}^{(\omega)}}. \tag{10.15}$$

Therefore, from (10.4)

$$n_2 = \frac{3}{8}\frac{\chi^{(3)(\omega)}}{n_{\text{L}}^{(\omega)}}. \tag{10.16}$$

Here, we have assumed that the nonlinear part of the absorption coefficient $\alpha^{(\omega)}$ is small and negligible. Typically,

$$n_2 \approx 2.6 \times 10^{-20} \quad (\text{m}^2\,\text{W}^{-1}) \tag{10.17}$$

for silica fibers. This value is quite small compared to most other nonlinear media. However, because of the small spot size ($\sim$10 μm) and the very low loss of the single mode silica fibers, the nonlinear interacting length is sufficiently long so that the nonlinear effects in optical fibers can be observed at relatively low power levels.

10.2 The Nonlinear Pulse Propagation Equation

It is extremely difficult to solve the nonlinear Maxwell equations. Fortunately, when we deal with the wave propagation problem in a nonlinear fiber, a number of approximations can be used to yield meaningful results. Using these approximations, we can obtain the nonlinear pulse propagation equation, which can be used to yield the evolution of an input pulse as it propagates along a nonlinear fiber.

Under the assumption of a weakly guiding fiber, the dominant single mode field may be approximated as a linearly polarized modal field with a Gaussian distribution (see Sect. 9.1). So, the time-dependent transverse electric field for the dominant mode can be expressed as follows:

$$E_y(\mathbf{r},t) = \frac{1}{2} E_y^{(t)}(\mathbf{r},t)\, e^{j\omega_0 t} + c.c., \tag{10.18}$$

where $E_y^{(t)}(\mathbf{r},t)$ is the slowly varying field amplitude of time, ω_0 is the optical carrier frequency, and c.c. means the complex conjugate terms. Because of the smallness of the nonlinear part of the relative dielectric constant, a perturbation technique can be used to derive the propagation equation for the envelope field. Let us define the following Fourier transform:

$$E_y^{(\omega)}(\mathbf{r},\omega) = \mathcal{F}\{E_y(\mathbf{r},t)\} = \int_{-\infty}^{\infty} E_y(\mathbf{r},t)\, e^{-j\omega t}\, dt. \tag{10.19}$$

Using (10.18), one has

$$\begin{aligned} E_y^{(\omega)}(\mathbf{r},\omega) &= \mathcal{F}\{E_y(\mathbf{r},t)\} = \mathcal{F}\left\{E_y^{(t)}(\mathbf{r},t)\, e^{j\omega_0 t}\right\} \\ &= \int_{-\infty}^{\infty} E_y^{(t)}(\mathbf{r},t)\, e^{j\omega_0 t} e^{-j\omega t}\, dt, \\ &= \int_{-\infty}^{\infty} E_y^{(t)}(\mathbf{r},t)\, e^{-j(\omega-\omega_0)t}\, dt, \end{aligned} \tag{10.20}$$

or

$$E_y^{(\omega)}(\mathbf{r},\omega-\omega_0) = \int_{-\infty}^{\infty} E_y^{(t)}(\mathbf{r},t)\, e^{-j(\omega-\omega_0)t}\, dt. \tag{10.21}$$

Its transform pair is

$$\begin{aligned} E_y^{(t)}(\mathbf{r},t) &= \mathcal{F}^{-1}\left\{E_y^{(\omega)}(\mathbf{r},\omega-\omega_0)\right\}, \\ &= \frac{1}{2\pi}\int_{-\infty}^{\infty} E_y^{(\omega)}(\mathbf{r},\omega-\omega_0)\, e^{j(\omega-\omega_0)t}\, d(\omega-\omega_0), \\ &= \frac{1}{2\pi}\int_{-\infty}^{\infty} E_y^{(\omega)}(\mathbf{r},\omega-\omega_0)\, e^{j(\omega-\omega_0)t}\, d\omega. \end{aligned} \tag{10.22}$$

We now summarize the approximations we must use to derive the pulse propagation equation for the field envelope:

(a) The nonlinear part of the index of refraction of silica is quite small; it is less than 10^{-6}.

(b) Polarization of the modal field is maintained throughout the field's propagation path.

(c) The optical fields in the fiber are quasi-monochromatic. This means that $\Delta\omega/\omega_0 \ll 1$, where $\Delta\omega$ is the spectral width and ω_0 is the operating carrier frequency. In other words, the pulse width is usually greater than 0.5 ps.

(d) The envelope field, $E_y^{(t)}(\mathbf{r}, t)$, is a slowly varying function of time, relative to the optical period.

(e) Nonlinear response of the induced susceptibility is instantaneous, and the third-harmonic term of $\chi^{(3)}$ is negligible, that is,

$$\epsilon_{\mathrm{NL}} = \frac{3}{4}\chi^{(3)(\omega)} \left| E_y^{(t)}(\mathbf{r}, t) \right|^2,$$

where

$$\epsilon = \epsilon_0 \left(\epsilon_{\mathrm{L}} + \epsilon_{\mathrm{NL}} \right),$$

$$\mathbf{D} = \epsilon \mathbf{E}.$$

(f) Perturbation theory applies.

The wave equation for a linearly polarized field is

$$\nabla^2 E_y(\mathbf{r}, t) = \mu_0 \frac{\partial^2 D_y(\mathbf{r}, t)}{\partial t^2}, \tag{10.23}$$

where $D_y(\mathbf{r}, t) = \epsilon E_y(\mathbf{r}, t)$ and $\epsilon = \epsilon_0 \left(\epsilon_{\mathrm{L}} + \epsilon_{\mathrm{NL}} \right)$. Taking the Fourier transform of (10.23) and using (10.19) yields

$$\nabla^2 E_y^{(\omega)}(\mathbf{r}, \omega) = -\mu_0 \epsilon^{(\omega)} \omega^2 E_y^{(\omega)}(\mathbf{r}, \omega). \tag{10.24}$$

Using (10.21), one has

$$\nabla^2 E_y^{(\omega)}(\mathbf{r}, \omega - \omega_0) = -\mu_0 \epsilon^{(\omega)} \omega^2 E_y^{(\omega)}(\mathbf{r}, \omega - \omega_0). \tag{10.25}$$

Here,

$$\epsilon^{(\omega)} = \epsilon_0 \left[\epsilon_{\rm L}^{(\omega)} + \epsilon_{\rm NL}^{(\omega)}\right] = \epsilon_0 \left[(1 + \chi^{(1)(\omega)}) + \frac{3}{4}\chi^{(3)(\omega)}\left|E_y\right|^2\right]. \tag{10.26}$$

Equation (10.25) will be solved by the separation of variables as follows. Let

$$E_y^{(\omega)} = F(x,y)\left[A^{(\omega)}(z,\omega-\omega_0)\,e^{-j\beta_0 z}\right], \tag{10.27}$$

where β_0 is a yet unknown constant. Substituting (10.27) in (10.25), one finds

$$\nabla_t^2 F + (\mu_0\epsilon^{(\omega)}\omega^2 - \beta^{(\omega)2})F \;=\; 0, \tag{10.28}$$

$$-2j\beta_0\frac{\partial A^{(\omega)}}{\partial z} + (\beta^{(\omega)2} - \beta_0^2) \;=\; 0, \tag{10.29}$$

where $\beta^{(\omega)}$ is the separation constant. The term $\partial^2 A^{(\omega)}/\partial z^2$ has been ignored due to the assumption that $A^{(\omega)}$ is a slowly varying function of z. According to the standard perturbation technique, the eigenfunction $F(x,y)$ of a perturbed system is assumed to be unchanged by the perturbation while the eigenvalue $\beta^{(\omega)}$ will be changed by $\Delta\beta^{(\omega)}$. Here the perturbation is the nonlinear part of $\epsilon^{(\omega)}$. The unperturbed eigenfunction $F(x,y)$ is simply the modal function for the fundamental HE_{11} or LP_{01} mode, which is obtained from (10.28) with $\epsilon^{(\omega)} = \epsilon_0\epsilon_{\rm L}^{(\omega)}$. This modal distribution may be approximated by a Gaussian, see (9.1)–(9.9). Since

$$\frac{\epsilon^{(\omega)}}{\epsilon_0} = \left(\epsilon_{\rm L}^{(\omega)} + \epsilon_{\rm NL}^{(\omega)}\right) = \mathrm{Re}\left(\epsilon_{\rm L}^{(\omega)}\right) + \Delta\left(\frac{\epsilon^{(\omega)}}{\epsilon_0}\right), \tag{10.30}$$

$$\Delta\left(\frac{\epsilon^{(\omega)}}{\epsilon_0}\right) = \epsilon_{\rm NL}^{(\omega)} - j\,\mathrm{Im}\left(\epsilon_{\rm L}^{(\omega)}\right), \tag{10.31}$$

and

$$\begin{aligned}\frac{\epsilon^{(\omega)}}{\epsilon_0} &= \left[n^{(\omega)}\right]^2 - \left[n_{\rm L}^{(\omega)} + \Delta n^{(\omega)}\right]^2,\\ &\approx n_{\rm L}^{(\omega)2} + 2n_{\rm L}^{(\omega)}\Delta n^{(\omega)}.\end{aligned} \tag{10.32}$$

So,

$$2n_{\rm L}^{(\omega)}\Delta n^{(\omega)} = \Delta\left(\frac{\epsilon^{(\omega)}}{\epsilon_0}\right), \tag{10.33}$$

$$\begin{aligned}\Delta n^{(\omega)} &= \frac{1}{2n_{\mathrm{L}}^{(\omega)}}\left[\frac{3}{4}\chi^{(3)(\omega)}\left|E\right|^2 - \frac{j\alpha^{(\omega)}n_{\mathrm{L}}^{(\omega)}}{k_0}\right],\\ &= \frac{3}{8}\chi^{(3)(\omega)}\left|E\right|^2 - \frac{j\alpha^{(\omega)}}{2k_0},\end{aligned} \tag{10.34}$$

where $k_0 = \omega\sqrt{\mu_0\epsilon_0}$. Substituting (10.32) in (10.28) and introducing

$$\beta^{(\omega)} = \beta^{(\omega)}_{\text{unperturbed}} + \Delta\beta^{(\omega)}, \tag{10.35}$$

we find

$$\Delta\beta^{(\omega)} = \frac{k_0\int_{-\infty}^{\infty}\int_{-\infty}^{\infty}\Delta n^{(\omega)}\left|F(x,y)\right|^2\,dxdy}{\int_{-\infty}^{\infty}\int_{-\infty}^{\infty}\left|F(x,y)\right|^2\,dxdy}. \tag{10.36}$$

Introducing the transform pair

$$\begin{aligned}A^{(t)}(z,t) &= \mathcal{F}^{-1}\left\{A^{(\omega)}(z,\omega-\omega_0)\right\},\\ &= \frac{1}{2\pi}\int_{-\infty}^{\infty}A^{(\omega)}(z,\omega-\omega_0)\,e^{j(\omega-\omega_0)t}\,d(\omega-\omega_0),\\ &= \frac{1}{2\pi}\int_{-\infty}^{\infty}A^{(\omega)}(z,\omega-\omega_0)\,e^{j(\omega-\omega_0)t}\,d\omega,\end{aligned} \tag{10.37}$$

$$A^{(\omega)}(z,\omega-\omega_0) = \mathcal{F}\left\{A^{(t)}(z,t)\right\} = \int_{-\infty}^{\infty}A^{(t)}(z,t)\,e^{-j(\omega-\omega_0)t}\,dt, \tag{10.38}$$

we obtain from (10.18)

$$\begin{aligned}E_y(\mathbf{r},t) &= \frac{1}{2}\mathcal{F}^{-1}\left\{E_y^{(t)}(\mathbf{r},t)\right\}\,e^{j\omega_0 t} + \text{c.c.},\\ &= \frac{1}{2}\mathcal{F}^{-1}\left\{F(x,y)A^{(\omega)}(z,\omega-\omega_0)\,e^{-j\beta_0 z}\right\}\,e^{j\omega_0 t} + \text{c.c.},\\ &= \frac{1}{2}F(x,y)\mathcal{F}^{-1}\left\{A^{(\omega)}(z,\omega-\omega_0)\right\}\,e^{-j(\beta_0 z-\omega_0 t)} + \text{c.c.},\\ &= \frac{1}{2}F(x,y)A^{(t)}(z,t)\,e^{-j(\beta_0 z-\omega_0 t)} + \text{c.c.}\end{aligned} \tag{10.39}$$

Here $A^{(t)}(z,t)$ is the slowly varying pulse envelope whose Fourier transform satisfies (10.29).

From (10.29), using the approximation

$$\beta^{(\omega)2} - \beta_0^2 = \left(\beta^{(\omega)} + \beta_0\right)\left(\beta^{(\omega)} - \beta_0\right) \approx 2\beta_0\left(\beta^{(\omega)} - \beta_0\right), \tag{10.40}$$

one obtains

$$-j\frac{\partial A^{(\omega)}}{\partial z} + \left(\beta^{(\omega)} - \beta_0\right)A^{(\omega)} = 0. \tag{10.41}$$

From (10.35)

$$\beta^{(\omega)} = \beta^{(\omega)}_{\text{unperturbed}} + \Delta\beta^{(\omega)}, \tag{10.42}$$

where

$$\beta^{(\omega)}_{\text{unperturbed}} = \beta_0 + (\omega - \omega_0)\,\beta_1 + \frac{1}{2}(\omega - \omega_0)^2\,\beta_2 + \cdots, \tag{10.43}$$

$$\beta_m = \left.\frac{d^m\beta^{(\omega)}_{\text{unperturbed}}}{d\omega^m}\right|_{\omega=\omega_0} \qquad m = 1, 2, 3, \ldots. \tag{10.44}$$

Equation (10.41) becomes

$$-j\frac{\partial A^{(\omega)}}{\partial z} + (\omega - \omega_0)\,\beta_1 A^{(\omega)} + \frac{1}{2}(\omega - \omega_0)^2\,\beta_2 A^{(\omega)} + \cdots + \Delta\beta A^{(\omega)} = 0. \tag{10.45}$$

Taking the inverse Fourier transform according to (10.37) gives

$$-j\frac{\partial A^{(t)}}{\partial z} - j\beta_1\frac{\partial A^{(t)}}{\partial t} - \frac{1}{2}\beta_2\frac{\partial^2 A^{(t)}}{\partial t^2} + \cdots + \mathcal{F}^{-1}\left\{\Delta\beta A^{(\omega)}\right\} = 0. \tag{10.46}$$

Using (10.34) and (10.36), one has

$$\mathcal{F}^{-1}\left\{\Delta\beta A^{(\omega)}\right\}$$

$$-A^{(t)}\left\{\frac{k_0\displaystyle\int_{-\infty}^{\infty}\int_{-\infty}^{\infty}\left[\frac{3}{8}\chi^{(3)(\omega)}\left|A^{(t)}\right|^2|F(x,y)|^4 - \frac{j\alpha^{(\omega)}}{2k_0}|F(x,y)|^2\right]dxdy}{\displaystyle\int_{-\infty}^{\infty}\int_{-\infty}^{\infty}|F(x,y)|^2\,dxdy}\right\}. \tag{10.47}$$

Following Agrawal [4], if we normalize the pulse amplitude $A^{(t)}$ such that $\left|A^{(t)}\right|^2$ represents the optical power, we may introduce an effective area A_{eff} defined as follows

$$A_{\text{eff}} = \frac{\left[\int_{-\infty}^{\infty}\int_{-\infty}^{\infty} |F(x,y)|^2 \, dxdy\right]^2}{\int_{-\infty}^{\infty}\int_{-\infty}^{\infty} |F(x,y)|^4 \, dxdy}. \tag{10.48}$$

Then (10.47) reads

$$\mathcal{F}^{-1}\left\{\Delta\beta A^{(\omega)}\right\} = \gamma\left|A^{(t)}\right|^2 A^{(t)} - \frac{j\alpha^{(\omega)} A^{(t)}}{2}, \tag{10.49}$$

where

$$\gamma = \frac{3}{8}\frac{\chi^{(3)} k_0}{A_{\text{eff}}} = \frac{n_2\omega_0}{cA_{\text{eff}}}. \tag{10.50}$$

Here A_{eff} is the effective core area in the range of 20–100 μm^2 in the 1.5 μm wavelength region, $n_2 \approx 3.2 \times 10^{-16}\,\text{cm}^2\,\text{W}^{-1}$ for silica fibers, ω_0 is the carrier frequency, and c is the velocity of light. Equation (10.46) becomes

$$\frac{\partial A^{(t)}}{\partial z} + \beta_1\frac{\partial A^{(t)}}{\partial t} - \frac{j\beta_2}{2}\frac{\partial^2 A^{(t)}}{\partial t^2} + \frac{\alpha^{(\omega)} A^{(t)}}{2} = -j\gamma\left|A^{(t)}\right|^2 A^{(t)}. \tag{10.51}$$

This is the pulse propagation equation we wish to derive. Here $A^{(t)}(z,t)$ is the slowly varying amplitude of the dominant mode, $\beta_1 = 1/v_{\text{g}}$, v_{g} is the group velocity, β_2 is the dispersion coefficient ($\beta_2 = dv_{\text{g}}^{-1}/d\omega$), α is the absorption coefficient, and γ is the nonlinear index coefficient in the range of 1–10 $\text{W}^{-1}\,\text{km}^{-1}$ for silica fibers.

Let us transform the pulse propagation equation to a frame of reference which is moving with the pulse at the group velocity, v_{g}. Introducing the normalizing coefficients

$$\begin{aligned} \tau &= (t - z/v_{\text{g}})/T_0, \\ \xi &= z/L_{\text{D}}, \\ L_{\text{D}} &= T_0^2/\left|\beta_2\right|, \end{aligned} \tag{10.52}$$

and setting

$$u(\xi,\tau) = \left[A^{(t)}(z,t)/\sqrt{P_0}\right] e^{\alpha L_{\text{D}}\xi/2}, \tag{10.53}$$

$$L_{\text{NL}} = 1/(\gamma P_0) \tag{10.54}$$

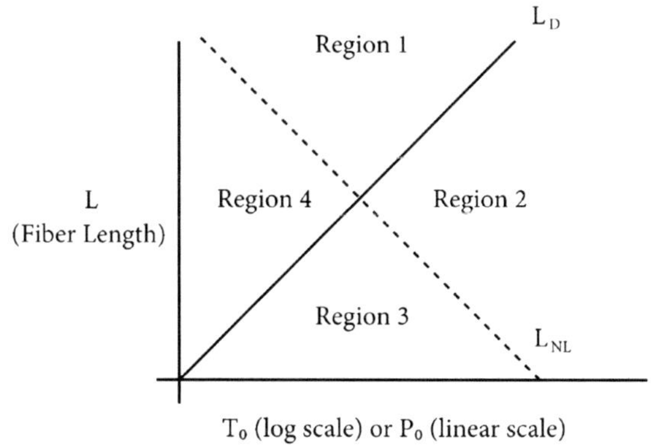

Figure 10.1. Various regions depicting the importance of dispersion effect and/or nonlinear effect

gives

$$-j\frac{\partial u}{\partial \xi}=\frac{\operatorname{sgn}(\beta_2)}{2}\frac{\partial^2 u}{\partial \tau^2}-\frac{L_D}{L_{NL}}e^{-\alpha L_D \xi}|u|^2 u, \tag{10.55}$$

where $\operatorname{sgn}(\beta_2)=\pm 1$ depending on the sign of β_2. We note that T_0 and P_0 are, respectively, the initial pulse width and the peak power of the initial pulse. The dispersion length L_D is the length scale in which dispersive effects become important, while the nonlinear length L_{NL} is the length scale in which nonlinear defects become important. This means when the fiber length L is such that $L \ll L_D$ and $L \ll L_{NL}$, then the propagating pulse is not significantly affected by the dispersion or the nonlinear effects. Only when $L \geq L_D$ or $L \geq L_{NL}$ will the dispersion or the nonlinear effects be important for the propagating pulses. A sketch of L_D vs. τ_0 for a given value of $|\beta_2|$ and that of L_{NL} vs. P_0 for a given value of γ are given in Fig. 10.1. One may identify four regions: (1) Region 1, $L > L_D$, $L \geq L_{NL}$, in which the dispersion effects and the nonlinear effects are significant; (2) Region 2, $L < L_D$, $L \geq L_{NL}$, in which only the nonlinear effects are important; (3) Region 3, $L < L_D$, $L < L_{NL}$, in which both the dispersion effects and the nonlinear effects may be ignored; and (4) Region 4, $L > L_D$, $L < L_{NL}$, in which only the dispersion effects are significant.

10.3 Solution of the Nonlinear Pulse Propagation Equation

Equation (10.55) is a nonlinear Schrödinger equation governing the pulse propagating behavior in a nonlinear silica fiber. Except for the special case of soliton pulses, the equation does not possess a closed-form solution. It must be solved numerically. One numerical method that has been used successfully is the forward-marching beam propagation method or the split-step Fourier method [14]. According to this method, the solutions may be advanced first using only the nonlinear part of (10.55). Next the solutions are allowed to advance using only the linear part of (10.55). This forward stepping process is repeated until the desired destination is reached. The Fourier transform is accomplished numerically via the well-known fast Fourier transform technique [15]. With this approach, the evolution of pulses on the dominant mode as they propagate down the fiber may be obtained.

In general, the effects of GVD (due to β_2 or L_{D}) and those of the nonlinear SPM (due to γ or L_{NL}) will be felt by the propagating pulses as described by the propagation equation (10.55). Let us introduce a parameter, N, defined as

$$N = \left(\frac{L_{\mathrm{D}}}{L_{\mathrm{NL}}}\right)^{1/2} = \left(\frac{\gamma P_0 T_0^2}{|\beta_2|}\right)^{1/2}, \tag{10.56}$$

which measures the relative importance of GVD and SPM effects. When $N \gg 1$, the GVD effects dominate, and when $N \ll 1$, the SPM effects dominate. For $N = 1$, the GVD effects and SPM effects are balanced in such a way that if $\mathrm{sgn}(\beta_2) = -1$ (the anomalous dispersion region), propagating solitons may result. These solitons (specially shaped pulses) may propagate along a lossless fiber without any degradation forever. More will be said later.

Substituting (10.56) in (10.55) yields

$$-j\,\frac{\partial u}{\partial \xi} = \mathrm{sgn}\,(\beta_2)\,\frac{1}{2}\,\frac{\partial^2 u}{\partial \tau^2} - N^2\,e^{-\alpha L_{\mathrm{D}}\xi}\,|u|^2\,u. \tag{10.57}$$

An example of how an unchirped Gaussian pulse evolves as it propagates down the fiber with $N = 1$ and $\alpha = 0$ is shown in Fig. 10.2. When $\beta_2 > 0$, as shown in Fig. 10.2a, the initial Gaussian pulse begins to degrade and flatten as the pulse travels down the fiber. On the other hand, when $\beta_2 < 0$, (the anomalous dispersion regime), as shown in Fig. 10.2b, the initial unchirped Gaussian pulse appears to evolve to a pulse shape that corresponds to a soliton pulse. After the formation of the soliton pulse, that soliton pulse remains unchanged as it continues to propagate down the fiber. This means that for the case $N = 1$, $\beta_2 < 0$, and for $\alpha = 0$, soliton is the stable pulse on this fiber. The situation is quite different when the initial Gaussian pulse has significant chirp. In that case, a stable soliton pulse will not be formed. Therefore, chirp in the initial pulse must be minimized in order to form a soliton pulse.

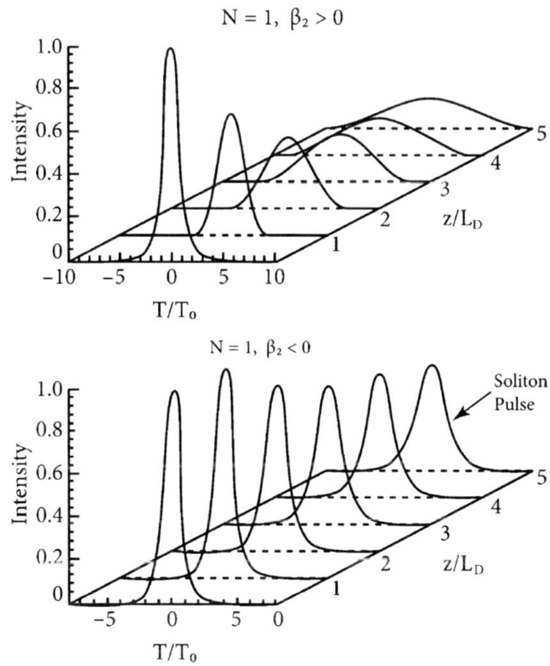

Figure 10.2. Evolution of pulse shapes over a distance of $5L_D$ for initially unchirped Gaussian pulse. The upper plot shows propagation in the normal dispersion region of the fiber with $\beta_2 > 0$ and $N = 1$. The lower plot shows propagation in the anomalous dispersion region with $\beta_2 < 0$ and $N = 1$. The pulse evolves to a soliton pulse [4]

10.4 Nonlinear Pulse Propagation for WDM Beams (Cross-Field Modulation Effects)

In a similar manner as the single beam case discussed in Sect. 10.2, we shall now provide the pulse propagation equations for the case where i optical pulses at different wavelengths are co-propagating in a single-mode fiber. For the sake of simplicity, we shall assume that these optical fields at different wavelengths are all linearly polarized in the y direction. Thus, the time dependent transverse electric field for the dominant mode can be written in the following form:

$$E_y(\mathbf{r},t) = \frac{1}{2}E_{y1}^{(t)}(\mathbf{r},t)\,e^{j\omega_1 t} + \frac{1}{2}E_{y2}^{(t)}(\mathbf{r},t)\,e^{j\omega_2 t} + \cdots$$

$$+ \frac{1}{2}E_{yi}^{(t)}(\mathbf{r},t)\,e^{j\omega_i t} + c.c. \tag{10.58}$$

$$(i = 1, 2, 3, \ldots I).$$

Here, $E_{yi}^{(t)}(\mathbf{r}, t)$ is the slowly varying pulse field amplitude of time on the ith beam, ω_i is the carrier frequency of the ith beam for the ith pulse, and I is the total number of beams. By substituting (10.58) in (10.3), and following through the procedures in the same way as was done for the single beam case, one may eventually arrive at the propagation equations for WDM beams, as was done by Agrawal [4].

Let us now consider this derivation from a heuristic and phenomenological point of view. Inspection of (10.51) shows that the nonlinear term in the propagation equation originates from the nonlinear index term, which in turn is caused by the intensity-induced polarization of the medium. This intensity refers to the total field intensity in the medium. When more than one field with different frequencies are present, the total intensity is the sum of all the intensities due to each separate field. This total field intensity will induce the nonlinear polarization, resulting in the nonlinear part of the index for the medium. This nonlinear index will prevail for all the beams. So, for each ith beam, from (10.34), the refractive index is

$$\Delta n_i \approx \frac{\epsilon_{\mathrm{NL}i}}{2n_{\mathrm{L}i}} \approx n_2 \left[|E_i|^2 + 2\sum_{m \neq i}^{I} |E_m|^2 \right] - \frac{j\alpha_i}{2k_0} \tag{10.59}$$

where $|E_i|^2$ is the intensity of the pulse on the ith beam and n_2 is the nonlinear coefficient for each beam. It has been assumed that this nonlinear coefficient is the same for all beams in the frequency spectrum of interest. The factor of 2 in front of the intensities due to all other beams other than the ith beam is not intuitively apparent. It may be traced back to the triple sum implied in (10.3) for the third term according to the more exact analysis [4].

It therefore follows that the pulse propagation equations for all the pulses on WDM beams are

$$\begin{aligned} &\frac{\partial A_i^{(t)}}{\partial z} + \beta_{1i} \frac{\partial A_i^{(t)}}{\partial t} - \frac{j\beta_{2i}}{2} \frac{\partial^2 A_i^{(t)}}{\partial t^2} + \frac{\alpha_i^{(\omega)} A_i^{(t)}}{2} \\ &\qquad = -j\gamma_i \left[\left|A_i^{(t)}\right|^2 + 2\sum_{m \neq i}^{I} \left|A_m^{(t)}\right|^2 \right] A_i^{(t)} \\ &\qquad\qquad (i = 1, 2, 3, \ldots, I). \end{aligned} \tag{10.60}$$

Here, for the ith wave, $A_i^{(t)}(z, t)$ is the slowly varying amplitude of the wave, $\beta_{1i} = 1/v_{\mathrm{g}i}$, where $v_{\mathrm{g}i}$ is the group velocity, β_{2i} is the dispersion coefficient ($\beta_{2i} = dv_{\mathrm{g}i}^{-1}/d\omega$), α_i the absorption coefficient, and

$$\gamma_i = \frac{n_2 \omega_i}{c A_{\text{eff}}} \tag{9.61}$$

is the nonlinear index coefficient with A_{eff} as the effective area for all beams and $n_2 \approx 3.2 \times 10^{-10}\,\text{m}^2\,\text{W}^{-1}$ for silica fibers, ω_i is the carrier frequency of the ith wave, and c is the speed of light in vacuum.

The fact that the nonlinear effects are twice as strong from the accompanying waves on different WDM wavelengths than from the self-wave is worth noting.

10.4.1 Self-Phase Modulation (SPM) and Cross-Phase Modulation (CPM)

Let us take a little detour and consider the case of I number of WDM plane waves co-propagating in a nonlinear medium characterized by the nonlinear index of refraction given by

$$n_i = n_{\text{L}i} + n_{\text{NL}i}, \tag{10.62}$$

where

$$n_{\text{L}i} = n_{\text{L}i}^{(\text{R})} - \frac{j\alpha_i}{2k_0}, \tag{10.63}$$

$$n_{\text{NL}i} = n_2 \left[|E_i|^2 + 2\sum_{m\neq i}^{I} |E_m|^2 \right]. \tag{10.64}$$

The zeroth order solution for these plane waves according to the perturbation theory is

$$\begin{aligned} \mathcal{E}_i(z,t) &\sim \frac{1}{2} e^{-j(\omega_i/c)(n_{\text{L}i}+n_{\text{NL}i})z} e^{j\omega_i t} + c.c., \\ &\sim \frac{1}{2} \exp\left(-j\frac{\omega_i n_{\text{L}i} z}{c}\right) \\ &\quad \exp\left[-j\frac{\omega_i}{c} n_2 \left(|E_i|^2 + 2\sum_{m\neq i}^{I} |E_m|^2 \right) z \right] e^{j\omega_i t} + c.c. \end{aligned} \tag{10.65}$$

The above expression shows that there is a self-induced phase shift, called SPM, as well as an induced phase shift due to the other co-propagating different wavelength field intensities experienced by the plane wave, called cross-phase modulation (CPM or XPM). The same nomenclatures will be used to describe the nonlinear effects in fiber.

10.4.2 Normalized Nonlinear Propagation Equations for WDM Beams

Introducing the normalizing coefficients

$$\tau = \frac{t - z/v_{g1}}{T_0},$$

$$d_{1i} = \frac{(v_{g1} - v_{gi})}{v_{g1} v_{gi}},$$

$$\xi = z/L_{D1},$$

$$L_{D1} = \frac{T_0^2}{|\beta_{21}|}, \tag{10.66}$$

and setting

$$u_i(\tau, \xi) = \frac{A_i(z,t)}{\sqrt{P_{0i}}} \exp\left(\alpha_i L_{Di} \xi / 2\right), \tag{10.67}$$

$$L_{NLi} = \frac{1}{\gamma_i P_{0i}},$$

$$L_{Di} = \frac{T_0^2}{|\beta_{2i}|} \tag{10.68}$$

gives

$$\begin{aligned} j\frac{\partial u_i}{\partial \xi} &= \frac{\mathrm{sgn}(\beta_{2i}) L_{D1}}{2 L_{Di}} \frac{\partial^2 u_i}{\partial \tau^2} + j \frac{d_{1i} L_{D1}}{T_0} \frac{\partial u_i}{\partial \tau} \\ &\quad - \frac{L_{D1}}{L_{NLi}} [\exp(-\alpha_i L_{D1} \xi) |u_i|^2 \\ &\quad + 2 \sum_{m \neq i}^{I} \exp(-\alpha_m L_{D1} \xi) |u_m|^2] u_i \end{aligned} \tag{10.69}$$

$$(i = 1, 2, 3, \ldots I).$$

Here, T_0 is the pulse width, P_{0i} is the incident optical power of the ith beam, and d_{1i}, the walk-off parameter between beam 1 and beam i, describes how fast a given pulse in beam i passes through the pulse in beam 1. In other words, the walk-off length is

$$L_{W(1i)} = \frac{T_0}{|d_{1i}|}. \tag{10.70}$$

So, $L_{W(1i)}$ is the distance for which the faster moving pulse (say in beam i) completely walked through the slower moving pulse in beam 1. The nonlinear interaction between these two optical pulses ceases to occur after a distance $L_{W(1i)}$. For CPM to take effect significantly, the group-velocity mismatch must be held to near zero.

Finding the analytic solution of (10.69), which is a set of simultaneously coupled nonlinear Schrödinger equations, is a formidable task. However, it may be solved numerically by the split-step Fourier method (forward marching beam propagation method) [14], which was used successfully earlier to solve the problem of propagation in complex fiber structures, such as fiber couplers, and to solve the thermal blooming problem for high energy laser beams [14]. According to this method, the solutions may be advanced first using only the nonlinear part of the equations, and then the solutions are allowed to advance using only the linear part of (10.69). This forward stepping process is repeated over and over again until the desired destination is reached. The Fourier transform is accomplished numerically via the well-known fast Fourier transform technique [15].

10.5 Soliton on a Single Beam

An ideal pulse to transmit one bit of information in a single mode optical fiber is a pulse that retains its shape and amplitude forever as it propagates in the fiber. This pulse is called a soliton. The existence of such a soliton was first shown theoretically by Hasegawa and Tappent in 1973 [1]. Then, in 1980, experimental demonstration of this soliton was carried out by Mollenauer et al. [1].

The equation that governs the solitons is the pulse propagation (10.57), with $\alpha = 0$,

$$-j\frac{\partial u}{\partial \xi} = \mathrm{sgn}\left(\beta_2\right)\frac{1}{2}\frac{\partial^2 u}{\partial \tau^2} - N^2\left|u\right|^2 u. \tag{10.71}$$

10.5.1 Bright Solitons

For the anomalous GVD case, $\mathrm{sgn}(\beta_2) = -1$, so (10.71) becomes

$$-j\frac{\partial u}{\partial \xi} = -\frac{1}{2}\frac{\partial^2 u}{\partial \tau^2} - N\left|u\right|^2 u. \tag{10.72}$$

This equation may be solved by the inverse scattering method or solved numerically by the split-step Fourier transform method. The inverse scattering method will not be covered here, but can be found elsewhere [16].

The nature of soliton is such that if the input pulse waveform at $\xi = 0$ is of a given soliton form, that waveform will repeat its shape and magnitude at certain time periods as the pulse propagates down the fiber. So, for $N = 1$, called the fundamental soliton case, the input pulse takes the following form:

$$u(0, \tau) = \text{sech}(\tau), \tag{10.73}$$

and the solution from (10.72) is

$$u(\xi, \tau) = \text{sech}(\tau)\, e^{-j\xi/2}. \tag{10.74}$$

Note that $\tau = (t - \beta z)/T_0$. This solution can be verified by substituting (10.74) back in (10.72). This means that if the input pulse is given by (10.73), this pulse will retain its shape and magnitude as it propagates along the fiber. The required condition $N = 1$ also provides a great deal of practical necessary requirements for the formation of the fundamental soliton. Since

$$N^2 = \frac{L_\text{D}}{L_\text{NL}} = \frac{\gamma P_0 T_0^2}{|\beta_2|}, \tag{10.75}$$

$N = 1$ means

$$L_\text{D} = L_\text{NL} \tag{10.76}$$

and

$$\frac{\gamma P_0 T_0^2}{|\beta_2|} = 1. \tag{10.77}$$

For example, if $T_0 = 10\,\text{ps}$, $\beta_2 = -2\,\text{ps}^2\,\text{km}$, $\gamma = 3\,\text{W}^{-1}\,\text{km}^{-1}$, then $P_0 = 6.7\,\text{mW}$, which is certainly obtainable.

For $N > 1$, higher order single beam solitons may form in a single mode fiber. In this case, the initial pulse must be

$$u(0, \tau) = N\,\text{sech}(\tau), \tag{10.78}$$

where N, the soliton order, is an integer and $u(\xi, \tau)$ must satisfy (10.72) for the $\text{sgn}(\beta_2) = -1$ case. For $N = 2$, the second-order soliton, the initial waveform is

$$u^{(2)}(0, \tau) = 2\,\text{sech}(\tau), \tag{10.79}$$

and for $\xi \geq 0$,

$$u^{(2)}(\xi, \tau) = \frac{4\left[\cosh(3\tau) + 3\, e^{-j4\xi}\cosh(\tau)\, e^{-j\xi/2}\right]}{\cosh(4\tau) + 4\cosh(2\tau) + 3(4\xi)}. \tag{10.80}$$

One notes that $\left|u^{(2)}(\xi,\tau)\right|^2$ is periodic in ξ with period $\xi = \pi/2$, which also turns out to be the period for all higher order solitons. The distinguishing feature for all higher order solitons ($N > 1$), in contrast with that of the fundamental soliton ($N = 1$), is that the shapes of all higher order solitons change as they propagate along the fiber but return to their initial shape periodically, while the shape of the fundamental soliton remains unchanged throughout its journey in the fiber.

10.5.2 Dark Solitons

For the normal GVD case, $\text{sgn}(\beta_2) = 1$, it appears that there can exist a dip in a uniform background whose shape is retained as it propagates along the fiber. This dip is called a dark soliton. The initial waveform for the fundamental dark soliton is

$$u_{\text{D}}(0,\tau) = \tanh(\tau), \tag{10.81}$$

while for $\xi \geq 0$,

$$u_{\text{D}}(\xi,\tau) = \tanh(\tau)\, e^{-j\xi}, \tag{10.82}$$

where u_{D} represent the field envelope for the dark soliton. The solution u_{D} satisfies the following equation:

$$-j\,\frac{\partial u_{\text{D}}}{\partial \xi} - \frac{1}{2}\frac{\partial^2 u_{\text{D}}}{\partial \tau^2} + \left|u_{\text{D}}\right|^2 u_{\text{D}} = 0. \tag{10.83}$$

It is pointed out that for $N > 1$ cases, there can exist gray solitons whose dips do not reach the zero level. Hence the name "gray" is used. A sketch of the bright, dark, and gray solitons is shown in Fig. 10.3.

10.6 Applications of Nonlinear Cross-Field Modulation (CPM) Effect

We note from the nonlinear pulse propagation equations (10.69) for WDM co-propagating beams that it is the nonlinear Kerr effect that produces the cross-field modulation effect, which couples the pulses on all co-propagating beams. The interaction of pulses on different beams via the nonlinear CPM effect is usually instrumental in destroying the integrity of solitons or other pulses on these WDM beams. Here we shall describe a few examples in which the CPM effect may be used to produce beneficial results. One of the most interesting applications is the formation of WDM solitons.

In the following examples, the governing equations are given by (10.69). They are a set of simultaneous coupled, nonlinear Schrödinger equations that may be solved numerically by the split-step Fourier method. With this approach, the evolution of all the pulses on all the co-propagting WDM beams as they propagate

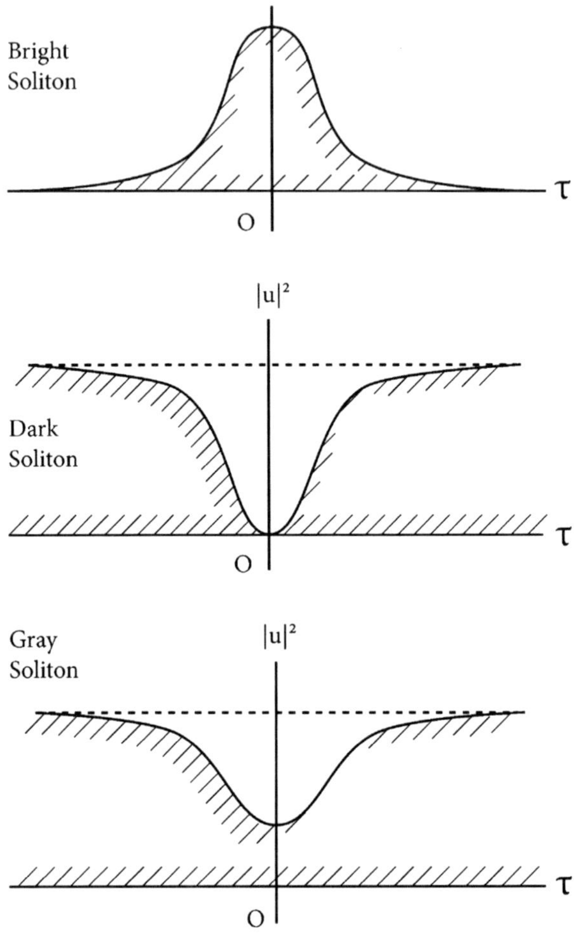

Figure 10.3. Sketches of bright, dark, and gray solitons

down the fiber may be obtained. It was through the numerical computations that the interesting results for these examples were found.

10.6.1 Pulse Shepherding Effect (Dynamic Control of In-Flight Pulses with a Shepherd Pulse) [9]

In a WDM system, the CPM effects caused by the nonlinearity of the optical fiber are unavoidable. These CPM effects occur when two or more optical beams co-propagate simultaneously and affect each other through the intensity dependence of the refractive index. This CPM phenomenon can be used to produce an interesting pulse shepherding effect [9]. It may be utilized to align the arrival time of pulses that are otherwise misaligned.

Consider now the evolution of two single pulses on two co-propagating beams whose operating wavelengths are separated by $\Delta\lambda = 4\,\text{nm}$. For this case, the four-wave-mixing effect is negligible. It is further assumed that the signal carrying pulses on each beam are separated by sufficiently large time intervals so that no interaction among succeeding pulses on the same beam occurs. The physical parameters chosen for the simulation correspond to an actual system that is of interest.

Parameter		
Length of fiber	L	50 km
Dispersion coefficient	β_2	$-1.61\,\text{ps}^2\,\text{km}^{-1}$
Operating wavelength of beam No. 1	λ_1	1.55 μm
Operating wavelength of beam No. 1	λ_2	1.546 μm
Nonlinear index coefficient	γ	$20\,\text{W}^{-1}\text{km}^{-1}$
Incident power of each beam	P_0	1 mW
Attenuation or absorption of each beam in fiber	α	$0.2\,\text{dB}\,\text{km}^{-1}$
Group velocity of beam	v_g	$2.051147 \times 10^8\,\text{m}\,\text{s}^{-1}$
Walk-off parameter of the beam No. 1 and beam No. j	$d_{1i} = v_{g1} - v_{gi}$	0 (no walk-off)
Pulse width	T_0	10 ps

With these values, the dispersion length L_D or the nonlinear length L_{NL}, which provides the length scale over which the dispersive or nonlinear effects for pulses on a single beam become important for pulse evolution along a fiber of length L, is $L_D = 62\,\text{km}$, or $L_{NL} - 50\,\text{km}$. Let us now consider the dynamic control of in-flight pulses with a shepherd pulse.

10.6.1.1 Without Shepherd Pulse

Shown in Fig. 10.4a is the evolution of two Gaussian pulses on two different wavelength beams as they propagate in this single mode fiber. These pulses are initially offset by 1/2 pulse width. It is seen that the pulses are affected by each other. Because of the nonlinear SPM and CPM effects, the pulses tend to attract each other. They appear to congregate towards a region of higher induced index of refraction. The leading pulse is pulled back while the trailing pulse is pushed forward so that these pulses tend to align with each other.

This observation is consistent with the earlier discovery of the self-focusing effect [17] where the induced higher refraction index region caused by the higher beam intensity tends to "attract" the propagating optical wave, resulting in the "focusing" of this optical wave. It is also consistent with the concept used to confine a thermally bloomed high-energy laser beam, where multiple surrounding beams are used to create an index environment in which the central main beam tends to

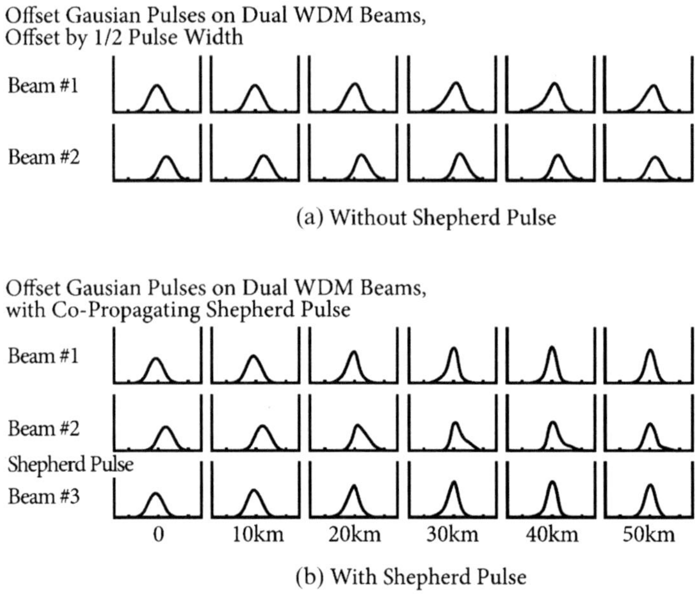

Figure 10.4. Evolution of two Gaussian pulses on two WDM beams separated by 1/2 pulse width: (**a**) without shepherd pulse on the third beam and (**b**) with shepherd pulse on the third beam [9]

expand less due to the lowering of the surrounding index of refraction caused by the heating from the surrounding beams [14]. It is also consistent with the "dragging" effect that occurs in weakly birefringent fibers [18].

It is expected, however, that when the two co-propagating pulses on two separate wavelength beams are separated by a sufficiently large distance, these two pulses will not interact with each other. This fact is demonstrated in Fig. 10.5a, where the two co-propagating pulses are separated by one pulse width. Each pulse propagates independently as if it were not aware of the presence of the other pulses. Thus, it appears that once these pulses are launched in this manner, the separation of these pulses cannot be altered, except through the introduction of a shepherd pulse as shown in the next section.

10.6.1.2 With Shepherd Pulse

It will be shown that the widely separated pulses on beam No. 1 and beam No. 2, as shown in Fig. 10.5a, can be brought significantly closer to each other by the launching of another pulse on a separate 1.542 μm wavelength beam with the proper magnitude and at the proper time. This pulse is called the shepherd pulse because of its shepherding behavior on the other pulse. In other words, it is possible to pull back the leading pulse and at the same time to push forward the trailing pulse to achieve near pulse alignment. This is shown in Fig. 10.5b.

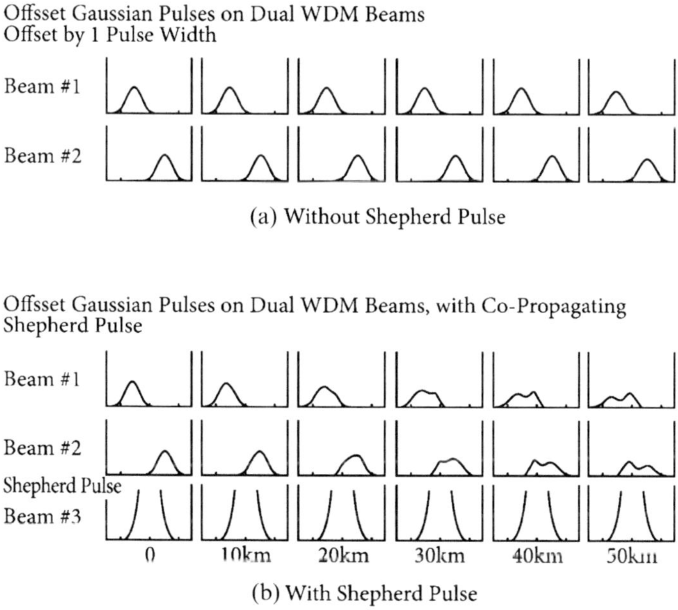

Figure 10.5. Evolution of two Gaussian pulses on two WDM beams separated by 1 pulse width: (**a**) without shepherd pulse on the third beam and (**b**) with shepherd pulse on the third beam [9]

The magnitude, shape, and the location of the shepherd pulse all contribute to the eventual success of this scheme to align these co propagating pulses. The fundamental phenomena that govern this scheme are the SPM, CPM, and GVD [4]. Computer simulation shows that a lower magnitude shepherd pulse does not possess sufficient attractive strength to pull the shepherded pulses together. For example, a magnitude 1 shepherd pulse, $\exp(-0.5\tau^2)$, situated in the middle of the shepherd pulses can bring these pulses only 10% closer to each other, while a magnitude 2 shepherd pulse, $2\exp(-0.5\tau^2)$, similarly situated can almost align these pulses (see Fig. 10.5b). It does not follow, however, that an even higher magnitude shepherd pulse can bring the shepherded pulses together sooner, because a magnitude 3 shepherd pulse's tremendous "pull" on the shepherded pulses tends to break up these pulses through the introduction of higher oscillations. There is a limit as to how strong the shepherd pulse can be.

Broadening the shepherd pulse, that is, using a $2\exp(-0.05\tau^2)$ pulse, only sharpens the shepherded pulses due to an increased apparent medium nonlinearity. The use of this broadened shepherd pulse can bring the shepherded pulses only 30% closer to each other, a far cry from the alignment achieved by the sharper shepherd pulse of $2\exp(-0.5\tau^2)$.

The next step, perhaps, is to use two shepherd pulses on two different wavelength beams to further enhance the shepherding effect. One, the $2\exp(-0.5\tau^2)$ shepherd pulse, on beam No. 3 at 1.542 μm may be used to pull the two shepherded pulses together, and the other $2\exp(-0.5\tau^2)$ shepherd pulse, on beam No. 4 at 1.538 μm may be used to sharpen the two shepherded pulses. When this simulation is done, it is discovered that the two shepherd pulses tend to break up the shepherded pulses into several oscillating pulses, an undesirable phenomenon.

The above computer simulation shows that there exists an optimum shepherd pulse with a certain magnitude, pulse width, pulse shape, and location with respect to the shepherded pulses that can provide the best alignment for these pulses. For the example with the physical parameters given here, the optimum shepherd pulse appears to be the $2\exp(-0.5\tau^2)$ pulse situated between the two shepherded pulses.

For the case of reduced separation of the pulses to be shepherded, as for the case in Fig. 10.4, a much more dramatic demonstration on the successful alignment achievable by a well designed shepherd pulse can be seen in Fig. 10.4b. Here, the Gaussian pulses on beam No. 1 and on beam No. 2 are offset by 1/2 pulse width. A Gaussian shepherd pulse of unit magnitude, aligned with the pulse on beam No. 1, is introduced on beam No. 3 whose wavelength is 1.542 μm. This wavelength is 4 nm from beam No. 2 and 8 nm from beam No. 1, assuring that the four wave mixing effect is negligible. It is observed that this shepherd pulse is capable of achieving excellent alignment of the wayward pulse (pulse on beam No. 2) with the reference pulse (pulse on beam No. 1)

Another case demonstrating the effectiveness of a shepherd pulse to control and align the shepherded pulses is shown in Fig. 10.6. Here, two Gaussian pulses on two different wavelength beams with wavelengths 1.55 and 1.546 μm, originating in an aligned position as shown in Fig. 10.6a, begin to separate from each other because of a slight difference in the group velocities for these two beams. Without the presence of a shepherd pulse, these beams will be approximately 1/2 pulse width apart at 50 km downstream as can be seen in Fig. 10.6a. With the shepherd pulse of $2\exp(-0.5\tau^2)$ on a third beam with wavelength 1.542 μm, originally aligned with the two shepherded pulses and propagating at the same velocity as the pulse on beam No. 1, at 50 km downstream, the shepherded pulses are still aligned as shown in Fig. 10.6b.

In conclusion, demanding "fast" walk-off of co-propagating beams from each other in order to avoid any deleterious walk-off effect among the beams and to minimize the interaction among these beams due to the nonlinear behavior of the fiber medium is unnecessary. On the contrary, it is found that, by requiring as little walk-off as possible, the "shepherding" effect among the various beams may be used to "herd" them together, resulting in the desirable characteristic of simultaneous arrival of co-propagating beams in a BPW (bit-parallel wavelength) system.

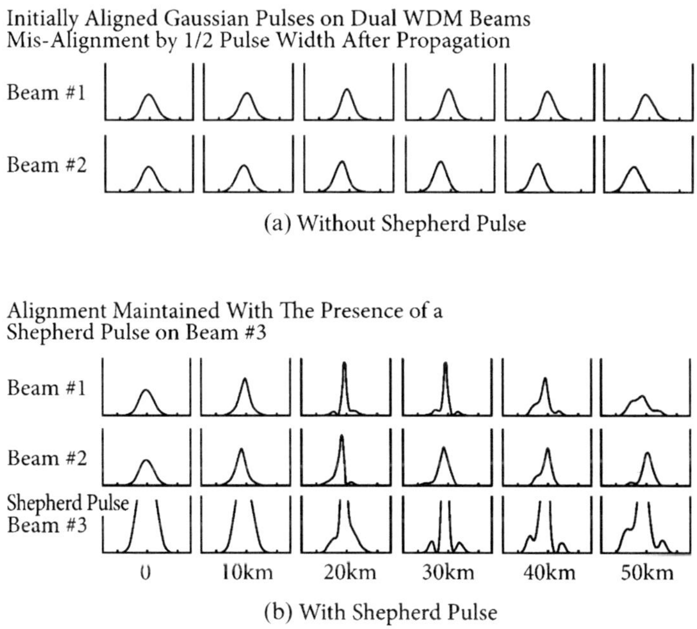

Figure 10.6. Evolution of two initially aligned Gaussian pulses on two WDM beams: (**a**) after propagation, separation occurs for pulses on beam No. 1 and beam No. 2 without shepherd pulse on the third beam, and (**b**) alignment maintained for pulses on beam No. 1 and beam No. 2 with shepherd pulse on the third beam [9]

10.6.2 Enhanced Pulse Compression in a Nonlinear Fiber by a WDM Optical Pulse [10]

Let us now demonstrate how the CPM effect can be used to produce a highly compressed pulse on a different wavelength beam [10]. The usual soliton-effect compressor [19], which makes use of higher-order solitons supported by fiber as a result of interplay between SPM and anomalous GVD, is well known. It is found that the interplay between CPM and GVD may also provide similar pulse compression effects. The significant difference is that pulse compression can take place for pulses on a different wavelength beam. This means that the high power pulse on one wavelength beam may be used to provide high compression to a low power pulse on another wavelength beam.

Now consider the evolution of two single soliton pulses on two co-propagating beams whose operating wavelengths are separated by $\Delta\lambda > 4\,\text{nm}$. For this case, the four wave mixing effect is negligible. Let us label the first pulse as the primary (P) pulse and the second pulse as the shepherd (S) pulse. The soliton number N_i for the pulse on the ith beam is defined as

$$N_i^2 = \frac{L_{\mathrm{D}i}}{L_{\mathrm{NL}i}}. \tag{10.84}$$

Furthermore, we assume that there is negligible walk-off, that is,

$$\begin{aligned} d_{1i} &= \text{the walk off parameter between beam No.1 and beam No. } i \\ &= v_{\mathrm{g1}} - v_{\mathrm{gi}} = 0, \end{aligned} \tag{10.85}$$

and there is no loss, that is,

$$\begin{aligned} \alpha_i &= \text{attenuation or absorption of pulses on beam } i \text{ in fiber} \\ &= 0. \end{aligned} \tag{10.86}$$

The neglect of fiber loss is justified since fiber lengths typically employed are only a small fraction of the absorption length ($\alpha_i L \ll 1$). Numerical simulation shows that significant pulse compression still exists for these interacting pulses.

10.6.2.1 Shepherding and Primary Pulses are all in the Anomalous Dispersion Region

For solitons propagating on a single beam in silica fibers, pulse compression is experienced when N, the soliton order, is larger than 1 [4]. This effect is due to the interaction of SPM and anomalous GVD during propagation. When two aligned pulses, one called the primary pulse and the other called the shepherd pulse, on two different wavelength beams co-propagate in a single-mode fiber, compression on both pulses occur due to the interaction of cross phase modulation of these two pulses and anomalous GVD during propagation.

Initial Pulse Widths are Identical

Computer simulation results are shown in Figs. 10.7–10.10 for co-propagating pulses with identical initial pulse widths. Both pulses are in the anomalous GVD regime. In Fig. 10.7 the maximum amount of compression experienced by both pulses, the primary (P) pulse and the shepherd (S) pulse, are plotted against the soliton order N_{s} for the shepherd pulse for various cases of primary pulse with the soliton order N_{p}. The amount of compression is expressed by the compression factor F_{c}, which is defined as [19]

$$F_{\mathrm{c}} = \frac{T_{\mathrm{FWHM}}}{T_{\mathrm{COMP}}}, \tag{10.87}$$

where the subscript FWHM means the full width at half maximum of the pulse, and the subscript COMP means the FWHM of the compressed pulse. It is seen that, in the absence of the shepherd pulse, that is, $N_{\mathrm{s}} = 0$, the primary pulse undergoes

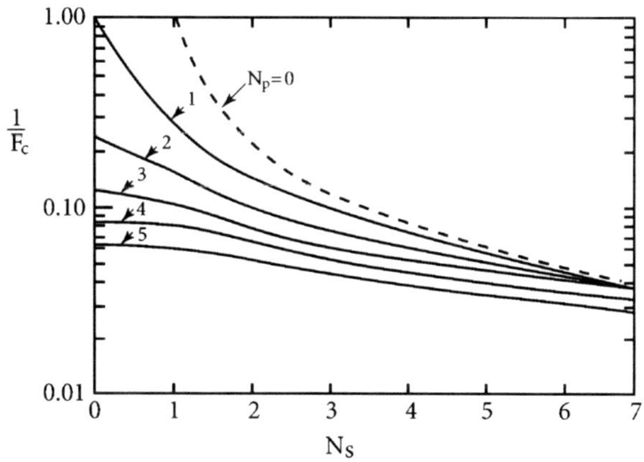

Figure 10.7. Compression factor (F_c) for various soliton values (N_p) of a primary pulse (P) as a function of soliton values (N_s) of a co-propagating shepherd pulse (S). The compression factor for the primary pulse is the same as the compression factor for the shepherd pulse. The initial pulse width for the primary pulse and that for the shepherd pulse is identical. The compression factor F_c is defined as the ratio between the full width at half maximum for the initial uncompressed pulse and that for the final compressed pulse [10]

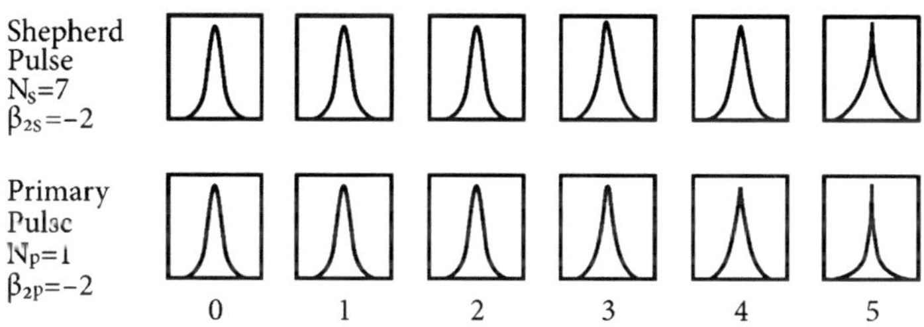

Figure 10.8. An illustration of the evolution of the shepherd pulse and the primary pulse for $N_s = 7$ and $N_p = 1$ and $z_{opt}/z_0 = 0.07$. Both pulses are in the anomalous dispersion region. The power amplitude $|u|^2$ is plotted in each frame. The highest power amplitude in each frame is normalized to unity. The initial power amplitude for the shepherd pulse is 49 ($N_s = 7$) and that of the primary pulse is 1 ($N_p = 1$). The final power amplitude for the shepherd pulse is 71.2 and that of the primary pulse is 2.15. The number along the horizontal abscissa refers to the normalized distance from the starting point of the fiber; in other words, when the normalized distance is 3, the distance is $3(z_{opt}/z_0)z_0/5$, where $z_0=(\pi/2)L_{Ds}$, and L_{Ds} is the dispersion length of the shepherd pulse, z_{opt} is the optimum fiber length in kilometer for the shepherd pulse when it experiences maximum pulse compression. Note that both pulses with different initial soliton numbers are similarly compressed, and the degree of compression for both pulses is higher than that experienced by each pulse when propagating alone. The dispersion coefficients β_{2s} and β_{2p} have units of ($ps^2\ km^{-1}$). All other numbers in the figure are dimensionless [10]

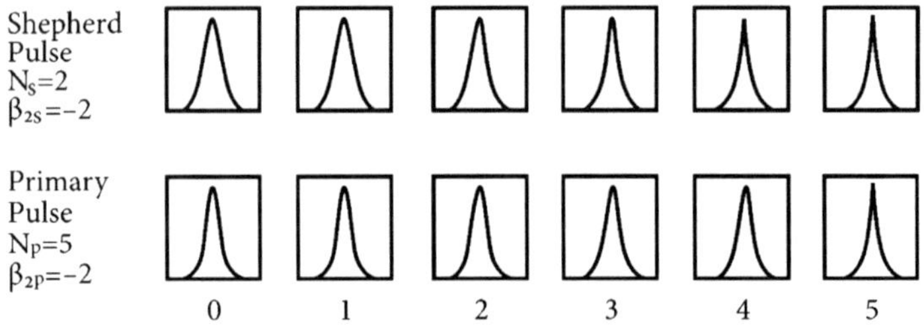

Figure 10.9. An illustration of the evolution of the shepherd pulse and the primary pulse for $N_\text{s} = 2$ and $N_\text{p} = 5$ and $z_\text{opt}/z_0 = 0.065$. Both pulses are in the anomalous dispersion region. The power amplitude $|u|^2$ is plotted in each frame. The highest power amplitude in each frame is normalized to unity. The initial power amplitude for the shepherd pulse is 4 ($N_\text{s} = 2$) and that of the primary pulse is 25 ($N_\text{p} = 5$). The final power amplitude for the shepherd pulse is 6.96 and that of the primary pulse is 35.1. The number along the horizontal abscissa refers to the normalized distance from the starting point of the fiber; in other words, when the normalized distance is 3, the distance is $3(z_\text{opt}/z_0)z_0/5$, where z_0=$(\pi/2)L_\text{Ds}$, and L_Ds is the dispersion length of the shepherd pulse, z_opt is the optimum fiber length in kilometer for the shepherd pulse when it experiences maximum pulse compression. Note that both pulses with different initial soliton numbers are similarly compressed, and the degree of compression for both pulses is higher than that experienced by each pulse when propagating alone. The dispersion coefficients $\beta_{2\text{s}}$ and $\beta_{2\text{p}}$ have units of (ps^2 km^{-1}). All other numbers in the figure are dimensionless [10]

the well-known soliton compression process for a single soliton number $N > 1$. As expected, the primary pulse retains its shape when $N_\text{p} = 1$. But, when a co-propagating shepherd pulse is present, both pulses undergo the same compression even if N_p is not equal to N_s or if $N_\text{s} < 1$ or if $N_\text{s} \ll 1$. Furthermore, the amount of compression is always larger than that achievable by a single stand-alone pulse.

For $N_\text{s} > N_\text{p}$, the shepherd pulse helps to compress the primary pulse further, especially when the soliton number for the primary pulse is near unity. For example, as N_s varies from 1 to 7, the pulse width of the $N_\text{p} = 1$ primary pulse can be compressed by the shepherd pulse by a factor of 27, while the pulse width of the $N_\text{p} = 2$ primary pulse will be compressed by a factor of 7. For an $N_\text{p} = 5$ primary pulse, its pulse width will be reduced by a factor on only 2.2 as N_s varies from 1 to 7. In other words, the weaker the intensity of the primary pulse, the more its pulse width will be compressed by the presence of a co-propagating high intensity shepherd pulse. Figure 10.8 gives an illustration of the evolution of the pulse shapes of the primary and shepherd pulse for the case where $N_\text{s} = 7$ and $N_\text{p} = 1$.

For $N_\text{s} < N_\text{p}$, the shepherd pulse still helps to compress the primary pulse further, but the effect is much more moderate. For example, as N_s varies from 0 to 2, the pulse width of the $N_\text{p} = 2$ primary pulse is compressed by a factor of 2.4,

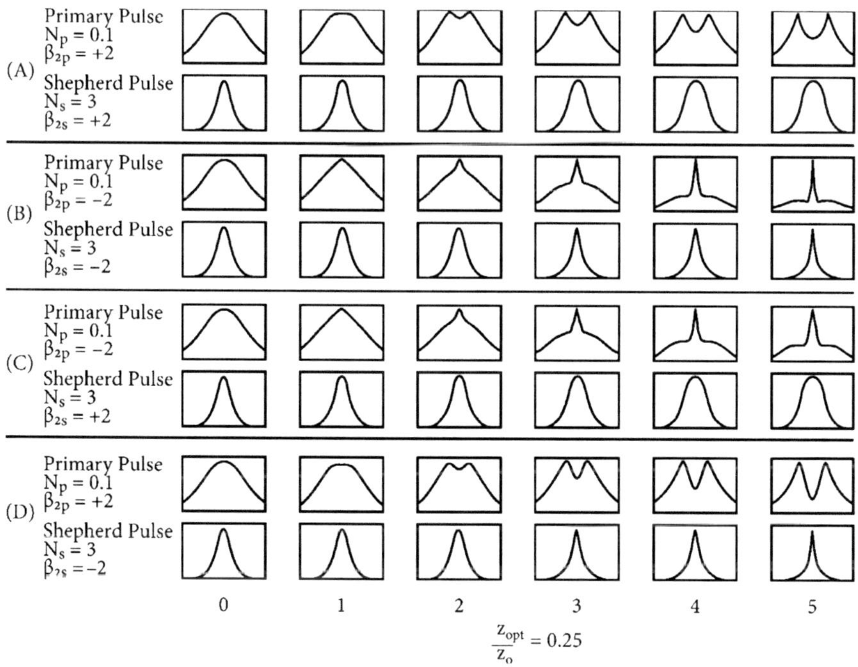

Figure 10.10. Evolution of two propagating pulses in different dispersion regions. The initial pulse amplitude of the primary pulse (pulse 1) is $N_p = 0.1$ and the initial pulse amplitude of the shepherd pulse (pulse 2) is $N_s = 3$. The initial pulse width of the primary pulse (pulse 1) is three times the initial pulse width of the shepherd pulse (pulse 2). (**A**) Primary pulse 1 and shepherd pulse are both in the normal dispersion region ($\beta_2 = +2$). (**B**) Primary pulse 1 and shepherd pulse 2 are both in the anomalous dispersion region ($\beta_2 = -2$). (**C**) Primary pulse 1 is in the anomalous dispersion region ($\beta_2 = -2$) and the shepherd pulse is in the normal dispersion region ($\beta_2 = +2$). (**D**) Primary pulse 1 is in the normal dispersion region ($\beta_2 = +2$) and the shepherd pulse is in the anomalous dispersion region ($\beta_2 = -2$). The power amplitude $|u|^2$ is plotted in each frame. The highest power amplitude in each frame is normalized to unity. The initial power amplitude for the shepherd pulse is 9 ($N_s = 3$) and that of the primary pulse is 0.01 ($N_p = 0.1$). The final power amplitudes for the shepherd pulse are (**A**) = 6.59, (**B**) = 14.8, (**C**) = 6.59, and (**D**) = 14.8 and those of the primary pulse are (**A**) = 0.0116, (**B**) = 0.0317, (**C**) = 0.0105, and (**D**) = 0.0121. The number along the horizontal abscissa refers to the normalized distance from the starting point of the fiber; in other words, when the normalized distance is 3, the distance is $3(z_{opt}/z_0)\ z_0/5$, where $z_0 = (\pi/2)L_{Ds}$, and L_{Ds} is the dispersion length of the shepherd pulse. z_{opt} is the optimum fiber length in kilometer for the shepherd pulse when it experiences maximum pulse compression. The dispersion coefficients β_{2s} and β_{2p} have units of ($ps^2\,km^{-1}$). All other numbers in the figure are dimensionless [10]

while the pulse width of the $N_p = 5$ primary pulse will be compressed by a factor of only 2 as N_s varies from 0 to 5. This means that to effectively enhance the pulse compression of a primary pulse, a higher intensity shepherd pulse must be used. Figure 10.9 shows the evolution of the pulse shapes of the primary and shepherd pulses for the case where $N_s = 2$ and $N_p = 5$.

It is known that a single pulse with $N < 1$, no pulse compression will occur. Hence, a $N_p < 1$ primary pulse traveling alone, or a $N_s < 1$ shepherd pulse traveling alone, will not experience any pulse compression. This is no longer true when these pulses co-propagate in the fiber. Even when $N_p + N_s < 1$, a slight pulse compression may still be observed for both the primary and secondary pulses. This is caused by the nonlinearity of the fiber medium. One also notes that when $N_p \ll 1$ and $N_s > 1$, pulse compression will be experienced by both the primary and shepherd pulses. The same degree of pulse compression will occur on the primary pulse even when $N_p \ll 1$. The degree of pulse compression for the primary or shepherd pulse is governed by the $N_s > 1$ shepherd pulse.

We have also calculated the normalized optimum fiber length z_{opt}/z_{0p} for the primary pulse as a function of N_s for various fixed values of N_p, where z_{opt} is the optimum length in kilometer for the primary or shepherd pulse when it experiences maximum pulse compression and z_{0p}=$(\pi/2)\, L_{Dp.}$ Here, L_{Dp}, the dispersion length for the primary pulse, is $T_0^2/\left|\beta_{2p}\right|$. We note that z_{opt} for the primary pulse occurs at the same location or very near the same location as that for the shepherd pulse. This means that the maximum pulse compression for the primary pulse and that for the shepherd pulse occur at the same location and at the same time. For high values of N_s, this normalized optimum fiber length can be much smaller than unity, indicating that the maximum pulse compression could occur at a length many times smaller than the dispersion length. Using, as an example, the following physical parameters:

Dispersion coefficient	β_2	$-2.0\,\text{ps}^2\,\text{km}^{-1}$
Operating wavelength of beam No. 1	λ_1	1,552 μm
Operating wavelength of beam No. 1	λ_2	1.548 μm
Nonlinear index coefficient	γ	$20\,\text{W}^{-1}\text{km}^{-1}$
Incident power of each beam	P_0	1 mW
Attenuation or absorption of each beam in fiber	α	$0\,\text{dB}\,\text{km}^{-1}$
Group velocity of beam	v_g	$2.051147 \times 10^8\,\text{m}\,\text{s}^{-1}$
Walk-off parameter of the beam No. 1	$v_{g1} - v_{gj}$	
and beam No. i	$v_{g1} - v_{gj}$	0 (no walk-off)
Pulse width	T_0	10 ps

we obtain

$$L_{Dp} = 50\,\text{km}.$$

Take the case of $N_p - 5$ and $N_s = 7$, onc finds $z_{opt}/z_{0p} = 0.04$. This means that maximum pulse compression can occur in a fiber with length of only 2.0 km long. For higher values of N_p and/or N_s, this length can be made even shorter.

Initial Pulse Widths are Not Identical

We also investigated the case where the pulse widths of the primary pulse and that of the shepherd pulse are not identical. Let us consider the case where a primary pulse has an initial intensity of $N_p = 1$and a shepherd pulse has an initial intensity of $N_s = 9$. It was assumed that the pulse width of the shepherd pulse is varied from the same to several times (3–5 times) wider than that of the primary pulse. The computer simulation shows that the primary pulse is similarly compressed for all the above cases. In other words, varying the pulse width of the shepherd pulse does not appear to affect the minimum pulse width achievable for the primary pulse, although the distance required to gain this minimum pulse width for the primary is increased as the pulse width of the shepherd pulse is increased. The amount of pulse compression for the primary pulse is governed by the intensity of the accompanying shepherd pulse. It is observed that for the broad shepherd pulse only the central portion of the shepherd pulse that overlaps the primary pulse is significantly affected and undergoes compression.

This simulation shows that the broader shepherd pulse with high intensity appears to enhance (or increase) the strength of the nonlinear coefficient of the fiber medium for the primary pulse, so as to enhance the pulse compression effect experienced by the primary pulse. This means that there is a way to increase the nonlinear effect of the medium dynamically through the addition of a broad, high intensity shepherd pulse. The amount of enhancement and the duration are controlled by the intensity and the pulse width of the shepherd pulse. The nonlinear effect of the medium is transferred to the primary pulse through the CPM effect.

Let us now investigate the case where the intensity of the narrow shepherd pulse is much higher than that of the broad primary pulse. In this simulation, the initial intensity of the narrow shepherd pulse is taken to be $N_s = 9$ and that of the broad primary pulse is $N_p < 1$. Both pulses undergo compression. The degree of compression is mostly governed by the high intensity narrow shepherd pulse. For example, at the maximum compression distance, the shepherd pulse is compressed by a factor of approximately 16, while a narrow pulse with the same compressed pulse width as that of the shepherd pulse appears to have been generated on top of the broad small intensity primary pulse, which appears as the pedestal for the narrow pulse.

We note here that what has been described above has practical significance. This scheme provides a practical pure optical way of generating very narrow bits on different wavelength streams for the bit-parallel data format.

10.6.2.2 The Shepherd Pulse is in the Normal Dispersion Region and the Primary Pulse is in the Anomalous Dispersion Regime

It is known that pulse compression of a single pulse in a fiber occurs because of the interaction of the nonlinear effect and the anomalous GVD effect [4]. This interaction also gives birth to the possible existence of a soliton pulse with $N = 1$. The above simulation results show that when a shepherd pulse is added as a co-propagating companion primary pulse, enhancement of pulse compression of the primary pulse is observed. It is of interest to learn if this pulse compression enhancement of the primary pulse still exists if the shepherd pulse is launched on a beam whose wavelength falls in the normal GVD regime. This computer experiment has been carried out. In this experiment, N_{p} is set to unity with $\beta_{2\mathrm{p}} = -2$, while N_{s} is set to 9 with $\beta_{2\mathrm{s}} = +2$. It is expected that without the shepherd pulse, the primary pulse is a soliton pulse, which will retain its shape without pulse compression or pulse spreading as it propagates down the fiber. Also, without the primary pulse, the high amplitude shepherd pulse in the normal dispersion regime is expected to propagate without experiencing pulse compression. When both of these pulses co-propagate on two separate beams, the pulse shepherding effect is observed, but no pulse compression is observed.

If N_{p} and N_{s} are both set equal to 9, the high amplitude of the primary pulse in the anomalous dispersion regime produces large pulse compression, but the degree of pulse compression (i.e., the narrowness of the compressed pulse) is not influenced by the presence of the high amplitude shepherd pulse in the normal dispersion regime. On the other hand, a very significant dip appears in the center of the shepherd pulse in the normal dispersion regime, breaking the original single shepherd pulse into two pulses. This is very different than the case where both primary and shepherd pulse are in the anomalous dispersion region. There, both pulses undergo compression.

10.6.2.3 The Shepherd Pulse and Primary Pulses are all in the Normal Dispersion Region

When both shepherd and primary pulses are in the normal dispersion region, no pulse compression occurs. Pulses tend to congregate toward the region of higher-induced index of refraction.

10.6.2.4 Additional Simulation Study on WDM Copropagating Pulses

The interaction between two separate pulses co-propagating on two different wavelength beams in a single-mode fiber is studied further. We note that the cross-phase modulation effect can be used effectively to provide another way to generate pulse compression in the anomalous dispersion region of a single-mode fiber. Because of the nonlinearity of the fiber medium, a slight pulse compression still occurs when the sum of the soliton numbers for the two beams is less than unity.

A more complex interaction is observed when one of the pulses is propagating in the normal dispersion region. The pulse in the normal dispersion region is seen to be broken up by the compression of the high soliton number pulse in the anomalous dispersion region. It also appears that if the pulse in the normal dispersion region is very broad compared with the high intensity narrow pulse in the anomalous dispersion region, a dark soliton like pulse can be generated on top of the broad pulse in the normal dispersion region, while the pulse in the anomalous dispersion region undergoes the usual pulse compression. Figure 10.10 is introduced to illustrate the evolution of the two propagating pulses when they exist in various different combinations of the dispersion regions.

It should be noted that the dispersion region in which the beam resides (i.e., where the beam wavelength resides) is all important in determining the behavior of the pulse on that beam even in the presence of a co-propagating pulse on a different wavelength beam. The co-propagating shepherd pulse, through the CPM effect due to the Kerr index nonlinearity, provides an additional phase retardation to the primary pulse as it travels down the fiber. In other words, an additional frequency chirp (in addition to that caused by SPM) is added to the primary pulse by the co-propagating shepherd pulse.

This "chirped" primary pulse is acted upon by the fiber's dispersion to yield the expected behavior. For example, if the primary pulse is on a beam whose wavelength is in the anomalous dispersion region (negative GVD region) and if the chirp caused by self-and cross-modulation effects is high enough, the leading half of the pulse containing the lowered frequencies will be retarded, while the trailing half, containing the higher frequencies, will be advanced, and the primary pulse will tend to collapse upon itself, resulting in pulse narrowing or pulse compression (see Figs. 10.10b,c).

On the other hand, if the primary pulse is on a beam whose wavelength is in the normal dispersion region (positive GVD region), the presence of co-propagating shepherd pulse on a different wavelength beam induces a dark-soliton-like behavior of the primary pulse, confirming the fact that the dispersive region in which the wavelength of the beam resides determines the propagation characteristic of that pulse. In contrast with the bright soliton case, a dark soliton possesses a nontrivial phase profile, which is a function of time, resulting in a rapid dip in the intensity of a broad pulse (see Figs. 10.10a,d).

The interaction of pulses on more than two beams has also been studied. As many as ten simultaneously propagating pulses on ten separate beams, with one carrying the shepherd pulse, were used. It was found that a single large amplitude shepherd pulse could similarly and simultaneously affect the other nine small amplitude pulses. The evolution of each of the small amplitude pulses depended mainly on the interaction of that pulse with the large amplitude shepherd pulse according to the manner discussed above for the two beam interaction case. Through CPM, co-propagating pulses on separate beams appear to share the nonlinear effect induced on any one of the pulses on separate beams

We find that for a WDM system, one shepherd pulse can cause the compression of all the other wavelength pulses, thereby improving their pulse widths as well as the separation of different pulses. Furthermore, since the longer wavelength pulses compress at rate different from the shorter wavelength pulses, we may conceivably give all pulses the same time width, which may make detection and discrimination easier to accomplish.

Here, we have shown a way to compress a bright or dark pulse. The nonlinear CPM effect is used to accomplish this on two or more co-propagating pulses on two or more WDM beams in a single-mode fiber. This numerical simulation shows that the effectiveness of compression is similar to that displayed by single higher-order soliton pulse propagating in a single beam. That this CPM effect can be used to compress pulses whose amplitudes are much less than unity (the traditional soliton number for a single beam) as long as a co-propagating pulse on a WDM beam undergoes compression should be noted.

10.6.3 Generation of Time-Aligned Picosecond Pulses on Wavelength-Division-Multiplexed Beams in a Nonlinear Fiber [11]

The difficulty in the generation of time-aligned pulses in the picosecond range on WDM beams is well recognized. Yet, these time-aligned pulses may be the backbone for the future ultra-high-speed bit-parallel communication system [12]. A way to generate these pulses is described here.

In spite of the intrinsically small value of the nonlinearity coefficient in fused silica, because of low loss and long interaction length, the nonlinear effects in optical fibers made with fused silica cannot be ignored even at relatively low power levels [20]. This nonlinear phenomenon in fibers has been used successfully to generate optical solitons [1] to compress optical pulses [4], to transfer energy from a pump wave to a Stokes wave through the Raman gain effect [4], to transfer energy from a pump wave to a counter propagating Stokes wave through the Brillouin gain effect [4], to produce four-wave mixing [4], to dynamically shepherd pulses [9], and to enhance pulse compression [10]. Here, we wish to describe one more: the generation of time-aligned pulses.

In a WDM fiber system, the CPM effects [4] caused by the nonlinearity of the optical fiber are unavoidable. These CPM effects occur when two or more optical beams co-propagate simultaneously and affect each other through the intensity dependence of the refractive index. This CPM phenomenon can be used to generate time-aligned data pulses [11].

10.6.3.1 Generation of Time-Aligned Pulses

A high-power, picosecond pulse, called the shepherd pulse, is launched on a given beam. A number of low-power beams that are selected based on the WDM format are launched without any signal pulses into a single-mode nonlinear fiber. These beams co-propagate with the beam carrying the shepherd pulse in this fiber. It will be shown first through numerical simulation results and then through experimental measurements that time-aligned pulses will appear on these low-power WDM beams. The nonlinear CPM effect in a single-mode fiber is instrumental in the generation of these time-aligned pulses. It is also required that the walk-off among all the beams be kept at a minimum acceptable value.

10.6.3.2 Computer Simulation Results

Computer-simulation results for the generation of three simultaneous pulses on three beams with separate wavelengths by a large amplitude shepherd pulse on the fourth beam are shown in Fig. 10.11. It is assumed that wavelength separation

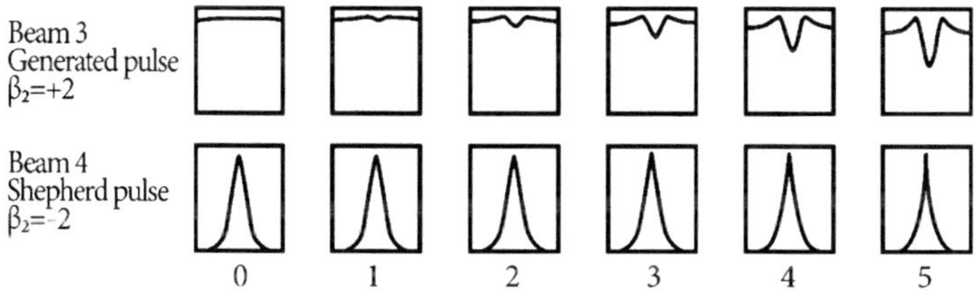

Figure 10.11. Pulse evolution picture for the generation of three simultaneous pulses on three beams with separate wavelengths by a large amplitude shepherd pulse on the fourth beam. For beam 1, $\beta_2 = -1.5$; for beam 2, $\beta_2 = -1.0$; for beam 3, $\beta_2 = +2.0$; and for beam 4, the shepherd pulse beam, $\beta_2 = -2.0$. The effect of different values of the dispersion coefficient on the induced pulses can be seen. In the negative β_2 region (the anomalous group-dispersion region of the fiber), the induced pulses are "bright" pulses, and in the positive β_2 region (the normal group-dispersion region of the fiber), the induced pulse is a "dark" pulse. The higher the $|\beta_2|$ value, the higher the amplitude of the induced pulse [11]

among the beams is larger than 5 nm and the shepherd pulse has a pulse width of 60 ps. Because of this wide wavelength separation of the beams as well as the width of the shepherd pulse, the four-wave-mixing effect is negligible for this case. Other parameters are the length of fiber $L = 20\,\text{km}$; the dispersion coefficient $|\beta_2| \leq 2\,\text{ps km}^{-1}$; the operating wavelength of beam 1, $\lambda_1 = 1.55\,\mu\text{m}$; the operating wavelength of beam 2, $\lambda_2 = 1.545\,\mu\text{m}$; the operating wavelength of beam 3, $\lambda_3 = 1.535\,\mu\text{m}$; the operating wavelength of beam 4, $\lambda_4 = 1.555\,\mu\text{m}$; the nonlinear index coefficient $\gamma = 20\,\text{W}^{-1}\,\text{km}^{-1}$; the attenuation or absorption of each beam in fiber, $\alpha = 0.2\,\text{dB km}^{-1}$; the group velocity of the beam, $v_g = 2.051147 \times 10^8\,\text{m s}^{-1}$; the walk-off parameter between the slowest beam and the fastest beam, $d_{12} < 3\,\text{ps km}^{-1}$; and the pulse width $T_0 = 60\,\text{ps}$. It has been assumed that the dispersion coefficients for these beams are the following: for beam 1, $\beta_2 = -1.5$; for beam 2, $\beta_2 = -1.0$; for beam 3, $\beta_2 = +2$; and for beam 4, the shepherd pulse beam, $\beta_2 = -2.0$. The effect of different values of the dispersion coefficient on the induced pulses can be seen from the resultant data. In the negative β_2 region (the anomalous GVD region), the induced pulses are "bright" pulses, and in the positive β_2 region (the normal GVD region), the induced pulse is a "dark" pulse, that is, a dip. The higher the $|\beta_2|$ value, the higher the amplitude of the induced pulse.

The simulation shows that large walk-off among the beams, that is, walk-off larger than a full shepherd pulse width within the dispersion length for the shepherd pulse, would destroy the capability of the shepherd pulse to generate time-aligned pulses on the co-propagating primary beams. It is for this reason that the selection of a proper fiber is of utmost importance. The following experiment will show that this demand, although rather stringent, can still be satisfied.

10.6.3.3 Experimental Setup and Results

Schematic block diagrams of two experimental setups are shown in Fig. 10.12. The pulse source is a tunable erbium-doped fiber-ring laser, producing a 100 MHz train of pulses 60 ps in width between the wavelength range of 1,530–1,560 nm. This erbium ring pulse is named the shepherd pulse operating at a peak power of higher than 200 mW. The primary sources are distribution feedback laser diodes at 1,535, 1,540, 1,545, and 1,557 nm operated under a dc bias well above threshold. This cw output from the primary laser diode source is about 1 mW, which is amplified through an erbium-doped fiber amplifier to around 33 mW.

Measurement of Fiber Characteristics

As indicated earlier, the selection of a proper fiber is of great importance in the successful generation of time-aligned pulses on co-propagating WDM beams. The Corning dispersion shifted (DS) fiber [21] is chosen to be the single-mode fiber

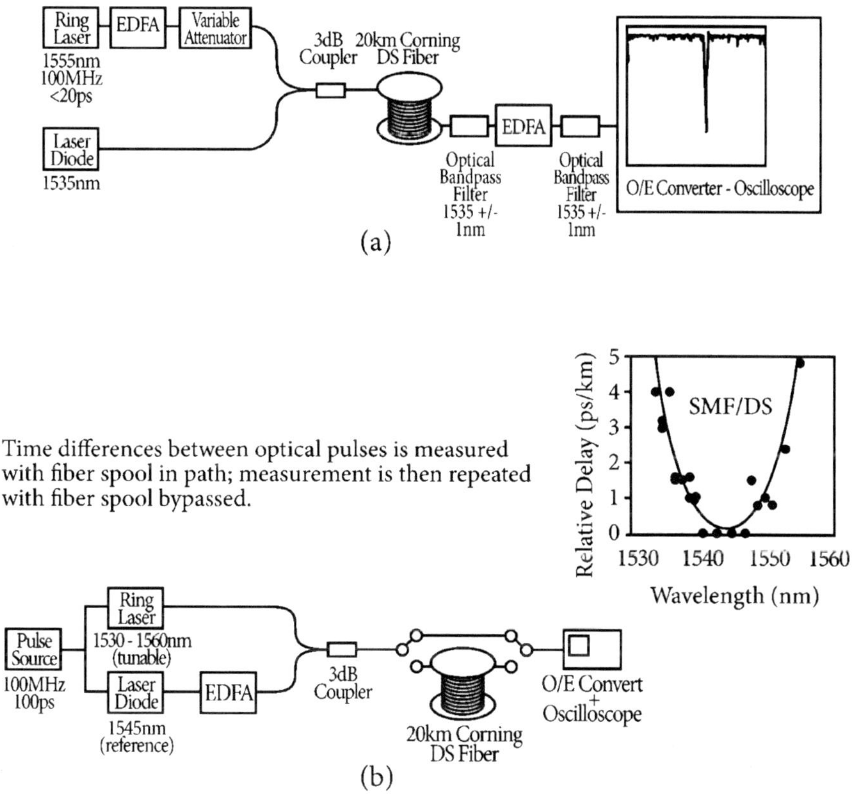

Figure 10.12. (**a**) A schematic block diagram for the experimental setup to measure the generated pulses. (**b**) A schematic block diagram for the experimental setup to measure the walk-off characteristics of the Corning DS fiber. The maximum walk-off for the wavelength 1,515–1,560 nm is less than 4 ps km^{-1} [11]

for the experiment because of its desirable dispersion and walk-off characteristics. To learn quantitatively the behavior of this fiber, the walk-off characteristics of this fiber are measured and displayed in Fig. 10.12b. It is seen that for the wavelength range of interest (1,535–1,560 nm), the dispersion coefficient β_2 varies between $+2$ and $-2\,\text{ps}^2\,\text{km}^{-1}$. The difference of group velocities as a function of the wavelength of the beams varies between 0 and 4 ps km^{-1}. An erbium-doped fiber amplifier (EDFA) is used to boost the power at the receiver.

Generation of Time-Aligned Pulses on Copropagating WDM Beams

Two sets of experiments were performed: The first set dealt with the generation of "dark" pulses on two or three semiconductor laser sources by a shepherd pulse on the ring laser; the second set dealt with the generation of "bright" pulses on a semiconductor laser source by a shepherd pulse on the ring laser. Signals from these sources (one or more signal sources from semiconductor lasers and one shepherd source from a ring laser) of different wavelengths are combined using two 2-to-1 fiber couplers. The combined output is sent through a 20-km spool of Corning DS fiber. At the output end of the fiber, an optical bandpass filter is used to reject the shepherd pulse signal from the ring laser. The signal from each laser diode is detected and viewed on an oscilloscope. A picture of this output is shown in Fig. 10.12a. A dip (dark pulse) or a rise (bright pulse) on the cw signal from a laser diode indicates the presence of a generated pulse as predicted by the computer-simulation result.

Generation of Dark Pulses

Systematic measurements are made for pulses generated on two or three primary cw beams from laser-diode sources due to the presence of a shepherd pulse on a beam from the ring laser.

Figure 10.13 shows the input and output pulses on the two primary cw beams with wavelengths 1,530 and 1,535 nm, and on the ring laser beam operating at 1,555 nm. It is seen that there were no pulses on the two primary beams at the input and there was a large shepherd pulse on the ring laser beam at the input.

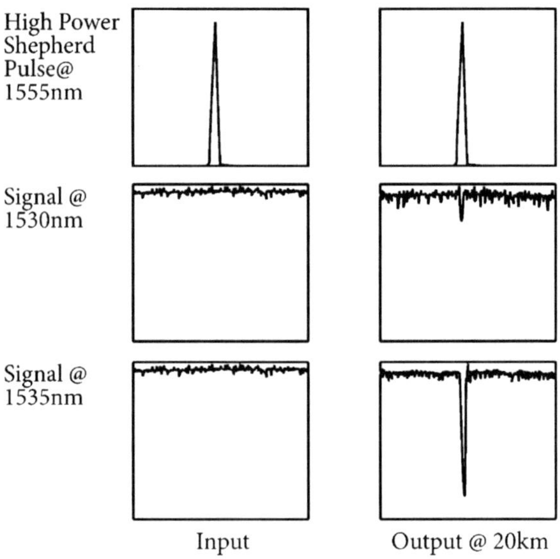

Figure 10.13. Oscilloscope picture for the generated pulses on two different wavelength primary beams by a co-propagating shepherd pulse on the third beam. "Dark" pulses are generated on the diode laser beams, since the operating wavelengths of these primary beams fall into the normal GVD region of the fiber [11]

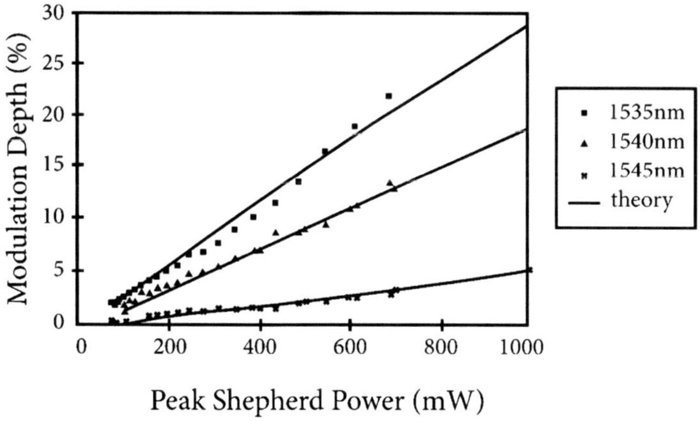

Figure 10.14. Measured and computed modulation depths of the generated "dark" pulses on three different wavelength beams as a function of the shepherd-pulse peak power [11]

After passing through the fiber, there appeared three time-aligned co-propagating output pulses on all three beams. As predicted by the theory, two dark pulses were generated on the primary diode laser beams because the operating wavelengths fell in the normal GVD region of the fiber and one was the shepherd pulse on the ring laser beam.

Shown in Fig. 10.14 is a plot of modulation depths on the cw beams as a function of the shepherd pulse peak power. Solid lines represent the computer-simulation results and the data points represent the measured results. Very close agreement is observed. It is noted that walk-off between the shepherd pulse and the generated primary pulses is less than 1 ps km^{-1}. For a fiber length of 20 km, the maximum pulse misalignment is 20 ps or 1/3 pulse width without the pulse-shepherding effect. It is expected that the presence of the unavoidable pulse-shepherding effect will diminish the pulse misalignment to a negligible level, as observed in the experimental results [12].

Generation of "Bright" Pulses

To verify the theoretical prediction that bright pulses can be generated if the operating wavelength of the primary laser diode falls in the anomalous GVD region of the fiber, the following experiment was performed: Two beams, one from the ring laser, carrying the shepherd pulse at 1,535 nm, and the other from the diode laser, carrying no pulse at 1,557 nm, were combined and sent through the Corning DS fiber. The input and output pulses on these beams are displayed in Fig. 10.15. As predicted by the theory, a bright pulse is generated on the diode laser beam because its operating wavelength falls in the anomalous GVD region of the fiber.

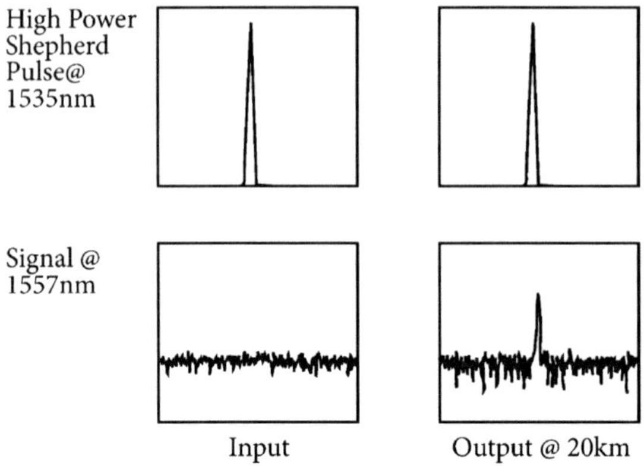

Figure 10.15. Oscilloscope picture for the generated pulse on a primary beam by a co-propagating shepherd pulse on the second beam. The "bright" pulse is generated on the diode laser beam, since the operating wavelength of the primary beam falls into the anomalous GDV region of the fiber [11]

These experiments show that through the use of a shepherd pulse on a co-propagating WDM beam, a simple way is found to generate time-aligned pulses on the other different wavelength beams in a nonlinear fiber. This was accomplished experimentally using a Corning single-mode-fiber-dispersion-shifted fiber. It should be noted that the success of this technique depends on the condition that the amount of walk-off or drifting between the large-amplitude shepherd pulse and the generated pulses must be less than half of the pulse width of the shepherd pulse for the entire pulse-generation interaction length of fiber. It should be noted that these generated pulses are *stable* pulses in the sense that they remain even after the shepherd pulse is channeled away. Successful generation of these time-aligned pulses on WDM beams is crucial to the realization of ultrahigh data-rate bit-parallel WDM single-fiber transmission system.

10.6.4 Bit Parallel WDM Solitons [8, 22]

The traditional way of transmitting solitons on WDM beams in a single-mode fiber is to make sure that soliton pulses on each WDM beam do not overlap with those on the other WDM beams. If overlapping occurs, rapid walk-off is designed in the system so that the overlapping time is kept at a minimum. So, basically this type of soliton propagation on WDM beams is not much different from the soliton propagation an a single beam case except new deleterious effects to destroy the integrity of a soliton are added due to the overlapping and walk-off interaction problem. Therefore, the CPM effect due to the presence of neighboring WDM beams is considered to be undesirable.

On the contrary, investigations presented in previous sections show that CPM effect may be used to gain constructive new approaches to solve problems of importance. The case-in-point is the discovery of bit parallel WDM solitons. Inspection of previous work on CPM effect shows that it tends to provide certain adhesion among co-propagating overlapping pulses even in the presence of walk-off. This stabilizing effect increases the tendency for co-propagating overlapping pulses to stay together. It was under this influence that the bit-parallel WDM solitons were found.

Let us start with an idealized fiber which is lossless (i.e., $\alpha_i = 0$ for all beams) and which possesses uniform GVD (i.e., $v_{gi} = v_g$ for all beams) within the wavelength range under investigation. The equations governing the propagation characteristics of signal pulses are

$$-j\frac{\partial u_i}{\partial \xi} = -\frac{1}{2}\frac{\partial^2 u_i}{\partial \tau^2}\frac{L_D}{L_{NL}}\left(|u_i|^2 + 2\sum_{m\neq i}^{I}|u_m|^2\right)u_i \qquad (10.88)$$

$$(i = 1, 2, 3, \ldots I)$$

The anomalous GVD case in which $\mathrm{sgn}(\beta_{2i}) = -1$ is considered. It is seen from the above equation that the summation term representing the cross phase modulation effect is twice as effective as the SPM effect for the same intensity. This observation also provides the idea that cross phase modulation may be used in conjunction with SPM on the WDM pulses to counteract the GVD effect; thus producing WDM solitons. Comparing the nonlinear term for a single beam with that for the co-propagating multiple WDM beams in (10.88) shows that if we choose the correct amplitudes for the initial pulses on WDM beams and retain the hyperbolic secant pulse form, it may be possible to construct a set of initial pulses that will propagate in the same manner as the single soliton pulse case, that is, undistorted and without change in shape for arbitrarily long distances. Let us choose the initial pulses as follows [8, 22]:

$$u_i(0, \tau) = [1 + 2(I - 1)]^{1/2}\,\mathrm{sech}(\tau) \qquad (10.89)$$

$$(i = 1, 2, 3, \ldots I),$$

where I is the number of WDM beams.

Using these initial pulse forms numerical simulation was carried out to solve (10.88). The split-step Fourier method was used. The fiber parameters used for the simulation are L, length of fiber, equal to 1,000 km; β_2, dispersion coefficient, equal to $-2\,\text{ps}^2\,\text{km}^{-1}$; γ, nonlinear index coefficient, equal to $20\,\text{W}^{-1}\text{km}^{-1}$; T_0, pulse width, equal to 10 ps; $L_D = 50\,\text{km}$; and $L_{NL} = 50\,\text{km}$. Four cases with $I = 1, 2, 3, 4$ were treated. The $I = 1$ case corresponds to the well-known single soliton case; here, the amplitude for the fundamental soliton is 1. For the two-beam case, the amplitude is $3^{-1/2} = 0.57735$. For the three-beam case, it is $5^{-1/2} = 0.4472136$. For the four-beam case, it is $7^{-1/2} = 0.37796447$. It is noted that the amplitude of the fundamental solitons on WDM multibeams becomes successively smaller as the number of beams is increased. This is because the nonlinear effect becomes more pronounced when more beams are present. Numerical simulation shows that after propagating 1,000 km through this fiber the original pulse shape for all these WDM pulses remains unchanged. It thus appears that the initial forms chosen for the pulses on WDM beams are the correct soliton forms for WDM beams.

An example on how bit-parallel WDM solitons may carry digital data streams on 4 WDM beams is demonstrated in Fig. 10.16. It shows how a bit-parallel word may be transmitted by this scheme. The key to form BP-WDM solitons is to set the magnitude of each bit to a predetermined value for a bit-parallel word based on the number of bits used to represent this word.

Preliminary results have shown that if the relative walk-off is kept below a certain value, all co-propagating pulses tend to stay together and co-propagate at an average group velocity. The shapes of these pulses may also be altered somewhat,

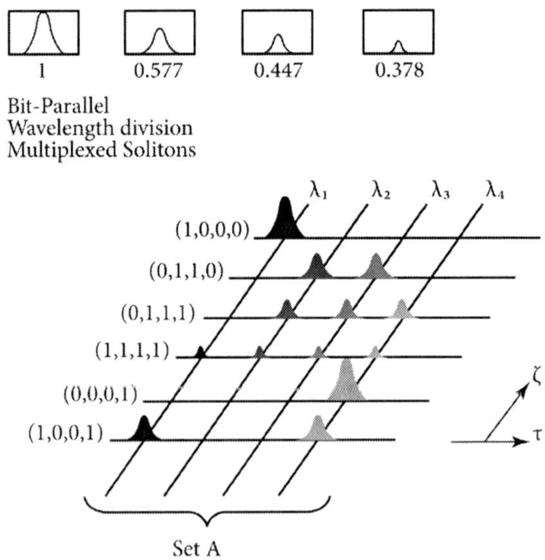

Figure 10.16. An example of bit-parallel WDM solitons in their information transmission

but the information carrying ability is maintained. This means the BP-WDM solitons are quite robust. Furthermore, because of the cohesive affects, due to CPM and pulse shepherding [8, 9, 23], some of the deleterious effects (such as Raman effect, the third order dispersion effect, and self-steepening effect) that may be present for the case of soliton propagation on a single beam may be diminished.

The existence of optical solitons on WDM beams in a fiber is not only of fundamental interest but also has enormous implications in the field of optical fiber communications. It is conceivable that multiple terabits of information can be sent through a single fiber in the bit-parallel WDM format [2] without degradation.

References

1. A. Hasegawa and T. Tappert, "Transmission of stationary nonlinear optical pulses in dispersive dielectric fibers. I. Anomalous dispersion," Appl. Phys. Lett. **23**, 142 (1973); L. F. Mollenauer, R. H. Stolen, and L. P. Gordon, "Experimental observation of picosecond pulse narrowing and solitons in optical fibers," Phys. Rev. Lett. **45**, 1095 (1980)
2. W. J. Tomlinson, "Wavelength multiplexing in multimode optical fibers," Appl. Opt. **16**, 2180 (1977); G. Winzer, "Wavelength multiplexing components – A review of single mode devices and their applications," J. Lightwave Tech. **2**, 369 (1984); H. Ishio, J. Minowa, and K. Nosu, "Review and status of wavelength division multiplexing technology and its application," J. Lightwave Tech. **2**, 448 (1984); L. A. Bergman, A. J. Mendez, and L. S. Lome, "Bit-parallel wavelength links for high performance computer networks," SPIE Critical Rev. **CR62**, 210 (1996)
3. G. I. Stegeman and M. Segev, "Optical spatial solitons and their interaction, universality, and diversity," Science **286**, 1518 (1999)
4. G. P. Agrawal, "Nonlinear Fiber Optics," 3rd edn., Elsevier, Singapore (2001)
5. D. Marcuse, "Pulse distortion in single-mode fibers," Appl. Opt. **19**, 1653 (1980); C. G. B. Garrett and D. E. McCumber, "Propagation of a gaussian light pulse through anomolous dispersive medium," Phy. Rev. A **1**, 305 (1970)
6. R, H. Stolen and C. Lin, "Self phase modulation in silica optical fibers," Phys. Rev. A **17**, 1448 (1978)
7. J. I. Gersten and J. Stone, "Measurement of cross phase modulation in coherent wavelength division multiplexing using injection lasers," Electron. Lett. **20**, 996 (1984); R. R. Alfano, P. L. Baldeck, F. Raccah, and P. P. Ho, "Cross phase modulation measured in optical fibers," Appl. Opt. **26**, 3491 (1987)
8. C. Yeh and L. A. Bergman, "Existence of optical solitons on wavelength division multiplexed beams," Phys. Rev. E **60**, 2306 (1999)
9. C. Yeh and L. A. Bergman, "Pulse shepherding in nonlinear fiber optics," J. Appl. Phys. **80**, 3174 (1996)

10. C. Yeh and L. A. Bergman, "Enhanced pulse compression in a nonlinear fiber by a wavelength division multiplexed optical pulse," Phys. Rev. E **57**, 2398 (1998)

11. C. Yeh, L. A. Bergman, J. Moroodian, and S. Monaco, "Generation of time-aligned picosecond pulses on WDM beams in a nonlinear fiber," Phys. Rev. E **57**, 6135 (1998)

12. L. A. Bergman, J. Moroodian, and C. Yeh, "An all optical long distance multi-Gbytes/s bit parallel WDM single fiber link," J. Lightwave Tech. **16**, 1577 (1998)

13. N. Bloembergen, "Nonlinear Optics," W. A. Benjamin, New York (1965)

14. C. Yeh, J. E. Pearson, and W. P. Brown, "Enhanced focal-plane irradiance in the presence of thermal blooming," Appl. Opt. **15**, 2913 (1976); J. E. Pearson, C. Yeh, and W. P. Brown, "Propagation of laser beams having an on-axis null in the presence of thermal blooming," J. Opt. Soc. Am. **66,** 1384 (1976); C. Yeh, L. Casperson, and B. Szejn, "Propagation of truncated gaussian beams in multimode fiber guides," J. Opt. Soc. Am. **68,** 989 (1978); C. Yeh, W. P. Brown, and B. Szejn, "Multimode inhomogeneous fiber couplers," Appl. Opt. **18**, 489 (1979)

15. J. W. Cooley and J. W. Tukey, "An algorithm for the machine calculation of complex Fourier series," Math. Comp. **19**, 297 (1965)

16. M. J. Ablowitz and P. A. Clarkson, "Solitons, Nonlinear Evolution Equations and Inverse Scattering," Cambridge University Press, New York (1991)

17. F. Shimizu, "Frequency broadening in liquid by a short light pulse," Phys. Rev. Lett. **19**, 1097 (1967)

18. C. R. Menyuk, "Stability of solitons in birefringent optical fibers. II. Arbitrary amplitudes," J. Opt. Soc. Am. B **5,** 392 (1988)

19. L. F. Mollenauer, R. H. Stolen, J. P. Gordon, and W. J. Tomlinson, "Extreme picosecond pulse narrowing by means of soliton effect in single-mode optical fibers," Opt. Lett. **8**, 289, (1983); H. Nakatsuka, D. Grischkowsky, and A. C. Balant, "Nonlinear picosecond pulse propagation through optical fibers with positive group velocity," Phys. Rev. Lett. **47**, 910 (1981); W. J. Tomlinson, R. H. Stolen, and C. V. Shank, "Compression of optical pulses chirped by self-phase modulation in fibers," J. Opt. Soc. Am. B **1**, 139 (1984); C. Headley III and G. P. Agrawal, "Simultaneous amplification and compression of picosecond optical pulses during Raman amplification," J. Opt. Soc. Am. B **10**, 2383 (1993)

20. E. P. Ippen,"Laser applications to Optics and Spectroscopy," Vol 2, Chap. 6, S. F. Jacobs, M. Sargent III, J. F. Scott, and M. O. Scully, eds., Addison Wesley, Reading, MA (1975)

21. Opto-Electronics Group, Corning Incorporated Report No. **MM26**, (1996)

22. C. Yeh and L. A Bergman, "Wavelength division multiplexed optical solitons," US Patent 6612, 743B1 (Sept. 2, 2003)

23. E. A. Ostrovskaya, Y. S. Kivshar, D. Mihalache, and J. C. Crasovan, "Multichannel soliton transmission and pulse shepherding in bit-parallel wavelength optical fiber links," IEEE J. Select. Topics Quantum Elect. **8**, 591 (2002)

11

ULTRA LOW-LOSS DIELECTRIC WAVEGUIDES

It is the extraordinary low-loss property of silica in the optical spectrum that enables the development of optical fibers as high capacity/long distance communication links. This low-loss behavior of silica is attained through the elimination of impurities, which are the causes of loss [1]. In the low frequency spectrum, lower than 30 GHz, conducting metals such as silver, copper, gold, and aluminum are generally adequate as structures for low-loss waveguides [2]. However, there remains an important region of the spectrum – from 30 to 3,000 GHz (the millimeter–submillimeter band) – where low-loss waveguides are not available. The main problem here in finding low-loss solids is no longer due to only one of the eliminating impurities, but is due to the presence of intrinsic vibration absorption bands [3]. The use of highly conducting materials is also precluded in this part of the spectrum owing to high skin-depth losses [4, 5]. In this chapter we show that a combination of material and waveguide geometry can circumvent these difficulties. For example, a ribbon-like waveguide structure with an aspect ratio of 10:1, fabricated from ceramic alumina (Coors 998 Alumina), can have an attenuation factor of less than 10 dB km^{-1} in the millimeter–submillimeter band. The attenuation is more than 100 times smaller than that of a typical ceramic (or other dielectric) circular rod waveguide. The main purpose of this chapter is to show that there may be another option, other than finding the ideal low-loss material, to construct a low-loss waveguide in the millimeter–submillimeter band.

11.1 Theoretical Foundation

According to the theory of waveguide propagation along a dielectric waveguide, the attenuation constant α of a dielectric waveguide surrounded by dry air is given by the following formula (see Sect. 3.5):

$$\alpha = 8.686\pi \frac{\epsilon_{r1} R \tan \delta_1}{\lambda_0} \qquad (\mathrm{dB\ m^{-1}}), \tag{11.1}$$

where

$$\epsilon_{r1} R = \frac{\epsilon_{r1} \displaystyle\int_{A_1} (\mathbf{E}_1 \cdot \mathbf{E}_1^*) \, dA}{\sqrt{\dfrac{\mu}{\epsilon_0}} \left[\displaystyle\int_{A_1} \mathbf{e}_z \cdot (\mathbf{E}_1 \times \mathbf{H}_1^*) \, dA + \int_{A_0} \mathbf{e}_z \cdot (\mathbf{E}_0 \times \mathbf{H}_0^*) \, dA \right]}. \tag{11.2}$$

Here, ϵ_{r1} and $\tan\delta_1$ are, respectively, the relative dielectric constant and the loss tangent of the dielectric core material, μ and ϵ_0 are, respectively, the permeability and permittivity of free space, λ_0 is the free space wavelength in meters, $\mathbf{e}_z$ is the unit vector in the direction of propagation, A_1 and A_0 are, respectively, the cross-sectional areas of the core and the cladding region, and $(\mathbf{E}_1, \mathbf{H}_1)$ and $(\mathbf{E}_0, \mathbf{H}_0)$ are, respectively, the modal electric and magnetic field vectors of the guided mode in the core region and in the cladding region. The factor $\epsilon_{r1} R$ is defined as the geometrical loss factor, and the attenuation α is directly proportional to it.

11.1.1 Normal Mode Solution

The normal modal fields for a given guiding structure are the eigen fields, which are obtained from the eigen-values and eigen-solutions of the wave equation corresponding to the appropriate boundary conditions (see Chaps. 2 and 3). The exact analytic solutions for a dielectric waveguiding structure are known only for a planar dielectric slab, for a circular dielectric cylinder (such as a typical optical fiber), and for an elliptical dielectric cylinder (see Chaps. 4–6). Numerical techniques, such as the finite-element technique [6], the finite difference time domain technique [7], or the beam propagation technique [8], must be used for other geometrical shapes. Only hybrid modes with all six field components can be supported by noncircular dielectric waveguides [9].

To find the normal mode solution for arbitrarily shaped dielectric waveguides, computational approaches based on the numerical solution of the Maxwell equations can be used (see Chap. 15 for detailed discussion).

11.1.2 Geometrical Loss Factor

Examination of the fundamental equation (11.1) governing the attenuation constant of a dominant mode guided by a simple solid dielectric waveguide surrounded by lossless air shows that it is dependent on the loss factor and the dielectric constant of the dielectric material and the geometrical cross-sectional shape of the guiding structure [10, 11] (see Fig. 11.1). Since the material loss factor and the dielectric constant of a solid are fixed, the only way to reduce the attenuation constant is to find the proper cross-sectional geometry of the waveguide. After performing a systematic study on a variety of geometries, it can be shown that a ribbon-shaped guide made with a low-loss, high dielectric constant material, such as ceramic alumina, can yield an attenuation constant for the dominant TM-like mode of less than 0.005 dB m^{-1}. (Two dominant modes with no cutoff frequency can be supported by this ceramic ribbon structure [9]: a TE-like dominant mode with most of its electric field aligned parallel to the major axis of the ribbon and a TM-like dominant mode with most of its electric field aligned parallel to the minor axis of the ribbon.) As a comparison, at operating frequency band around 100 GHz, the attenuation constant is 1.3 dB m^{-1} for the typical cylindrical Teflon dielectric waveguide, 2.4 dB m^{-1} for the metallic rectangular waveguide, and 3 dB m^{-1} for the microstripline [12]. This remarkable low-loss behavior of a ceramic ribbon guiding the dominant TM-like mode is shown in Fig. 11.1. The loss behavior of a ceramic ribbon guiding TE-like mode is also shown. Several important conclusions can be drawn from the results given in Fig. 11.1:

- Much lower geometrical loss factors are obtained for high aspect ratio ribbon waveguide with high dielectric constant. For example, when the normalized cross-sectional area, $A\left(\epsilon_{r1}-1\right)/\lambda_0^2$, is 0.4, the geometrical loss factor (as well as the attenuation factor) for this ribbon supporting the TM-like mode is about 140 times smaller than that for a circular rod with the same cross-sectional area supporting the dominant HE_{11} mode.

- To achieve these dramatically lower geometrical loss factors, the guiding structure must be of ribbon shape with high aspect ratio as well as high dielectric constant, supporting the dominant TM-like mode.

- In the low-loss region, that is, $\epsilon_{r1}R < 0.05$, the geometrical loss factor curve for the 10:1 ribbon with dielectric constant $\epsilon_{r1} = 10$ supporting the TM-like mode is much flatter than that for the circular rod supporting the HE_{11} mode. This indicates that the geometrical loss factor for the ribbon is insensitive to small deviations of the normalized cross-sectional area of the ribbon, whereas the geometrical loss factor for the circular rod is very sensitive to size changes in the rod. This means that TM-like mode on the ribbon is a very stable mode, not easily disturbed by any geometrical imperfections.

- Separation of the geometrical loss curves for TE-like mode and TM-like mode becomes larger for larger dielectric constant of the guiding ribbon. Furthermore, there is a definite relationship between the geometrical loss curves for TE-like mode and that for the TM-like mode. These facts are significant, because they can be used to devise a fundamentally new way to measure the super low-loss characteristics of TM-like mode guided along low-loss high dielectric constant ribbons.

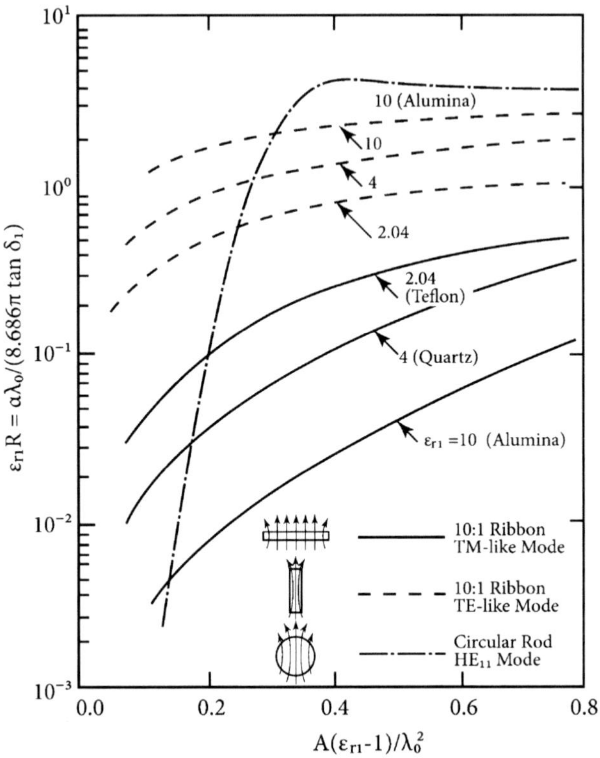

Figure 11.1. The geometrical loss factor $\epsilon_{r1}R$ as a function of the normalized cross-sectional area $A(\epsilon_{r1}-1)/\lambda_0^2$. Here, A is the cross-sectional area of the waveguide, ϵ_{r1} is the relative dielectric constant of the dielectric guide, and λ_0 is the free-space wavelength. Dielectric ribbons with an aspect ratio of 10 and an alumina circular rod are considered. For the ribbon case, the geometrical loss factors for the dominant TM-like (low-loss) and TE-like (high-loss) modes are obtained for three different dielectric materials: alumina with $\epsilon_{r1} = 10$, quartz with $\epsilon_{r1} = 4$, and teflon with $\epsilon_{r1} = 2.04$. The case for the alumina circular rod supporting the dominant HE_{11} mode is shown for comparison purposes. The alumina ribbon supporting the dominant TM-like mode provides the most dramatic reduction in the geometrical loss factor when compared with that for the alumina circular rod. Sketches of the transverse electric field lines for the TE-like, TM-like, and HE_{11} modes are also shown

- Inspection of the expression for the geometrical loss factor, (11.2), shows that the numerator term representing the electric field intensity within the dielectric waveguide governs the magnitude of the geometrical loss factor. To minimize this factor, the electric field intensity must be chosen to be as small as possible over the cross-sectional area of the dielectric guide. It is noted that the TM-like mode on a ribbon structure provides precisely this behavior while the opposite is true for the TE-like mode on this structure. Thus, the TE-like mode yields much higher geometrical loss factor than TM-like mode.

11.1.3 Relationship Between Geometrical Loss Factors for TE-Like Mode and for TM-Like Mode

Let us now investigate the relationship between the geometrical loss factors for the TE-like mode and TM-like mode on a high aspect ratio ribbon. The ratio r_α is defined as follows:

$$r_\alpha = \frac{(\epsilon_{r1}R)_{\text{TE-like}}}{(\epsilon_{r1}R)_{\text{TM-like}}}, \tag{11.3}$$

$$r_\alpha = \frac{\alpha_{\text{TE-like}}}{\alpha_{\text{TM-like}}}. \tag{11.4}$$

This ratio is displayed in Fig. 11.2. A large difference is seen for the attenuation constants of TE-like and TM-like modes when the dielectric constant of the ribbon is 10 or higher. This relationship can be used to measure the very low attenuation factor for the TM-like mode on a ribbon.

11.1.4 External Field Decay Consideration

Having established theoretically the fact that the geometrical factor $\epsilon_{r1}R$ can be significantly reduced with the choice of high dielectric constant core material for the waveguide as well as the choice of a thin ribbon geometry, it is important to answer the question of whether even higher dielectric constant material can provide even smaller (or better) $\epsilon_{r1}R$ factor. To answer this question, it is important to note that, for an open waveguide structure such as the dielectric waveguide, the field extent (or field decaying rate) away from the guiding structure represents a very important measure on how well the structure "guides" the wave. So, one must rephrase the question: *For a given (same) field decaying rate away from the surface of the dielectric core, will a higher dielectric constant material yield a smaller $\epsilon_{r1}R$?* Figure 11.3 is developed to address this question. In this figure, $\epsilon_{r1}R$ is plotted as a function of the dielectric constant of the core for a fixed decaying rate of unity normalized to the free-space wavelength for the exterior field from the core surface. This decaying rate means that the exterior field would have reduced by 1 e^{-1} at one free-space wavelength from the guiding surface. The following experiment will show that the guided field is still very well bonded to the guiding structure at this decaying rate. It is seen that $\epsilon_{r1}R$ initially reduces very rapidly as the dielectric constant is increased, reaching a minimum at around 10–14. Then, $\epsilon_{r1}R$ increases very gently for further increase in the dielectric constant.

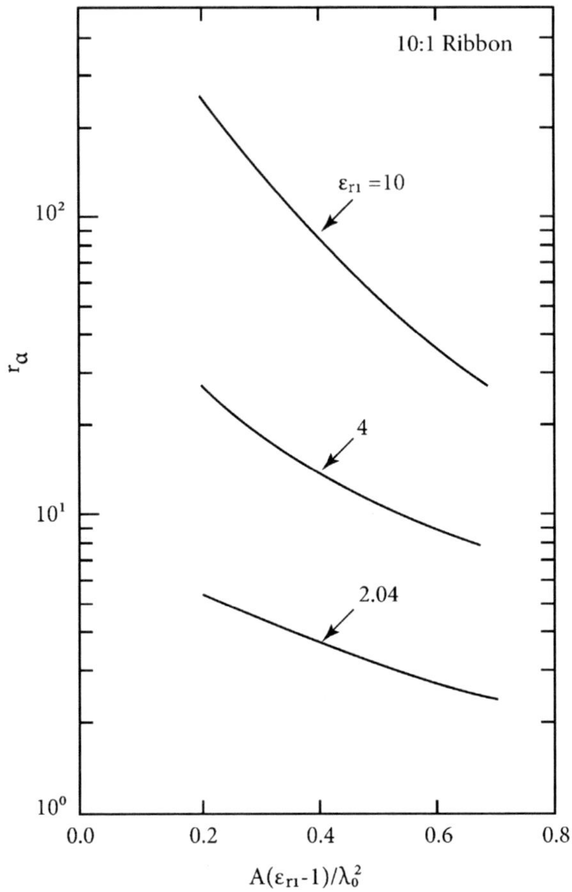

Figure 11.2. The ratio r_α as a function of the normalized cross-sectional area $A(\epsilon_{r1}-1)/\lambda_0^2$ for a 10:1 ribbon waveguide for three different ribbon materials with $\epsilon_{r1} = 10$, 4, and 2.04. The ratio r_α is defined as $(\epsilon_{r1}R)_{\text{TE}} \,/\, (\epsilon_{r1}R)_{\text{TM}}$ or $\alpha_{\text{TE}}/\alpha_{\text{TM}}$, and A is the cross-sectional area of the waveguide, ϵ_{r1} is the relative dielectric constant of the dielectric guide, and λ_0 is the free-space wavelength. It is seen that for high relative dielectric constant ribbon material, such as $\epsilon_{r1} = 10$, the ratio may be quite high, implying that the attenuation for the dominant TM-like mode and that for the dominant TE-like mode can be different by two orders of magnitude. For example, when the normalized area $A(\epsilon_{r1} - 1)/\lambda_0^2$ is 0.4, r_α is about 88, implying that the attenuation constant for the TM-like mode can be 88 times smaller than that of the TE-like mode on the same ribbon structure

Hence, core material with a relative dielectric constant of around 10 appears to be the preferred choice. For smaller normalized decaying rate (i.e., for weaker "guidance"), the curve in Fig. 11.3 moves lower and shifts slightly towards higher dielectric constant. This implies that $\epsilon_{r1}R$ can be made even smaller, but at the expense

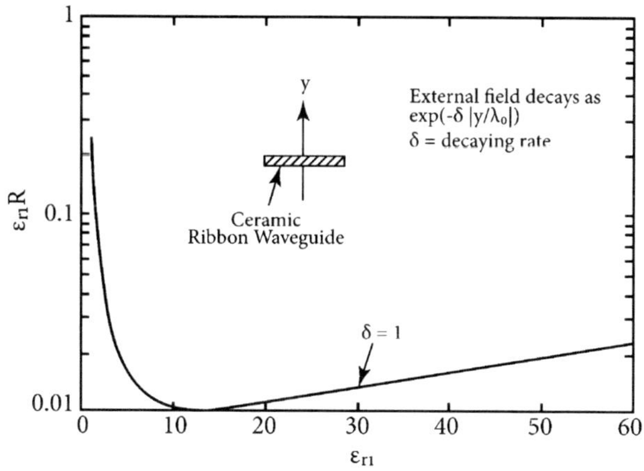

Figure 11.3. The geometrical loss factor $\epsilon_{r1}R$ as a function of the relative dielectric constant ϵ_{r1} of the 10:1 ribbon waveguide for a fixed decaying rate of unity normalized to the free-space wavelength for the exterior field from the core surface of the guide. It is seen that the geometrical loss factor is close to a minimum near $\epsilon_{r1} = 10$. Thus the choice of material with $\epsilon_{r1} \approx 10$ is near optimum

of much weaker guidance. In that case, somewhat higher dielectric constant material should be used to obtain the lowest value of $\epsilon_{r1}R$. Keeping the normalized decaying rate higher than unity, the choice of a low-loss dielectric with a relative dielectric constant of approximately 10 appears to be close to optimum.

11.2 Experimental Verification

The cavity resonator method [13] can be used to verify experimentally the very low attenuation constant of the ceramic ribbon waveguide. By measuring the Q of the alumina-ribbon resonator as shown in Sect. 5.8, we may obtain the attenuation constant of the dielectric waveguide α in (dB m^{-1}) according to the following formula [14] (see Chap. 3):

$$\alpha = 8.686 \frac{v_\mathrm{p}}{v_\mathrm{g}} \frac{\beta}{2Q}, \tag{11.5}$$

where β is the propagation constant of the mode under consideration and v_p and v_g are the phase and the group velocity, respectively, of that mode. Several factors affect the accuracy and sensitivity of this technique: the alignment of the dielectric waveguide with the coupling holes, the alignment of the parallel plates, the uniformity or the straightness of the dielectric waveguide, the coupling or radiation losses and the metallic wall losses of the plates. Previous work [13] shows that the maximum Q that can be reliably measured by this technique in the K_a band (26.5 – 40 GHz) is approximately 30,000. This limits the smallest value of the measured attenuation constant to around 0.1 dB m^{-1}, whereas one expects the ceramic ribbon to exhibit a value around a tenth of that.

To circumvent this difficulty, an indirect way to measure this ultra-low attenuation was developed. As seen from (11.1)–(11.3), there is a definite relationship between the attenuation constants or Q values for the TM-like and TE-like modes on a dielectric ribbon:

$$\begin{aligned} r_\alpha &= \frac{\alpha_{\text{TE-like}}}{\alpha_{\text{TM-like}}} = \frac{Q_{\text{TE-like}}}{Q_{\text{TM-like}}} \frac{v_{g\text{TE-like}}}{v_{g\text{TM-like}}}, \\ &= \frac{(\epsilon_{r1} R)_{\text{TE-like}}}{(\epsilon_{r1} R)_{\text{TM-like}}}, \end{aligned} \tag{11.6}$$

where $v_{g\text{TE-like}}$ and $v_{g\text{TM-like}}$ are the group velocity of the TE-like and TM-like modes, respectively, and $Q_{\text{TE-like}}$ and $Q_{\text{TM-like}}$ are the Q values for the TE-like and TM-like modes, respectively. Measured results are displayed in Fig. 11.4 as data points and calculated theoretical results are given as curves. Excellent agreement between measured values and theoretical values affirms the correctness of the derived theoretical ratio ($Q_{\text{TM-like}}/Q_{\text{TE-like}}$). This relationship can now be used reliably to obtain $Q_{\text{TM-like}}$ when $Q_{\text{TE-like}}$ is known.

Using this technique, we measured the Q and the attenuation constant for the TM-like mode on ultra-low-loss alumina ribbon. Measured results are summarized in Table 11.1. The measured data points as well as the calculated results are displayed in Fig. 11.5. Excellent agreement between the experimental data and theoretical results can be seen. The importance of the geometry of the guide is apparent from these results. At 30 GHz, the alumina rectangular ribbon with aspect ratio of 10 and a loss tangent of 1.59×10^{-4} can support a low-loss TM-like mode with an attenuation constant at 0.0098 dB m^{-1}, which is 120 times less than that for the dominant mode on an alumina circular rod with similar cross-sectional area and a loss tangent of 0.0001. It is also 61 times less than that for the dominant mode in a standard metallic rectangular waveguide (WR-28). Significant improvement to less than 0.006 dB m^{-1} can be obtained for the ribbon if the same alumina material that was used for the circular rod is used.

To demonstrate the viability of the alumina ribbon as an actual transmission line, we have performed the following experiment. Two horns, one transmitting and one receiving, are separated by a free-space distance of 86 cm. A 120 ps pulse is emitted from the transmitting horn and is received by the receiving horn as pulse B after traversing through this free-space distance of 86 cm. We performed another experiment, where a 91-cm-long ceramic-ribbon waveguide with cross-sectional

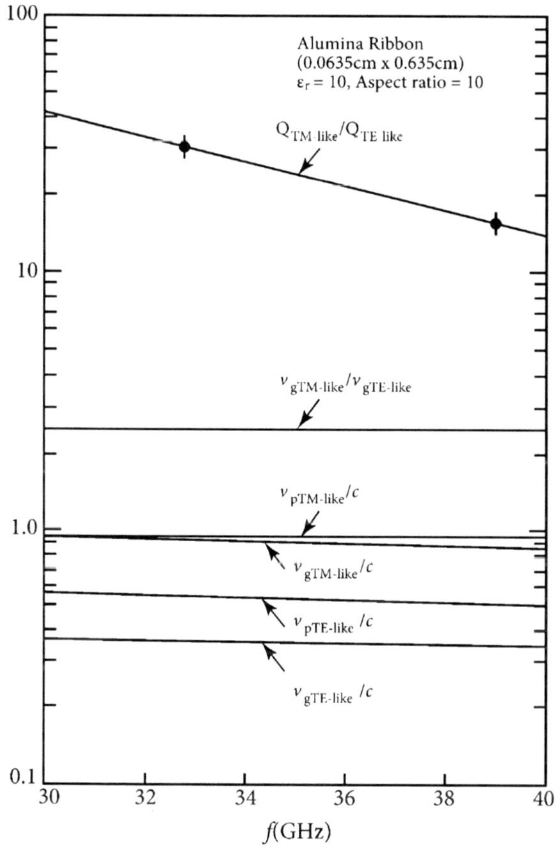

Figure 11.1. Ratios of $Q_{\text{TM-like}}/Q_{\text{TE-like}}$, $v_{\text{gTM-like}}/v_{\text{gTE-like}}$, $v_{\text{pTM-like}}/c$, $v_{\text{gTM-like}}/c$, $v_{\text{pTE-like}}/c$, and $v_{\text{gTE-like}}/c$ vs. frequency f (in GHz). Values taken for a 10:1 aspect ratio alumina ribbon with dimensions $0.0635 \times 0.635\,\text{cm}^2$ and relative dielectric constant $\epsilon_{r1} = 10$ supporting the TM-like or the TE-like mode. Here, $Q_{\text{TM-like}}$, $v_{\text{gTM-like}}$, $v_{\text{pTM-like}}$, and $Q_{\text{TE-like}}$, $v_{\text{gTE-like}}$, $v_{\text{pTE-like}}$ are the resonant Q, the group velocity, and the phase velocity of the TM-like and TE-like modes, respectively. The velocity of light in vacuum is c. The ratio $Q_{\text{TM-like}}/Q_{\text{TE-like}}$ can vary from 42 at 20 GHz to 14 at 40 GHz. This means that for a low-loss alumina ribbon with a loss tangent of 5×10^{-5}, if $Q_{\text{TE-like}}$ is measured as 22,760 at 30 GHz (this Q value is well within the measurement capability of our apparatus), then $Q_{\text{TM-like}}$ must be 955,900 (this Q value is well beyond the measurement capability of any known room temperature resonant-cavity apparatus). To ensure that this measurement capability is not exceeded, so that $Q_{\text{TE-like}}$ and $Q_{\text{TM-like}}$ may both be measured directly in order to verify the calculated ratio ($Q_{\text{TM-like}}/Q_{\text{TE-like}}$), low-loss alumina ribbon was made high-loss by coating the ribbon with a thin layer of high-loss dried India ink. Measured results are shown as data points and calculated results are shown as curves. Excellent agreement is found [11]

Table 11.1 Measured Q, attenuation constant, and loss tangent for ceramic waveguides[1]

Batch	F, GHz	$Q_{\text{TE-like}}$	$Q_{\text{TM-like}}$	α_{TM} (dB m^{-1})	$\tan\delta_1$ ($\times 10^{-4}$)
1	38.60	3,860±300	64,700±4,800	0.062± 0.005	2.8±0.21
1	32.80	3,920±290	121,700±4,800	0.026±0.002	2.8±0.21
2	38.89	6,480±490	103,700±4,800	0.035±0.003	1.59±0.12
2	32.98	7,233±540	216,990±16,300	0.014±0.001	1.59±0.12
3	39.96	10,948±450	10,948±450	1.45±0.09	1.0± 0.04
3	30.03	11,117±500	11,117±500	1.17±0.08	1.0±0.04

area of $0.0635 \times 0.635\,\text{cm}^2$ was inserted between the horns. Specially designed transitions are used to maximize launching and receiving efficiencies. A launching (or receiving) efficiency of 84% (or loss of less than 0.825 dB) at 39.86 GHz has been measured for an exponential launching horn. The same 120 ps pulse is now sent through this ceramic-ribbon waveguide structure. The received pulse is labeled pulse A. The received pulses for these two cases are displayed in Fig. 11.6. The received signal through the ceramic-ribbon waveguide is at least 21 dB greater than that through free space, providing clear evidence that the signal can easily be guided by the ceramic ribbon.

The ceramic-ribbon waveguide is an open structure surrounded by dry air. How to support such a structure is an important consideration. One viable supporting structure is made with plastic fishing-line, that is, nylon thread. Thin nylon (low dielectric constant) threads, spaced at least 10 cm apart, are strung across wooden rails separated by 5 cm (far enough apart so that the exterior guiding field at the ribbon edges has decayed to negligible value). Ceramic-ribbon waveguide can simply be laid on top of the nylon threads along the middle of the rails. The nylon threads can easily support the ceramic ribbon. Any perturbation caused by the nylon-thread support on the propagation characteristics of the guided TM-like mode on the ceramic ribbon waveguide is not detectable.

[1]Three batches of alumina samples were obtained from Coors Ceramic Company (Golden, Colorado, USA). Batch 1, made from Coors' Superstrate 996S20-71 (99.6%, Hirel Thin Film Substrate), contains ribbons with dimensions $0.0635 \times 0.635 \times 11.43\,\text{cm}^3$. Batch 2, made from Coors' extruded 998 Alumina (99.8% alumina) rectangular rod, contains ribbons with dimensions $0.0635 \times 0.635 \times 91\,\text{cm}^3$. Batch 3 is Coors' extruded 998 Alumina (99.8% alumina) circular rod with dimensions 0.244 cm (diameter) $\times$ 91 cm (length). Numerous repeated measurements are made on these samples at various frequencies within the frequency band from 30-40 GHz, using the waveguide resonator technique. As Batch 3 contains a 91-cm-long circular alumina rod, the guiding property of the dominant mode is independent of the orientation of the transverse electric field. Thus, $Q_{\text{TE-like}} = Q_{\text{TM-like}} = Q_{\text{HE}}$ for the circular rod. Here Q_{HE} is the Q value for the HE_{11} mode on a circular dielectric rod [11]

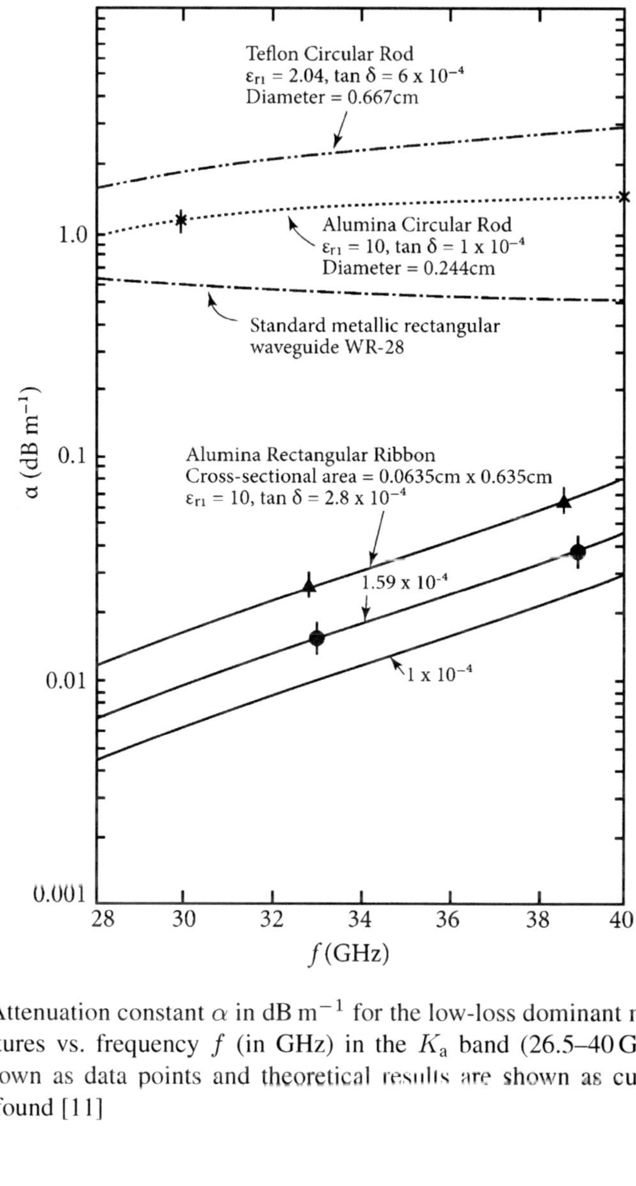

Figure 11.5. Attenuation constant α in dB m^{-1} for the low-loss dominant mode in various guiding structures vs. frequency f (in GHz) in the K_a band (26.5–40 GHz). Measured results are shown as data points and theoretical results are shown as curves. Excellent agreement is found [11]

Another practical problem is how to join sections of ceramic-ribbon waveguides. A "shiplap" joint may be used to provide a strong bond between two ends of ceramic ribbon. A quarter-inch length of the jointing ribbon end is ground to a thickness of 0.0317 cm, which is half the original thickness of the ribbon. The jointing end of the other ceramic ribbon is prepared similarly. The ends are lapped together, aligned, and then glued with "super glue," resulting in a strong bond. We found no measurable loss attributable to the joint.

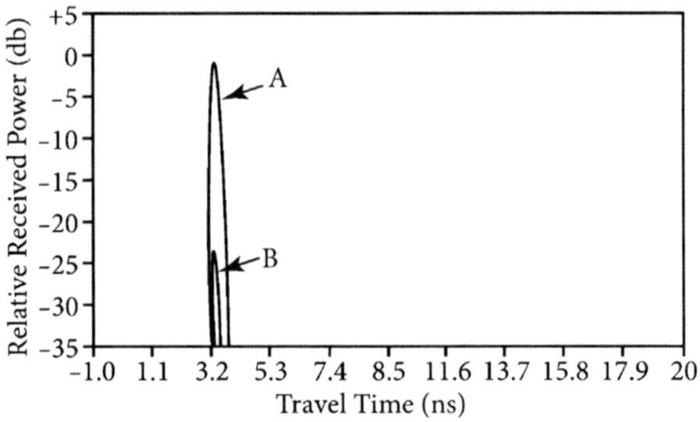

Figure 11.6. Transmission by ceramic ribbon. Comparison between the received pulses for a 120 ps pulse transmitted between horns linked by the alumina ribbon waveguide (pulse A) and that linked by free-space (pulse B). The slight delay of the arrival of pulse A indicates that, owing to wave guidance by the alumina ribbon, pulse A is being guided at the group velocity of the TM-like mode on this structure. The guided group velocity is slower than c, the free-space group velocity [11]

11.3 An Example of Low-Loss Terahertz Ribbon Waveguide [15]

It is recalled that in the terahertz frequency range, all presently-known solid dielectric materials, even without impurities, as well as metallic materials are quite lossy. We now apply the considerations given in previous sections to the case of low-loss guidance for terahertz signals.

The significance of the development of the ribbon-based transmission system can be seen in Fig. 11.7. There the attenuation constant in dB m^{-1} is plotted as a function of operating frequency from 30 to 3,000 GHz for several commonly used single-mode traditional waveguides and for the ribbon waveguide. These single-mode traditional waveguides are the rectangular metallic waveguide, the circular dielectric guide with the same cross-sectional area as the dielectric ribbon, and the microstrip line. It is seen that at low frequencies (around 30 GHz), the rectangular metallic waveguide, quartz circular rod, and microstrip line are acceptable guides having attenuation constants ranging from 0.4 to 8 dB m^{-1}. The high loss of the microstrip line is tolerated since only short lengths are normally used. On the other hand, the silicon dielectric ribbon exhibits an attenuation constant of only 0.014 dB m^{-1}. As operating frequency f increases, attenuation also increases. For the rectangular metallic waveguide and the microstrip line, the attenuation increases at a rate proportional to approximately $f^{3/2}$. For the quartz circular rod and the silicon dielectric ribbon, it increases at a rate proportional to approximately f. At 300 GHz the attenuation for the microstrip line has reached approximately 150 dB m^{-1} (an onerous figure indeed), while that for the

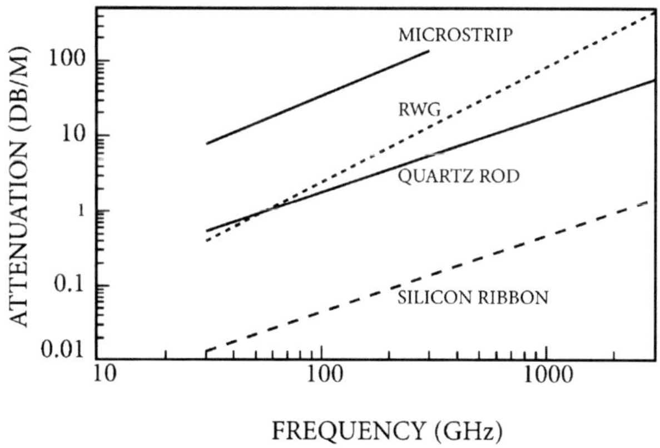

Figure 11.7. Typical performance comparison between several conventional waveguide structures and the high dielectric constant (Si) ribbon waveguide for the frequency range from 30 GHz to 3 THz. Note that the waveguide losses of typical conventional waveguides can be as much as 100 times larger than those of the ribbon waveguide in this spectrum [15]

rectangular metallic waveguide is 15 dB m^{-1}, and the quartz circular rod has a loss of 5.5 dB m^{-1}. At 300 GHz the attenuation for the silicon dielectric ribbon is approximately 0.15 dB m^{-1}, a figure significantly below that for the rectangular metallic waveguide at one tenth the frequency. At 3 THz the attenuation for the rectangular metallic waveguide has risen to more than 400 dB m^{-1} while that for the quartz circular rod has risen to more than 50 dB m^{-1}. These are extremely unattractive figures, rendering these waveguides impractical. Of course, the microstrip line is also impractical. It appears that the only remaining candidate at 3 THz is the silicon or other low-loss high dielectric constant ribbon. There the attenuation for the ribbon is approximately 1.5 dB m^{-1}, which is a workable figure.

Let us now address the important issues that involve using the ribbon waveguide in actual circuits. These issues include how to transport terahertz signals around bends (corners) without excessive radiation loss, how to efficiently couple terahertz power to and from conventional waveguides and microstrip structures, and how to design specific circuit elements. Typical low-loss terahertz circuits or interconnects may contain not only the basic straight sections of low-loss dielectric ribbons but also the other necessary circuit components, such as couplers, bends and curves, branches and combiners, and filters. Many of these passive circuit components require a tightly confined mode. We show that this is accomplished by using polymer-coated high dielectric constant ribbon structure.

From the discussion above, one expects that very low-loss guiding of terahertz band signals can be obtained with a ribbon-shaped guiding structure made with a moderately low-loss high dielectric constant material. However, since more than

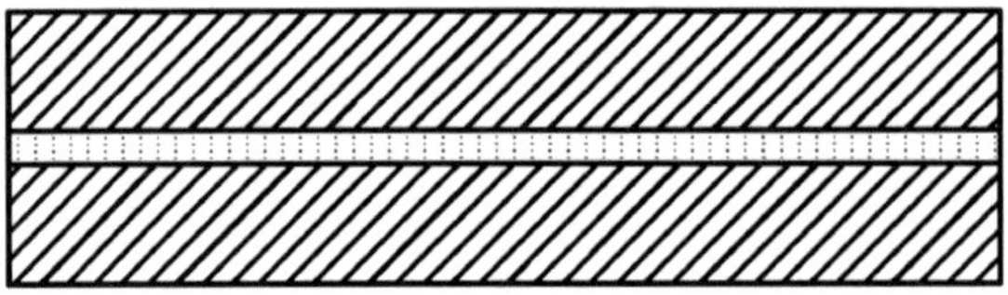

Figure 11.8. Longitudinal cross-sectional geometry of a polymer-coated high dielectric constant ribbon. The thickness and the width of the high dielectric constant ribbon are, respectively, approximately $0.0635\lambda_0 \times 0.635\lambda_0$. The thickness of the polymer coating is $\sim 0.25\lambda_0$ and the width is the same as the high dielectric constant ribbon. The relative dielectric constant of the ribbon is 10 while that of the polymer is 2.04

90% of the guided power is carried in the lossless dry air region outside the ribbon structure, it is expected that a significant amount of guided power will be radiated and lost when the guiding structure encounters sharp curves or corners. To remedy this problem, a variant of the described ribbon guide is envisioned wherein a thick polymer coating is added to the dielectric substrate in the vicinity of the bend or radiating region. A sketch of the structure is shown in Fig. 11.8. The polymer coating is introduced to provide tighter confinement of the guided power near the high dielectric constant ribbon. When a 10:1 aspect ratio, high dielectric constant ribbon ($0.625\lambda_0 \times 0.0625\lambda_0$) is coated on both sides with $1/4\lambda_0$ thick Teflon layer, more than 90% of the guided power for the dominant TM-like mode is contained within the boundary of the coated waveguide. This means that this coated ribbon can guide the dominant mode around corners, can divide or combine the power into or out of multiple waveguides, or can keep the mode confined in regions where radiation is likely. The calculated attenuation constant for the dominant TM-like mode in the Teflon-coated alumina ribbon is approximately 0.2 dB mm^{-1} at 3 THz (or 0.02 dB/free-space wavelength), an acceptable value for short-distance propagation. It thus appears that the Teflon-coated alumina ribbon can be used (with appropriate input–output matching) as an intermediate section for the design of all the components mentioned above. For long straight-run distances, a bare high dielectric constant ribbon with an attenuation constant of 0.0006 dB mm^{-1} at 3 THz should still be used.

A summary of the analyses and uses of the polymer-coated ribbon waveguide follows:

1. The distribution of the guided power (Poynting's vector) for the dominant $_{e}HE_{11}$ mode (TM-like mode) as a function of the distance away from the center major axis of the coated guide is shown in Fig. 11.9. Several cases with various Teflon-coating thicknesses are shown. It is seen that there exists a large discontinuity in the power distribution at the boundary between the low dielectric constant Teflon medium and the high dielectric constant alumina. The reason is that the

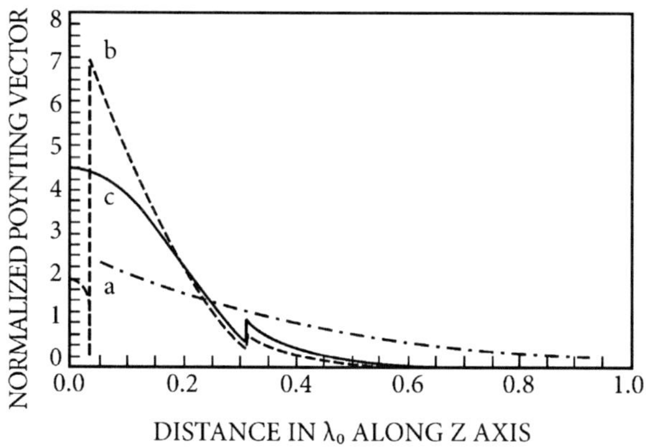

Figure 11.9. Normalized power intensity in a polymer-coated high dielectric constant ribbon supporting the dominant TM-like mode. Three cases are shown. For case (**a**), the guiding structure is a bare $0.635\lambda_0$ (width) $\times\, 0.0635\lambda_0$ (height) high dielectric constant ribbon surrounded by dry air; for case (**b**), it is the same high dielectric constant ribbon coated on the top and bottom surfaces of the ribbon with a layer of Teflon $0.25\lambda_0$ thick; for case (**c**), it is a plain Teflon ribbon with dimensions $0.635\lambda_0$ (width) $\times\, 0.635\lambda_0$ (height) surrounded by dry air. Here, $\epsilon_{r2} = 2.06$ and $\epsilon_{r1} = 10$ are, respectively, the relative dielectric constants of Teflon and the high dielectric constant material, and λ_0 is the free-space wavelength. The distributions of guided power for the three cases are as follows: (**a**) 1.04% in the high dielectric constant material and 98.96% in the air region; (**b**) 5.75% in the high dielectric constant material, 87.8% in the Teflon material, and 6.45% in the air region; (**c**) 69.23% in the Teflon material and 10.77% in the air region. Note that for case (**b**) more than 93% of the guided power is contained within the coated waveguided structure [15]

normal displacement vector (**D**) is continuous at the boundary and therefore the normal electric vector has a large discontinuity at the boundary when the dielectric constant differs greatly at the interface, resulting in a large discontinuity in the distribution of the Poynting's vector (power density) across the boundary. Specifically, the normal electric vector undergoes a jump at the boundary that is equal to the ratio of the inner to outer relative dielectric constants of the two dielectrics involved. For the thin high dielectric constant ribbon case, the power density within the high dielectric constant region is quite small while the power density just outside the high dielectric constant boundary is very large. This distinctive hole-in-the-middle power distribution for this high dielectric constant ribbon must be considered in the design of an efficient coupler for this ribbon structure. When a thin layer (of around $1/4\lambda_0$) of the polymer is coated on the high dielectric constant ribbon, more than 90% of the guided power is contained within the coated guiding structure. This fact is very important because this means that this coated ribbon can turn corners without suffering significant radiation losses.

Figure 11.10. Relative intensity (lighter shades indicate higher intensity) envelope of the transverse E-field in the upper half plane for a bare hight dielectric constant waveguide (alumina) and a Teflon-coated high dielectric constant waveguide. A sketch of the structure is shown in Fig. 11.8. The high dielectric constant ribbon is $0.6\lambda_0$ wide and $0.06\lambda_0$ thick. For the left plot there is no polymer coat, and for the right plot the Teflon coat is $0.26\lambda_0$ thick. Note that the field is more confined to the structure for the polymer coated waveguide [15]

2. A plot of the relative intensity envelope of the transverse E field in the upper half plane for a bare alumina ribbon waveguide and a Teflon-coated alumina waveguide is shown in Fig. 11.10. The alumina is $0.6\lambda_0$ wide and $0.06\lambda_0$ thick. On the left plot there is no polymer coat. On the right plot the Teflon coat is $0.26\lambda_0$ thick. Most of the guided power is confined within the coated guiding structure when the Teflon thickness is about $1/4\lambda_0$. When the thickness is reduced to zero, most of the power is confined within a distance of $1\lambda_0$ from the surface of the high dielectric constant ribbon in free space. Although there is a dip in the transverse field in the high dielectric constant material, the high dielectric constant ribbon is very thin and the dip may not be clearly seen in Fig. 11.10.

3. Possible applications are as follows. A polymer-coated high dielectric constant ribbon waveguide may be used: (1) as a transition region for (a) turning corners, (b) for making waveguide splits, (c) for shielding the guided wave from extraneous obstacles or interference, and (d) for providing efficient excitation of a desired mode; and (2) as a fundamental circuit element for designing low-pass, high-pass, or bandpass filters, couplers, patch antennas, array antenna elements, etc. Various high dielectric constant semiconductor materials such as GaAs, InP, or Si may be used to construct ribbon or polymer-coated ribbon waveguides. The additional advantage is that active elements may now be constructed from and/or on these ribbon or coated-ribbon waveguides, enabling the natural integration of passive and/or active elements and waveguiding structures. Such low-loss transmission media do not exist at this time for common commercial circuits at millimeter wavelengths, submillimeter wavelengths, or at terahertz frequencies. Components that would benefit from the use of Si, GaAsG, or InP ribbon waveguides include amplifiers, filters, up–down converters, oscillators, radiating antennas, phase shifters, millimeter wave monolithic integrated circuits (MMIC), or terahertz interconnects. As the fre-

quency increase, the semiconductor-composed ribbon waveguide circuits become more desirable and can provide a whole new category of rf components and subsystems. They also solve major interconnect problems that now limit multichip MMICs at millimeter wavelengths. Furthermore, they also provide high-power carrying capabilities.

In conclusion, we have shown how the lack of viable low-loss transmission systems in the terahertz spectrum may be overcome with the use of newly developed ribbon-based waveguides [11] and transitions [15]. The propensity of the fields to lie outside the guide substrate was shown to be problematic due to loose fundamental mode coupling and hence increased radiation around bends, transitions, or discontinuities. This problem has been overcome by adding a low dielectric constant polymer coating to the top and bottom of the ribbon guide in the immediate vicinity of the radiating regions. The extra dielectric loss by the coating is minimized by keeping the coated length short (<10 wavelengths). Several new transitions into and out of the coated and uncoated ribbon guide have been analyzed in [15]. These gave mismatch losses below 0.25 dB per transition, which is acceptable for small numbers of transition regions within a circuit. The same polymer-coated ribbon structures that are used to confine the dominant mode around bends can be used to make low-loss transitions into and out of a low dielectric microstrip line or can be used as filter elements on a straight ribbon waveguide. It is possible to realize a ribbon guide at terahertz frequencies with silicon being one of the materials of choice. Although we have presented the basis for the realization of ribbon guide circuit elements for terahertz applications, there are still material and fabrication issues that must be solved before these components can find their way into practical circuits. One application is to build flexible sensor and/or source heads for some of the terahertz imaging instruments [16]. At this time to our knowledge no other low-loss guide media exist at these wavelengths [17]. The concepts presented here may lead to many more component and transition realizations and to practical use of the ribbon guide in the near future.

Just as the low-loss optical fiber has become the backbone of high-speed communication links, the low-loss dielectric ribbon waveguide [15] and other subwavelength fiber waveguides [18, 19] may become the backbone of ultra-high-speed circuits, replacing metal-based microstrip lines or waveguides. Armed with this low-loss terahertz interconnects medium, we can find applications in medical diagnostics, radar, mechanically flexible sensors and sources, short- or long-distance communication links, MMIC interconnects, and perhaps even micromachined on-chip transmission line networks for ultrafast computers and/or processors.

References

1. F. P. Kapron, D. B. Keck, and R. D. Maurer, "Radiation losses in glass optical waveguides," Appl. Phys. Lett. **17**, 423 (1970); K. C. Kao and G. A. Hockam, "Dielectric fiber surface waveguides for optical frequencies," Proc. IEEE **133**, 1151 (1966)
2. F. E. Terman, "Radio Engineers' Handbook," McGraw-Hill, New York (1943)
3. M. N. Afsar and K. J. Button, "Millimeter wave dielectric measurement of materials," Proc. IEEE **73**, 131 (1985); R. Birch, J. D. Dromey, and J. Lisurf, "The optical constants of some common low-loss polymers between 4 and 40 cm^{-1} ,"Infrared Phys. **21**, 225 (1981); M. N. Afsar, "Precision dielectric measurements of nonpolar polymers in the millimeter wavelength range," IEEE Trans. Microw. Theory Tech. **33**, 1410 (1985)
4. C. Yeh, "American Institute of Physics Handbook," 3rd edn., D. E. Gray, ed., McGraw-Hill, New York (1972)
5. S. Ramo, J. R. Whinnery, and T. Van Duzer, "Fields and Waves in Communication Electronics," 2nd edn., Wiley, New York (1984)
6. C. Yeh, K. Ha, S. B. Dong, and W. P. Brown, "Single-mode optical waveguides," Appl. Opt. **18**, 1490 (1979)
7. K. S. Yee, "Numerical solution of initial boundary value problems involving Maxwell's equations in isotropic media," IEEE Trans. Antenn. Propag. **14**, 302 (1966); A. Taflove, "Computational Electrodynamics – The Finite-Difference Time-Domain Method," Artech House, Norwood, MA (1995)
8. C. Yeh, L. Casperson, and B. Szejn, "Propagation of truncated gaussian beams in multimode fiber guides," J. Opt. Soc. Am. **68**, 989 (1978); C. Yeh and F. Manshadi, "On weakly guiding single-mode optical waveguides," J. Lightwave Tech. **3**, 199 (1985)
9. C. Yeh, "Elliptical dielectric waveguides," J. Appl. Phys. **33**, 3235 (1962); C. Yeh, "Attenuation in dielectric elliptical cylinders," IEEE Trans. Antenna Prop. **11**, 177 (1963)
10. C. Yeh, F. I. Shimabukuro, and J. Chu, "Ultra-low-loss dielectric ribbon waveguide for millimeter/submillimeter waves," J. Appl. Phys. **54**, 1183 (1989)
11. C. Yeh, F. Shimabukuro, P. Stanton, V. Jamnejad, W. Imbriale, and F. Manshadi, "Communication at millimeter-submillimeter wavelengths using a ceramic ribbon," Nature **404**, 584 (2000)
12. S. K. Koul, "Millimeter Wave and Optical Dielectric Integrated Guides and Circuits," Series in Microwave and Optical Engineering, Wiley, New York (1997); T. C. Edwards, "Foundation for Microstrip Circuit Design," Wiley, New York (1981)
13. F. Shimabukuro and C. Yeh, "Attenuation measurement of very low-loss dielectric waveguides by the cavity resonator method applicable in the millimeter/submillimeter wavelength range," IEEE Trans. Microw. Theory Tech. **36**, 1160 (1988)
14. C. Yeh, "A relation between α and Q," Proc. IRE **50**, 2145 (1962)

15. C. Yeh, F. Shimabukuro, and P. H. Siegel, "Low-loss terahertz waveguides," Appl. Opt. **28**, 5937 (2005)
16. P. H. Siegel, Private communication (2005)
17. P. H. Siegel, "Terahertz technology," IEEE Trans. Microw. Theory Tech. **MTT-50**, 910 (2002); D. van der Weider, "Application and outlook or electronic terahertz technology," Opt. Photon. News **14**, 48 (2003); J. Mullins, "Using unusable frequencies," IEEE Spectr. **39**, 22 (2002); G. Gallot, S. P. Jamison, R. W. McGowan, and D. Grischkowsky, "Terahertz waveguides," J. Opt. Soc. Am. B **17**, 851 (2000); J. F. Roux, F. Aquistapace, F. Garet, L. Duvillaret, and J. L. Coutaz, "Grating-assisted coupling of terahertz waves into a dielectric waveguide studies by terahertz time-domain spectroscopy," Appl. Opt. **41**, 6507 (2002); G. L. Carr, M. C. Martin, W. C. Mckinney, K. Jordan, G. R. Neill, and G. P. Williams, "High power terahertz radiation from relativistic electrons," Nature **420**, 153 (2002); R. Mendis and D. Grischkowsky, "Plastic ribbon THz waveguides," J. Appl. Phys. **88**, 4449 (2000); K. Wang and D. Mittelman, "Metal wires for terahertz waveguides," Nature **432**, 376 (2004)
18. L. J. Chen, H. W. Chen, T. F. Kao, J. Y. Lu, and C. K. Sun, "Low-loss subwavelength plastic fiber for terahertz wave guiding," Opt. Lett. **31**, 308 (2006)
19. H. W. Chen, Y. T. Li, J. L. Kuo, J. Y. Lu, L. J. Chen, C. L. Pan, and C. K. Sun, "Investigation on spectral loss characteristics of subwavelength terahertz fibers," Opt. Lett. **32**, 1017 (2007)

12

PLASMON (SUBWAVELENGTH) WAVEGUIDES

A surface plasmon-polariton wave (SPP) is a guided surface wave along the interface of a metal film or a metal substrate and a surrounding dielectric material. This type of wave is formed by the coupling of an electromagnetic wave to oscillating free electrons at the surface of a conductor. The first pioneering investigation on this phenomenon was carried by Ritchie in the 1950s [1]. Subsequently, for the next half a century, the studies of SPPs and their applications have expanded from the diagnostics of metal surface qualities to a broad spectrum of scientific endeavors in physics, chemistry, and biology [2–10]. Furthermore, the subwavelength model size and very slow group velocity over an unusually large bandwidth of the SPP wave may provide exciting applications to the field of nanophotonics and microfabrication [4, 6, 10].

Typically, the material property of a nonmagnetic metallic film or a metallic substrate that enables the existence of a SPP wave is characterized by a permittivity, ϵ_m, and a permeability, μ_0, with

$$\begin{aligned}\epsilon_m &= \epsilon_0\left[1 - \frac{\omega_\mathrm{p}^2}{\omega^2 - j\nu\omega}\right],\\ &= \epsilon_0\left[1 - \frac{\omega_\mathrm{p}^2}{\omega^2 + \nu^2} - j\frac{\omega_\mathrm{p}^2\nu}{\omega(\omega^2 + \nu^2)}\right],\end{aligned} \tag{12.1}$$

where ω is the excitation frequency, ω_p is the bulk electron plasma frequency, and ν is the effective electron collision frequency. At optical wavelengths, $\omega^2 + \nu^2 < \omega_\mathrm{p}^2$ for most metals, making the real part of ϵ_m a negative value. As will be shown later, this negative real-part of ϵ_m for the metal and the positive real-part of the surrounding dielectric are crucial in allowing the existence of the SPP waves at the interface.

It is known that the Zenneck wave (a surface guided wave) also exists along the interface between a lossy conducting medium and a dielectric medium. According to Stratton [11], the lossy conducting medium may be characterized by its complex dielectric constant at dc for the frequency range far away from the violet or ultraviolet region, which is the region where the resonant frequencies of the bound electrons of metallic atoms are known. So,

$$\epsilon_m^{(\mathrm{dc})} = \epsilon_0 \left(1 - j \frac{\sigma'}{\omega \epsilon_0} \right), \tag{12.2}$$

where $\sigma' = \sigma_{\mathrm{dc}} = \left(\epsilon_0 \omega_{\mathrm{p}}^2\right)/\nu$, σ_{dc} is the dc conductivity. This equation can be derived from (12.1) if we consider the range

$$\frac{\omega}{\nu} \ll 1. \tag{12.3}$$

Substituting (12.3) in (12.1) yields

$$\begin{aligned} \epsilon_m &= \epsilon_0 \left[1 - \frac{\sigma_{\mathrm{dc}}}{\epsilon_0 \nu} - j \frac{\sigma_{\mathrm{dc}}}{\omega \epsilon_0} \right], \\ &\sim -j \frac{\sigma_{\mathrm{dc}}}{\omega} \quad \text{if } \omega \to 0. \end{aligned} \tag{12.4}$$

This is the permittivity that we use when we consider the Zenneck wave case.

At optical frequencies, the unavoidable intrinsic loss of conducting materials due to inelastic scattering mechanisms means that SPP waves are inherently lossy. Nevertheless, the useful propagation distance for applications in subwavelength aperture arrays, thin-film guided wave spectroscopy, and nonlinear interactions due to surface field enhancement can still be shorter than the acceptable loss length.

Here, starting from the basic electromagnetic equations, as given in Chap. 4, fundamentals of SPP waves will be obtained and discussed.

12.1 TM Wave Guidance Along a Metallic Substrate

The geometry of this problem is shown in Fig. 12.1. A metallic substrate with permittivity ϵ_m is immersed in a dielectric medium with permittivity ϵ_1. The nonzero components of a TM wave in 2D are E_z, E_x, and H_y. The propagation direction is z and the propagation constant and frequency are β and ω, respectively. For harmonic fields, the factor $e^{-j\beta z + j\omega t}$will be assumed and suppressed. From Sect. 4.1, the TM field components are the following:

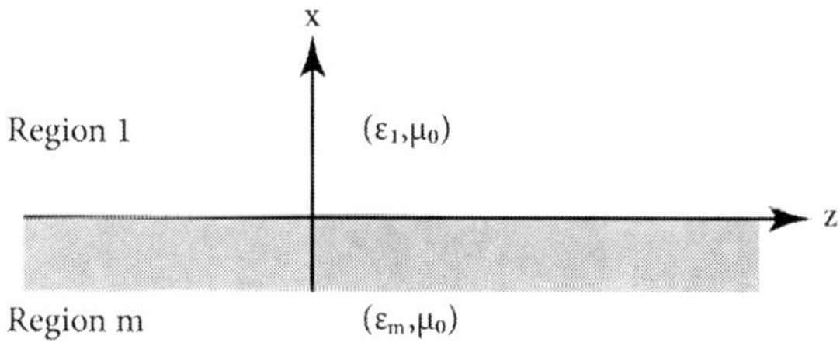

Figure 12.1. Geometry of the problem

In region m (the metallic substrate region), $x < 0$

$$E_z^{(m)} = A_m\, e^{p_m x}, \tag{12.5}$$

$$E_x^{(m)} = \frac{j\beta}{p_m^2}\frac{\partial E_z^{(m)}}{\partial x} = \frac{j\beta}{p_m} A_m\, e^{p_m x}, \tag{12.6}$$

$$H_y^{(m)} = \frac{j\omega\epsilon_m}{p_m^2}\frac{\partial E_z^{(m)}}{\partial x} = \frac{j\omega\epsilon_m}{p_m} A_m\, e^{p_m x}, \tag{12.7}$$

where A_m is an arbitrary constant,

$$p_m^2 = \beta^2 - k_m^2, \tag{12.8}$$

with

$$k_m^2 = \omega^2 \mu_0 \epsilon_m \tag{12.9}$$

and μ_0 is the free space permeability. The function $e^{p_m x}$ is chosen such that all fields will decay exponentially to zero as $x \to -\infty$. So, $\mathrm{Re}(p_m)$ must be positive. If β is found to have a positive real part, then this TM wave can exist along the interface of the metallic substrate

In region 1 (the dielectric region), $x > 0$,

$$E_z^{(1)} = A_1\, e^{-p_1 x}, \tag{12.10}$$

$$E_x^{(1)} = \frac{j\beta}{p_1^2}\frac{\partial E_z^{(1)}}{\partial x} = -\frac{j\beta}{p_1} A_1\, e^{-p_1 x}, \tag{12.11}$$

$$H_y^{(1)} = \frac{j\omega\epsilon_1}{p_1^2}\frac{\partial E_z^{(1)}}{\partial x} = -\frac{j\omega\epsilon_1}{p_1} A_1\, e^{-p_1 x}, \tag{12.12}$$

where A_1 is an arbitrary constant, and

$$p_1^2 = \beta^2 - k_1^2, \tag{12.13}$$

$$k_1^2 = \omega^2 \mu_0 \epsilon_1. \tag{12.14}$$

The function $e^{-p_1 x}$ is chosen such that all fields will decay exponentially to zero at $x \to \infty$. So $\mathrm{Re}(p_1)$ must be positive. Matching the tangential fields at the boundary surface $x = 0$ yields

$$A_m = A_1, \tag{12.15}$$

$$\frac{j\omega\epsilon_m}{p_m} A_m = -\frac{j\omega\epsilon_1}{p_1} A_1. \tag{12.16}$$

The nontrivial solution is

$$\frac{\epsilon_m}{p_m} + \frac{\epsilon_1}{p_1} = 0. \tag{12.17}$$

Subtracting (12.8) from (12.13) yields

$$p_m^2 - p_1^2 = k_1^2 - k_m^2. \tag{12.18}$$

Substituting (12.18) in (12.17) and simplifying gives

$$p_m^2 \left(1 - \frac{\epsilon_1^2}{\epsilon_m^2}\right) = k_1^2 - k_m^2. \tag{12.19}$$

From (12.8) we have

$$\beta^2 = p_m^2 + k_m^2. \tag{12.20}$$

Substituting (12.19) in (12.20) yields

$$\beta^2 = k_m^2 + \frac{k_1^2 - k_m^2}{\left(1 - \frac{\epsilon_1^2}{\epsilon_m^2}\right)} = \frac{k_1^2 \epsilon_m}{\epsilon_m + \epsilon_1} = \left(\frac{\omega}{c}\right)^2 \frac{\epsilon_1 \epsilon_m}{\epsilon_m + \epsilon_1}. \tag{12.21}$$

So,

$$\beta = \frac{\omega}{c} \sqrt{\frac{\epsilon_1 \epsilon_m}{\epsilon_m + \epsilon_1}}, \tag{12.22}$$

$$p_1 = \frac{\omega}{c} \sqrt{\frac{-\epsilon_1^2}{\epsilon_0 (\epsilon_m + \epsilon_1)}}, \tag{12.23}$$

$$p_m = \frac{\omega}{c}\sqrt{\frac{-\epsilon_m^2}{\epsilon_0\left(\epsilon_m + \epsilon_1\right)}}. \tag{12.24}$$

Let us consider the following cases:

(a) ϵ_1/ϵ_0 and ϵ_m/ϵ_0 are both real and positive

For this case, β is real, indicating a propagating wave, but p_1 and p_m are both imaginary, indicating a nondecaying function of x for the field components, violating the required condition for a guided surface wave. The same conclusion was reached in Sect. 4.7.

(b) ϵ_1/ϵ_0 is real and positive and ϵ_m/ϵ_0 is real and negative

From (12.22) we note that if ϵ_1 is real and positive, $\epsilon_m/\epsilon_0 < 0$, and $\epsilon_m/\epsilon_0 < \epsilon_1/\epsilon_0$, the quantity under the square root is negative, making β an imaginary value. This means that there is no propagating wave. On the other hand, if $\epsilon_m/\epsilon_0 < 0$ and $|\epsilon_m/\epsilon_0| > \epsilon_1/\epsilon_0$, β is real, indicating a propagating wave. Under this condition, one can also show from (12.23) and (12.24) that p_m and p_1 are both real and positive, indicating decaying field away from the interface. A sketch of the ω-β diagram is shown in Fig. 12.2.

(c) ϵ_1/ϵ_0 is real and positive and ϵ_m/ϵ_0 is complex

Using

$$\frac{\epsilon_m}{\epsilon_0} = \epsilon_{rm}^r - j\epsilon_{rm}^i, \tag{12.25}$$

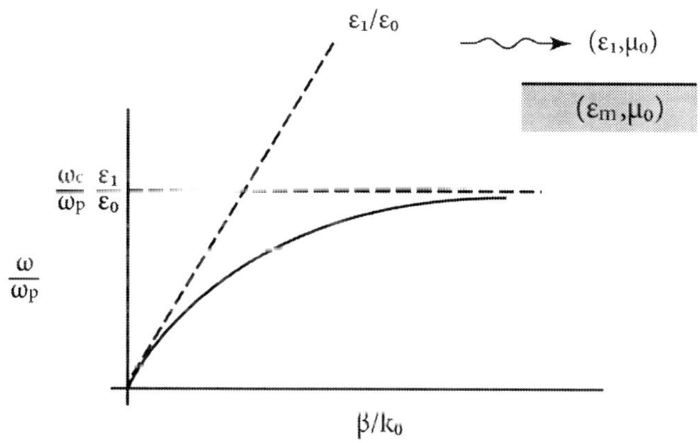

Figure 12.2. A sketch of the ω-β diagram for case where $\epsilon_m < 0$, $|\epsilon_m| > \epsilon_1$, and ϵ_1 is real and positive. ω_c is the high frequency cutoff, given by $\omega_c = \omega_p/\sqrt{1+\epsilon_1/\epsilon_0}$. ϵ_1 and ϵ_m are both real

where $\epsilon_{rm}^r = \mathrm{Re}(\epsilon_m/\epsilon_0)$ and $\epsilon_m^i = \mathrm{Im}(\epsilon_m/\epsilon_0)$, (12.22) becomes

$$\beta = \frac{\omega}{c}\left[\frac{\left(\epsilon_{rm}^r - j\epsilon_{rm}^i\right)\epsilon_{r1}}{\left(\epsilon_{r1} + \epsilon_{rm}^r\right) - j\epsilon_{rm}^i}\right]^{1/2} = \frac{\omega}{c}a(1 - jb), \tag{12.26}$$

where

$$a = \left[\epsilon_{r1}\frac{\epsilon_{rm}^r\left(\epsilon_{r1} + \epsilon_{rm}^r\right) + \epsilon_{rm}^{i2}}{\left(\epsilon_{r1} + \epsilon_{rm}^r\right)^2 + \epsilon_{rm}^{i2}}\right]^{1/2}, \tag{12.27}$$

$$b = \frac{\epsilon_{r1}\epsilon_{rm}^i}{\epsilon_{rm}^r\left(\epsilon_{r1} + \epsilon_{rm}^r\right) + \epsilon_{rm}^{i2}}. \tag{12.28}$$

Equation (12.26) simplifies as follows:

Case 1 If $|b| \ll 1$

$$\beta \approx \frac{\omega}{c}a\left(1 - \frac{jb}{2}\right) = \frac{\omega}{c}a - j\frac{\omega ab}{2c}. \tag{12.29}$$

Case 2 If $|b| \gg 1$

$$\beta \approx \frac{\omega}{c}a(-jb)^{1/2} = \frac{\omega}{c}a\sqrt{b}\left(e^{-j\pi/4}\right) = \frac{\omega}{c}a\left(\frac{b}{2}\right)^{1/2}(1 - j). \tag{12.30}$$

Case 3 If $|b| \approx 1$,

$$\beta \approx \frac{\omega}{c}a(1 + b^2)^{1/2}\left[\cos\left(\frac{\tan^{-1} b}{2}\right) - j\sin\left(\frac{\tan^{-1} b}{2}\right)\right]. \tag{12.31}$$

From (12.22)–(12.24), we note that if $\left|\epsilon_{rm}^i\right| < \left|\epsilon_{rm}^r\right|$ and $\epsilon_{rm}^r < 0$, when $\mathrm{Re}(\beta) \neq 0$, indicating a propagating wave, $\mathrm{Re}(p_m)$ and $\mathrm{Re}(p_1)$ will be > 0, indicating decaying fields away from the interface. When these conditions are satisfied, SPP modes can exist.

Let us review a few realistic cases:

Interface of Ag and SiO_2 *at 633 nm*

In this case, the complex dielectric constant of Ag at $\lambda = 633$ nm is

$$\frac{\epsilon_m}{\epsilon_0} = -19 - j0.53 \tag{12.32}$$

and that of SiO_2 is $\epsilon_1/\epsilon_0 = 4$. From (12.28) we find

$$b = 0.0074, \tag{12.33}$$

which is $\ll 1$. So, this belongs to case 1 above. Therefore,

$$\beta = \frac{\omega}{c}a - j\frac{\omega ab}{2c}$$

and $\mathrm{Re}(p_m) > 0$ and $\mathrm{Re}(p_1) > 0$, indicating a propagating SPP wave.

Interface of Al and SiO_2 at 633 nm

Here the complex dielectric constant of Al at $\lambda = 633$ nm is

$$\frac{\epsilon_m}{\epsilon_0} = -29.8 - j11.6 \tag{12.34}$$

and that of SiO_2 is $\epsilon_1/\epsilon_0 = 4$. From (12.28) we find

$$b = 0.051, \tag{12.35}$$

which is again $\ll 1$. So case 1 applies. An SPP wave can exist.

Interface of Al and SiO_2 at 0.4 THz

Here, the complex dielectric constant of Al at 0.4 THz is

$$\frac{\epsilon_m}{\epsilon_0} = -3.3 \times 10^4 - j1.6 \times 10^6 \tag{12.36}$$

and that of SiO_2 is $\epsilon_1/\epsilon_0 = 4$. From (12.28) we find

$$b \simeq 2.5 \times 10^{-6}, \tag{12.37}$$

which is again $|b| \ll 1$. So case 1 applies. An SPP wave can exist.

It is seen from the above examples that an SPP wave appears to be a rather robust wave that can exist in a broad range of wavelengths along different metallic interfaces.

As another illustration, the ω-β diagram for the SPP on a metal substrate with or without a coated layer of dielectric material is given in Fig. 12.3 [10].

12.2 TM Wave Guidance Along a Metallic Film [12, 13]

The geometry of this problem is shown in Fig. 12.4. A two-dimensional metallic film of thickness h with dielectric constant ϵ_m is immersed in a dielectric medium whose dielectric constant above the film is ϵ_3 and below the film is ϵ_1.

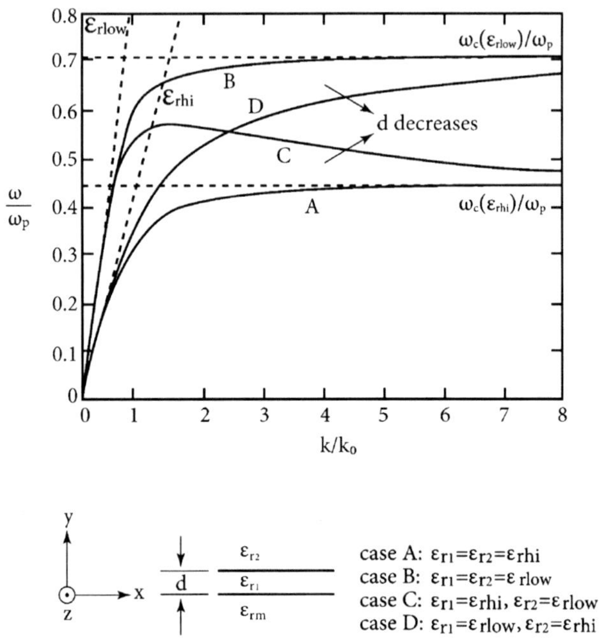

Figure 12.3. ω-k diagrams (*solid curves*) for conventional layered surface-plasmon structures (Cases A-D) with $\epsilon_{\rm rhi} = 4$ and $\epsilon_{\rm rlow} = 1$ (*air*). Layer thicknesses $d/\lambda_p = 0.015$, 0.02, and 0.025 are used for C and D (*solid + dotted curves*). The light lines $\omega/\omega_p = \sqrt{\epsilon/\epsilon_0}\,k/k_0$ (*vertical dashed lines*) and the cutoff frequencies $\omega_c(\epsilon)/\omega_p = 1/\sqrt{1+\epsilon/\epsilon_0}$ (*horizontal dashed lines*) are shown [10]

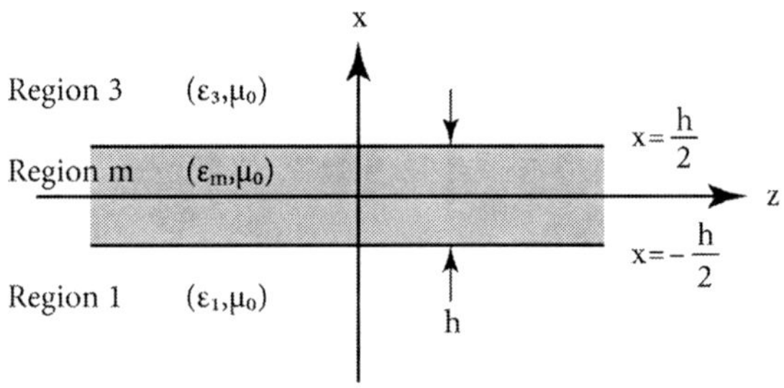

Figure 12.4. Thin film geometry

We are interested in the possible propagation of TM waves along the film. The TM waves have components E_z, E_x, and H_y with a z dependent propagation factor $e^{-j\beta z+j\omega t}$, which will be suppressed. Here, β is the propagation constant of the wave and ω is the harmonic frequency. From Sect. 4.1, the 2D TM fields with $\partial/\partial y = 0$ satisfy the following equations:

$$E_x = -\frac{j\beta}{p^2}\frac{\partial E_z}{\partial x}, \tag{12.38}$$

$$H_y = -\frac{j\omega\epsilon}{p^2}\frac{\partial E_z}{\partial x}, \tag{12.39}$$

$$\left(\frac{\partial^2}{\partial x^2} + p^2\right) E_z = 0, \tag{12.40}$$

$$p^2 = \omega^2\mu\epsilon - \beta^2, \tag{12.41}$$

where ϵ and μ are the permittivity and permeability of the medium in which the equations apply. Using the appropriate solutions that may satisfy the required boundary conditions and radiation condition, one finds the following expressions for the fields tangential to the field surfaces:

In region 1, $(x \leq -h/2)$

$$E_z^{(1)} = A_1\, e^{p_1 x}, \tag{12.42}$$

$$H_y^{(1)} = -\frac{j\omega\epsilon_1}{p_1} A_1\, e^{p_1 x}, \tag{12.43}$$

$$p_1^2 = \beta^2 - \omega^2\mu\epsilon_1; \tag{12.44}$$

In region m, $(-h/2 \leq x \leq h/2)$,

$$E_z^{(m)} = A_2 \cosh(p_m x) + A_3 \sinh(p_m x), \tag{12.45}$$

$$H_y^{(m)} = -\frac{j\omega\epsilon_m}{p_m}\left[A_2 \sinh(p_m x) + A_3 \cosh(p_m x)\right], \tag{12.46}$$

$$p_m^2 = \beta^2 - \omega^2\mu\epsilon_m; \tag{12.47}$$

In region 3, $(x \geq h/2)$,

$$E_z^{(3)} = A_4\, e^{-p_3 x}, \tag{12.48}$$

$$H_y^{(3)} = -\frac{j\omega\epsilon_3}{p_3} A_4\, e^{-p_3 x}, \tag{12.49}$$

$$p_3^2 = \beta^2 - \omega^2\mu\epsilon_3, \tag{12.50}$$

where A_1, A_2, A_3, and A_4 are arbitrary constants. Matching the boundary conditions at $x = \mp h/2$ yields

$$A_1\, e^{-hp_1/2} = A_2 \cosh\left(\frac{p_m h}{2}\right) - A_3 \sinh\left(\frac{p_m h}{2}\right), \tag{12.51}$$

$$\frac{j\omega\epsilon_1}{p_1} A_1\, e^{-hp_1/2} = \frac{j\omega\epsilon_m}{p_m}\left[-A_2 \sinh\left(\frac{p_m h}{2}\right) + A_3 \cosh\left(\frac{p_m h}{2}\right)\right], \tag{12.52}$$

$$A_2 \cosh\left(\frac{p_m h}{2}\right) + A_3 \sinh\left(\frac{p_m h}{2}\right) = A_4\, e^{-p_3 h/2}, \tag{12.53}$$

$$\frac{j\omega\epsilon_m}{p_m}\left[A_2 \sinh\left(\frac{p_m h}{2}\right) + A_3 \cosh\left(\frac{p_m h}{2}\right)\right] = -\frac{j\omega\epsilon_3}{p_3} A_4\, e^{-p_3 h/2}. \tag{12.54}$$

For a nontrivial solution, the determinant of the above linear equations must be set to zero. This yields

$$\tanh(p_m h)\,(1 + s_1 s_3) = -(s_1 + s_3), \tag{12.55}$$

with

$$s_1 = \frac{p_1}{p_m}\frac{\epsilon_m}{\epsilon_1}, \tag{12.56}$$

$$s_3 = \frac{p_3}{p_m}\frac{\epsilon_m}{\epsilon_3}. \tag{12.57}$$

Equation (12.55) is the dispersion relation for the TM wave from which the propagation characteristics and the ω-β diagram may be obtained.

Let us investigate the following special cases :

(a) If $\epsilon_1 = \epsilon_3$, the dispersion relation (12.55) reduces to

$$\tanh(p_m h) = -\frac{2s_1}{1 + s_1^2}. \tag{12.58}$$

It may be further reduced to

$$\left[\tanh\left(\frac{p_m h}{2}\right) + s_1\right]\left[s_1 \tanh\left(\frac{p_m h}{2}\right) + 1\right] = 0 \tag{12.59}$$

or

$$\tanh\left(\frac{p_m h}{2}\right) = -s_1 \tag{12.60}$$

and

$$\tanh\left(\frac{p_m h}{2}\right) = -\frac{1}{s_1}. \tag{12.61}$$

One may recognize that (12.60) represents the dispersion relation for a symmetric (or even) mode having $A_3 = 0$ in (12.45), while (12.61) represents the dispersion relation for an antisymmetric (or odd) mode, having $A_2 = 0$ in (12.45). It has been found that the symmetric mode possesses smaller attenuation constant than the antisymmetric mode for thin metal films [13].

(b) If $h \to \infty$, then $\tanh(p_m h) \to 1$ in (12.55). The dispersion relation is reduced to

$$(1 + s_1 s_3) = -(s_1 + s_3) \tag{12.62}$$

or

$$(1 + s_1)(1 + s_3) = 0. \tag{12.63}$$

Thus, one has

$$s_1 = -1, \tag{12.64}$$

$$s_3 = -1. \tag{12.65}$$

These are the dispersion relations for the half-space problem treated in Sect. 12.1. It is understood that if the thickness is too large, that is, larger than the skin depth in the metal substrate, the fields along the upper interface and those along the lower interface are decoupled, resulting in two separate interface problems.

As an illustration, for SPP wave propagation along a metal film, we shall consider the following case [13]:

$$\epsilon_m/\epsilon_0 = -29.8 - j11.6 \qquad \text{(aluminum film)},$$

$$\epsilon_1/\epsilon_0 = 2.137,$$

$$\epsilon_3/\epsilon_0 = 1.69,$$

$$\lambda_0 = 6.33\,\text{nm}.$$

Using (12.55), the real and imaginary parts of the complex propagation constant β are plotted against the thickness of the film in Fig. 12.5. Both symmetric and antisymmetric modes are shown. Strictly speaking, if the guiding structure is not symmetric about the y-z plane, that is, $\epsilon_1 \neq \epsilon_3$, no purely symmetric or antisymmetric mode can exist. In other words, A_2 and A_3 in (12.45) must both be nonzero in order to satisfy the boundary conditions. Nevertheless, there can exist modes that are symmetric-like modes, whose fields are mostly symmetric about the y-z

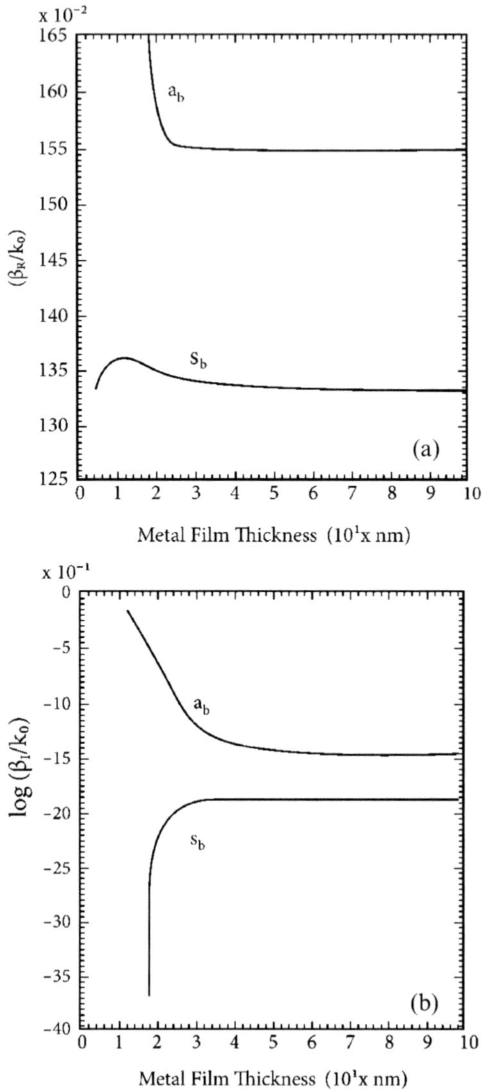

Figure 12.5. (*Top*) Normalized real part of the complex propagation constant of the SPP modes vs. the metal film thickness (h). (*Bottom*) Normalized imaginary part of the complex propagation constant of the SPP modes vs. the metal film thickness (h). $n_1 = 1.462$, $n_3 = 1.3$, $\epsilon_m/\epsilon_0 = -29.8 - j11.6$, and $\lambda_0 = 6.33$ nm [13]

plane, or antisymmetric-like modes, whose fields are mostly antisymmetric about the y-z plane. It is seen from Fig. 12.5 that the symmetric mode (s_b) possesses much lower attenuation than the antisymmetric mode (a_b). Therefore, the symmetric SPP mode is called the long range SPP wave of LR-SPP. It is of interest to

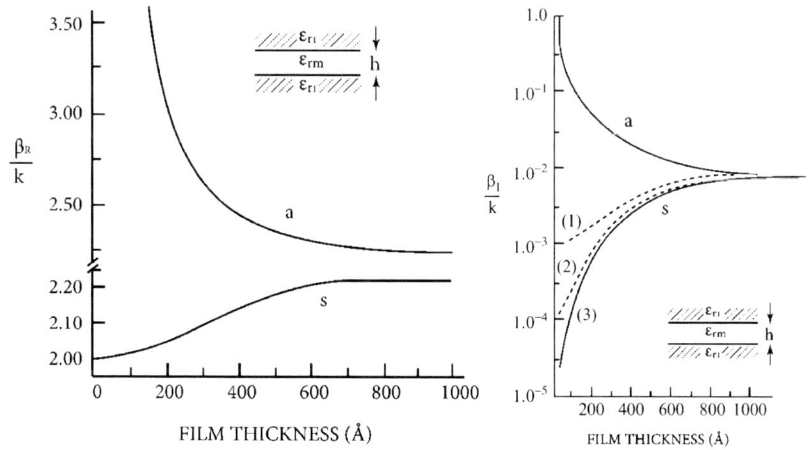

Figure 12.6. (*Left*) Normalized real part of the propagation wave vector β as a function of film thickness for the symmetric structure with $(\epsilon_{r1})^{1/2} = 2.0$. a and s refer to the antisymmetric and symmetric waves, respectively. (*Right*) Normalized imaginary part of the propagation wave vector β as a function of film thickness for the symmetric structure with $(\epsilon_{r1})^{1/2} = 2.0$. a and s refer to the antisymmetric and symmetric waves, respectively. (1) $(\epsilon_{r1})^{1/2} = 2.0 - j0.001$, (2) $(\epsilon_{r1})^{1/2} = 2.0 - j0.0001$, and (3) $(\epsilon_{r1})^{1/2} = 2.0$, $\epsilon_{rm} = -19 - j0.53$ [12]

note that at $h = 19.39$ nm, the s_b mode changes from an attenuation type (which can be excited) to a growing type (which cannot be excited). Therefore, no solution is displayed for the range where the wave cannot be excited.

Another illustration deals with the case of a metal film immersed in a single dielectric medium, that is, a symmetric structure. Here, the dielectric medium has a dielectric constant of ϵ_1. Computed results for the symmetric and antisymmetric modes are shown in Fig. 12.6. [12].

A more refined model for the metal (such as Ag) whose optical constant was obtained experimentally rather than being represented as a damped free electron gas was used by Dionne et al. to treat the SPP problem [14]. They found that unlike the free electron gas model, the surface plasmon wave vector remains finite at resonances with the antisymmetric field plasmon converging to a pure photon mode for very thin films. Furthermore, allowed excitation modes are found to exist between the bound and the radiative branches of the dispersion curves.

12.3 Wave Guidance by Metal Ribbons

Analytical solutions are only available for the problem of wave guidance by metal films, as have been demonstrated in the previous sections [12–14]. Although, in principle, by approximating the metal ribbon by a flat elliptical cylinder one may

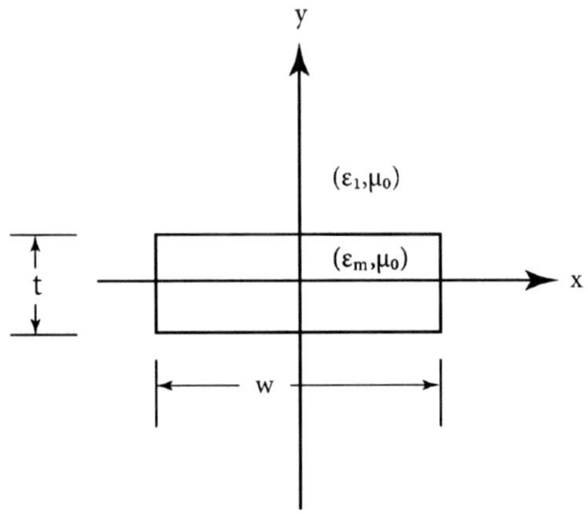

Figure 12.7. The geometry of a metal ribbon

solve the problem in terms of Mathieu-type functions, as shown in Chap. 6, however, the complexity of the solutions as well as the need to use Mathieu functions with complex arguments, which have not been well studied, render the solutions less than useful. This is where the use of numerical techniques to find the desired solutions is most appropriate. Berini [15] used the method of lines, which had been well developed [16], to find the needed solutions for the metal ribbon problem (see Fig. 12.7). Here we shall only describe some of his results as illustrations. We note that all six components of the field must be present for the metal ribbon structure.

The following symbols will be used to designate various modes on a metal ribbon:

a	denotes asymmetric transverse electric field distribution,
s	denotes symmetric electric field distribution,
Subscript b	denotes bound modes,
Subscript m	denotes number of field extrema along x axis,
w	denotes the width of the ribbon,
t	denotes the thickness of the ribbon,
s_{b}	denotes symmetric bound mode on a metal film with $w = \infty$,
a_{b}	denotes asymmetric bound mode on a metal film with $w = \infty$,
sa_{b}^{m} or aa_{b}^{m}	denotes modes on a ribbon guide that have dominant E_y field exhibiting asymmetry with respect to the x-axis; the first letter refers to whether the transverse electric field is asymmetric or symmetric with respect to the y-axis,
ss_{b}^{m} or sa_{b}^{m}	denotes mode on a ribbon that have dominant E_y field exhibiting symmetry with respect to the x-axis.

To generate the numerical data, the following parameters are used:

$$\lambda_0 = 0.633\,\text{nm},$$

$$\epsilon_m/\epsilon_0 = -19 - j0.53,$$

$$\epsilon_1/\epsilon_0 = 4,$$

$$k_0 = \omega\sqrt{\mu_0\epsilon_0} = \frac{\omega}{c},\quad c = \text{speed of light in vacuum},$$

$$\beta = \beta^r - j\alpha = \text{complex propagation constant.}$$

It is seen from Figs. 12.8 and 12.9 that the basic propagation behavior of the film guide and the ribbon guide remains relatively unchanged [15]. Again the symmetric mode is the low-loss mode. Figure 12.10 is introduced to show the contour plot of $\mathrm{Re}(S_z)$, where S_z is the z-directed Poynting's vector. One notes that a wider guide provides better guidance and lower loss, since the fields are distributed over a wider area. Also evident in Fig. 12.10 is the edge effect, that is, field density is increased at the edges of the guide in accordance with the edge condition given in Sect. 2.5.

We note that the fundamental mode is the ss_b^0 mode, which is the ${}_\mathrm{e}\mathrm{HE}_{11}$ mode discussed earlier in Chap. 6 on elliptical waveguides. The ss_b^0 mode evolves to the TEM wave as the thickness of the ribbon approaches zero. This is also the only long-range mode with attenuation $\alpha \sim 10 - 0.1\,\mathrm{dB\,m^{-1}}$ [15]. Furthermore, a_b and s_b modes do not have a cutoff thickness. For $w/t > 1$, the E_y field component dominates and for $w/t < 1$, the E_x field component dominates. Also, higher order modes on a ribbon have a cutoff width.

12.4 SPP Waves Along Cylindrical Structures

12.4.1 TM Waves

Let us consider the propagation of SPP waves along a circular cylindrical structure, shown in Fig. 12.11.

There are three distinct regions: region 1 with ϵ_1 and $r \leq a$; region 2 with ϵ_2 and $a \leq r \leq b$; and region 3 with ϵ_3 and $r \geq b$. The SPP waves are assumed to be propagating in the z-direction with a propagation factor $e^{-j\beta z + j\omega t}$, where β is the propagation constant. The field components tangential to the cylindrical surfaces for the lowest order TM mode are the following:

In region 1 ($r \leq a$)

$$E_z^{(1)} = A_1 I_0(p_1 r), \tag{12.66}$$

$$H_\theta^{(1)} = \frac{j\omega\epsilon_1}{p_1} A_1 I_0'(p_1 r), \tag{12.67}$$

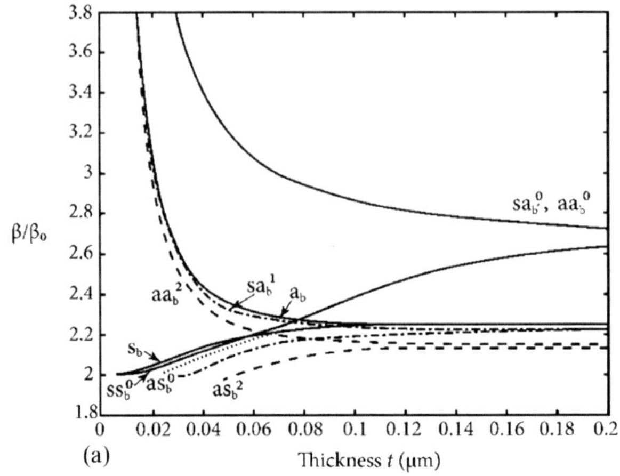

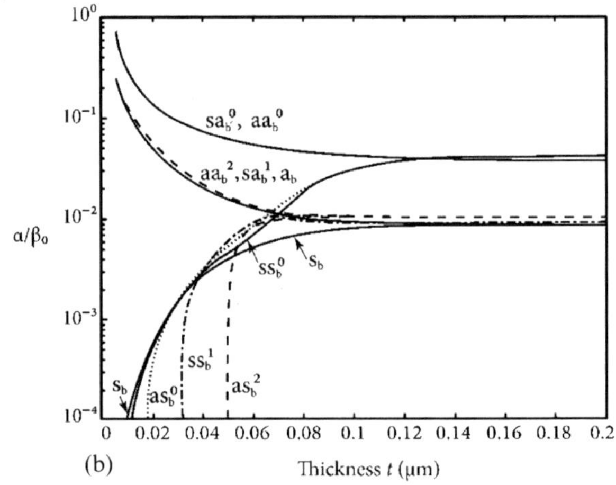

Figure 12.8. Dispersion characteristics: (*top*) normalized phase constant, and (*bottom*) normalized attenuation constant as a function of thickness of the first eight modes supported by a metal film wave guide of width $w = 1$ μm. The a_b and s_b modes supported for the case $w = \infty$ are shown for comparison [15]

$$p_1^2 = \beta^2 - k_1^2, \tag{12.68}$$

$$k_1^2 = \omega^2 \mu_0 \epsilon_1. \tag{12.69}$$

In region 2 ($a \le r \le b$)

$$E_z^{(2)} = A_2 I_0(p_2 r) + A_3 K_0(p_2 r), \tag{12.70}$$

$$H_\theta^{(2)} = \frac{j\omega\epsilon_2}{p_2}\left[A_2 I_0'(p_2 r) + A_3 K_0'(p_2 r)\right], \tag{12.71}$$

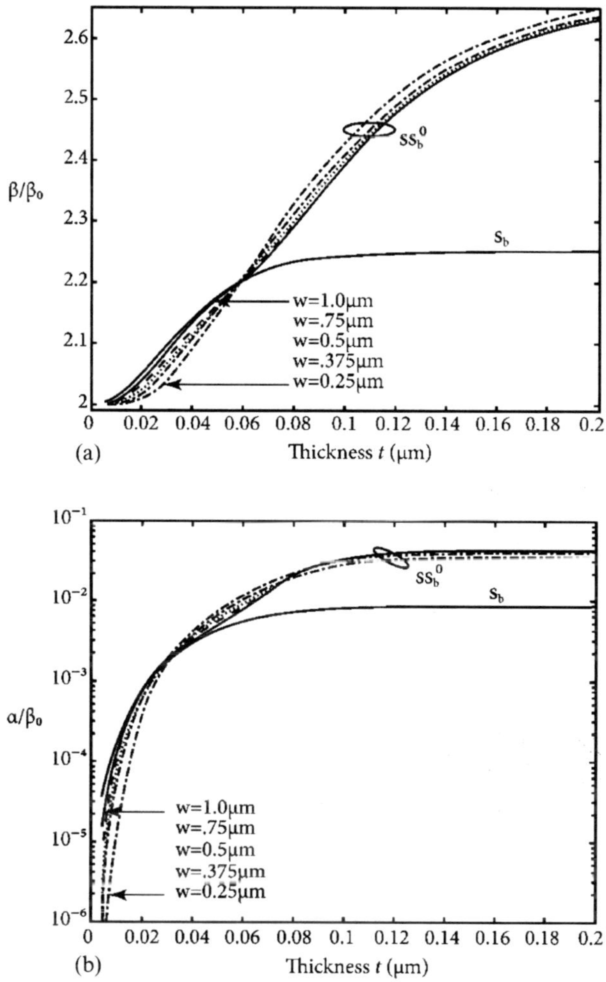

Figure 12.9. Dispersion characteristics: (*top*) normalized phase constant and (*bottom*) normalized attenuation constant as a function of thickness of the s_b mode supported by metal film wave guide of various widths. The s_b mode supported for the case $w = \infty$ is shown for comparison [15]

$$p_2^2 = \beta^2 - k_2^2, \tag{12.72}$$

$$k_2^2 = \omega^2 \mu_0 \epsilon_2 \tag{12.73}$$

In region 3 ($r \geq b$)

$$E_z^{(3)} = A_4 K_0(p_3 r), \tag{12.74}$$

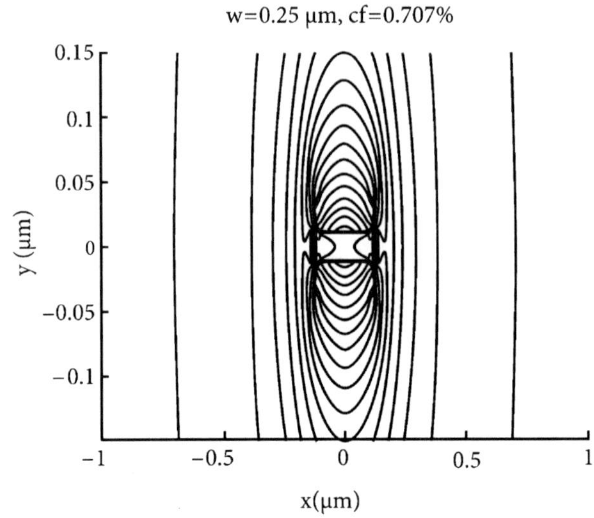

Figure 12.10. Contour plot of $\mathrm{Re}\{S_z\}$ associated with the ss_b^0 mode for metal film guides of thickness $t = 20\,\mathrm{nm}$ and $w = 0.25\,\mu\mathrm{m}$. The power confinement factor cf is also given in all cases with the area of the wave guide core A_c taken as the area of the metal region. In all cases, the outline of the metal film is shown as the rectangular dashed contour [15]

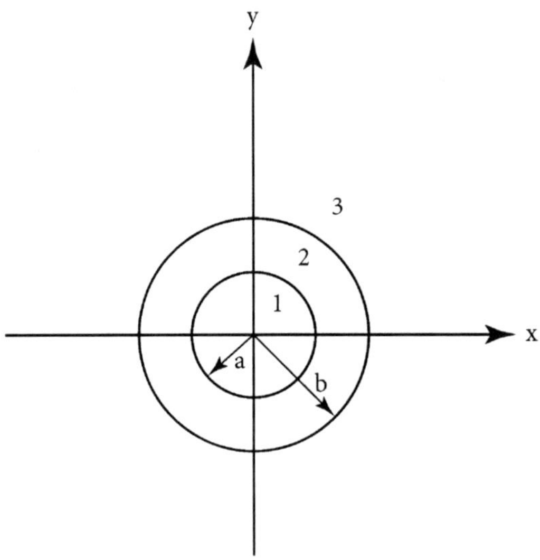

Figure 12.11. Geometry of the cylindrical structure

$$H_\theta^{(3)} = \frac{j\omega\epsilon_3}{p_3} A_4 K_0'(p_3 r), \tag{12.75}$$

$$p_3^2 = \beta^2 - k_3^2, \tag{12.76}$$

$$k_3^2 = \omega^2 \mu_0 \epsilon_3. \tag{12.77}$$

Here A_1, A_2, A_3, and A_4 are arbitrary constants, I_0 and K_0 are the modified Bessel functions of order zero, and the prime on these functions signifies the derivative of the function with respect to its argument. These modified Bessel functions are so chosen that the radiation condition is satisfied. If one of the dielectric constants is complex, the propagation constant will also be complex, indicating attenuation of the SPP wave.

Matching the tangential electric and magnetic fields at the boundary $r = a$ and at the boundary $r = b$, one obtains

$$A_1 I_0(p_1 a) = A_2 I_0(p_2 a) + A_3 K_0(p_2 a), \tag{12.78}$$

$$\frac{j\omega\epsilon_1}{p_1} A_1 I_0'(p_1 a) = \frac{j\omega\epsilon_2}{p_2} \left[A_2 I_0'(p_2 a) + A_3 K_0'(p_2 a)\right], \tag{12.79}$$

$$A_2 I_0(p_2 b) + A_3 K_0(p_2 b) = A_4 K_0(p_3 b), \tag{12.80}$$

$$\frac{j\omega\epsilon_2}{p_2} \left[A_2 I_0'(p_2 b) + A_3 K_0'(p_2 b)\right] = \frac{j\omega\epsilon_3}{p_3} A_4 K_0'(p_3 b). \tag{12.81}$$

Setting the determinant of the above four linear simultaneous equations to zero yields

$$AB - CD = 0, \tag{12.82}$$

where

$$A = \frac{I_0(p_2 a)}{I_0(p_1 a)} - \frac{\epsilon_2 p_1}{\epsilon_1 p_2} \frac{I_0'(p_2 a)}{I_0'(p_1 a)}, \tag{12.83}$$

$$B = \frac{\epsilon_2 p_3}{\epsilon_3 p_2} \frac{K_0'(p_2 b)}{K_0'(p_3 b)} - \frac{K_0(p_2 b)}{K_0(p_3 b)}, \tag{12.84}$$

$$C = \frac{K_0(p_2 a)}{I_0(p_1 a)} - \frac{\epsilon_2 p_1}{\epsilon_1 p_2} \frac{K_0'(p_2 a)}{I_0'(p_1 a)}, \tag{12.85}$$

$$D = \frac{\epsilon_2 p_3}{\epsilon_3 p_2} \frac{I_0'(p_2 b)}{K_0'(p_3 b)} - \frac{I_0(p_2 b)}{K_0(p_3 b)}. \tag{12.86}$$

Equation (12.82) is the dispersion relation for the SPP modes propagating along the multilayered cylindrical structure for which the ω-β diagram may be derived.

Several special cases are of interest.

(a) If $\epsilon_1 = \epsilon_m$, the permittivity of a metal, and ϵ_2 and ϵ_3 are both nonmetal dielectrics, the solution represents that of a metal cylinder surrounded by a layer of dielectric material immersed in another dielectric medium

(b) If $\epsilon_2 = \epsilon_m$, the permittivity of a metal, and ϵ_1 and ϵ_3 are both nonmetal dielectrics, the solution represents that of a cylindrical metal film immersed in dielectrics.

(c) If $\epsilon_1 = \epsilon_3 = \epsilon_m$, the permittivity of a metal, and ϵ_2 is a nonmetal dielectric, the solution represents that of a cylindrical slot waveguide.

Let us now look at a few limiting cases:

The following formulas for modified Bessel functions are useful. If $\xi \to \infty$, where ξ is the argument of the modified Bessel function,

$$I_0(\xi) \sim \frac{e^{\xi}}{\sqrt{2\pi\xi}}, \tag{12.87a}$$

$$I_0'(\xi) \sim \frac{e^{\xi}}{\sqrt{2\pi\xi}}, \tag{12.87b}$$

$$K_0(\xi) \sim \sqrt{\frac{\pi}{2\xi}} e^{-\xi}, \tag{12.87c}$$

$$K_0'(\xi) \sim -\sqrt{\frac{\pi}{2\xi}} e^{-\xi}. \tag{12.87d}$$

If $\xi \to 0$,

$$I_0(\xi) \sim 1 + \frac{1}{4}\xi^2, \tag{12.88a}$$

$$I_0'(\xi) \sim \frac{1}{2}\xi, \tag{12.88b}$$

$$K_0(\xi) \sim -\left[\ln\left(\frac{\xi}{2}\right) + \gamma\right], \tag{12.88c}$$

$$K_0'(\xi) \sim -\frac{1}{\xi}. \tag{12.88d}$$

(d) If $b \to \infty$, the case reduces to either the situation where a metal cylinder with radius a is immersed in a dielectric with $\epsilon_1 = \epsilon_m$, the permittivity of a metal, and ϵ_2, permittivity of a dielectric, or the situation where a metal medium with $\epsilon_2 = \epsilon_m$ has a cylindrical hole filled with dielectric material, with ϵ_1 as its permittivity. Using (12.87), (12.84) and (12.86) become

$$\begin{aligned} B \underset{b\to\infty}{\simeq} & \frac{\epsilon_2 p_3}{\epsilon_3 p_2}\left(\frac{e^{-p_2 b}e^{p_3 b}}{\sqrt{p_2 p_3}}\right) - \frac{e^{-p_2 b}e^{p_3 b}}{\sqrt{p_2 p_3}}, \\ \simeq & \left[\left(\frac{\epsilon_2 p_3}{\epsilon_3 p_2}\frac{1}{\sqrt{p_2 p_3}} - \frac{1}{\sqrt{p_2 p_3}}\right)\right] e^{-(p_2-p_3)b}, \\ \simeq 0 & \quad \text{if } p_2 > p_3, \quad (b\to\infty), \end{aligned} \tag{12.89}$$

$$\begin{aligned} D \underset{(b\to\infty)}{\simeq} & \frac{\epsilon_2 p_3}{\epsilon_3 p_2}\left(\frac{e^{p_2 b}}{\sqrt{2\pi p_2 b}}\right)\left[-\frac{e^{p_3 b}}{(\pi/2k_3 b)^{1/2}}\right] \\ & - \left(\frac{e^{p_2 b}}{\sqrt{2\pi p_2 b}}\right)\left[\frac{e^{p_3 b}}{(\pi/2k_3 b)^{1/2}}\right], \\ \simeq \infty & \qquad (\text{if } \; b\to\infty). \end{aligned} \tag{12.90}$$

Hence, from (12.82), $A\cdot 0 - C\cdot\infty = 0$. Since $A \neq 0$, C must be zero. Or,

$$\frac{K_0(p_2 a)}{I_0(p_1 a)} - \frac{\epsilon_2 p_1}{\epsilon_1 p_2}\frac{K_0'(p_2 a)}{I_0'(p_1 a)} = 0. \tag{12.91}$$

This is the dispersion relation for a cylinder of radius a supporting the SPP wave.

If a is very large, the problem should reduce to the classic case of a metal substrate bordering a dielectric half-space. For $a\to\infty$, (12.91) becomes

$$\begin{aligned} & e^{-p_2 a}\left(\frac{\pi}{2p_2 a}\right)^{1/2} e^{-p_2 a}\sqrt{2\pi p_1 a} \\ & \qquad + \frac{\epsilon_2 p_1}{\epsilon_1 p_2} e^{-p_2 a}\left(\frac{\pi}{2p_2 a}\right)^{1/2} e^{-p_1 a}\sqrt{2\pi p_1 a} = 0. \end{aligned} \tag{12.92}$$

Simplifying yields

$$1 + \frac{\epsilon_2 p_1}{\epsilon_1 p_2} = 0. \tag{12.93}$$

This is the same dispersion equation as (12.17) for the classic planar SPP wave if one replaces ϵ_2 and p_2 by ϵ_m and p_m, respectively.

If a is very small and $\epsilon_1 = \epsilon_m$, the problem reduces to that of SPP wave propagation along a metal nanowire. Substituting (12.88) in (12.91) gives

$$(p_2 a)^2 \ln\left(\frac{p_2 a}{2}\right) = 2\frac{\epsilon_2}{\epsilon_m}. \tag{12.94}$$

This is the dispersion relation for a SPP wave along a nanowire immersed in a dielectric medium of dielectric constant ϵ_2.

If a is very small and $\epsilon_2 = \epsilon_m$, the problem reduces to that of SPP wave passing through a small hole filled with dielectric ϵ_1 in a metal screen. Equation (12.91) becomes

$$(p_m a)^2 \ln\left(\frac{p_m a}{2}\right) = 2\frac{\epsilon_m}{\epsilon_1}. \tag{12.95}$$

This is the dispersion relation for a SPP wave passing through a small hole filled with dielectric ϵ_1. It is understood that the cutoff frequency for a normal waveguide mode for this small hole is much higher than the operating frequency for the SPP wave.

Let us return to the problem of an SPP wave propagation along a metal cylinder of arbitrary radius a in a dielectric medium. Assuming

$$\frac{\epsilon_m}{\epsilon_0} = 1 - \frac{\omega_\mathrm{p}^2}{\omega^2} \tag{12.96}$$

and $\epsilon_2/\epsilon_0 = 1$, the ω-β diagram for this wave is given in Fig. 12.12 [17–19]. It is seen that there is no low frequency cutoff.

The dispersion relation for SPP wave propagation in a dielectric cylinder enclosed by a metal wall is also given by (12.91), provided that $\epsilon_2 = \epsilon_m$, the metal permittivity, and ϵ_1 is the permittivity of a dielectric material. It is anticipated that the usual metallic cylinder filled with dielectric has a low frequency cutoff and no high frequency cutoff, while the SPP wave in such a structure has no low frequency cutoff.

(e) If $b = a + \delta$, where δ is very small, the problem reduces to the case of a thin cylindrical metal film with $\epsilon_2 = \epsilon_m$, the metal permittivity. Referring back to (12.82) we note that only B (12.84) and D (12.86) are functions of b. Using a Taylor series expansion, one may write

$$B(b) = B(a) + \delta B'(a) + \cdots, \tag{12.97}$$

$$D(b) = D(a) + \delta D'(a) + \cdots, \tag{12.98}$$

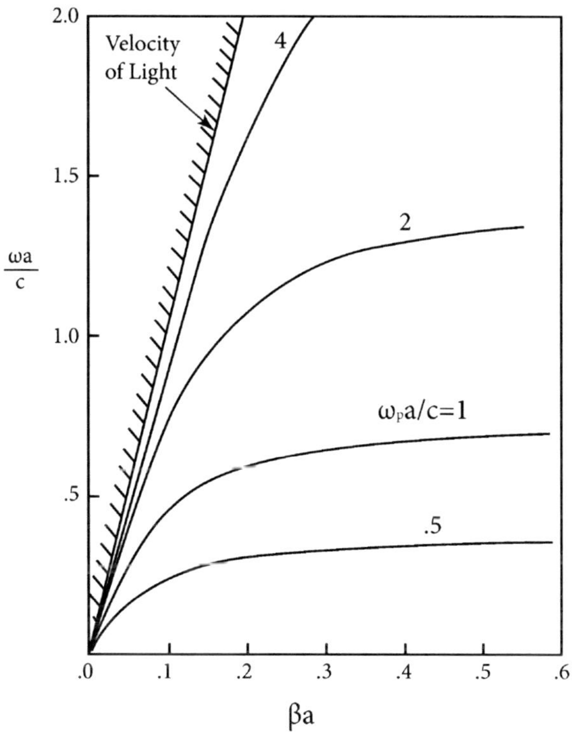

Figure 12.12. ω-β diagram of a plasma column for various values of $\omega_p a/c$ [18]

where the prime signifies the derivative with respect to r. Equation (12.82) becomes

$$(AB - CD)|_{b=a} + \delta\left(AB' - CD'\right)\Big|_{b=a} = 0. \tag{12.99}$$

The ω-β diagram for the SPP wave along a thin cylindrical metal film can be obtained from (12.99)

12.4.2 HE Waves

For noncircularly symmetric waves guided in a circular cylindrical hole, field components of HE waves must be used to satisfy the boundary conditions. For a subwavelength cylindrical hole in an optically thick metallic film with the metal described by a plasmonic model, propagating modes near the surface plasmon frequency, regardless of how small the hole is, always exist. It is understood that this statement is only true if the plasmonic model is valid. For a sufficiently small hole, that is, <100 nm, this plasmonic model must be modified. Let us now consider the dispersion relation for waves supported by this subwavelength metallic hole in an optically metallic film. From Chap. 5, (5.50), we find the dispersion for HE waves:

$$\left[\frac{J_n'(p_1 a)}{p_1 J_n(p_1 a)} + \frac{K_n'(q_m a)}{q_m K_n(q_m a)}\right]\left[\frac{k_1^2 J_n'(p_1 a)}{p_1 J_n(p_1 a)} + \frac{k_m^2 K_n'(q_m a)}{q_m K_n(q_m a)}\right]$$

$$= \left(\frac{\beta n}{a}\right)^2 \left(\frac{1}{p_1^2} + \frac{1}{q_m^2}\right)^2, \quad (12.100)$$

with

$$p_1^2 = k_1^2 - \beta^2,$$

$$q_m^2 = \beta^2 - k_m^2,$$

$$k_1^2 = \omega^2 \mu_0 \epsilon_1,$$

$$k_m^2 = \omega^2 \mu_0 \epsilon_m,$$

$$\frac{\epsilon_m}{\epsilon_0} = 1 - \frac{\omega_p^2}{\omega^2 - j\nu\omega},$$

$$n = 0, 1, 2, 3, \ldots.$$

The propagation factor $e^{-j\beta z + j\omega t}$ is assumed. The Drude plasmonic model for metal as given by (12.1) was used for metal.

The above dispersion relation can be solved in the same way as described in Chap. 5, when we assume that the plasma collision frequency is zero. The ω-β diagram and the cutoff frequencies are shown for the dominant HE_{11} mode in Fig. 12.13 [20]. For the sake of comparison, the same diagrams for the TE_{11} modes for a perfectly conducting circular waveguide are also shown. It is seen that plasmonic waveguide can sustain a propagating mode at a significantly longer cutoff wavelength than is to be expected from a perfectly conducting hollow waveguide.

12.5 Nanofibers (Subwavelength Guiding Structures)

Another subwavelength guiding structure is a nanofiber guide. Photonic nanofibers are dielectric fibers with sub-micron (tens to hundreds of nanometers) transverse dimensions [21]. These dimensions are considered subwavelength sizes since

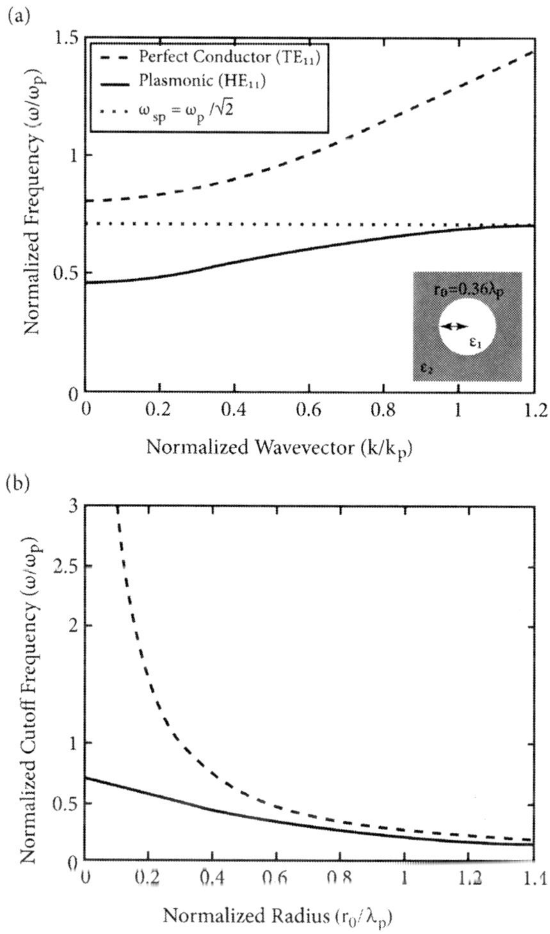

Figure 12.13. (**a**) Dispersion relation of the lowest-order waveguide mode of a cylindrical hole in metal. The dashed line corresponds to the TE_{11} mode in a perfect conductor. The solid line corresponds to the HE_{11} mode in a Drude plasmonic metal. The radius of the hole is $0.36\lambda_p$, where λ_p is the plasmonic wavelength in the Drude model. (**b**) Cutoff frequency of the lowest-order waveguide vs. hole radius r_0 for perfect conductor (TE_{11}, *dashed line*) and plasmonic (HE_{11}, *solid line*) models [20]

the operating free-space wavelengths are usually larger. Typically these photonic nanofibers are made with silica glass, SF8 glass, chalcogenide glass, silicon, InGaAs or AlGaAs. They are either surrounded by air or low-index insulators, resulting in large core to cladding index contrast, usually >1.35. The modal solutions for the guided waves along these structures cannot be obtained using the simpler scalar-wave approximation discussed in Sect. 2.9 due to the presence of high index contrast. They must be obtained from the exact vector wave equation.

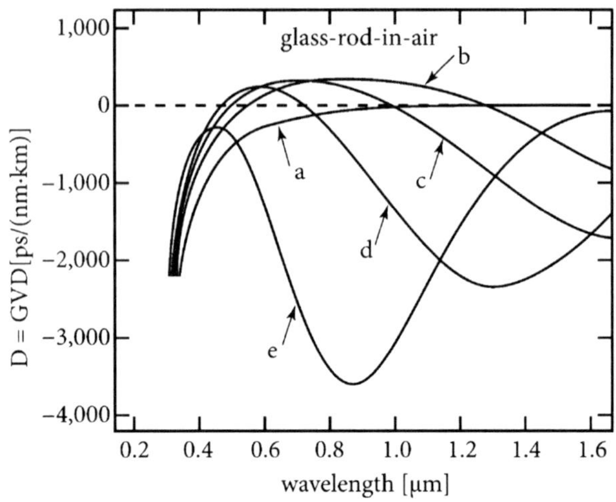

Figure 12.14. Net group-velocity dispersion of silica glass photonic nanowires. As the core is reduced in size, the zero-GVD points shift to shorter wavelengths. The overall behavior of the net-GVD is characterized primarily by the waveguide-GVD and is consistent among nanowires of different materials although they occur at smaller (larger) characteristic dimensions for larger (smaller) index contrasts. Core dimensions are (**a**) bulk silica; (**b**) 1 μm; (**c**) 800 nm; (**d**) 600 nm; and (**e**) 400 nm [22]

As an illustration, let us consider the following two examples [22]: (a) A circular cylindrical fiber silica glass fiber in air supporting the dominant HE_{11} mode; and (b) A rectangular cylindrical silicon with aspect ratio of 1:1.5 supporting the ${}_0HE_{11}$ (TE-like) mode embedded in a SiO_2 cladding. Plots of the total GVD (waveguide and material dispersion) as a function of wavelength for different fiber sizes are shown in Figs. 12.14 and 12.15. It is seen that GVD (see Chap. 10) can vary from anomalous to normal dispersion by core size changes. Because of the small sizes of the core, A_{eff} (see Chap. 10) becomes quite small. Thus, the nonlinear coefficient γ (see Chap. 10) can be quite large. Combination of large enhancement of nonlinear coefficient and varied GVD can make photonic nanowires ideally suited for many nonlinear optical applications.

It should be noted that the use of subwavelength structures to guide electromagnetic waves are not new. Most dielectric waveguides used to guide microwaves belong to this family [23]. For example, the use of flexible Teflon wires to guide centimeter or millimeter waves has been in existence since the 1960s [23]. Furthermore, any radiation loss due to the bending with bending radii $>10\lambda_0$ can usually be tolerated. We can also use the rigorous analysis shown in Chaps. 5 and 6 or several numerical methods (finite element and finite difference) based on the exact Maxwell equations (Chap. 15) to analyze the wave propagation characteristics of nanofibers.

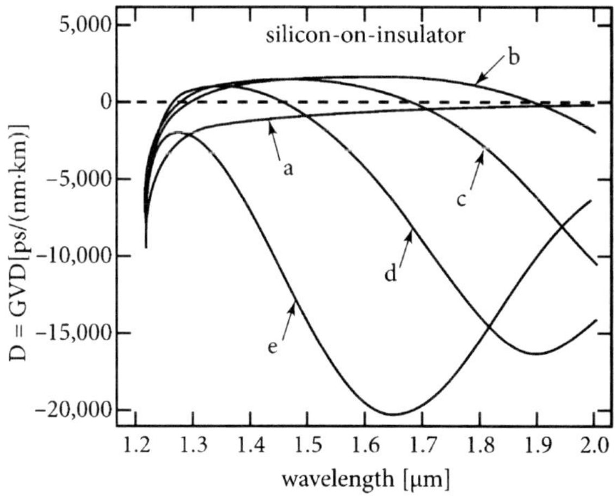

Figure 12.15. Net-group velocity dispersion of silicon photonic nanowires of various dimensions. The core sizes are (**a**) bulk silicon; (**b**) $300 \times 450\,\text{nm}^2$; (**c**) $266 \times 400\,\text{nm}^2$; (**d**) $233 \times 350\,\text{nm}^2$; and (**e**) $200 \times 350\,\text{nm}^2$ [22]

12.6 Conclusions and Discussion

Formal analysis of wave guidance, named surface plasmon-polariton (SPP), along the interface of a metal substrate and along a metal film has been presented. The "noble metals" model was used to characterize the electrical properties of the metals. In other words, the properties of the metals resemble those of a cold plasma with an electron gas. Gold, silver, copper, and aluminum belong to this noble metals family. A more accurate description of the electrical properties of silver and SiO_2 can be found from the following empirical formulas: [14]

Ag relative dielectric constant ($\epsilon_r = \epsilon_r' + \epsilon_r''$) fitting parameters, Johnson and Christy data:

$$\epsilon_r' = \begin{cases} \text{cubic spline} & \lambda < 500\,\text{nm} \\ a_0 + a_1\lambda + a_2\lambda^2 + a_3\lambda^3 + a_4\lambda^4 & \lambda \geq 500\,\text{nm}, \end{cases}$$

$$\epsilon_r'' = \begin{cases} \text{cubic spline} & \lambda < 400\,\text{nm} \\ b_0 + b_1\lambda + b_2\lambda^2 + b_3\lambda^3 + b_4\lambda^4 & \lambda \geq 400\,\text{nm}, \end{cases}$$

where

$$a_0 = 29.34, \qquad b_0 = -1.753,$$

$$a_1 = -0.11028, \quad b_1 = 0.009962,$$

$$a_2 = 1.1218 \times 10^{-4}, \qquad b_2 = -1.696 \times 10^{-5},$$

$$a_3 = -1.08164 \times 10^{-7}, \quad b_3 = 1.178 \times 10^{-8},$$

$$a_4 = 2.44496 \times 10^{-11}, \quad b_4 = -2.334 \times 10^{-12}.$$

Ag relative dielectric constant ($\epsilon_r = \epsilon_r' + \epsilon_r''$) fitting parameters, Palik data:

$$\epsilon_r' = \begin{cases} \text{cubic spline} & \lambda < 350\,\text{nm} \\ c_0 + c_1\lambda + a_2\lambda^2 & \lambda \geq 350\,\text{nm}, \end{cases}$$

$$\epsilon_r'' = \begin{cases} \text{cubic spline} & \lambda < 350\,\text{nm} \\ d_0 + d_1\lambda + d_2\lambda^2 & \lambda \geq 350\,\text{nm}, \end{cases}$$

$$c_0 = 10.314, \qquad d_0 = 1.0481,$$

$$c_1 = -0.026295, \qquad d_1 = -0.003259,$$

$$c_2 = -2.51 \times 10^{-5}, \quad d_2 = 5.3387 \times 10^{-6}.$$

SiO_2 relative dielectric constant (ϵ_{r2}) fitting parameters, Palik data:

$$\epsilon_{r2} = \begin{cases} e_0 + e_1\lambda + e_2\lambda^2 + e_3\lambda^3 & \lambda \leq 612\,\text{nm} \\ f_0 + f_1\lambda + f_2\lambda^2 + f_3\lambda^3 & \lambda \geq 612\,\text{nm}, \end{cases}$$

where

$$e_0 = 2.806, \qquad f_0 = 2.222,$$

$$e_1 = -0.00347, \qquad f_1 = -2.46178 \times 10^{-4},$$

$$e_2 = 6.10502 \times 10^{-6}, \quad f_2 = -1.71928 \times 10^{-7},$$

$$e_3 = -3.68627 \times 10^{-9}, \quad f_3 = -4.49923 \times 10^{-11}.$$

The importance of the SPP wave is in its applications. Since surface plasmons with the same frequency as the exciting electromagnetic wave possess much shorter guide wavelength (as small as 8% of that of the free space electromagnetic wave), this phenomenon would allow the plasmons to travel along nanoscale subwavelength wires. These nanoscale wires can form interconnects for ultra small and ultra fast transistors and chips. A new family of ultra-small, nanoscale devices can also be envisioned, such as faster computers and processors, higher resolution microscopes, more efficient light-emitting diodes (LEDs), and more sensitive chemical and biological detectors [4, 5, 14].

References

1. R. H. Ritchie, "Plasma losses by fast electrons in thin films," Phys. Rev. **106**, 874 (1957)
2. H. Raether, "Surface Plasmons," Springer, Berlin Heidelberg New York (1988)
3. J. R. Sambles, G. Bradbery, and F. Z. Yang, "Optical-excitation of surface plasmons - An introduction," Contemp. Phys. **32**, 173 (1991)
4. W. L. Barnes, A. Deseux, and T. W. Ebbesen, "Surface plasmon subwavelength optics," Nature **424**, 824 (2003)
5. H. A. Atwater, "The promise of plasmonics," Sci. Am. **296**, 56 (2007)
6. E. Ozbay, "Plasmonics: Merging photonics and electronics at nanoscale dimensions," Science **311**, 189 (2006)
7. S. A. Maier, "Plasmonics: Fundamentals and Applications," Springer, Berlin Heidelberg New York (2007)
8. S. A. Maier and H. A. Atwater, "Plasmonics: Localization and guiding of electromagnetic energy in metal/dielectric structures," J. Appl. Phys. **99**, 011101 (2005)
9. D. Qu, D. Grischkowsky, and W. Zhang, "Terahertz transmission properties of thin, subwavelength metallic hole arrays," Opt. Lett. **29**, 896 (2004); T. Jeon and D. Grischkowsky, "THz Zenneck surface wave (THz surface plasmon) propagation on a metal sheet," Appl. Phys. Lett. **88**, 06113 (2003)
10. A. Karalis, E. Lidorikis, M. Ibanescu, J. D. Joannopoulos, and M. Soljaĉiĉ, "Surface-plasmon-assisted guiding of broadband slow and subwavelength light in air," Phys. Rev. Lett. **95**, 06390 (2005)
11. J. A. Stratton, "Electromagnetic Theory," McGraw-Hill, New York (1941)
12. J. J. Burke, G. I. Stegeman, and T. Tamir, "Surface-polariton-like wave guided by thin lossy metal films," Phys. Rev. B **33**, 5186 (1986); E. N. Economou, "Surface plasmons in thin films," Phys. Rev. **182**, 539 (1969)
13. M. N. Zervas, "Surface plasmon-polariton waves guided by thin metal films," Opt. Lett. **16**, 720 (1991)
14. J. A. Dionne, L. Sweatlock, H. A. Atwater, and A. Polman, "Planar metal plasmon waveguides – Frequency dependent dispersion, propagation, localization and loss beyond the free electron model," Phys. Rev. B **72**, 075405 (2005)
15. P. Berini, "Plasmon-polariton waves guided by thin lossy metal films of finite width: Bound modes of symmetric structures," Phys. Rev. B **61**, 10484 (2000)

16. R. Pregla and W. Pascher, "Numerical Techniques for Microwave and Millimeter-Wave Passive Structures," T. Itoh, ed., Wiley, New York (1989)

17. F. I. Shimabukuro, "A study of dispersion in plasmas," Ph. D. Thesis, California Inst. of Technol. (1962)

18. A. W. Trivelpiece and R. W. Gould, "Space charge waves in cylindrical plasma columns," J. Appl. Phys. **3**, 1784 (1959)

19. C. Yeh, "Wave propagation on a moving plasma column," J. Appl. Phys. **39**, 6112 (1968)

20. P. B. Catrysse, H. Shin, and S. Fan, "Propagating modes in subwavelength cylindrical holes," J. Vac. Sci. Tech. **23**(6), 2675 (2005)

21. L. Tong, R. R. Gattass, J. B. Ashcom, S. He, J. Lou, M. Shen, I. Maxwell, and E. Mazur, "Subwavelength-diameter silica wires for low-loss optical wave guiding," Nature **426**, 816 (2003)

22. M. A. Foster, A. C. Turner, M. Lipson, and A. L. Gaeta, "Nonlinear optics in photonic nanowires," Opt. Exp. **16**, 1300 (2008)

23. R. E. Collin, "Field Theory of Guided Waves," McGraw-Hill, New York (1960)

13

PHOTONIC CRYSTAL WAVEGUIDES

The wave propagation in or along periodic structures such as photonic crystal waveguides produces many interesting features that are absent for uniform structures. These features can then provide large variety of applications ranging from integrated optics [1–3], microwave filters and gratings [4], and DFB lasers [4] in the early 1970s to the study of photonic band gap structures and photonic crystals in the late 1980s [5–7] and artificial metamaterial in the middle 2000s, to name a few. The first comprehensive study of wave-propagation in periodic structures was done by Brillouin [8] in 1953. Tamir et al. in 1964 [9] provided the complete exact solution to the case of TE wave propagation in a sinusoidally stratified (periodic) dielectric medium. In 1965, Yeh et al. [10] gave the complete exact solution to the case of TM wave propagation in a sinusoidally stratified (periodic) dielectric medium. Casey et al. [11] also found the exact solution for the case of wave-propagation in a sinusoidally varying plasma medium. Elachi and Yeh in 1973 [1] gave a formal solution to the problem of wave propagation along a periodic dielectric waveguide.

13.1 Fundamental Properties of Guided Waves in Periodic Structures

The distinguishing characteristic of a wave guided by a periodic structure is that its guided field at any plane normal to z (the direction of propagation) along the propagating path is identical to the field at another normal plane, which is a distance, d, away except for a propagation factor $e^{-j\beta d}$. Here, d is the periodicity length of the periodic medium and β is the propagation constant. In other words,

$$\mathbf{E}\left(x, y, z+d\right) = e^{-j\beta d}\mathbf{E}\left(x, y, z\right), \tag{13.1}$$

$$\mathbf{H}\left(x, y, z+d\right) = e^{-j\beta d}\mathbf{H}\left(x, y, z\right). \tag{13.2}$$

This is a statement of the Floquet theorem [8]. Another way of stating the above is

$$\mathbf{E}\left(x,y,z\right)=e^{-j\beta d}\mathbf{E}_p\left(x,y,z\right), \tag{13.3}$$

$$\mathbf{H}\left(x,y,z\right)=e^{-j\beta d}\mathbf{H}_p\left(x,y,z\right), \tag{13.4}$$

where $\mathbf{E}_p, \mathbf{H}_p$ are periodic functions of z with period d or

$$\mathbf{E}_p\left(x,y,z\right)=\mathbf{E}_p\left(x,y,z-d\right), \tag{13.5}$$

$$\mathbf{H}_p\left(x,y,z\right)=\mathbf{H}_p\left(x,y,z-d\right). \tag{13.6}$$

Expanding the periodic functions by Fourier series yields

$$\mathbf{E}_p\left(x,y,z\right)=\sum_{n=-\infty}^{n=\infty}a_n\left(x,y\right)e^{-j2n\pi z/d}, \tag{13.7}$$

$$\mathbf{H}_p(x,y,z)=\sum_{n=-\infty}^{n=\infty}b_n\left(x,y\right)e^{-j2n\pi z/d}, \tag{13.8}$$

where a_n and b_n are vector functions of x and y. Substituting (13.7) and (13.8) in (13.3) and (13.4) gives

$$\mathbf{E}\left(x,y,z\right)=\sum_{n=-\infty}^{n=\infty}a_n\left(x,y\right)e^{-j\beta_n z}, \tag{13.9}$$

$$\mathbf{H}\left(x,y,z\right)=\sum_{n=-\infty}^{n=\infty}b_n\left(x,y\right)e^{-j\beta_n z}, \tag{13.10}$$

where

$$\beta_n=\beta+\frac{2n\pi}{d}, \tag{13.11}$$

$$n=0,\ \pm 1,\ \pm 2,\ldots.$$

Equations (13.9) and (13.10) are called the Floquet spatial harmonic expansions. This means we should expect all fields in any periodic medium to vary according to (13.9) and (13.10). The eigenvalue equation for β for a given problem dealing with periodic structures always yields solutions, $\beta_n=\beta+2n\pi/d$, in addition to the fundamental solution. Hence a complete ω-β (Brillouin) diagram will display ω as a periodic function of β.

13.2 Stop-Band and Pass-Band Property

An examination of (13.9) and (13.10) shows that there exist forward propagating waves (with $\beta_n > 0$) and backward propagating waves (with $\beta_n < 0$). These waves may interact to produce the stop-band and pass-band behavior for the guided waves. It is defined that pass-band refers to the frequency band in which the resultant propagation constant of the total guided wave is a real root of the dispersion relation, while stop-band refers to the frequency band in which no real root of the dispersion relation can be found, implying the nonexistence of any guided wave. This stop-band region is sometimes called the radiation region. (This definition is in agreement with one given earlier for the existence of surface waves in Chap. 3.)

The essence of the propagation characteristics of guided waves along a periodic dielectric structure can be gained from the consideration of a dielectric slab (2D) waveguide structure with small periodic index variation along the propagation direction [1]. The dielectric slab immersed in free-space has a thickness of $2a$ and a dielectric variation of

$$\epsilon(z) = \epsilon_1 \left[1 + \eta \cos \frac{2\pi z}{d} \right], \tag{13.12}$$

where $\epsilon_1 > \epsilon_0$, $|\eta| < 1$, and d is the periodicity of the dielectric variation (see Fig. 13.1). We are interested in the characteristics of waves guided by this structure. Let us first consider the case where $\eta = 0$, that is, the uniform slab case with $\epsilon = \epsilon_1$. From Chap. 4, we may obtain the ω-β diagram for the dominant mode, which is now displayed in Fig. 13.2a. It is seen that the ω-β curve for the dominant guided mode is bounded between two straight lines representing the dispersionless

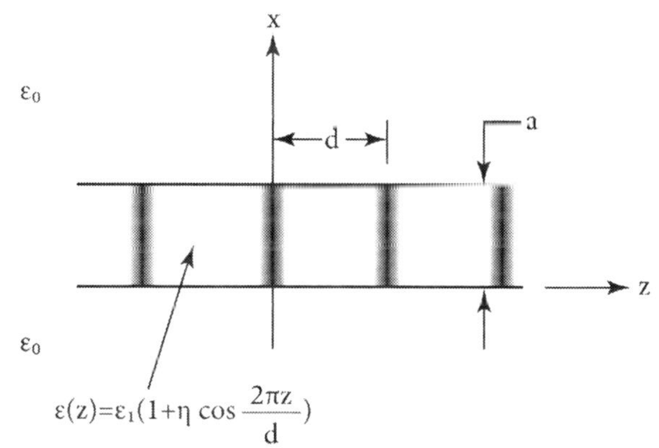

Figure 13.1. Slab waveguide with sinusoidally varying dielectric medium in the z-direction

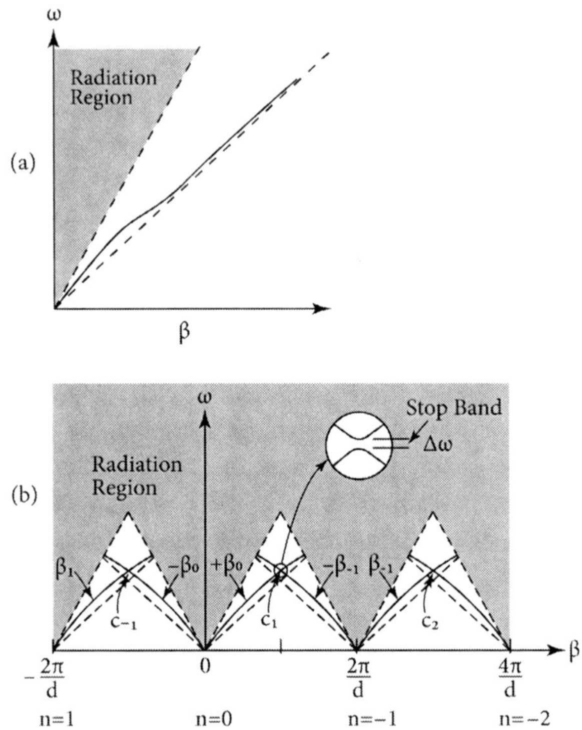

Figure 13.2. ω-β diagram

ω-β curves for a plane wave propagating in an infinite medium. The straight line with the steeper slope refers to the free space medium with permittivity ϵ_0, while the other straight line refers to the infinite dielectric medium with permittivity ϵ_1. Only between these two straight lines will the roots of the dispersion relation for waves guided by the slab dielectric waveguide be real. Hence, guided waves can be present only when the roots are real. Curves in the second quadrant of Fig. 13.2a refers to negative (backward) propagating wave where β is negative.

Let us now investigate the situation where $|\eta|$ is small but not zero. This means that the dielectric slab has a small periodic variation in its permittivity along the direction of the guided wave. The presence of this dielectric periodicity dictates that the Floquet condition (13.9)-(13.11) discussed earlier must be satisfied by the guided wave. In other words,

$$\begin{aligned} \beta_n &= \beta_0 + \frac{2\pi n}{d}, \\ n &= 0, \ \pm 1, \ \pm 2, \ldots, \end{aligned} \tag{13.13}$$

where β_0 is the propagation constant of the slab with unperturbed dielectric constant ϵ_1. So, for each n, the relationship between ω and β_n can be obtained by shifting the ω vs. β diagram of the unperturbed case by $2\pi n/d$ along the β axis as shown in Fig. 13.2b. The solid curves represent the allowed propagation with real values of β. Our discussion will be limited to the region where β_n is real. The $n = 0$, ω vs. β_0 curve has two branches – one refers to the forward propagating wave with positive group velocity and the other refers to the backward propagating wave with negative group velocity. Similar behavior is observed for the ω-β diagram for the $n = -1$ mode, and so forth, for the other $\pm n$ modes.

We note that there exist many cross-over points as indicated by c_1, $c_2 \ldots c_{-1}$, c_{-2}. There, the forward propagating wave and the backward propagating wave possess the same phase velocity. This means that wave coupling phenomenon can occur and there is energy exchange between the forward propagating wave and the backward propagating wave. resulting in the modifications of the dispersion relation. As shown in the inset, the ω vs. β curves are split due to the coupling effect. In the split region between the two ω-β curves, no real value solution of β can be found. Therefore, within the $\Delta\omega$ band, no guided wave is permitted to propagate along the axial z direction. This $\Delta\omega$ is called the stop band. This is a direct consequence of the presence of the periodic medium/structure [1].

13.3 Dielectric-Rod Array Waveguide

It is of interest to study the case where the periodic dielectric variation is very strong. This was done by Fan et al. [12] who used computer simulations to treat the case of a waveguide composed of an array of parallel dielectric-square rods immersed in free-space as shown in Fig. 13.3. The cross section of the rod is $0.3\,d \times 0.3\,d$, where d is the periodicity of the array. The relative dielectric constant of the rod is 13. They calculated the ω-β diagrams for two types of dominant modes[1]: TE (E_y, H_x, H_z) and TM (H_y, E_x, E_z) modes. The wave is assumed to propagate in the z direction as $e^{-j\beta z + j\omega t}$. The completed results are displayed in Fig. 13.4. It is seen that there is a large stop band for the TE dominant mode. This is because for the selected parameters, the fields for the TE mode is much more closely coupled to the dielectric structure than that for the TM mode.

Also, in contrast to the weak periodic case, the stop band for the present case is much larger, indicating strong influence of the large periodic variations.

[1]The definition for modes in Fan et al.'s paper [12] is different than the conventional definition used in this book. Their TM [TE] modes, defined in terms of the 2D plane normal, correspond, respectively, to our conventional TE [TM] modes, defined with respect to the direction of propagation of guided waves.

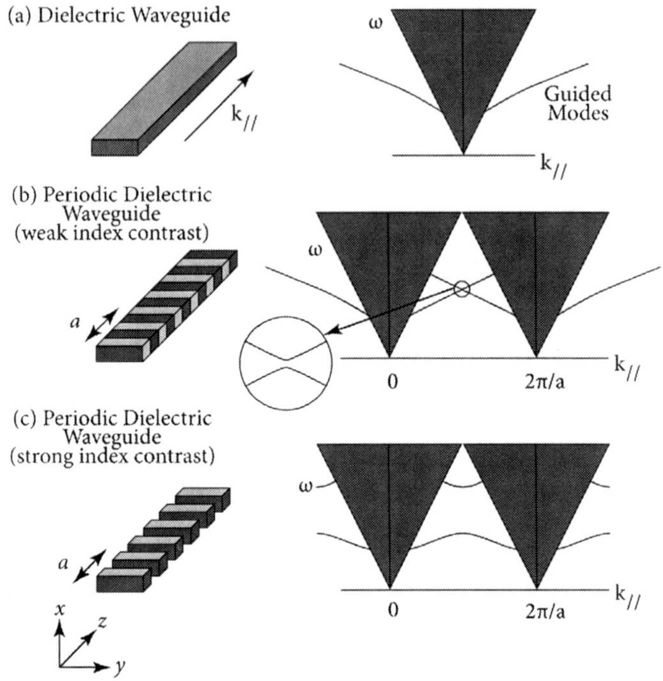

Figure 13.3. Schematic of the band structures for waveguides with different dielectric constant contrast along the guide. (**a**) Uniform dielectric waveguide, with the band structure shown on the right hand side. (**b**) Periodic dielectric waveguide with weak index contrast. The origin of the gap is shown on the right-hand side. (**c**) Periodic dielectric waveguide with strong index contrast. The expected band structure is sketched on the right-hand side [12]

13.4 Band Gap and Waveguide Bends

One of the most intriguing aspect of wave propagation in periodic structures is the presence of stop band and pass band characteristics. Bloch and Brillouin have studied this effect many years ago [9]. Certainly, it has also been known that any defects in the periodic medium would destroy or affect this stop-band, pass-band behavior, at least in the regions where the defects are. But, it was not until 1987 that Yablonovitch [6] and John [5] realized the possibility of using these defects as well as controlled periodicity in the material to manipulate and influence the nature of wave modes in this material. For example, a point defect could act like a microcavity, a line defect like a waveguide, and a planar defect like a perfect mirror.

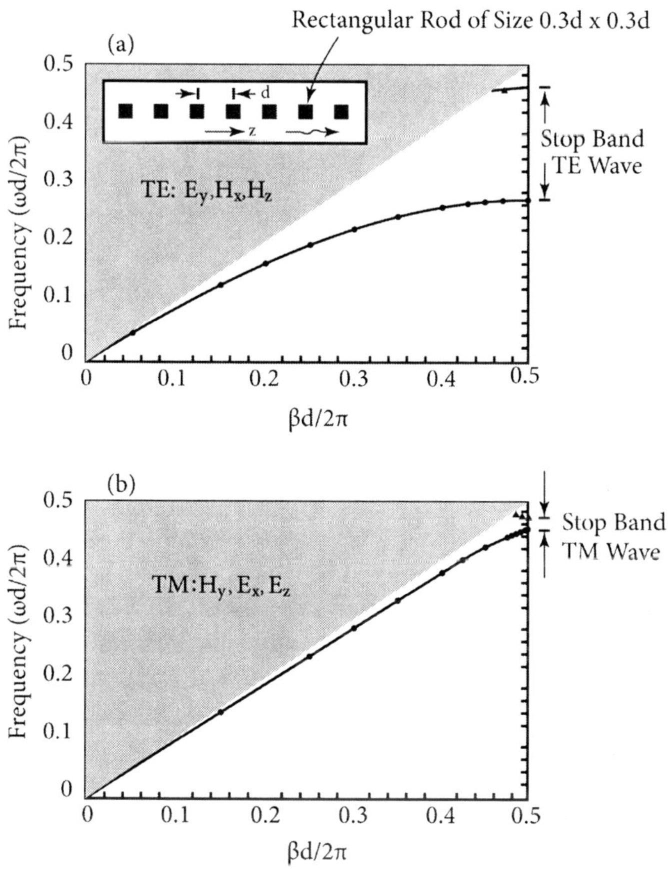

Figure 13.4. ω-β diagram for (*top*) TE (E_y, H_x, H_z) and (*bottom*) TM (H_y, E_x, E_z) modes for rod array waveguide shown in Fig. 13.3. The relative dielectric constant of the rods is 13 and the surrounding medium is air [12]

It is recalled that pass band means the range of frequencies in which real roots (real β's) can be found from the dispersion relation while no real root exists in the stop band. However, complex roots with complex β's do exist. They become physical if the periodicity of the medium is broken by the presence of a defect. Stop band is also named bandgap. The name "photonic crystal" was first coined by Yablonovitch [6] to represent a solid periodic medium. To illustrate the idea of using defects to guide light waves in photonic crystal, we shall follow the work of Mekis et al. [13]. In their computer simulation study, they used arrays of GaAs rods of circular cross section, with an index of refraction of 3.4 and radius of $0.3\,d$, where d is the periodicity of the array lattice. For the TE mode with (E_y, H_x, H_z),

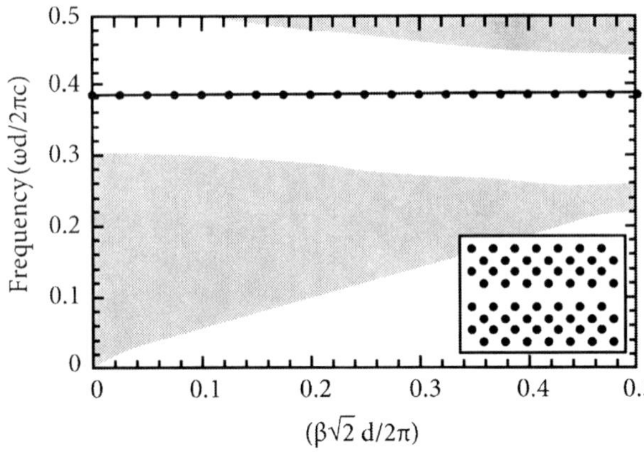

Figure 13.5. Dispersion relations for a PBG waveguide shown in the inset. The gray areas are the projected band structure of the perfect crystal. The filled circles correspond to even modes [13]

where z is the direction of propagation of the wave, the band gap is centered at frequency $\omega = 0.37 \times 2\pi c/d$, where c is the speed of light. If $d = 0.57\,\mu\text{m}$, $\lambda = 2\pi c/\omega = 1.54\,\mu\text{m}$. The results are shown in Fig. 13.5 for the defect case where one row of rods is removed. The ω-β diagrams show that one propagating mode may exist. The mode confined within the defect has fields that decay exponentially away from the defect and propagate along the defect.

Mekis et al. [13] have also obtained the results of the simulation for a guided mode passing through a 90° bend. They showed that 98% of the power in the guided mode that goes in one end comes out of the other.

13.5 Photonic Bandgap Fiber

The photonic bandgap waveguide discussed in previous sections are all single mode or low-order mode guides. Confinement of this guided mode depends on its binding behavior with the surrounding periodic structure. In other words, some of the guided field must penetrate into the periodic structure, which is a material structure that possesses absorption losses as well as scattering losses.

Shown in the left figure of Fig. 13.6 is a new type of fiber, called the hollow core photonic bandgap fiber, proposed by Knight in 1996 [14]. He showed that by operating the beam signal in the bandgap region of the periodic tube structure, the wall of the hollow core could minimize the finite-aperture diffraction expansion of a gaussian-like beam in the air core. Thereby, the beam can be guided by this

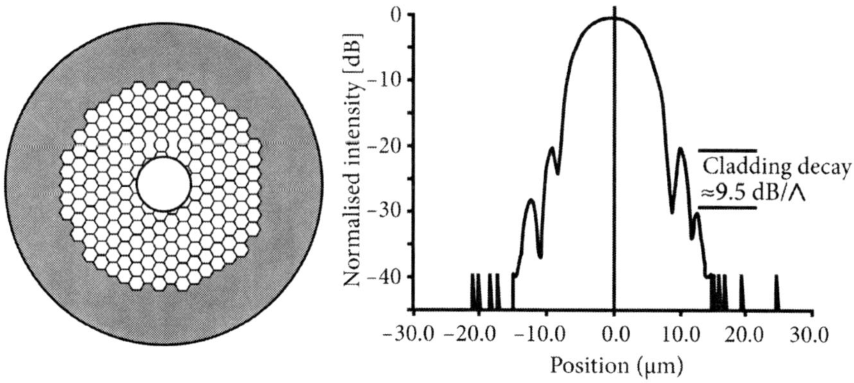

Figure 13.6. (*Left*) A sketch of the hollow-core photonic bandgap fiber. (*Right*) Normalized near field intensity distribution at the end of 500 m of this fiber. The core boundary is at approximately 9 μm from the center. About 98.3% of the power is located in the hollow core. Λ is the cladding period [14]

hollow core with very little losses. As shown in the plot on the right side of Fig. 13.6 more than 98% of the beam power is located in the hollow core.

It is worthwhile to point out that there is very little bending losses for this bandgap fiber in the optical spectrum, since the bending radius is usually much larger than the wavelength; even a 1 cm bending radius is more than 1,000λ. On the other hand, if we wish to use the same design for a terahertz signal, the 1 cm bending radius would be only 10λ. For this case, excessive mode coupling at the bend would introduce unacceptable bending losses [15].

13.6 Analytical Study of Surface Wave Propagation Along a Periodic Structure

In this section we discuss the analytical formulation of the guided wave propagation along a periodic structure. A rigorous analytic solution will be found for a canonical problem on the propagation along an open, periodic dielectric slab [1, 16, 17]. Formal solutions for TE and TM wave propagation in a periodic dielectric medium have been found by Tamir et al. in 1964 [9] and by Yeh et al. in 1965 [10]. Here, we make use of these solutions to solve the periodic dielectric slab problem.

The geometry is shown in Fig. 13.1. The slab with thickness a is immersed in free space. It has a permittivity variation of

$$\epsilon(z) = \epsilon_0 \sum_n s_n \exp\left(-\frac{j2n\pi z}{d}\right), \tag{13.14}$$

where s_n are known coefficients and d is the period of periodic medium. For wave propagating in the z-direction along a 2D slab, all fields are independent of y. From Sect. 8.2.2, we find that the 2D fields for the TE and TM waves in a periodic dielectric medium can be derived from the following equations:

TE Waves

$$\mathbf{E}^{(\mathrm{TE})}(x,z) = \mathbf{e}_y \mathrm{F}^{(\mathrm{TE})}(x,z), \tag{13.15}$$

$$\mathbf{H}^{(\mathrm{TE})}(x,z) = \frac{j}{\omega\mu_0}\boldsymbol{\nabla}_{x,z} \times \mathbf{E}^{(\mathrm{TE})}(x,z). \tag{13.16}$$

TM Waves

$$\mathbf{H}^{(\mathrm{TM})}(x,z) = \mathbf{e}_y\, \epsilon^{1/2}(z)\, \mathrm{F}^{(\mathrm{TM})}(x,z), \tag{13.17}$$

$$\mathbf{E}^{(\mathrm{TM})}(x,z) = -\frac{j}{\omega\epsilon(z)}\boldsymbol{\nabla}_{x,z} \times \mathbf{H}^{(\mathrm{TM})}(x,z). \tag{13.18}$$

Here $\mathbf{e}_y$ is a unit vector in the y-direction, and $\boldsymbol{\nabla}_{x,z}$ is a 2D del operator

$$\boldsymbol{\nabla}_{x,z} = \mathbf{e}_x \frac{\partial}{\partial x} + \mathbf{e}_z \frac{\partial}{\partial z} \tag{13.19}$$

and $\mathrm{F}^{(\mathrm{TE}),(\mathrm{TM})}(x,z)$ satisfies

$$\left(\boldsymbol{\nabla}^2_{x,z} + k^2_{\mathrm{TE,TM}}(z)\right) \mathrm{F}^{(\mathrm{TE}),(\mathrm{TM})}(x,z) = 0, \tag{13.20}$$

where

$$k^2_{\mathrm{TE}}(z) = k_0^2 \epsilon(z), \tag{13.21}$$

$$k^2_{\mathrm{TM}}(z) = k_0^2\epsilon(z) - \frac{3}{4}\left[\frac{1}{\epsilon(z)}\frac{\partial \epsilon(z)}{\partial z}\right]^2 + \frac{1}{2\epsilon(z)}\frac{\partial^2\epsilon(z)}{\partial z^2}, \tag{13.22}$$

$$k_0^2 = \omega^2\mu_0\epsilon_0.$$

Since $\epsilon(z)$ is a periodic function of z, we may represent $k^2_{\mathrm{TE,TM}}(z)$ by a Fourier series

$$k^2_{\mathrm{TE,TM}}(z) = k_0^2 \sum_n \left[p_n^{(\mathrm{TE}),(\mathrm{TM})}\right]^2 \exp\left(-\frac{j2n\pi z}{d}\right), \tag{13.23}$$

where $p_n^{(\mathrm{TE}),(\mathrm{TM})}$ are known coefficients.

Let us assume that the solution of (13.20) takes the following form:

$$\mathrm{F}^{(\mathrm{TE}),(\mathrm{TM})}(x, z) = \sum_n q_n^{(\mathrm{TE}),(\mathrm{TM})}(x) \exp\left(-j\beta_n^{(\mathrm{TE}),(\mathrm{TM})} z\right), \tag{13.24}$$

where

$$\beta_n^{(\mathrm{TE}),(\mathrm{TM})} = \beta_0^{(\mathrm{TE}),(\mathrm{TM})} + \frac{2n\pi}{d}, \tag{13.25}$$

$$(n = 0, \ \pm 1, \ \pm 2, \ldots)$$

and $\beta_0^{(\mathrm{TE}),(\mathrm{TM})}$ are the propagation constants. Substituting (13.24) in (13.20) yields

$$\frac{d^2\underline{q}(x)}{dx^2} = -\underline{\underline{\mathrm{P}}}\,\underline{q}(x), \tag{13.26}$$

where $\underline{q}$ is a column vector and $\underline{\underline{\mathrm{P}}}$ is a constant square matrix:

$$\underline{q}(x) = \begin{bmatrix} \cdot \\ \cdot \\ \cdot \\ q_2(x) \\ q_1(x) \\ q_0(x) \\ q_{-1}(x) \\ q_{-2}(x) \\ \cdot \\ \cdot \\ \cdot \end{bmatrix}, \tag{13.27a}$$

$$\underline{\underline{\mathrm{P}}} = \begin{bmatrix} \cdot & \cdot & \cdot & \cdot & \cdot & \cdot & \cdot & \cdot & \cdot & \cdot & \cdot \\ \cdot & \cdot & \cdot & \cdot & \cdot & \cdot & \cdot & \cdot & \cdot & \cdot & \cdot \\ \cdot & \cdot & \cdot & \cdot & \cdot & \cdot & \cdot & \cdot & \cdot & \cdot & \cdot \\ \cdot & \cdot & \cdot & \mathrm{P}_{-22} & \mathrm{P}_{-12} & \mathrm{P}_{02} & \mathrm{P}_{12} & \mathrm{P}_{22} & \cdot & \cdot & \cdot \\ \cdot & \cdot & \cdot & \mathrm{P}_{-21} & \mathrm{P}_{-11} & \mathrm{P}_{01} & \mathrm{P}_{11} & \mathrm{P}_{21} & \cdot & \cdot & \cdot \\ \cdot & \cdot & \cdot & \mathrm{P}_{-20} & \mathrm{P}_{-10} & \mathrm{P}_{00} & \mathrm{P}_{10} & \mathrm{P}_{20} & \cdot & \cdot & \cdot \\ \cdot & \cdot & \cdot & \mathrm{P}_{-2-1} & \mathrm{P}_{-1-1} & \mathrm{P}_{0-1} & \mathrm{P}_{1-1} & \mathrm{P}_{2-1} & \cdot & \cdot & \cdot \\ \cdot & \cdot & \cdot & \mathrm{P}_{-2-2} & \mathrm{P}_{-1-2} & \mathrm{P}_{0-2} & \mathrm{P}_{1-2} & \mathrm{P}_{2-2} & \cdot & \cdot & \cdot \\ \cdot & \cdot & \cdot & \cdot & \cdot & \cdot & \cdot & \cdot & \cdot & \cdot & \cdot \\ \cdot & \cdot & \cdot & \cdot & \cdot & \cdot & \cdot & \cdot & \cdot & \cdot & \cdot \\ \cdot & \cdot & \cdot & \cdot & \cdot & \cdot & \cdot & \cdot & \cdot & \cdot & \cdot \end{bmatrix}, \tag{13.27b}$$

with

$$\mathrm{P}_{nl} = k_0^2\, p_{n-l} - \beta_n^2\, \delta_{nl}. \tag{13.28}$$

The symbols (TE),(TM) are implied in the above equations and $\delta_{nl} = 1$ if $n = l$ and 0 if $n \neq l$. Equation (13.26) is a set of coupled second order differential equations with constant coefficients whose solutions are

$$\underline{q}(x) = \underline{\mathrm{c}}\, e^{\pm j\kappa x}, \tag{13.29}$$

where κ is a constant and $\underline{\mathrm{c}}$ is a constant vector, which is independent of x. substituting (13.29) in (13.26) gives

$$\underline{\underline{\mathrm{P}}}\,\underline{\mathrm{c}} = \kappa^2\, \underline{\mathrm{c}}. \tag{13.30}$$

We note that κ^2 is an eigenvalue of the matrix $\underline{\underline{\mathrm{P}}}$. Thus the characteristic equation is given by

$$\det\left[\underline{\underline{\mathrm{P}}} - \kappa^2\, \underline{\underline{\mathrm{I}}}\right] = 0, \tag{13.31}$$

where $\underline{\underline{\mathrm{I}}}$ is a unit matrix. The eigenvalues of (13.31) are designated by κ_m^2 and the corresponding eigenvectors are $\underline{\mathrm{c}}^{(m)}$ (with elements $\mathrm{c}_n^{(m)}$, $n = 0, \pm 1, \pm 2, \cdot\cdot$). Rewriting (13.29) gives

$$\underline{q}^{(m)+}(x) \;=\; \underline{\mathrm{c}}\, e^{-j\kappa_m x}, \tag{13.32a}$$

$$\underline{q}^{(m)-}(x) \;=\; \underline{\mathrm{c}}\, e^{j\kappa_m x}. \tag{13.32b}$$

Substituting (13.32) back in (13.24), we obtain

$$\mathrm{F}^{(\mathrm{TE}),(\mathrm{TM})} = \sum_m \left\{ \begin{array}{c} \exp\left[-j\kappa_m^{(\mathrm{TE}),(\mathrm{TM})}\right] \\ \exp\left[j\kappa_m^{(\mathrm{TE}),(\mathrm{TM})}\right] \end{array} \right\} \sum_n \mathrm{c}_n^{(m)(\mathrm{TE}),(\mathrm{TM})} \exp\left(-j\beta_n^{(\mathrm{TE}),(\mathrm{TM})} z\right), \tag{13.33}$$

with

$$\beta_n^{(\mathrm{TE}),(\mathrm{TM})} = \beta^{(\mathrm{TE}),(\mathrm{TM})} + \frac{2n\pi}{d},$$

$$n = 0, \ \pm 1, \ \pm 2, \ldots.$$

The ratio $\mathrm{c}_n^{(m)}/\mathrm{c}_0^{(m)}$ can be found from the eigen-equation for (13.30).

Let us now consider the dielectric slab with periodic medium problem as shown in Fig. 13.1. TE waves will be treated first.

Region 1 $(0 < x < a)$ *(the periodic dielectric region)*

Using (13.33), (13.15), and (13.16), the fields tangential to the slab boundary can be obtained:

$$E_y^{(1)} = \sum_m \left[A_1^{(m)} e^{-j\kappa_m x} + A_2^{(m)} e^{j\kappa_m x}\right] \sum_n v_n^{(m)} e^{-j\beta_n z}, \tag{13.34}$$

$$\begin{aligned} H_z^{(1)} &= \frac{j}{\omega\mu_0} \frac{\partial F_y^{(1)}}{\partial x} \\ &= \sum_m \left[A_1^{(m)} e^{-j\kappa_m x} - A_2^{(m)} e^{j\kappa_m x}\right] \sum_n \frac{\kappa_m}{\omega\mu_0} v_n^{(m)} e^{-j\beta_n z}, \end{aligned} \tag{13.35}$$

where $A_1^{(m)}$ and $A_2^{(m)}$ are unknown constants and $v_n^{(m)} = \mathrm{c}_n^{(m)\mathrm{TE}}/\mathrm{c}_0^{(m)\mathrm{TE}}$.

Region 0 $(x > a)$ *(The free space region)*

$$F_y^{(0)} = \sum_n A_3^{(n)} e^{-k_{xn} x} e^{-j\beta_n z}, \tag{13.36}$$

$$H_z^{(0)} = \frac{j}{\omega\mu_0} \frac{\partial F_y^{(0)}}{\partial x} = \sum_n \frac{-jk_{xn}}{\omega\mu_0} A_3^{(n)} e^{-k_{xn} x} e^{-j\beta_n z}, \tag{13.37}$$

where $A_3^{(m)}$ are unknown constants and

$$\begin{aligned}\beta_n^2 &= k_0^2 + k_{xn}^2, \\ \beta_n &= \beta_0 + \frac{2\pi n}{d}, \\ k_0^2 &= \omega^2\mu_0\epsilon_0.\end{aligned} \tag{13.38}$$

Region 2 $(x < 0)$ *(the free space region)*

$$E_y^{(2)} = \sum_n A_4^{(n)}\, e^{k_{xn}x} e^{-j\beta_n z}, \tag{13.39}$$

$$H_z^{(2)} = \frac{j}{\omega\mu_0}\frac{\partial E_y^{(2)}}{\partial x} = \sum_n \frac{-jk_{xn}}{\omega\mu_0} A_4^{(n)}\, e^{k_{xn}x} e^{-j\beta_n z}, \tag{13.40}$$

where $A_4^{(n)}$ are unknown constants.

Satisfying the continuity conditions for the tangential electric and magnetic fields at the boundaries at $x = a$ and at $x = 0$ gives

$$\sum_m \left[A_1^{(m)}\, e^{-j\kappa_m a} + A_2^{(m)}\, e^{j\kappa_m a}\right] \sum_n v_n^{(m)}\, e^{-j\beta_n z} = \sum_n A_3^{(n)}\, e^{-k_{xn}a} e^{-j\beta_n z}, \tag{13.41}$$

$$\sum_m \left[A_1^{(m)}\, e^{-j\kappa_m a} - A_2^{(m)}\, e^{j\kappa_m a}\right] \sum_n \frac{\kappa_m}{\omega\mu_0} v_n^{(m)}\, e^{-j\beta_n z} = \sum_n \frac{-jk_{xn}}{\omega\mu_0} A_3^{(n)}\, e^{-k_{xn}a} e^{-j\beta_n z}, \tag{13.42}$$

$$\sum_m \left[A_1^{(m)} + A_2^{(m)}\right] \sum_n v_n^{(m)}\, e^{-j\beta_n z} = \sum_n A_4^{(n)}\, e^{-j\beta_n z}, \tag{13.43}$$

$$\sum_m \left[A_1^{(m)} - A_2^{(m)}\right] \sum_n \frac{\kappa_m}{\omega\mu_0} v_n^{(m)}\, e^{-j\beta_n z} = \sum_n \frac{jk_{xn}}{\omega\mu_0} A_4^{(n)}\, e^{-j\beta_n z}. \tag{13.44}$$

Matching the coefficients with the same dependence yields

$$\sum_m \left[A_1^{(m)}\, e^{-j\kappa_m a} + A_2^{(m)}\, e^{j\kappa_m a}\right] v_n^{(m)} = A_3^{(n)}\, e^{-k_{xn} a}, \tag{13.45}$$

$$\sum_m \left[A_1^{(m)}\, e^{-j\kappa_m a} - A_2^{(m)}\, e^{j\kappa_m a}\right] \left(\frac{j\kappa_m}{k_{xn}}\right) v_n^{(m)} = A_3^{(n)}\, e^{-k_{xn} a}, \tag{13.46}$$

$$\sum_m \left[A_1^{(m)} + A_2^{(m)}\right] v_n^{(m)} = A_4^{(n)}, \tag{13.47}$$

$$\sum_m \left[A_1^{(m)} - A_2^{(m)}\right] \left(\frac{-j\kappa_m}{k_{xn}}\right) v_n^{(m)} = A_4^{(n)}, \tag{13.48}$$

$$(n = 0, \pm 1, \pm 2, \ldots).$$

Eliminating $A_3^{(n)}$ and $A_4^{(n)}$ gives

$$\sum_m A_1^{(m)}\, e^{-j\kappa_m a} \left(1 - \frac{j\kappa_m}{k_{xn}}\right) v_n^{(m)} + A_2^{(m)}\, e^{j\kappa_m a} \left(1 + \frac{j\kappa_m}{k_{xn}}\right) v_n^{(m)} = 0, \tag{13.49}$$

$$\sum_m A_1^{(m)} \left(1 + \frac{j\kappa_m}{k_{xn}}\right) v_n^{(m)} + A_2^{(m)} \left(1 - \frac{j\kappa_m}{k_{xn}}\right) v_n^{(m)} = 0, \tag{13.50}$$

or

$$\sum_m \left[A_1^{(m)} a_m^{(n)} + A_2^{(m)} b_m^{(n)}\right] = 0, \tag{13.51}$$

$$\sum_m \left[A_1^{(m)} e_m^{(n)} + A_2^{(m)} f_m^{(n)}\right] = 0, \tag{13.52}$$

$$(n = 0, \pm 1, +2, \ldots),$$

with

$$f_m^{(n)} = \left(1 - \frac{j\kappa_m}{k_{xn}}\right) v_n^{(m)}, \tag{13.53}$$

$$e_m^{(n)} = \left(1 + \frac{j\kappa_m}{k_{xn}}\right) v_n^{(m)}, \tag{13.54}$$

$$a_m^{(n)} = e^{-j\kappa_m a} f_m^{(n)}, \tag{13.55}$$

$$b_m^{(n)} = e^{j\kappa_m a} e_m^{(n)}. \tag{13.56}$$

Equations (13.51) and (13.52) are two coupled systems of linear equations that relate the two sets of unknown amplitudes $A_1^{(m)}$ and $A_2^{(m)}$. Setting the determinant of these linear equations to zero yields the dispersion relation for the TE waves guided by this dielectric slab with longitudinally periodic dielectrics.

Let us consider the TM case.

Region 1 $(0 < x < a)$ *(the periodic dielectric region)*

Using (13.33), (13.17), and (13.18), we can derive the fields tangential to the slab boundary:

$$\begin{aligned} H_y^{(1)} &= \epsilon^{1/2}(z) \sum_m \left[B_1^{(m)} e^{-j\kappa_m x} + B_2^{(m)} e^{j\kappa_m x} \right] \sum_n u_n^{(m)} e^{-j\beta_n z}, \\ &= \sum_m \left[B_1^{(m)} e^{-j\kappa_m x} + B_2^{(m)} e^{j\kappa_m x} \right] \sum_n w_n^{(m)} e^{-j\beta_n z}, \end{aligned} \tag{13.57}$$

$$\begin{aligned} E_z^{(1)} &= -\frac{j}{\omega\epsilon(z)} \frac{\partial H_z^{(1)}}{\partial x}, \\ &= \sum_m \left[B_1^{(m)} e^{-j\kappa_m x} - B_2^{(m)} e^{j\kappa_m x} \right] \left(-\frac{\kappa_m}{\omega\epsilon_0} \right) \sum_n s_n^{(m)} e^{-j\beta_n z}, \end{aligned} \tag{13.58}$$

where

$$\epsilon^{1/2}(z) = \sum_n g_n e^{-j2\pi n z/d}, \tag{13.59}$$

$$\frac{1}{\epsilon(z)} = \frac{1}{\epsilon_0} \sum_n g'_n e^{-j2\pi n z/d}, \tag{13.60}$$

$$w_n^{(m)} = \sum_r g_{n-r} u_n^{(m)}, \tag{13.61}$$

$$s_n^{(m)} = \sum_r g'_{n-r} w_n^{(m)}. \tag{13.62}$$

Here, g_n and g'_n are known expansion coefficients, $B_1^{(m)}$ and $B_2^{(m)}$ are unknown constants, and $u_n^{(m)} = \mathrm{c}_n^{(m)\mathrm{TM}} / \mathrm{c}_0^{(m)\mathrm{TM}}$.

Region 0 $(x > a)$

$$H_y^{(0)} = \sum_n B_3^{(n)} \, e^{-k_{xn}x} e^{-j\beta_n z}, \tag{13.63}$$

$$E_z^{(0)} = \frac{-j}{\omega\epsilon_0} \frac{\partial H_y^{(0)}}{\partial x} = \sum_n \frac{-jk_{xn}}{\omega\epsilon_0} B_3^{(n)} \, e^{-k_{xn}x} e^{-j\beta_n z}, \tag{13.64}$$

where $B_3^{(m)}$ are unknown constants.

Region 2 $(x < 0)$

$$H_y^{(0)} = \sum_n B_4^{(n)} \, e^{k_{xn}x} e^{-j\beta_n z}, \tag{13.65}$$

$$E_z^{(0)} = \frac{-j}{\omega\epsilon_0} \frac{\partial H_y^{(0)}}{\partial x} = \sum_n \frac{-jk_{xn}}{\omega\epsilon_0} B_4^{(n)} \, e^{-k_{xn}x} e^{-j\beta_n z}, \tag{13.66}$$

where $B_4^{(n)}$ are unknown constants.

In a similar way as was done before for the TE case, the resultant equations for the TM case are

$$\sum_m \left[B_1^{(m)} a_m^{(n)\prime} + B_2^{(m)} b_m^{(n)\prime} \right] = 0, \tag{13.67}$$

$$\sum_m \left[B_1^{(m)} e_m^{(n)\prime} + B_2^{(m)} f_m^{(n)\prime} \right] = 0, \tag{13.68}$$

$$(n = 0, \pm 1, \pm 2, \ldots),$$

with

$$f_m^{(n)\prime} = \left(w_n^{(m)} + \frac{\kappa_m}{jk_{xn}} s_n^{(m)} \right), \tag{13.69}$$

$$e_m^{(n)\prime} = \left(w_n^{(m)} - \frac{\kappa_m}{jk_{xn}} s_n^{(m)} \right), \tag{13.70}$$

$$a_m^{(n)\prime} = e^{-j\kappa_m a} f_m^{(n)\prime}, \tag{13.71}$$

$$e_m^{(n)\prime} = e^{j\kappa_m a} e_m^{(n)\prime}. \tag{13.72}$$

Again, (13.67) and (13.68) are two coupled systems of linear equations that relate the two sets of unknown amplitudes $B_1^{(m)}$ and $B_2^{(m)}$. Setting the determinant of these linear equations to zero yields the dispersion relation for the TM waves guided by this dielectric slab with longitudinally periodic dielectrics.

The ω-β diagram can be obtained from the dispersion relation as follows:

1. Given a periodic dielectric profile means $p_n^{(\mathrm{TE}),(\mathrm{TM})}$ in (13.23) are known for any given combination of $k_0 = \sqrt{\mu_0 \epsilon_0}$ and β_0, the eigenvalues κ_m and the eigenvectors with elements $\mathrm{c}_n^{(m)}$ can be determined from (13.30).

2. In all of the above computations, successive approximation is used to truncate all infinite matrices.

3. The information gained from (1) on κ_m and $\mathrm{c}_n^{(m)}$ is used in the dispersion relation to calculate β_0 for a given k_0.

The above process is by no means simple [16, 17]. A better way to obtain the ω-β diagram for this slab problem or other problems involving periodic structures is perhaps to use numerical techniques, such as the FDTD technique discussed in Sect. 15.4 or other approximate techniques such as the perturbation technique or the coupled-mode approach. [1]

Finally, we wish to add that the worthiness of this exact approach is in its ability to delineate the salient features of guided waves on periodic structures, that is, the presence of space-harmonics (even for very weak periodic variations), the decoupling of TE and TM waves on a 2D structure, and the coupling of TE and TM waves on a 3D structure, such as a dielectric ribbon with periodic medium.

References

1. C. Elachi and C. Yeh, "Periodic structures in integrated optics," J. Appl. Phys. **44**, 3146 (1973); K. Sakuda and A. Yariv, "Analyses of optical propagation in a corrugated dielectric waveguide," Opt. Commun. **8**, 1 (1973)
2. D. Marcuse, ed., "Integrated Optics," IEEE Press, New York (1973)
3. R. E. Collin, "Foundations for Microwave Engineering," McGraw-Hill, New York (1966)
4. H. Kogelnik and C. V. Shank, "Stimulated emission in a periodic structure," Appl. Phys. Lett. **18**, 152 (1971); H. Kogelnik and C. V. Shank, "Coupled-wave theory of distributed feedback lasers," J. Appl. Phys. **43**, 2325 (1972)
5. S. John, "Strong localization of photon in certain disordered dielectric superlattices," Phys. Rev. Lett. **58**, 2486 (1987)

6. E. Yablonovitch, "Inhibited spontaneous emission in solid state physics and electronics," Phys. Rev. Lett. **58**, 2059 (1987)

7. J. Joannopoulos, E. Meade, and J. Winn, "Photonic Crystals," Princeton Press, New Jersey (1995)

8. L. Brillouin, "Wave Propagation in Periodic Structures," Dover, New York (1953)

9. T. Tamir, H. C. Wang, and A. A. Oliner, "Wave-propagation in sinusoidally stratified dielectric media," IEEE Trans. Microw. Theory Tech. **MTT-12**, 323 (1964)

10. C. Yeh, K. F. Casey, and Z. A. Kaprielian, "Transverse magnetic wave propagation in sinusoidally stratified dielectric media," IEEE Trans. Microw. Theory Tech. **MTT-13**, 297 (1965)

11. K. F. Casey, J. R. Mathes, and C. Yeh, "Wave propagation in sinusoidally stratified plasma media," J. Math. Phys. **10**, 891 (1969)

12. S. Fan, J. N. Winn, A. Devenyi, J. C. Chen, R. D. Meade, and J. D. Joannopoulos, "Guided and defect modes in periodic dielectric waveguides," J. Opt. Soc. Am. B **12**, 1267 (1995)

13. A. Mekis, S. Fan, and J. D. Joannopoulos, "Bound states in photonic crystal waveguides and waveguide bends," Phys. Rev. B **58**, 4809 (1998)

14. J. C. Knight, T. A. Birks, P. J. Russell, and D. M. Atkins, "All-silica single-mode optical fiber with photonic crystal cladding," Opt. Lett. **21**, 1547 (1996); B. J. Managan, L. Farr, A. Langford, P. J. Roberts, D. P. Williams, F. Couny, M. Lawman, M. Mason, S. Coupland, R. Flea, H. Sabert, T. A. Birks, J. C. Knight, and P. S. T. Russell, "Low loss (1.7 dB/km) hollow core photonic band-gap film," PDP24, Opt. Soc. Am. (2004)

15. C. Yeh, F. I. Shimabukuro, and P. H. Siegel, Private communication (2005)

16. S. T. Peng, T. Tamir, and H. L. Bertoni, "Theory of periodic dielectric waveguides," IEEE Trans. Microw. Theory Tech. **MTT-23**, 123 (1975)

17. M. Akbari, M. Shahabadi, and K. Schuenemann, "A rigorous two dimensional field analysis of DFB structures," Prog. Electromagn. Res. **22**, 197 (1999)

14

METAMATERIAL AND OTHER WAVEGUIDES

Surface waves are quite robust. In addition to the many surface wave structures discussed in previous chapters, we find the existence of a variety of other structures that can support surface waves. A few of these additional structures will be discussed here. They include the moving medium guide, the anisotropic medium guide, and the metamaterial waveguide.

14.1 Moving Dielectric Waveguides [1, 2]

Applying the theory of special relativity, Lorentz transformation, and Minkowski theory to electromagnetics, we can solve the problem of wave propagation along moving dielectric and moving plasma waveguides.

14.1.1 Relativity, Lorentz Transformation, and Minkowski Transformation

It is worthwhile to cite the relativity postulates by Einstein [3–5]:

Postulate 1: When properly formulated, the laws of physics are invariant to a transformation from one reference system to another moving with a linear, uniform relative velocity. (Stratton, p. 74) [5].

Postulate 2: The velocity of propagation of an electromagnetic disturbance in free space is a universal constant, c, which is independent of the reference system. (Stratton, p. 75) [5].

Based on these postulates, the Lorentz transformations between an inertial frame $S(\mathbf{r}, t)$ and another inertial frame $S'(\mathbf{r}', t')$, which is moving at a uniform velocity $\mathbf{v}$ with respect to S, can be written in the general form.

$$\mathbf{r}' = \mathbf{r} - \gamma \mathbf{v} t + (\gamma - 1)\frac{\mathbf{r} \cdot \mathbf{v}}{v^2}\mathbf{v}, \tag{14.1}$$

$$t' = \gamma\left(t - \frac{\mathbf{r} \cdot \mathbf{v}}{c^2}\right), \tag{14.2}$$

where

$$\gamma = (1-\beta^2)^{-1/2}, \qquad \beta = \frac{v}{c},$$

$$\mathbf{r} = x\mathbf{e}_x + y\mathbf{e}_y + z\mathbf{e}_z,$$

$$c = \text{velocity of light in vacuum.} \tag{14.3}$$

To assure the covariance of the Maxwell equations between S and S' systems, the following Minkowski transformations for the field vectors must be used:

$$\mathbf{E}' = \gamma\left(\mathbf{E} + \mathbf{v}\times\mathbf{B}\right) + (1-\gamma)\frac{\mathbf{E}\cdot\mathbf{v}}{v^2}\mathbf{v}, \tag{14.4}$$

$$\mathbf{B}' = \gamma\left(\mathbf{B} - \frac{1}{c^2}\mathbf{v}\times\mathbf{E}\right) + (1-\gamma)\frac{\mathbf{B}\cdot\mathbf{v}}{v^2}\mathbf{v}, \tag{14.5}$$

$$\mathbf{D}' = \gamma\left(\mathbf{D} + \frac{1}{c^2}\mathbf{v}\times\mathbf{H}\right) + (1-\gamma)\frac{\mathbf{D}\cdot\mathbf{v}}{v^2}\mathbf{v}, \tag{14.6}$$

$$\mathbf{H}' = \gamma\left(\mathbf{H} - \mathbf{v}\times\mathbf{D}\right) + (1-\gamma)\frac{\mathbf{H}\cdot\mathbf{v}}{v^2}\mathbf{v}. \tag{14.7}$$

Another consequence of these postulates is the birth of the principle of phase invariance, that is,

$$-\mathbf{k}'\cdot\mathbf{r}' + \omega' t' = -\mathbf{k}\cdot\mathbf{r} + \omega t, \tag{14.8}$$

or, using the Lorentz transforms, we obtain

$$\mathbf{k}' = \mathbf{k} - \gamma\frac{\mathbf{v}\omega}{c^2} + (\gamma-1)\frac{\mathbf{k}\cdot\mathbf{v}}{v^2}\mathbf{v}, \tag{14.9}$$

$$\omega' = \gamma\left(\omega - \mathbf{v}\cdot\mathbf{k}\right). \tag{14.10}$$

Let us now apply the above transformations to treat the moving media problem.

14.1.2 Reflection and Transmission of Electromagnetic Waves by a Moving Plasma Medium [6]

Let us consider the case of the reflection and transmission of waves by a moving dispersive dielectric half-space or slab. Specifically, the dispersive medium is assumed to be a homogeneous cold plasma. The permittivity of a cold plasma is

$$\epsilon_{\mathrm{p}} = \epsilon_0\left[1 - \left(\frac{\omega_{\mathrm{p}}^2}{\omega'^2}\right)\right], \tag{14.11}$$

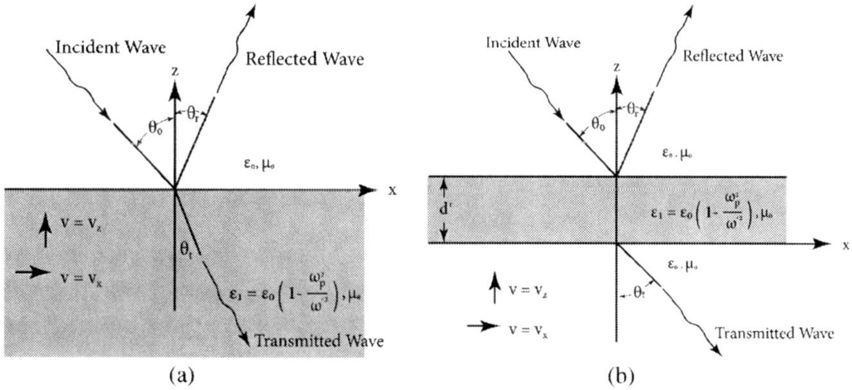

Figure 14.1. Geometry of the problem. (**a**) Moving plasma half space; (**b**) moving plasma slab

where ϵ_0 is the free-space permittivity, ω_p is the plasma frequency, and ω' is the frequency in the the moving system S', which is stationary with respect to the uniformly moving medium.

A harmonic plane wave in the free-space region with its electric vector polarized in the y-direction is assumed to be incident on a moving semi-infinite plasma half-space or on a moving plasma slab of thickness d (see Fig. 14.1). In the observer's system S, which is stationary with respect to the free space, the incident plane wave is

$$E_y^{(i)} = E_0\, e^{-j(k_x x + k_z z)}\, e^{j\omega t}, \tag{14.12}$$

$$B_y^{(i)} = 0, \tag{14.13}$$

where E_0 and ω are, respectively, the amplitude and the frequency of the incident wave, $k_x = k_0 \sin\theta_0$, $k_z = -k_0 \cos\theta_0$, $k_0 = \omega\,(\mu_0\epsilon_0)^{1/2}$. θ_0 is the angle between the propagation vector and the positive z axis in the x-z plane. The reflected wave and the transmitted wave, in the observer's system S, takes the following forms:

For the reflected wave

$$E_y^{(r)} = A_r \exp\left[-j\left(k_x^{(r)} x - k_z^{(r)} z\right)\right] \exp\left[j\omega^{(r)} t\right], \tag{14.14}$$

$$B_y^{(r)} = 0. \tag{14.15}$$

For the transmitted wave

$$E_y^{(t)} = G_t \exp\left[-j\left(k_x^{(t)} x - k_z^{(t)} z\right)\right] \exp\left[j\omega^{(t)} t\right], \tag{14.16}$$

$$B_y^{(t)} = 0. \tag{14.17}$$

A_r, G_t, $k_x^{(r)}$, $k_x^{(t)}$, $k_z^{(r)}$, $k_z^{(t)}$, $\omega^{(r)}$, and $\omega^{(t)}$ are given later. Making use of the principle of phase invariance of plane waves, the covariance of the Maxwell equations and the Lorentz transformations, and satisfying the boundary conditions, one obtains the following relations [3, 4, 6]:

(a) For a moving half-space, if it is moving uniformly with a velocity v_x in the positive x direction,

$$\omega^{(r)} = \omega^{(t)} = \omega, \tag{14.18a}$$

$$k_x^{(r)} = k_x^{(t)} = k_x = k_0 \sin\theta_0, \tag{14.18b}$$

$$k_z^{(r)} = k_z = -k_0 \cos\theta_0, \tag{14.18c}$$

$$k_z^{(t)} = -k_0 \left(\cos^2\theta_0 - \frac{\omega_p^2}{\omega^2}\right)^{1/2}, \tag{14.18d}$$

$$A_r = E_0 \frac{\cos\theta_0 + \dfrac{k_z^{(t)}}{k_0}}{\cos\theta_0 - \dfrac{k_z^{(t)}}{k_0}}, \tag{14.18e}$$

$$G_t = E_0 \frac{2\cos\theta_0}{\cos\theta_0 - \dfrac{k_z^{(t)}}{k_0}}; \tag{14.18f}$$

if the half-space is moving uniformly with a velocity v_z in the positive z-direction,

$$\omega^{(r)} = \omega\gamma_z^2\left[\left(1+\beta_z^2\right) + 2\beta_z \cos\theta_0\right], \tag{14.19a}$$

$$k_x^{(r)} = k_x = k_0 \sin\theta_0, \tag{14.19b}$$

$$k_z^{(r)} = -k_0\gamma_z^2\left[2\beta_z + \cos\theta_0\left(1+\beta_z^2\right)\right], \tag{14.19c}$$

$$\omega^{(t)} = \omega\gamma_z^2 \left[(1 + \beta_z \cos\theta_0) - \beta_z Q\right], \tag{14.19d}$$

$$k_z^{(t)} = k_0\gamma_z^2 \left[\beta_z (1 + \beta_z \cos\theta_0) - Q\right], \tag{14.19e}$$

$$A_r = E_0 \frac{\omega^{(r)}}{\omega} \left[\frac{\cos\theta_0 + \beta_z - Q}{\cos\theta_0 + \beta_z + Q}\right], \tag{14.19f}$$

$$G_t = E_0 \left[\frac{2\dfrac{\omega^{(t)}}{\omega}(\cos\theta_0 + \beta_z)}{\cos\theta_0 + \beta_z + Q}\right], \tag{14.19g}$$

with $\beta_z = v_z/c$, $\gamma_z = 1/(1 \quad \beta_z^2)$, and

$$Q = \frac{1}{\gamma_z}\left[\gamma_z^2 (1 + \beta_z \cos\theta_0)^2 - \frac{\omega_\mathrm{p}^2}{\omega^2} - \sin^2\theta_0\right]^{1/2}. \tag{14.20}$$

(b) For a moving slab of thickness d, if it is moving uniformly with a velocity v_x in the positive x direction,

$$\omega^{(r)} = \omega^{(t)} = \omega, \tag{14.21a}$$

$$k_x^{(r)} = k_x^{(t)} = k_x = k_0 \sin\theta_0, \tag{14.21b}$$

$$k_z^{(r)} = k_z^{(t)} = k_z = -k_0 \cos\theta_0, \tag{14.21c}$$

$$A_r = E_0 \left\{\frac{j\left(\dfrac{\omega_\mathrm{p}^2}{\omega^2}\right)\sin(\eta_x k_0 d)\exp(2jk_0 d\cos\theta_0)}{2\eta_x \cos\theta_0 \cos(\eta_x k_0 d) + j(\eta_x^2 + \cos^2\theta_0)\sin(\eta_x k_0 d)}\right\}, \tag{14.21d}$$

$$G_t = E_0 \left\{\frac{2\eta_x \cos\theta_0 \exp(jk_0 d\cos\theta_0)}{2\eta_x \cos\theta_0 \cos(\eta_x k_0 d) + j(\eta_x^2 + \cos^2\theta_0)\sin(\eta_x k_0 d)}\right\}, \tag{14.21e}$$

with

$$\eta_x = \left(\cos^2\theta_0 - \frac{\omega_{\mathrm{p}}^2}{\omega^2}\right)^{1/2}; \tag{14.21f}$$

if the slab is moving uniformly with a velocity v_z in the positive z direction,

$$\omega^{(r)} = \omega\gamma_z^2\left[1+\beta_z^2+2\beta_z\cos\theta_0\right], \tag{14.22a}$$

$$k_x^{(r)} = k_x^{(t)} = k_0\sin\theta_0, \tag{14.22b}$$

$$k_z^{(r)} = -k_0\gamma_z^2\left[2\beta_z+\cos\theta_0\left(1+\beta_z^2\right)\right], \tag{14.22c}$$

$$\omega^{(t)} = \omega, \tag{14.22d}$$

$$k_z^{(t)} = -k_0\cos\theta_0, \tag{14.22e}$$

$$A_r = E_0\left\{\frac{j\gamma_z^2\left(\dfrac{\omega_{\mathrm{p}}^2}{\omega^2}\right)\mathrm{C1}\,\sin(\eta_z k_0 d)\exp\left[2jk_0 d\,\gamma_z\mathrm{C2}\right]}{2\eta_z\gamma_z\,\mathrm{C2}\,\cos(\eta_z k_0 d)+j\left[\eta_z^2+\gamma_z^2\,\mathrm{C2}^2\right]\sin(\eta_z k_0 d)}\right\}, \tag{14.22f}$$

$$G_t = E_0\left\{\frac{2\eta_z\gamma_z\,\mathrm{C2}\exp\left[j\gamma_z k_0 d\,\mathrm{C2}\right]}{2\eta_z\gamma_z\,\mathrm{C2}\,\cos(\eta_z k_0 d)+j\left[\eta_z^2+\gamma_z^2\,\mathrm{C2}^2\right]\sin(\eta_z k_0 d)}\right\}, \tag{14.22g}$$

with

$$\eta_z = \left[\gamma_z^2(1+\beta_z\cos\theta_0)^2 - \left(\frac{\omega_{\mathrm{p}}^2}{\omega^2}\right) - \sin^2\theta_0\right]^{1/2}, \tag{14.22h}$$

$$\mathrm{C1} = 1+\beta_z^2+2\beta_z\cos\theta_0,$$

$$\mathrm{C2} = \cos\theta_0+\beta_z.$$

It is rather surprising to note from (14.18) and (14.21) that all reflected and transmitted fields for both the plasma half-space case and the plasma slab case are independent to the velocity of the plasma medium, which is moving in a direction parallel to the interface.

Of special interest are the reflection coefficient for the half-space case and the reflection and transmission coefficient for the slab case. The reflection coefficient and the transmission coefficient are defined, respectively, by the relations

$$R = \mathbf{n} \cdot \mathbf{S}_r / \mathbf{n} \cdot \mathbf{S}_i \tag{14.23}$$

and

$$T = \mathbf{n} \cdot \mathbf{S}_t / \mathbf{n} \cdot \mathbf{S}_i, \tag{14.24}$$

where $\mathbf{n}$ is the unit vector normal to the interface and

$$\mathbf{S}_i = \frac{1}{2}\left[\mathbf{E}^{(i)} \times \mathbf{H}^{*(i)}\right], \tag{14.25}$$

$$\mathbf{S}_r = \frac{1}{2}\left[\mathbf{E}^{(r)} \times \mathbf{H}^{*(r)}\right], \tag{14.26}$$

$$\mathbf{S}_t = \frac{1}{2}\left[\mathbf{E}^{(t)} \times \mathbf{H}^{*(t)}\right]. \tag{14.27}$$

The * signifies the complex conjugate of the function. The more interesting case of a plasma medium moving in a direction normal to its interface is now considered in detail. The reflection coefficient for a plasma half space moving in the z direction is [1]

$$R_z^{(\mathrm{HS})} = \frac{A_r A_r^*}{E_0^2} \frac{k_z^{(r)}}{k_0 \cos\theta_0} \frac{\omega}{\omega^{(r)}}, \tag{14.28}$$

where A_r, $k_z^{(r)}$, and $\omega^{(r)}$ are given in (14.19). The reflection and transmission coefficient for a plasma slab moving in the z-direction are, respectively,

$$R_z^{(\mathrm{S})} = \frac{A_r A_r^*}{E_0^2} \frac{k_z^{(r)}}{k_0 \cos\theta_0} \frac{\omega}{\omega^{(r)}} \tag{14.29}$$

[1]The definitions given by (14.23) and (14.24) for the reflection and transmission coefficients are meaningful only for harmonic dependent fields (i.e., all fields have $e^{j\Omega t}$ dependence, where Ω is real). Although the incident wave as viewed from the S system has harmonic dependence, the frequency of the transmitted wave in the moving plasma medium as viewed from the S system [i.e., $\omega^{(t)}$ in (14.19d)], in general, may be complex. Consequently, the transmission coefficient given by (14.24) is not applicable to the case in which the plasma half-space is moving towards or away from the incident wave.

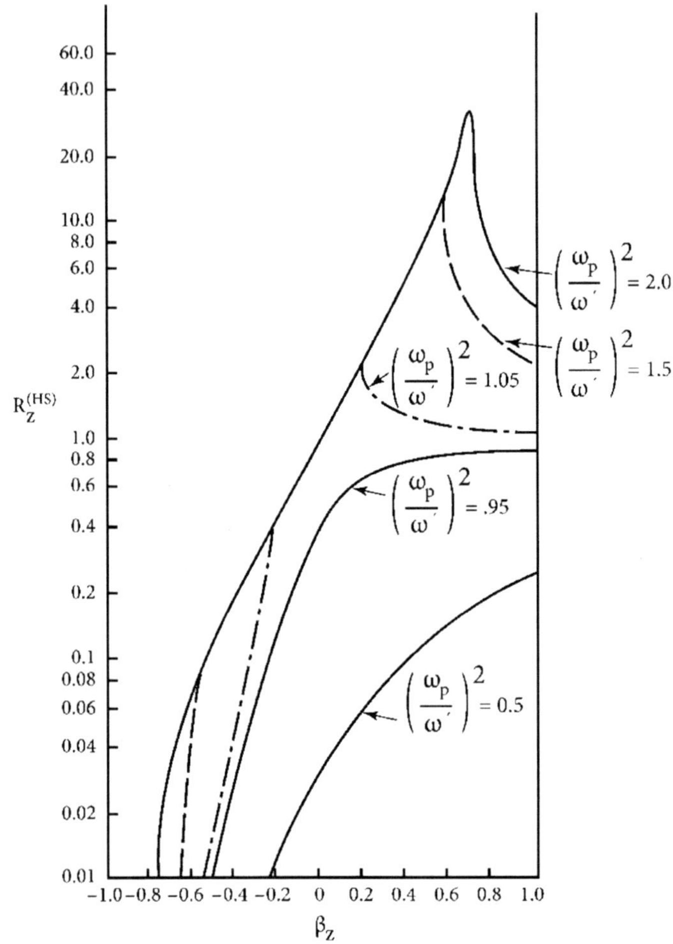

Figure 14.2. Reflection coefficient for the moving half-space case [6]

and

$$T_z^{(S)} = \frac{G_t G_t^*}{E_0^2}, \tag{14.30}$$

where A_r, $k_z^{(r)}$, $\omega^{(r)}$, and G_t are given in (14.22). It can be seen from the above equations that the reflection and transmission coefficients are rather complicated functions of the angle of incidence, the velocity, and the plasma frequencies of the medium. To have a qualitative idea of how the reflection and transmission coefficients vary as a function of the velocity of the moving medium, we shall consider the limiting case of normal incidence.

Figure 14.2 shows the reflection coefficient $R_z^{(HS)}$ for the plasma half-space as a function of β_z for various values of ω_p/ω. Unlike the nondispersive dielectric half-space case, the reflection coefficient no longer increases monotonically as β_z increases for all values of ω_p/ω. For the case $\omega_p/\omega > 1$, $R_z^{(HS)}$ reaches a maximum as β_z increases from -1.0 and then decays to the value $(\omega_p/\omega)^4$. Furthermore, if $\omega_p/\omega > 1$ and $\beta_z > 1$, $R_z^{(HS)}$ is always greater than unity. In other words, when the plasma half-space is moving towards the incident wave, the reflected energy is greater than the energy of the incident wave. Another interesting feature of the $\omega_p/\omega > 1$ case is that a portion of the $R_z^{(HS)}$ vs. β_z curve is identical for different values of ω_p/ω. For example, the portion of the curve between $\beta_z = -0.5$ and $\beta_z = 0.55$ is the same for $(\omega_p/\omega)^2 = 1.5$ case and for $(\omega_p/\omega)^2 > 1.5$ case. If $\omega_p/\omega < 1$, $R_z^{(HS)}$ increases monotonically from zero to $(\omega_p/\omega)^4$ as β_z varies from -1 to +1.

The reflection coefficient $R_z^{(S)}$ and the transmission coefficient $T_z^{(S)}$ for a moving plasma slab are plotted, respectively, in Figs. 14.3 and 14.4 as a function of the velocity of the slab. It is assumed that $k_0 d = \pi/4$, where d is the thickness of the slab in the reference frame, which is at rest with the plasma slab, and k_0 is the free-space wave number. The reflection coefficient as well as the transmission co-

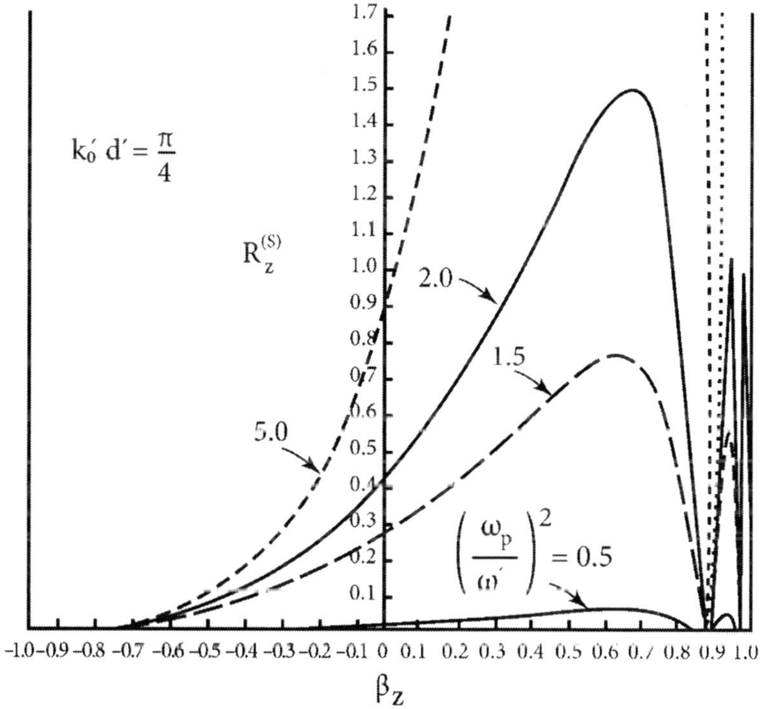

Figure 14.3. Reflection coefficient for the moving slab case [6]

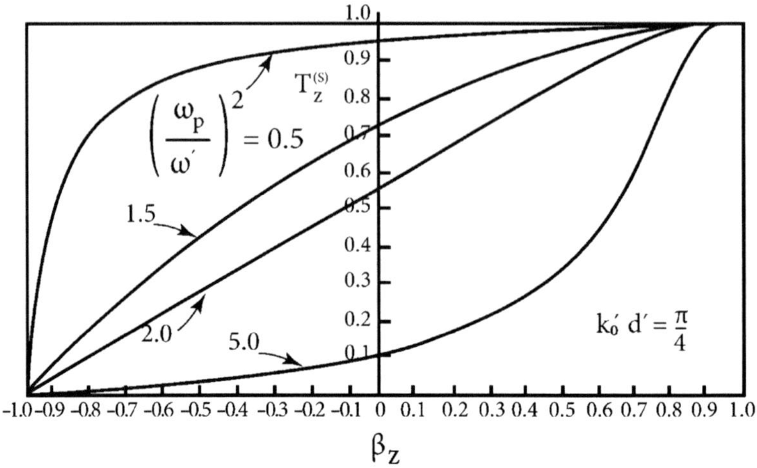

Figure 14.4. Transmission coefficient for the moving slab case [6]

efficient decrease monotonically to zero for all values of ω_p/ω as β_z changes from 0 to -1. As the velocity of the slab moving towards the incident wave increases, that is, as β_z increases from 0, the reflection coefficient increases and reaches a maximum and then starts to oscillate between zero and certain constant. The envelope of the oscillating function is a decaying function; it approaches $(\omega_p/\omega)^4$ as β_z approaches $+1$. The maximum value of $R_z^{(S)}$ increases as the plasma density increases. The transmission coefficient increases monotonically as β_z increases from 0 to about 0.9. As β_z increase from 0.9 to 1.0, very small ripples are observed as $T_z^{(S)}$ approaches 1.0. Also noted is the fact that $T_z^{(S)}$ approaches unity faster as the plasma becomes less dense. It is further noted that $R_z^{(S)} + T_z^{(S)} = 1$ at $\beta_z = 0$, as expected. However, if $\beta_z \neq 0$, then $R_z^{(S)} + T_z^{(S)} \neq 1$ in general.

14.1.3 Mode Propagation Along Moving Dielectric Slabs [1]

An isotropic dielectric slab having a permittivity of ϵ_1, a permeability of μ_0, and a conductivity of zero is assumed to be moving in direction parallel to the direction of propagation of the surface waves. Two types of modes may be guided along a moving dielectric slab: TE and TM modes. The TE modes have a single component of electric field E_y and magnetic-field components H_x and H_z, while the TM modes have a single component of magnetic field H_y and electric-field components E_x and E_z. The dielectric slab is assumed to have a thickness of $2d$.

14.1.3.1 TE Modes

In the moving system S', which is stationary with respect to the uniformly moving dielectric slab, the electric-field component of the TE modes takes the form (see Chap. 4)

$$E'_{y'} = \begin{cases} A' \exp\{-(p' |x'| - d) - j\beta' z' + j\omega' t'\} & |x'| \geq d, \\ A' \sec(h'd) \cos(h'x') \exp(-j\beta' z' + j\omega' t') & |x'| \leq d \end{cases} \tag{14.31}$$

for the even solution (i.e., $E'_{y'}$ is symmetric with respect to the plane $x' = 0$) and the form

$$E'_{y'} = \begin{cases} A' \exp\left[-p' \left(x' - d\right) - j\beta' z' + j\omega' t'\right] & x' \geq d, \\ A' \csc(h'd) \sin(h'x') \exp(-j\beta' z' + j\omega' t') & |x'| \leq d, \\ -A' \exp\left[p' \left(x' + d\right) - j\beta' z' + j\omega' t'\right] & x' \leq -d \end{cases} \tag{14.32}$$

for the odd solution (i.e., $E'_{y'}$ is asymmetric with respect to the plane $x' = 0$). It is noted that

$$\left(\beta' d\right)^2 = \omega'^2 \mu_0 \epsilon_1 d^2 - \left(h' d\right)^2, \tag{14.33}$$

$$\left(\beta' d\right)^2 = \omega'^2 \mu_0 \epsilon_0 d^2 + \left(p' d\right)^2. \tag{14.34}$$

Matching the tangential electric and magnetic fields at the boundaries, $x' = \pm d$, we obtain

$$p' d = h' d \tan\left(h' d\right) \tag{14.35}$$

for the even solution, and

$$p' d = -h' d \cot\left(h' d\right) \tag{14.36}$$

for the odd solution. Combining (14.33) and (14.34) gives

$$\left(p' d\right)^2 + \left(h' d\right)^2 = \left(\frac{\epsilon_1}{\epsilon_0} \quad 1\right) \omega'^2 \mu_0 \epsilon_0 d^2. \tag{14.37}$$

Making use of Minkowski's transformation theory and the Lorentz transformations [7], we have the following expressions for the electric-field component of the TE modes in the observer's system S:

For the even solution

$$E_y = \begin{cases} A \exp\{-(p'|x| - d) - j\beta z + j\omega t\} & |x| \geq d, \\ A \sec(h'd) \cos(h'x) \exp(-j\beta z + j\omega t) & |x| \leq d, \end{cases} \tag{14.38}$$

with

$$p'd = h'd \tan\left(h'd\right) \tag{14.39}$$

$$\left(p'd\right)^2 + \left(h'd\right)^2 = \left[\frac{\epsilon_1}{\epsilon_0} - 1\right] \left(k_0 d^2\right) \gamma^2 \left[1 + \frac{v_z}{c} \frac{1}{k_0 d} \left(k_0^2 d^2 + p'^2 d^2\right)^{1/2}\right]^2, \tag{14.40}$$

$$\left(\beta d\right)^2 = \left(k_0 d\right)^2 + \left(p'd\right)^2, \tag{14.41}$$

where $k_0 = \omega \left(\mu_0 \epsilon_0\right)^{1/2}$, c is the speed of light in vacuum, $\gamma^2 = 1/\left[1 - \left(v_z/c\right)^2\right]$, A is an arbitrary constant, and v_z is the velocity of the moving dielectric slab.

For the odd solution

$$E_y = \begin{cases} A \exp\left[-p'\left(x - d\right) - j\beta z + j\omega t\right] & x \geq d, \\ A \csc(h'd) \sin(h'x') \exp(-j\beta z + j\omega t) & |x| \leq d, \\ -A \exp\left[p'\left(x + d\right) - j\beta z + j\omega t\right] & x \leq -d, \end{cases} \tag{14.42}$$

with

$$p'd = -h'd \cot\left(h'd\right) \tag{14.43}$$

as well as (14.40) and (14.41). The propagation constant βd for the even solution or for the odd solution can be obtained, respectively, by solving (14.39)–(14.41), or (14.40), (14.41), and (14.43).

14.1.3.2 TM Modes

Similar analysis may be carried out for TM modes. In the observer's system, S, the magnetic field component of the TM modes take the following form (see Chap. 4):

For the even solution

$$H_y = \begin{cases} B \cos(h'x) \exp\left(-j\beta z + j\omega t\right) & |x| \leq d, \\ B \cos(h'd) \exp\left[-p'\left(|x| - d\right) - j\beta z + j\omega t\right] & |x'| \geq d, \end{cases} \tag{14.44}$$

with

$$\frac{\epsilon_1}{\epsilon_0} p'd = h'd \tan\left(h'd\right), \tag{14.45}$$

$$\left(p'd\right)^2 + \left(h'd\right)^2 = \left[\frac{\epsilon_1}{\epsilon_0} - 1\right] \left(k_0 d\right)^2 \gamma^2 \left[1 + \frac{v_z}{c} \frac{1}{k_0 d} \left(k_0^2 d^2 + p'^2 d^2\right)^{1/2}\right]^2, \tag{14.46}$$

$$\left(\beta d\right)^2 = \left(k_0 d\right)^2 + \left(p'd\right)^2, \tag{14.47}$$

where B is an arbitrary constant.

For the odd solution

$$H_y = \begin{cases} B \exp\left[-p'\left(x-d\right) - j\beta z + j\omega t\right] & x \geq d, \\ B \csc(h'd) \sin(h'x) \exp(-j\beta z + j\omega t) & -d \leq x \leq d, \\ -B \exp\left[p'\left(x+d\right) - j\beta z + j\omega t\right] & x \leq -d, \end{cases} \tag{14.48}$$

with

$$\frac{\epsilon_1}{\epsilon_0} p'd = -h'd \cot\left(h'd\right) \tag{14.49}$$

as well as (14.46) and (14.47). Again the propagation constant βd for the even solution or for the odd solution can be obtained, respectively, by solving (14.45)–(14.47) or (14.46), (14.47), and (14.49).

14.1.4 Mode Propagation Along a Moving Dielectric Cylinder

Unlike the dielectric slab case, the modes cannot, in general, be separated into TE or TM modes. All modes with angular dependence are a combination of a TE and TM mode, and are classified as hybrid HE modes. This is still true when the dielectric cylinder moves in the axial direction. In the observer's system S, the axial components of the electric and magnetic fields are

$$E_z = \begin{cases} f_m^{(1)}\left(r\right) \exp\left(-jm\theta - j\beta z + j\omega t\right) & r < a, \\ f_m^{(2)}\left(r\right) \exp\left(-jm\theta - j\beta z + j\omega t\right) & r > a, \end{cases} \tag{14.50}$$

$$H_z = \begin{cases} g_m^{(1)}(r)\exp(-jm\theta - j\beta z + j\omega t) & r < a, \\ g_m^{(2)}(r)\exp(-jm\theta - j\beta z + j\omega t) & r > a, \end{cases} \tag{14.51}$$

where $f_m^{(1),(2)}(r)$ and $g_m^{(1),(2)}(r)$ are appropriate combinations of Bessel or modified Bessel functions. Using a similar technique, as described for the moving dielectric-slab case, we may obtain the characteristic equation for the moving-dielectric-cylinder case

$$\left[\frac{1}{h'd}\frac{J_m'(h'd)}{J_m(h'd)} + \frac{1}{p'd}\frac{K_m'(p'd)}{K_m(p'd)}\right]\left[\frac{1}{h'd}\frac{J_m'(h'd)}{J_m(h'd)} + \frac{\epsilon_0}{\epsilon_1}\frac{1}{p'd}\frac{K_m'(p'd)}{K_m(p'd)}\right]$$

$$= m^2 \frac{\left[(h'd)^2 + (p'd)^2\right]\left[(p'd)^2 + (h'd)^2\frac{\epsilon_0}{\epsilon_1}\right]}{(h'd)^4(p'd)^4}, \tag{14.52}$$

$$(p'd)^2 + (h'd)^2 = \left(\frac{\epsilon_1}{\epsilon_0} - 1\right)(k_0 d)^2\gamma^2 \left[1 + \frac{v_z}{c}\frac{1}{k_0 d}\left((k_0 d)^2 + (p'd)^2\right)^{1/2}\right], \tag{14.53a}$$

$$(\beta d)^2 = (k_0 d)^2 + (p'd)^2. \tag{14.53b}$$

Here d represents the radius of the dielectric cylinder and $J_m(\xi)$and $K_m(\xi)$ are, respectively, the Bessel function and the modified Bessel function of order m and argument ξ. The prime on the Bessel function or the modified Bessel function indicates the derivative of the function with respect to its argument.

It is of interest to note from the above analysis for both kinds of mode structure that as far as the characteristic equations are concerned, the velocity of the moving dielectric enters into only one of the characteristic equations [i.e., (14.46) or (14.53a)] and that equation is the same for mode propagation along a moving-dielectric slab or along a moving-dielectric cylinder. Consequently, the computation of the characteristic roots for the modes along a moving-dielectric structure is greatly simplified. For example, we suppose that we wish to find the propagation constant β of a hybrid HE$_{mn}$ mode for a moving-dielectric rod. Solution of the transcendental (14.52) provides a plot of $p'd$ vs. $h'd$ for fixed values of m and ϵ_1/ϵ_0. Solution of the algebraic (14.53a) provides another plot of $p'd$ vs. $h'd$ for fixed values of $k_0 d$, ϵ_1/ϵ_0, and v_z/c. The intersection of these two plots gives the root for (14.52) and (14.53a) for fixed values of $k_0 d$, ϵ_1/ϵ_0, and v_z/c. Substituting

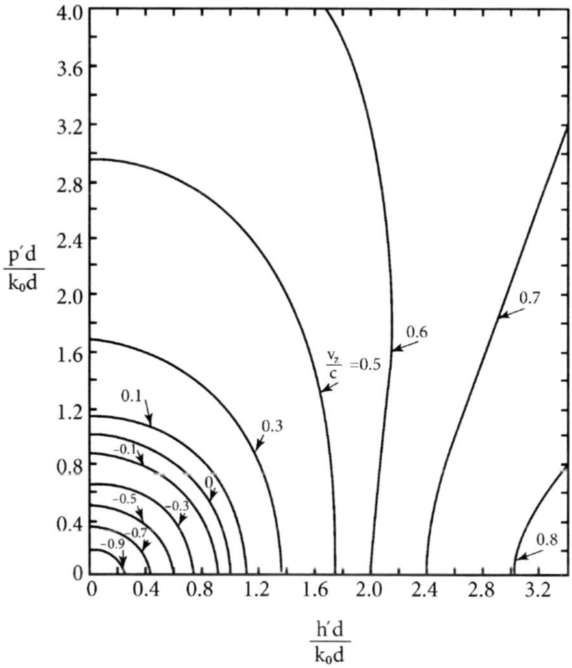

Figure 14.5. $(h'd/k_0d)$ vs. $(p'd/k_0d)$ for various v_z/c with $\epsilon_1/\epsilon_0 = 2.0$, that is, a plot of (14.53a) [1]

the value of $p'd$ corresponding to the root in (14.53b) gives the propagation constant βd. It is important to note that the much more difficult plot for the transcendental (14.52) is independent of the velocity of the dielectric rod, while the simpler plot for the algebraic (14.53a), which is also valid for the dielectric-slab problem, is a function of v_z/c. Hence the solution for the transcendental equation for the stationary-dielectric problem may be used directly for the moving-dielectric problem.

A set of $(p'd/k_0d)$ vs. $(h'd/k_0d)$ curves computed from (14.53a) for various values of v_z/c with $\epsilon_1/\epsilon_0 = 2.0$ is given in Fig. 14.5. This set of curves is independent of k_0d and is symmetrical about $(p'd/k_0d)$ axis as well as $(h'd/k_0d)$ axis. At $v_z/c = 0$, as expected, the curve is a circular arc. For $v_z/c < 0$, that is, when the dielectric material is moving in a direction that is opposite to the direction of propagation of the mode, the curves look like elliptical arcs. As v_z/c approaches -1, the size of the arc approaches 0. As v_z/c increases from 0, the size of the arc grows. For $v_z > 0.5$, the curves no longer look like the elliptical arcs. When $p'd/k_0d = 0$, these curves intercept the $h'd/k_0d$ axis at $[(\epsilon_1/\epsilon_0 - 1)\,(1 + (v_z/c)\,/\,(1 - v_z/c)^{1/2}]$. When $h'd/k_0d = 0$ and for $v_z/c \leq (\epsilon_0/\epsilon_1)^{1/2}$, these curves intercept the $p'd/k_0d$ axis at $\{[\epsilon_1/\epsilon_0 - 1] \times [1 - (v_z/c)^2]\}^{1/2}/\{1 - (v_z/c)[\epsilon_1/\epsilon_0]^{1/2}\}$. For $v_z/c > (\epsilon_0/\epsilon_1)^{1/2}$, the curves

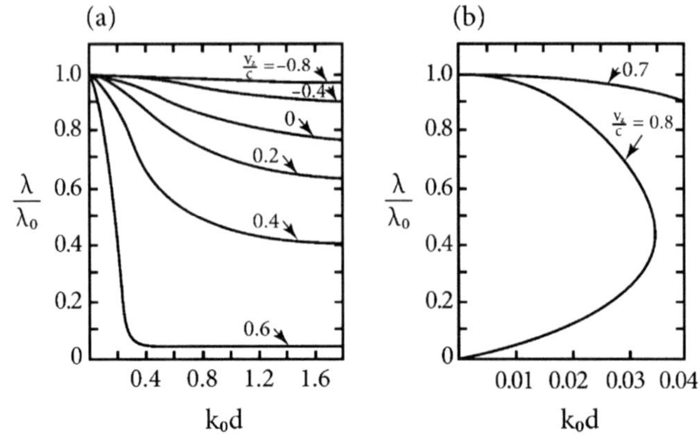

Figure 14.6. Normalized guide wavelength λ/λ_0 as a function of k_0d for the dominant TE even mode along a moving dielectric slab with thickness $2d$ and $\epsilon_1/\epsilon_0 = 2.0$. Note that the scale for k_0d in (**b**) is expanded

will not intercept the $p'd/k_0d$ axis and the asymptotes of these curves are $(h'd/k_0d) = \pm\left\{[(\epsilon_1/\epsilon_0)(v_z/c) - 1]/[1-(v_z/c)^2]\right\}^{1/2}(p'd/k_0d)$.

Figure 14.6 is introduced to illustrate how the normalized guide wavelength λ/λ_0 for the dominant $\mathrm{TE}^{(\mathrm{even})}$ mode along a moving-dielectric slab of thickness $2d$ varies as a function of k_0d for various values of v_z/c. It is assumed that $\epsilon_1/\epsilon_0 = 2.0$. At $v_z/c = 0$, as expected, λ/λ_0 is unity for $k_0d = 0$ and λ/λ_0 approaches asymptotically to $[\epsilon_1/\epsilon_0]^{-1/2}$as k_0d becomes infinite. However, when $v_z/c < 0$, λ/λ_0 varies from 1 to $\{1-(v_z/c)[\epsilon_1/\epsilon_0]^{1/2)}\}/\{[\epsilon_1/\epsilon_0]^{1/2} - v_z/c\}$, which is larger than $[\epsilon_1/\epsilon_0]^{-1/2}$ for negative values of v_z/c. When $v_z/c = -1$, $\lambda/\lambda_0 = 1$ for all values of k_0d. In other words, when the dielectric medium is moving opposite to the direction or mode propagation, the guide wavelength of the mode becomes closer to the free-space wavelength or the mode becomes less attached to the guiding structure. On the other hand, when the dielectric medium is moving in the same direction as the mode (i.e., $v_z/c > 0$), the guide wavelength approaches $\{1-(v_z/c)[\epsilon_1/\epsilon_0]^{1/2}\}\{[\epsilon_1/\epsilon_0]^{1/2} - v_z/c\}$, which is smaller than $[\epsilon_1/\epsilon_0]^{1/2}$ as k_0d increases or the mode clings closer to the guiding structure. When $v_z/c = [\epsilon_1/\epsilon_0]^{-1/2}$, λ/λ_0 varies from 1 to 0 as k_0d varies from 0 to ∞. As v_z/c goes beyond $[\epsilon_1/\epsilon_0]^{-1/2}$, the characteristics of the λ/λ_0 vs. k_0d curve are altered significantly as can be seen in Fig. 14.6b. For example, when $v_z/c = 0.8$, the guiding structure can support not only the usual forward mode but also a backward mode within the region $0 \le k_0d < 0.035$. Beyond $k_0d = 0.035$, the structure can no longer support this dominant mode, that is, there exists a high-frequency cutoff for this mode. This is a unique feature of the moving nondispersive dielectric structure.

We note from this example that the movement of the dielectric medium introduces many interesting features that cannot be found for the stationary-dielectric problem. It is expected that these unique features will also be present for mode propagation along a moving-dielectric rod.

14.1.5 Wave Propagation on a Moving Plasma Column

A circular plasma column of radius a is assumed to be surrounded by a free-space region. The plasma medium having a permittivity of

$$\epsilon_{\mathrm{p}} = \epsilon_0 \left(1 - \frac{\omega_{\mathrm{p}}^2}{\omega'^2}\right) \tag{14.54}$$

is assumed to be moving uniformly with a velocity v_z in the axial direction (i.e., the z direction), which is parallel to the direction of propagation of the surface waves. In (14.54), ϵ_0 is the free-space permittivity, ω_{p} is the plasma frequency, and ω' is the frequency in the moving system S', which is stationary with respect to the plasma medium. We are interested in the propagation characteristics of surface waves along this structure. It can be shown that in the S' system the plasma structure can support two types of surface modes: a TM circularly symmetric mode and a hybrid HE angularly dependent mode. The axial components of electric and magnetic fields in the moving system S' are

$$\begin{aligned} E'_z &= A'_m I_m(p'r') \exp\left(-jm\theta' - j\beta' z' + j\omega' t'\right) \quad r < a, \\ &= B'_m K_m(q'r') \exp\left(-jm\theta' - j\beta' z' + j\omega' t'\right) \quad r > a, \end{aligned} \tag{14.55}$$

$$\begin{aligned} H'_z &= C'_m I_m(p'r') \exp\left(-jm\theta' - j\beta' z' + j\omega' t'\right) \quad r < a, \\ &= D'_m K_m(q'r') \exp\left(-jm\theta' - j\beta' z' + j\omega' t'\right) \quad r > a, \end{aligned} \tag{14.56}$$

with

$$(\beta' a)^2 = (p'a)^2 + \left(\frac{\omega' a}{c}\right)^2 \left(1 - \frac{\omega_{\mathrm{p}}^2}{\omega'^2}\right), \tag{14.57}$$

$$(\beta' a)^2 = (q'a)^2 + \left(\frac{\omega' a}{c}\right)^2, \tag{14.58}$$

where I_m and K_m are modified Bessel functions and A'_m, B'_m, C'_m, and D'_m are arbitrary constants. Matching the tangential electric and magnetic fields at the boundary $r' = a$, one obtains

$$\left[\left(\frac{K_m'(q'a)}{q'aK_m(q'a)}\right)-\left(\frac{I_m'(p'a)}{p'aI_m(p'a)}\right)\right]\left[\left(\frac{K_m'(q'a)}{q'aK_m(p'a)}\right)\right.$$

$$\left.-\left(1-\frac{\omega_{\mathrm{p}}^2}{\omega'^2}\right)\left(\frac{I_m'(p'a)}{p'aI_m(p'a)}\right)\right]=m^2\left[\frac{\left(p'^2a^2-q'^2a^2\right)}{(p'a)^4(q'a)^4}\right]$$

$$\left\{(p'a)^2-\left[1-\frac{\omega_{\mathrm{p}}^2}{\omega'^2}\right](q'a)^2\right\}, \quad (14.59)$$

where the prime on the modified Bessel function indicates the derivative of the function with respect to its argument. Making use of Lorentz transformations [3], we have

$$\beta=\gamma\left[\beta'+\omega'\left(v_z/c^2\right)\right], \quad (14.60a)$$

$$\omega=\gamma\left(\omega'+\beta' v_z\right), \quad (14.60b)$$

$$(\beta a)^2=\frac{(p'a)^2+\left(\frac{a^2}{c^2}\right)\left[\gamma^2\left(\omega-\beta v_z\right)^2-\omega_{\mathrm{p}}^2\right]}{\gamma^2\left[1-\frac{\omega}{\beta}\frac{v_z}{c^2}\right]^2}, \quad (14.60c)$$

$$(\beta a)^2=(q'a)^2+\left(\frac{\omega a}{c}\right)^2. \quad (14.60d)$$

The dispersion relation is still given by (14.59), with

$$\frac{\omega_{\mathrm{p}}^2}{\omega'^2}=\frac{\omega_{\mathrm{p}}^2/\omega^2}{\gamma^2\left\{1-\frac{v_z}{c}\left[1+\frac{q'^2a^2}{\left(\omega_{\mathrm{p}}^2a^2/c^2\right)}\frac{\omega_{\mathrm{p}}^2}{\omega^2}\right]^{1/2}\right\}^2}, \quad (14.61)$$

$$p'^2a^2=q'^2a^2+\frac{\omega_{\mathrm{p}}^2a^2}{c^2}, \quad (14.62)$$

$$\beta^2a^2=q'^2a^2+\left(\frac{\omega^2}{\omega_{\mathrm{p}}^2}\right)\left(\frac{\omega_{\mathrm{p}}^2a^2}{c^2}\right), \quad (14.63)$$

where β is the propagation constant of the surface waves.

It is of interest to investigate analytically the properties of the dispersion relation for the extreme values of βa. At cutoff $q'^2a^2=0$; in other words, the surface wave is very loosely bounded to the guiding structure near cutoff, and below the cutoff frequency of a certain mode, the guiding structure can no longer support this mode. By substituting $q'a \to 0$ into (14.59), (14.61), (14.62), and (14.63), one can show that there exists no cutoff frequency for the circularly symmetric TM mode $(m = 0)$ and for the lowest-order hybrid mode $(m = 1)$. On the other hand, there exists cutoff frequencies for the higher-order hybrid modes $(m \geq 2)$. The cutoff frequencies are governed by the following equation:

$$\left(\frac{\omega^2}{\omega_\mathrm{p}^2}\right)_{\text{cutoff}} \cong \left(\frac{1 + v_z/c}{1 - v_z/c}\right) \left(\frac{\omega^2}{\omega_\mathrm{p}^2}\right)_{\text{cutoff for } v_z = 0 \text{ case}} \tag{14.64}$$

and

$$\left(\frac{\omega^2}{\omega_\mathrm{p}^2}\right)_{\text{cutoff for } v_z = 0 \text{ case}} = \left\{2 + \frac{P I_m(P)}{(m-1)\, I_{m-1}(P)}\right\}^{-1},$$

with

$$P = \frac{\omega_\mathrm{p} a}{c}.$$

It can be seen that when the plasma is moving in the direction of wave propagation, the cutoff frequencies for the moving plasma case are higher than those for the stationary plasma case, and when the plasma is moving in the opposite direction of wave propagation, the cutoff frequencies for the moving plasma case are lower than those for the stationary plasma case.

Another interesting property for the guided wave along a moving plasma column can be obtained from the dispersion relation. Unlike the stationary plasma waveguide case, in which all guided modes have an upper cutoff frequency, that is, $\omega = \omega_\mathrm{p}/\sqrt{2}$, all guided modes propagating along a moving plasma possess no upper cutoff frequency, provided that the plasma medium is moving in the same direction of the propagating modes. When the plasma medium is moving in the opposite direction of propagating modes, backward wave appears as the propagation constant βa increases and there exists a maximum value for βa. This point can best be indicated by the following numerical examples.

To obtain the propagation constant βa of a particular surface wave mode, m, it is necessary to first solve the transcendental (14.59) with (14.61) and (14.62) for the root $q'a$ for fixed values of ω/ω_p and $\omega_\mathrm{p} a/c$ and then substitute the resultant values for $q'a$, ω/ω_p, and $\omega_\mathrm{p} a/c$ in (14.63). Figures 14.7 and 14.8 are introduced, respectively, to illustrate how the propagation constants βa for the dominant circularly symmetric TM mode $(m = 0)$ and for the dipolar hybrid HE mode $(m = 1)$ vary as a function of ω/ω_p for various values of v_z/c, with $\omega_\mathrm{p} a/c = 1.0$. It can be seen from these figures that the general behavior of the ω/ω_p vs. βa curves for the $m = 0$ mode and the $m = 1$ mode are quite similar. At $v_z/c = 0$, as expected,

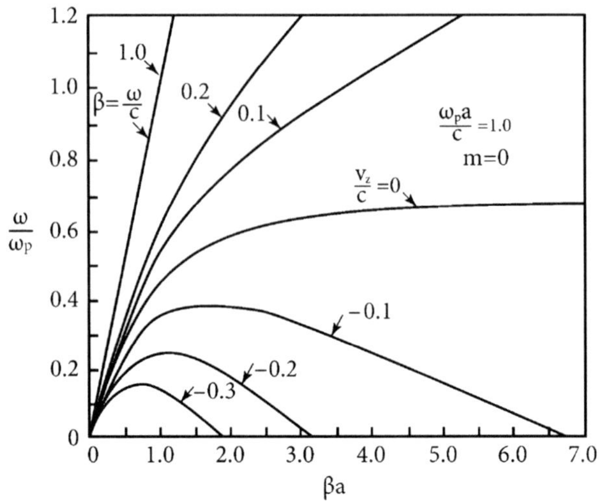

Figure 14.7. The ω-β diagram for the circularly symmetric TM mode ($m = 0$) propagating along a moving plasma column [2]

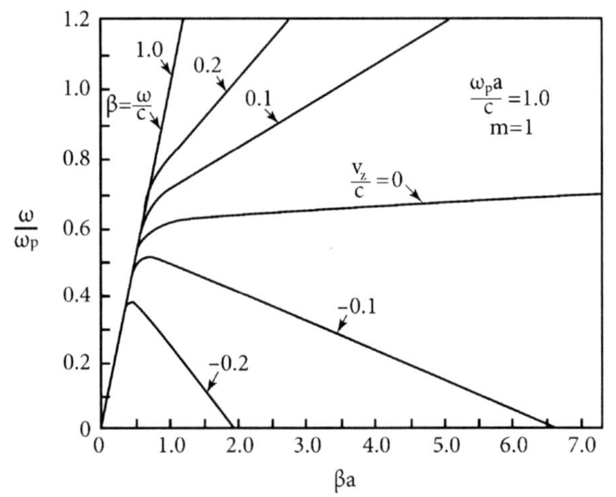

Figure 14.8. The ω-β diagram for the lowest-order hybrid HE mode ($m = 1$) propagating along a moving plasma column [2]

βa is zero for $\omega/\omega_p = 0$ and βa becomes infinite as ω/ω_p approaches asymptotically $1/\sqrt{2}$. However, when $v_z/c > 0$, the high frequency cutoff for $m = 0, 1$ mode disappears; βa increases without bounds as ω/ω_p increases. When $v_z/c = 1$, β becomes ω/c. In other words, when the plasma medium is moving in the direction of wave propagation, the propagation constant becomes closer to the free-space propagation constant or the surface wave becomes less attached to the plasma column. On the other hand, when the plasma medium is moving opposite to the direction of propagation of the surface wave (i.e., $v_z/c < 0$), in addition to the presence of a high-frequency cutoff, it is noted that the guiding structure can not only support a forward wave but also a backward wave. Furthermore, as v_z/c decreases from 0, the high-frequency cutoff decreases and when $v_z/c = 1.0$, no surface wave can exist along this structure. One also notes that the phase velocity for the backward wave is less than that for the forward wave and that below a certain frequency the group velocity for the backward wave is almost independent of frequency and approaches the speed of the moving plasma. The value for $\omega_{\mathrm{p}}a/c$ is so chosen that the appearance of backward wave is due solely to the movement of the plasma medium and not due to the intrinsic property of a stationary plasma column as a backward wave structure.

In conclusion, we note from this specific example that the movement of the plasma medium introduces many features, which cannot be found for the stationary plasma problem. It is expected that these unique features will also be present for the surface wave propagation along structures of different geometrical shapes, such as slabs. Furthermore, the case of the wave propagation along a moving plasma column with an impressed dc magnetic field in the axial direction can also be treated in a similar manner.

14.2 Anisotropic Medium Waveguides

Solutions are available for two special families of anisotropic medium surface waveguides; one, whose permittivity and permeability are

$$\epsilon = \epsilon_1, \tag{14.65}$$

$$\underline{\underline{\mu}} = \begin{bmatrix} \mu & -j\kappa & 0 \\ j\kappa & \mu & 0 \\ 0 & 0 & \mu_z \end{bmatrix}, \tag{14.66}$$

where ϵ_1, μ, and κ are given constants, represents a ferrite medium with an impressed static magnetic field in the axial direction, the propagation direction of the guided wave, and the other, whose permittivity and permeability are:

$$\underline{\underline{\epsilon}} = \begin{bmatrix} \epsilon_1 & j\epsilon_2 & 0 \\ j\epsilon_2 & \epsilon_1 & 0 \\ 0 & 0 & \epsilon_3 \end{bmatrix}, \tag{14.67}$$

$$\mu = \mu_0, \tag{14.68}$$

where ϵ_1, ϵ_2, ϵ_3, and μ_0, are given constants, represents a plasma medium with an impressed static magnetic field in the axial direction, the propagation direction of the guided wave. Let us derive first the wave equations for the anisotropic ferrite medium represented by (14.65) and (14.66).

Putting (14.65) and (14.66) into the constitutive equations yields

$$\mathbf{D} = \epsilon_1 \mathbf{E}, \tag{14.69}$$

$$\begin{bmatrix} B_x \\ B_y \\ B_z \end{bmatrix} = \begin{bmatrix} \mu H_x - j\kappa H_y \\ j\kappa H_x + \mu H_y \\ \mu_z H_z \end{bmatrix}, \tag{14.70}$$

using (14.69) and (14.70) and assuming all field components to vary as $e^{jwt - j\beta z}$, which is suppressed, we obtain the Maxwell equations as follows:

$$\frac{\partial E_z}{\partial y} + j\beta E_y = -j\omega\mu H_x - \omega\kappa H_y, \tag{14.71}$$

$$-\frac{\partial E_z}{\partial x} - j\beta E_x = \omega\kappa H_x - j\omega\mu H_y, \tag{14.72}$$

$$\frac{\partial E_y}{\partial x} - \frac{\partial E_x}{\partial y} = -j\omega\mu_z H_z, \tag{14.73}$$

$$\frac{\partial H_z}{\partial y} + j\beta H_y = j\omega\epsilon_1 E_x, \tag{14.74}$$

$$-\frac{\partial H_z}{\partial x} - j\beta H_x = j\omega\epsilon_1 E_y, \tag{14.75}$$

$$\frac{\partial H_y}{\partial x} - \frac{\partial H_x}{\partial y} = j\omega\epsilon_1 E_z. \tag{14.76}$$

Substituting E_x from (14.74) in (14.72) results in an equation in H_x, H_y, $\partial E_z/\partial x$, and $\partial H_z/\partial y$; substituting E_y from (14.75) in (14.71) results in an equation in H_x, H_y, $\partial E_z/\partial y$, and $\partial H_z/\partial x$. These two simultaneous equations are then solved for H_x and H_y in terms of $\partial E_z/\partial x$, $\partial E_z/\partial y$, $\partial H_z/\partial x$, and $\partial H_z/\partial y$. Finally, these results for H_x and H_y are substituted in (14.76), resulting in the following equation:

$$\nabla_{\mathrm{t}}^2 E_z + a_1 \nabla_{\mathrm{t}}^2 H_z + a_2 E_z = 0, \tag{14.77}$$

where ∇_{t} is the transverse del operator given by $\mathbf{e}_x\,\partial/\partial x + \mathbf{e}_y\,\partial/\partial y$, ∇_{t}^2 is given by $\partial^2/\partial x^2 + \partial^2/\partial y^2$, and a_1 and a_2 are known constants. Equations (14.71) and (14.74) are solved for H_x and H_y in terms of E_x, E_y, $\partial E_z/\partial y$, and $\partial E_z/\partial x$. Substituting the resultant H_x in (14.75) yields an equation in E_x, E_y, $\partial E_z/\partial y$, $\partial E_z/\partial x$, and $\partial H_z/\partial x$, and substituting the resultant H_y in (14.74) yields an equation in E_x, E_y, $\partial E_z/\partial y$, $\partial E_z/\partial x$, and $\partial H_z/\partial y$. From these two equations, we solve for E_x and E_y in terms of $\partial E_z/\partial x$, $\partial E_z/\partial y$, $\partial H_z/\partial x$, and $\partial H_z/\partial y$. The resultant E_x and E_y are substituted in (14.73) yielding

$$\nabla_{\mathrm{t}}^2 E_z + a_3 \nabla_{\mathrm{t}}^2 H_z + a_4 E_z = 0, \tag{14.78}$$

where a_3 and a_4 are known constants. Equations (14.77) and (14.78) can now be combined to give the following sought-after equations:

$$\nabla_{\mathrm{t}}^2 E_z + b_1 E_z + b_2 H_z = 0, \tag{14.79}$$

$$\nabla_{\mathrm{t}}^2 H_z + b_3 H_z + b_4 E_z = 0, \tag{14.80}$$

where

$$b_1 = p^2 - k_2^2 \frac{\kappa}{\mu}, \tag{14.81}$$

$$b_2 = \frac{j\omega\beta\mu_z\kappa}{\mu}, \tag{14.82}$$

$$b_3 = p^2 \frac{\mu_z}{\mu}, \tag{14.83}$$

$$b_4 = -\frac{j\omega\beta\epsilon_1\kappa}{\mu}, \tag{14.84}$$

$$p^2 = \omega^2 \epsilon_1 \mu - \beta^2, \tag{14.85}$$

$$k_2^2 = \omega^2 \epsilon_1 \kappa, \tag{14.86}$$

$$k^2 = \omega^2 \epsilon_1 \mu. \tag{14.87}$$

Following Kales [8], we can transform (14.79) and (14.80) into a pair of wave equations as follows:

$$E_z = s_1 u_1 + s_2 u_2, \tag{14.88}$$

$$H_z = \frac{(s_1 - b_1)\, s_1}{b_2} u_1 + \frac{(s_2 - b_1)}{b_2} u_2, \tag{14.89}$$

$$\mathbf{E}_\mathrm{t} = -j\beta\nabla_\mathrm{t}\,(u_1 + u_2) + \frac{\mu}{\beta\kappa}\mathbf{e}_z \times \nabla_\mathrm{t}\left[(s_1 - b_1)\, u_1 + (s_2 - b_1)\, u_2\right], \tag{14.90}$$

$$\mathbf{H}_\mathrm{t} = \frac{1}{\omega\kappa}\nabla_\mathrm{t}\left[\left(p^2 - s_1\right) u_1 + \left(p^2 - s_2\right) u_2\right] - j\omega\epsilon_1\mathbf{e}_z \times \nabla_\mathrm{t}\,(u_1 + u_2)\,, \tag{14.91}$$

where s_1 and s_2 are the roots of

$$(b_1 - s)\,(b_3 - s) - b_2 b_4 = 0 \tag{14.92}$$

and

$$\nabla_t^2 u_1 + s_1 u_1 = 0, \tag{14.93}$$

$$\nabla_t^2 u_2 + s_2 u_2 = 0. \tag{14.94}$$

The case for a cold plasma medium with an impressed axial static magnetic field can be treated in a similar manner, except the constitutive relations are given by

$$\mathbf{D} = \underline{\underline{\epsilon}} \cdot \mathbf{E}, \tag{14.95}$$

$$\mathbf{B} = \mu\,\mathbf{H}, \tag{14.96}$$

where $\underline{\underline{\epsilon}}$ and μ are given by (14.67) and (14.68). The form of the solution for the axial fields is the same as (14.88) and (14.89), except the constants s_1, s_2, b_1, and b_2 will be different. They must be derived from the Maxwell equations with the constitutive relations (14.95) and (14.96). The transverse fields may also be derived from the Maxwell equations.

Let us now consider the example of wave propagation along a ferrite cylinder with radius a, situated in free-space. A static magnetic field is assumed to be impressed along the axial direction. We are interested in deriving the characteristic equation for the hybrid $n = \pm 1$ modes.

The relevant solutions are
For region 1 $(r \leq a)$

$$u_1 = A_1 J_n \left(q_1 r\right) e^{jn\theta}, \tag{14.97}$$

$$u_2 = A_2 J_n \left(q_2 r\right) e^{jn\theta}, \tag{14.98}$$

$$n = 0, \ \pm 1, \ \pm 2, \ldots,$$

where $q_1 = \sqrt{s_1}$ and $q_2 = \sqrt{s_2}$, s_1, and s_2 are roots of (14.92). From (14.88)–(14.91), we obtain the field components

$$E_z^{(1)} = \left[s_1 A_1 J_n \left(q_1 r\right) + s_2 A_2 J_n \left(q_2 r\right)\right] e^{jn\theta}, \tag{14.99}$$

$$H_z^{(1)} = \left[\frac{s_1 - b_1}{b_2} A_1 J_n \left(q_1 r\right) + \frac{s_2 - b_1}{b_2} A_2 J_n \left(q_2 r\right)\right] e^{jn\theta}, \tag{14.100}$$

$$E_r^{(1)} = \left\{A_1 \left[-j\beta q_1 J_n'(q_1 r) - j\frac{\mu n}{\beta\kappa}\left(s_1 - b_1\right)\frac{1}{r} J_n\left(q_1 r\right)\right]\right.$$

$$\left. + A_2 \left[-j\beta q_2 J_n'\left(q_2 r\right) - j\frac{\mu n}{\beta\kappa}\left(s_2 - b_1\right)\frac{1}{r} J_n\left(q_2 r\right)\right]\right\} e^{jn\theta}, \tag{14.101}$$

$$E_\theta^{(1)} = \left\{A_1 \left[\frac{\beta n}{r} J_n\left(q_1 r\right) + \frac{\mu}{\beta\kappa}\left(s_1 - b_1\right) q_1 J_n'\left(q_1 r\right)\right]\right.$$

$$\left. + A_2 \left[\frac{\beta n}{r} J_n\left(q_2 r\right) + \frac{\mu}{\beta\kappa}\left(s_2 - b_1\right) q_2 J_n'\left(q_2 r\right)\right]\right\} e^{jn\theta}, \tag{14.102}$$

$$H_r^{(1)} = \left\{A_1 \left[\frac{\left(p^2 - s_1\right) q_1}{\omega\kappa} J_n'(q_1 r) - \frac{\omega\varepsilon_1 n}{r} J_n\left(q_1 r\right)\right]\right.$$

$$\left. + A_2 \left[\frac{\left(p^2 - s_2\right) q_2}{\omega\kappa} J_n'(q_2 r) - \frac{\omega\varepsilon_1 n}{r} J_n\left(q_2 r\right)\right]\right\} e^{jn\theta}, \tag{14.103}$$

$$H_{\theta}^{(1)} = \left\{ A_1 \left[\frac{jn(p^2 - s_1)}{r\omega\kappa} J_n(q_1 r) - j\omega\epsilon_1 q_1 J_n'(q_1 r) \right] \right.$$

$$\left. + A_2 \left[\frac{jn(p^2 - s_2)}{r\omega\kappa} J_n(q_2 r) - j\omega\epsilon_1 q_2 J_n'(q_2 r) \right] \right\} e^{jn\theta}, \quad (14.104)$$

where the prime signifies the derivative with respect to the argument.

For region 0 ($r \geq a$)

$$E_z^{(0)} = A_3 K_n(q_0 r)\, e^{jn\theta}, \quad (14.105)$$

$$H_z^0 = A_4 K_n(q_0 r)\, e^{jn\theta}, \quad (14.106)$$

$$E_r^{(0)} = A_3 \frac{j\beta}{q_0} K_n'(q_0 r) - A_4 \frac{\omega\mu_0 n}{q_0^2 r} K_n(q_0 r), \quad (14.107)$$

$$E_\theta^{(0)} = A_3 \frac{-\beta n}{q_0^2 r} K_n(q_0 r) - A_4 \frac{-j\omega\mu_0}{q_0} K_n'(q_0 r), \quad (14.108)$$

$$H_r^{(0)} = A_3 \frac{\omega\epsilon_0 n}{q_0^2 r} K_n(q_0 r) + A_4 \frac{j\beta}{q_0} K_n'(q_0 r), \quad (14.109)$$

$$H_\theta^{(0)} = A_3 \frac{j\omega\epsilon_0}{q_0} K_n'(q_0 r) - A_4 \frac{\beta n}{q_0^2 r} K_n(q_0 r), \quad (14.110)$$

with $q_0^2 = \beta^2 - k_0^2$, $k_0^2 = \omega^2\mu_0\epsilon_0$, and $K_n(q_0 r)$ is the modified Bessel function of the second kind of order n and argument $q_0 r$. Here, A_1, A_2, A_3, and A_4 are arbitrary constants.

Matching the tangential fields at $r = a$ yields

$$\begin{bmatrix} d_{11} & d_{12} & d_{13} & d_{14} \\ d_{21} & d_{22} & d_{23} & d_{24} \\ d_{31} & d_{32} & d_{33} & d_{34} \\ d_{41} & d_{42} & d_{43} & d_{44} \end{bmatrix} \begin{bmatrix} A_1 \\ A_2 \\ A_3 \\ A_4 \end{bmatrix} = 0, \quad (14.111)$$

where

$$d_{11} = s_1 J_n(q_1 a), \quad (14.112a)$$

$$d_{12} = s_2 J_n(q_2 a), \quad (14.112b)$$

$$d_{13} = -K_n(q_0 a), \quad (14.112c)$$

$$d_{14} = 0, \quad (14.112d)$$

$$d_{21} = \frac{s_1 - b_1}{b_2} J_n(q_1 a), \tag{14.112e}$$

$$d_{22} = \frac{s_2 - b_1}{b_2} J_n(q_2 a), \tag{14.112f}$$

$$d_{23} = 0, \tag{14.112g}$$

$$d_{24} = -K_n(q_0 a), \tag{14.112h}$$

$$d_{31} = \frac{\beta n}{a} J_n(q_1 a) + \frac{\mu}{\beta \kappa}(s_1 - b_1) q_1 J_n'(q_1 a), \tag{14.112i}$$

$$d_{32} = \frac{\beta n}{a} J_n(q_2 a) + \frac{\mu}{\beta \kappa}(s_2 - b_1) q_2 J_n'(q_2 a), \tag{14.112j}$$

$$d_{33} = \frac{\beta n}{q_0^2 a} K_n(q_0 a), \tag{14.112k}$$

$$d_{34} = \frac{-j\omega\mu_0}{q_0} K_n'(q_0 a), \tag{14.112l}$$

$$d_{41} = \frac{jn(p^2 - s_1)}{\omega\kappa a} J_n(q_1 a) - j\omega\epsilon_1 q_1 J_n'(q_1 a), \tag{14.112m}$$

$$d_{42} = \frac{jn(p^2 - s_2)}{\omega\kappa a} J_n(q_2 a) - j\omega\epsilon_1 q_2 J_n'(q_2 a), \tag{14.112n}$$

$$d_{43} = \frac{-j\omega\epsilon_0}{q_0} K_n'(q_0 a), \tag{14.112o}$$

$$d_{44} = \frac{\beta n}{q_0^2 a} K_n(q_0 a). \tag{14.112p}$$

Setting the determinant of the simultaneous equation to zero gives the characteristic equation for the hybrid HE_{nm} mode. This equation can be solved numerically. It is anticipated that, similar to the case of wave propagation in a circular finite rod in an enclosed conducting circular waveguide, "Faraday Effect" will also be exhibited for the $n = \pm 1$ modes here.

14.3 Metamaterial Artificial Dielectric Waveguides

Metamaterial is a material that possesses negative permittivity ($\epsilon < 0$) and/or negative permeability ($\mu < 0$). There is no known naturally occurring material for which ϵ and μ are simultaneously negative at the same frequency. However, they can be realized artificially. Waves in this material medium were first studied by Veselago in 1967 [9]. Then, Pendry et al. in 1999 [10] and Shelby et al. in 2001

[11] demonstrated experimentally the existence of this metamaterial. This metamaterial is also called the left-handed material with negative refractive index $(n < 0)$. Further noted is that in order for the condition $(\epsilon < 0)$ and $(\mu < 0)$ to exist, ϵ and μ must be dispersive [12].

14.3.1 Some Special Properties of Metamaterial [12]

Here we discuss a few special properties of metamaterial

14.3.1.1 If $\epsilon < 0$ and $\mu < 0$, Then $n < 0$

We prove that if ϵ is negative and μ is negative, the index of refraction n must also be negative. For all physical material ϵ, μ, or n must be a complex quantity, and so

$$\epsilon/\epsilon_0 = r_1\, e^{j\theta_1} = r_1 \cos\theta_1 + jr_1 \sin\theta_1, \tag{14.113}$$

$$\mu/\mu_0 = r_2\, e^{j\theta_2} = r_2 \cos\theta_2 + jr_2 \sin\theta_2, \tag{14.114}$$

where r_1, r_2, θ_1, and θ_2 are all real numbers. By definition,

$$\begin{aligned} n &= \left(\frac{\epsilon\mu}{\epsilon_0\mu_0}\right)^{1/2} = \sqrt{r_1 r_2}\, e^{j(\theta_1+\theta_2)/2} \\ &= \sqrt{r_1 r_2}\left[\cos\left(\frac{\theta_1+\theta_2}{2}\right) + j\,\sin\left(\frac{\theta_1+\theta_2}{2}\right)\right]. \end{aligned} \tag{14.115}$$

Here ϵ_0 and μ_0 are the free-space permittivity and permeability. For a passive medium, $\mathrm{Im}\,(n)$ must be positive, where Im means the imaginary part of, that is,

$$\sin\left(\frac{\theta_1+\theta_2}{2}\right) > 0 \text{ or } 0 \le \frac{1}{2}(\theta_1+\theta_2) < \pi. \tag{14.116}$$

If we require $\mathrm{Re}\,(\epsilon)$ and $\mathrm{Re}\,(\mu)$ to be negative, where Re means the real part of, then, from (14.113) and (14.114),

$$\cos\theta_1 < 0 \text{ or } \frac{\pi}{2} < \theta_1 < \frac{3\pi}{2}, \tag{14.117}$$

$$\cos\theta_2 < 0 \text{ or } \frac{\pi}{2} < \theta_2 < \frac{3\pi}{2}. \tag{14.118}$$

Substituting (14.117) and (14.118) in (14.116) yields

$$\frac{\pi}{2} < \frac{1}{2}(\theta_1+\theta_2) < \pi. \tag{14.119}$$

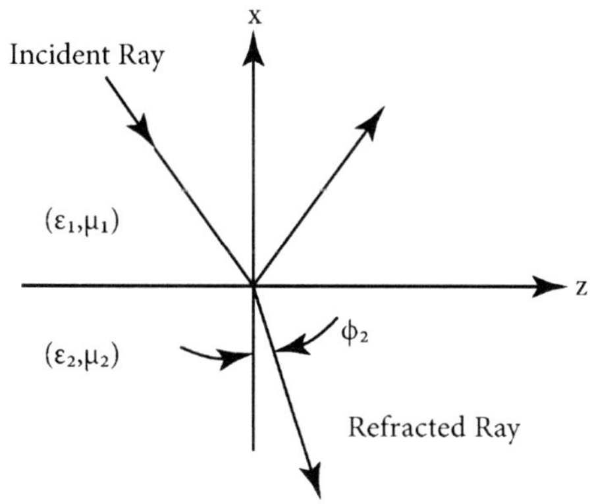

Figure 14.9. Traditional Snell's law for $n_1 > 0$ and $n_2 > 0$. ϕ_2 is positive

From (14.115), this means $\cos\left[(\theta_1 + \theta_2)/2\right]$ is negative, or, $\mathrm{Re}\,(n) < 0$, that is, the index of refraction must be negative. The fact that ϵ, μ, and n must be complex is consistent with the causality condition.

14.3.1.2 Snell's Law for $n < 0$

Given an interface shown in Fig. 14.9; The upper medium $(x > 0)$, medium 1 with (ϵ_1, μ_1), and the lower medium $(x < 0)$, medium 2 with (ϵ_2, μ_2). A plane wave (ray) is incident as shown from medium 1. According to the Snell's law

$$n_1 \sin\phi_1 = n_2 \sin\phi_2, \tag{14.120}$$

where ϕ_1 is the angle of incidence and ϕ_2 is the angle of refraction and $n_1 = \sqrt{(\epsilon_1\mu_1)/(\epsilon_0\mu_0)}$ and $n_2 = \sqrt{(\epsilon_2\mu_2)/(\epsilon_0\mu_0)}$. If $n_1 > 0$ and $n_2 > 0$, in the traditional way, ϕ_2 is always positive. If $n_1 > 0$ and $n_2 < 0$, as in the case of metamaterial medium, $\sin\phi_2 < 0$ or $\phi_2 < 0$. This situation is shown in Fig. 14.10. The ray is bent in a different direction.

14.3.1.3 Poynting's Vector and Wave Vector in Metamaterial

For a propagating plane wave, we have

$$\mathbf{E} = \mathbf{E}_0\, e^{-j\mathbf{k}.\mathbf{r}+j\omega t}, \tag{14.121}$$

$$\mathbf{H} = \mathbf{H}_0\, e^{-j\mathbf{k}.\mathbf{r}+j\omega t}, \tag{14.122}$$

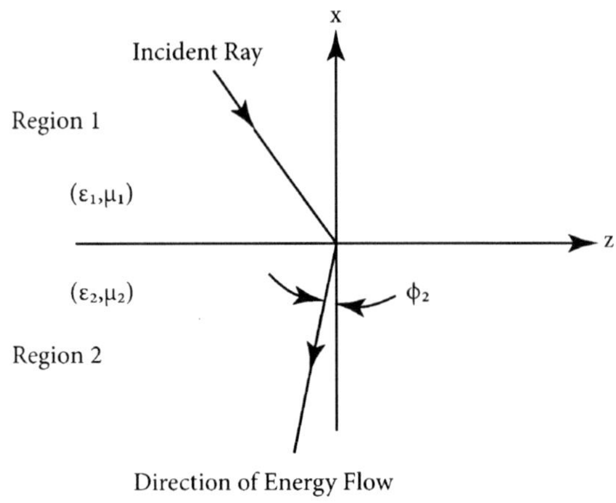

Figure 14.10. Metamaterial Snell's law for $n_1 > 0$ and $n_2 < 0$. Here, $\phi_2 < 0$

where $\mathbf{k}$ is the wave vector, E_0 and H_0 are the electric and magnetic vectors of the plane wave. From the Maxwell equations, for $\epsilon > 0,\ \mu > 0$

$$\mathbf{k} \times \mathbf{E}_0 = \omega\mu\mathbf{H}_0,$$

$$\mathbf{k} \times \mathbf{H}_0 = -\omega\epsilon\mathbf{E}_0, \tag{14.123}$$

$$\mathbf{S} = \mathbf{E}_0 \times \mathbf{H}_0 = \mathrm{S}_0\mathbf{k}.$$

For $\epsilon < 0,\ \mu < 0$

$$\mathbf{k} \times \mathbf{E}_0 = -\omega\left|\mu\right|\mathbf{H}_0,$$

$$\mathbf{k} \times \mathbf{H}_0 = \omega\left|\epsilon\right|\mathbf{E}_0, \tag{14.124}$$

$$\mathbf{S} = \mathbf{E}_0 \times \mathbf{H}_0 = -\mathrm{S}_0\mathbf{k}.$$

The above results show that in a traditional medium $(\epsilon,\ \mu > 0)$, the wave vector $\mathbf{k}$ and the vectors $\mathbf{E}$, $\mathbf{H}$ form a right-handed triad and the Poynting's vector points in the same direction as that of the wave vector, while in a metamaterial medium $(\epsilon,\ \mu < 0)$, the wave vector $\mathbf{k}$ and the vectors $\mathbf{E}$, $\mathbf{H}$ form a left-handed triad and the Poynting's vector points in an opposite direction as that of the wave vector (see Fig. 14.11).

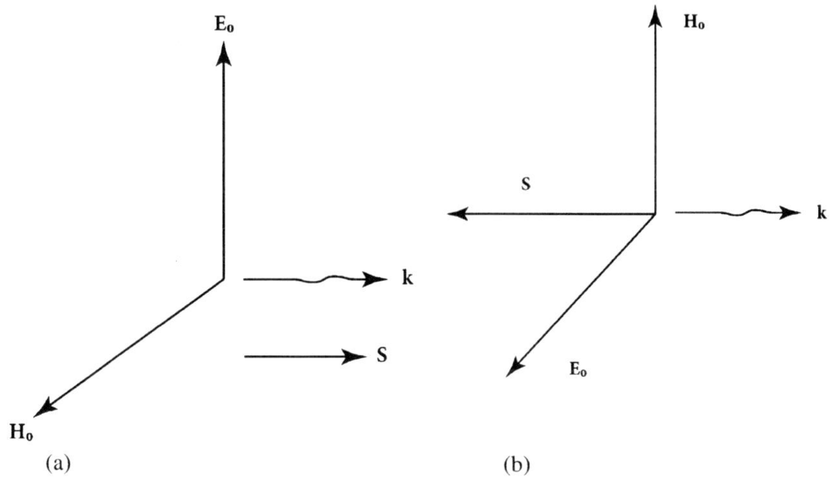

Figure 14.11. Poynting vector of plane wave in (**a**) traditional medium and (**b**) metamaterial medium

Let us now consider the field energy density of this plane wave in the metamaterial medium, from (2.79) and (2.81).

$$U = \frac{1}{4}\left[\epsilon \left|E_0\right|^2 + \mu \left|H_0\right|^2\right]. \tag{14.125}$$

Substituting $\epsilon = -\left|\epsilon\right|$ and $\mu = -\left|\mu\right|$ into (14.125) gives

$$U = \frac{1}{4}\left[-\left|\epsilon\right|\left|E_0\right|^2 - \left|\mu\right|\left|H_0\right|^2\right], \tag{14.126}$$

which is a negative quantity. This means that if the metamaterial medium is neither dispersive nor lossy, the total field energy would be negative, which is not allowed. Therefore, the condition $(\epsilon < 0)$ and $(\mu < 0)$ is not permitted if $(\epsilon,\ \mu)$ are not dispersive nor absorptive.

14.3.1.4 Fresnel Formulas

By carrying out the usual analysis of the problem of a plane wave obliquely incident onto the interface of two media characterized by $(\epsilon_{1,}\mu_1)$ and $(\epsilon_{2,}\mu_2)$ (see Fig. 14.9), we can obtain the following formulas for the transmitted (refracted) and reflected field. It has been assumed that the incident electric field is polarized perpendicular to the plane of incidence, that is, $\mathbf{E}^{(i)} = E^{(i)}\mathbf{e}_y$.

$$\frac{E_0^{(t)}}{E_0^{(i)}} = \frac{\dfrac{2n_1 \cos\phi_1}{\mu_1}}{\dfrac{n_1 \cos\phi_1}{\mu_1} + \dfrac{n_2}{\mu_2}\left(1 - \dfrac{n_1^2}{n_2^2}\sin^2\phi_1\right)^{1/2}} \tag{14.127}$$

and

$$\frac{E_0^{(r)}}{E_0^{(i)}} = \frac{\dfrac{n_1 \cos\phi_1}{\mu_1} - \dfrac{n_2}{\mu_2}\left(1 - \dfrac{n_1^2}{n_2^2}\sin^2\phi_1\right)^{1/2}}{\dfrac{n_1 \cos\phi_1}{\mu_1} + \dfrac{n_2}{\mu_2}\left(1 - \dfrac{n_1^2}{n_2^2}\sin^2\phi_1\right)^{1/2}}, \tag{14.128}$$

where $E_0^{(t)}$, $E_0^{(r)}$, and $E_0^{(i)}$ are, respectively, the amplitudes of the transmitted, reflected, and incident electric fields. The above formulas are valid for positive as well as negative index media, provided that $\left(n_1^2/n_2^2\right)\sin^2\phi_1 < 1$. For $\left(n_1^2/n_2^2\right)\sin^2\phi_1 > 1$, the expression

$$\left(1 - \frac{n_1^2}{n_2^2}\sin^2\phi_1\right)^{1/2} = \pm j\left(\frac{n_1^2}{n_2^2}\sin^2\phi_1 - 1\right)^{1/2}. \tag{14.129}$$

The $-$ sign is chosen because the transmitted field must not diverge at infinity for $n_2 > 0$. The + sign is chosen for $n_2 < 0$.

It is of interest to note that if $n_1 > 0$ and $n_2 < 0$ and if $\epsilon_2 = -\epsilon_1$ and $\mu_2 = -\mu_1$, then

$$\frac{E_0^{(r)}}{E_0^{(i)}} = 0. \tag{14.130}$$

We can also show that the same is true for the case where the incident plane wave is polarized in the plane of incidence. This means that no matter what the angle of incidence or what the field polarization is, there is no reflected field for this case. An interesting scenario can be envisioned:

Given a slab of negative index material with thickness d and $\epsilon = -\epsilon_0$, $\mu = -\mu_0$ as shown in Fig. 14.12, a point source is located at a distance $l < d$ from the surface of the slab. Two focal points exist, one inside the slab, F_1, and one outside the slab, F_2. According to (14.128), there is no reflected ray at either surface 1 or surface 2. The presence of F_2 on the other side of the slab means that an image of the original object at A will be formed at F_2. This means that the slab acts as a planar lens with no aberration [13].

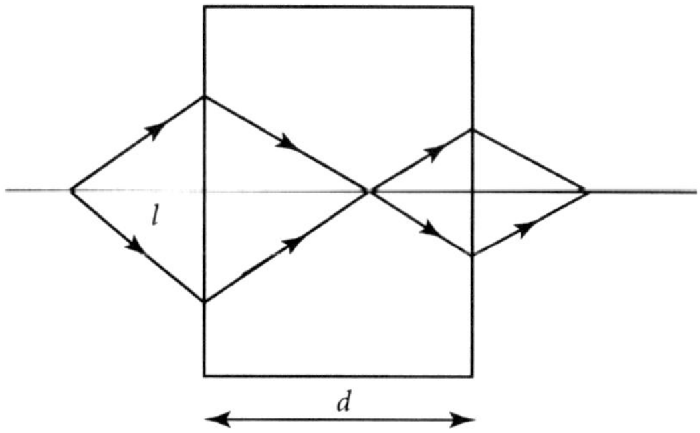

Figure 14.12. A planar slab of negative-index material of width d will produce two focal points, one inside the material $< d$ and the other outside when a point source is placed a distance $l < d$ from it. With $\epsilon = -\epsilon_0$ and $\mu = -\mu_0$, there are no reflected waves when the lens is in vacuum, and a real image appears at a distance $> d$ from the object

14.3.1.5 Formation of Metamaterials

Since there are no naturally occurring materials that possesses both negative permittivity and negative permeability simultaneously, they must be made artificially.

It has been shown by Pendry [10] that, at least in the microwave range, a composite structure consisting of periodically spaced thin wires and split ring resonators can have negative permittivity and negative permeability, resulting in a negative index of refraction. The artificial permittivity and permeability for this composite structure may be represented as follows:

$$\epsilon(\omega) = \epsilon_0 \left(1 - \frac{\omega_p^2}{\omega^2}\right), \tag{14.131}$$

$$\mu(\omega) = \mu_0 \left(1 - \frac{F_1\omega^2}{\omega^2 - \omega_0^2}\right), \tag{14.132}$$

where ω_p, F_1, and ω_0 are known constants, which are functions of the size, the geometry (shape), and the spacing of the structure. The existence of $\epsilon < 0$ and $\mu < 0$ over a common frequency range by this composite structure has been verified by experiments [14].

14.3.1.6 Cloaking with Metamaterial [14, 15]

Cloaking means that an object can be made invisible. In other words, by cloaking an object, an observer can see through the object and see the scene behind the now invisible object. Metamaterial can provide this cloaking possibility. This concept is

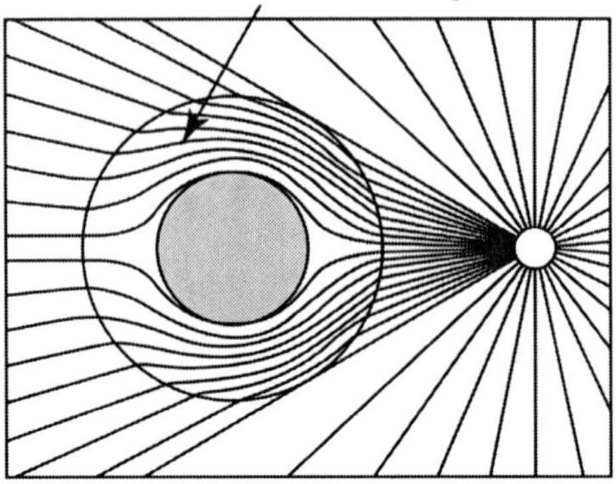

Figure 14.13. The rays go around the inner region and then go straight. An observer at the left of the cloak would see the point source

demonstrated in Fig. 14.13. It is seen that the rays can go around the metamaterial shell region and emerge as undisturbed rays. An observer at the left of cloaked object can see the point source and not the uncloaked object.

It has been shown theoretically by Pendry [16] that the concealed region of space must incorporate metamaterial that possesses the following characteristics: permittivity and permeability must be anisotropic and they must be equal.

Attempts to realize Pendry's cloaking idea have been initiated with encouraging results [14, 15].

14.3.2 Metamaterial Surface Waveguides [17–23]

Let us now consider the problem of surface wave propagation along a slab of metamaterial with thickness $2d$ and constitutive parameters (ϵ_2, μ_2) (see Fig. 14.14). The slab is immersed in a medium with parameters (ϵ_1, μ_1). Here, (ϵ_2, μ_2) may be both negative or positive. Following the same approach as given in Chap. 4, we can obtain the following results: all fields have the factor $e^{-j\beta z + j\omega t}$ attached and suppressed.

TE Modes

(a) The even (symmetric) case

Axial Fields:

$$\begin{aligned} H_z &= A_1 \sin p_2 x \quad (-d \le x \le d), \\ H_z &= A_1' \, e^{-q_1|x|} \quad (|x| \ge d), \end{aligned} \tag{14.133}$$

$$E_z = 0 \qquad \text{(for all } x\text{)} . \tag{14.134}$$

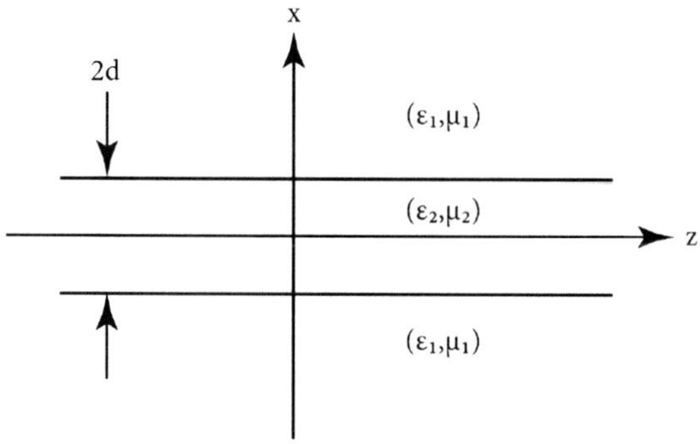

Figure 14.14. Geometry of metamaterial slab waveguide

Dispersion relation:

$$\tan(p_2 d) = \frac{\mu_2}{\mu_1}\frac{q_1 d}{p_2 d} \qquad (p_2 d \text{ is real}), \tag{14.135}$$

$$\tanh(\alpha_2 d) = -\frac{\mu_2}{\mu_1}\frac{q_1 d}{\alpha_2 d} \qquad (p_2 d = -j\alpha_2 d, p_2 d \text{ is imaginary}). \tag{14.136}$$

(b) The odd (antisymmetric) case

Axial Fields:

$$\begin{aligned} H_z &= A_2 \cos p_2 x \quad (-d \le x \le d), \\ H_z &= A_2' \, e^{-q_1|x|} \quad (|x| \ge d), \end{aligned} \tag{14.137}$$

$$E_z = 0 \qquad (\text{for all } x)\,. \tag{14.138}$$

Dispersion Relation:

$$\cot(p_2 d) = -\frac{\mu_2}{\mu_1}\frac{q_1 d}{p_2 d} \qquad (p_2 d \text{ is real}), \tag{14.139}$$

$$\coth(\alpha_2 d) = -\frac{\mu_2}{\mu_1}\frac{q_1 d}{\alpha_2 d} \qquad (p_2 d = -j\alpha_2 d,\ p_2 d \text{ is imaginary}). \tag{14.140}$$

TM mode

(a) The even (symmetric) case

$$\begin{aligned} E_z &= A_3 \sin p_2 x \quad (-d \le x \le d), \\ E_z &= A_3' \, e^{-q_1|x|} \quad (|x| \ge d), \end{aligned} \tag{14.141}$$

$$H_z = 0 \qquad (\text{for all } x)\,. \tag{14.142}$$

Dispersion Relation:

$$\tan(p_2 d) = \frac{\epsilon_2}{\epsilon_1}\frac{q_1 d}{p_2 d} \qquad (p_2 d \text{ is real}), \tag{14.143}$$

$$\tanh(\alpha_2 d) = -\frac{\epsilon_2}{\epsilon_1}\frac{q_1 d}{\alpha_2 d} \quad (p_2 d = -j\alpha_2 d,\ p_2 d \text{ is imaginary}). \tag{14.144}$$

(b) The odd (antisymmetric) case
Axial Fields:

$$\begin{aligned} E_z &= A_4 \cos p_2 x \quad (-d \le x \le d), \\ E_z &= A_4' \, e^{-q_1|x|} \quad (|x| \ge d), \end{aligned} \tag{14.145}$$

$$H_z = 0 \qquad (\text{for all } x)\,. \tag{14.146}$$

Dispersion Relation:

$$\cot(p_2 d) = -\frac{\epsilon_2}{\epsilon_1}\frac{q_1 d}{p_2 d} \quad (p_2 d \text{ is real}), \tag{14.147}$$

$$\coth(\alpha_2 d) = -\frac{\epsilon_2}{\epsilon_1}\frac{q_1 d}{\alpha_2 d} \quad (p_2 d = -j\alpha_2 d,\ p_2 d \text{ is imaginary}). \tag{14.148}$$

Here, $A_{1,2,3,4}$ and $A'_{1,2,3,4}$ are arbitrary constants. All transverse fields can be derived from (4.1)–(4.6) with the appropriate constants. The factors $q_1 d$, $p_2 d$ for each mode are related to βd, ω/c, $\epsilon_1\mu_1$, and $\epsilon_2\mu_2$ by the following relations:

$$(q_1 d)^2 = (\beta d)^2 - d^2\frac{\omega^2}{c^2}\epsilon_1\mu_1, \tag{14.149}$$

$$(p_2 d)^2 = d^2\frac{\omega^2}{c^2}\epsilon_2\mu_2 - (\beta d)^2 \quad (p_2 d \text{ is real}), \tag{14.150}$$

$$(\alpha_2 d)^2 = (\beta d)^2 - d^2\frac{\omega^2}{c^2}\epsilon_2\mu_2 \quad (p_2 d = j\alpha_2 d,\ p_2 d \text{ is imaginary}), \tag{14.151}$$

$$(q_1 d)^2 + (p_2 d)^2 = d^2\frac{\omega^2}{c^2}(\epsilon_2\mu_2 - \epsilon_1\mu_1) \quad (p_2 d \text{ is real}), \tag{14.152}$$

$$(q_1 d)^2 - (\alpha_2 d)^2 = d^2\frac{\omega^2}{c^2}(\epsilon_2\mu_2 - \epsilon_1\mu_1) \quad (p_2 d = j\alpha_2 d, p_2 d \text{ is imaginary}). \tag{14.153}$$

where $\omega/c = k_0$ is the free-space wave number, c is the speed of light in vacuum. To find the ω-β diagram for each mode, we must find the roots from the dispersion relation for a given $k_0 d = \omega d/c$. This scheme has been discussed in Chap. 4.

We note that the expressions (14.133)–(14.153) are valid for any values of ϵ_1, ϵ_2, μ_1, μ_2, and $(p_2 d)$ may be purely real or purely imaginary. It is important to point out that there is a great distinction in the guiding characteristics between the traditional surface wave structure for which ϵ_1, ϵ_2, μ_1, and μ_2 are all > 0 and the metamaterial surface wave structure for which $\epsilon_2 < 0$ and $\mu_2 < 0$, while $\epsilon_1 > 0$ and $\mu_1 > 0$. For the traditional case, (14.135), (14.139), (14.143), (14.147), (14.149), (14.150), and (14.152) apply and $p_2 d$ must be real to yield guided waves. For the metamaterial case, in addition to the above equations, (14.136), (14.140), (14.144), (14.148), (14.149), (14.151), and (14.153) also apply and $p_2 d$ may be purely real or purely imaginary to yield guided waves.

The case where $\epsilon_2 < 0$ and $\mu_2 < 0$ (the metamaterial slab case) will now be considered.

Case A $\quad \epsilon_2/\epsilon_0 = -2.5,\ \mu_2/\mu_0 = -0.5,\ \epsilon_1/\epsilon_0 = 1,\ \mu_1/\mu_0 = 1$

Figure 14.15 is introduced to illustrate how we may obtain the roots (solutions for the dispersion relations) for a metamaterial slab waveguide. As an example, we have chosen the TE odd modes and TM even modes. The $q_1 d$ vs. the $p_2 d$ curves for dispersion relation (14.139) and the $q_1 d$ vs. the $\alpha_2 d$ curves for dispersion relation (14.140) are shown in Fig. 14.15 (top). The accompanying (14.140) and (14.153) are also plotted in the figure for a given frequency ω/c. The medium parameters are assumed to be given. They are $\epsilon_2/\epsilon_0 = -2.5$, $\mu_2/\mu_0 = -0.5$, $\epsilon_1/\epsilon_0 = 1.0$, and $\mu_1/\mu_0 = 1.0$. The roots are the intersections of the curves for (14.139) and (14.152) and for (14.140) and (14.153). Similar results for the TM even modes are shown in Fig. 14.15 (bottom). Only positive values of $q_1 d$ are of interest in order to satisfy the radiation condition. The ω-β diagram, that is, β/k_0 vs. $k_0 d$ can be obtained using (14.150) or (14.151), where appropriate. Results are shown in Fig. 14.16 for the TE_1 mode. It is seen that this mode possesses a cutoff frequency and that there exists a small band of frequency around $k_0 d = 3$ where two types of waves can propagate: one with a negative group velocity and one with a positive group velocity.

Case B $\quad \epsilon_2/\epsilon_0 = -2,\ \mu_2/\mu_0 = -1.5,\ \epsilon_1/\epsilon_0 = 1,\ \mu_1/\mu_0 = 1$

The ω-β diagrams for TM_1, TM_2, TE_0, and TE_1 modes for this case are displayed in Fig. 14.17. The double solutions behavior is again observed in these figures. To observe the total modal power flow for TM and TE modes, Fig. 14.18 is introduced.

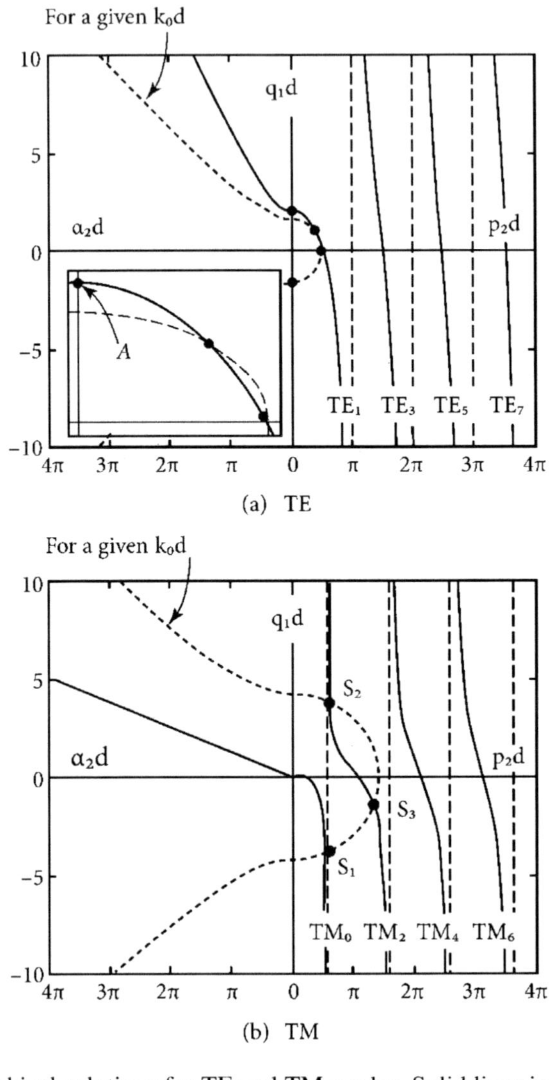

Figure 14.15. Graphical solutions for TE and TM modes. Solid lines in the first and fourth quadrants represent (14.139) or (14.143); solid lines in the second quadrant represent (14.140) or (14.144); dashed lines in the first and fourth quadrants represents (14.152); dashed lines in the second and third quadrants represents (14.153). The medium parameters are $\epsilon_2 = -2.5$, $\mu_2 = -0.5$, $\epsilon_1 = 1.0$, and $\mu_1 = 1.0$. Intersections of the dashed and solid curves represent the roots (or the solutions) of the dispersion relations. $k_0 d = (\omega/c)d$ [23]

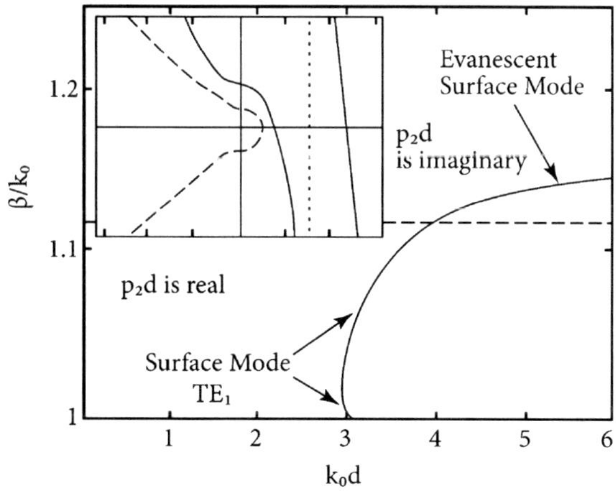

Figure 14.16. Dispersion curves for TE surface modes [23]

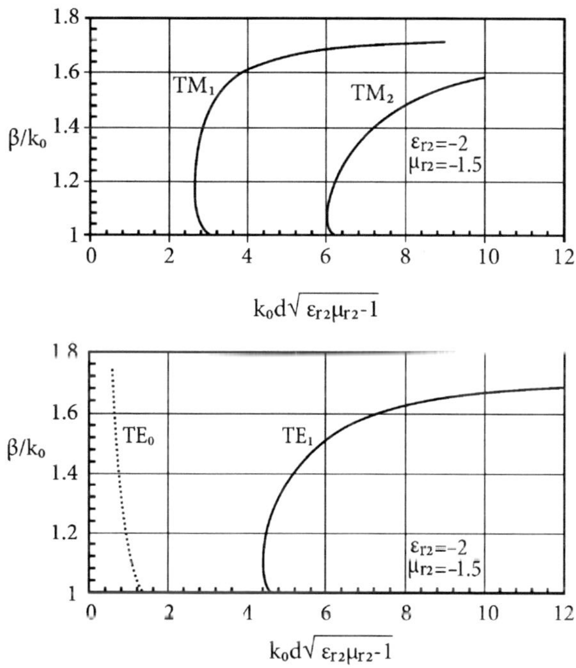

Figure 14.17. Dispersion curves for (*top*) TM modes and (*bottom*) TE modes [21]. Here, $V = k_0 d\sqrt{\epsilon_{r2}\mu_{r2} - 1}$. The TM_1 mode starts at some minimum, $V = V_{\min}$. Then for $V_{\min} < V < \pi$, there are two solutions for the TM_1 mode. One has a cutoff at $V = \pi$ and a negative group velocity (*lower branch*). The other exists at any value of $V > V_{\min}$ (*upper branch*). It is a tightly bound mode for large V. Similar behavior can be found for the TM_2 and TE_1 modes. The TE_0 mode is the dominant mode since it can exist in the range $1/|\mu_{r2}| < V < \pi/2$ (with $|\mu_{r2}| > 2/\pi$) [21]

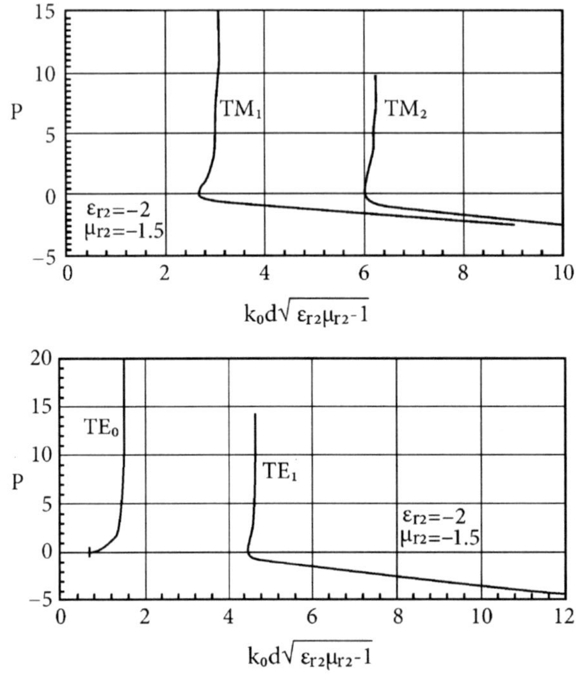

Figure 14.18. Total power flow of (*top*) TM modes and (*bottom*) TE modes vs. $V = k_0 d\sqrt{\epsilon_{r2}\mu_{r2} - 1}$. Note the TE_0 mode can carry forward (positive) power. One branch of the TE_1, TM_1, or TM_2 modes can carry a forward power with a cutoff V and another branch can carry a backward power with no high frequency cutoff [21]

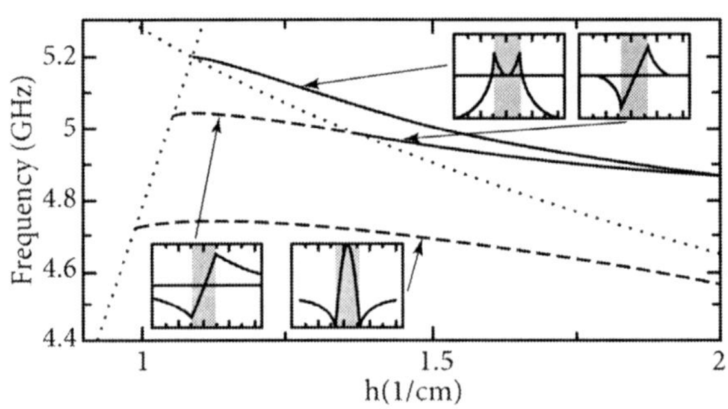

Figure 14.19. Frequency dispersion curves for three lower-order guided modes of the metamaterial slab waveguide (slab thickness = 4 cm). Insets show the characteristic mode profiles. Dashed curves refer to the region where $p_2 d$ is real. Solid curves refer to the region where $p_2 d$ is imaginary [18]

Case C ϵ_2, μ_2 *are dispersive and* $\epsilon_1/\epsilon_0 = 1, \mu_1/\mu_0 = 1$

From our previous discussion in Sect. 14.3.1.5, metamaterial can be realized only if the dispersive nature of ϵ_2 and μ_2 is taken into account. Using (14.131)–(14.132) for $\epsilon_2(\omega)$ and $\mu_2(\omega)$ with $\omega_p/2\pi = 10\,\text{GHz}$, $\omega_0/2\pi = 4\,\text{GHz}$, $F_1 = 0.56$, and a slab of thickness 4 cm, the ω-β diagrams for the first three modes are obtained and displayed in Fig. 14.19.

In conclusion, we note that wave propagation along structures with negative permittivity and negative permeability material provides many interesting features that are not present when the material possesses positive ϵ and μ. Here we have only provided a glimpse of what is to come based on the fundamental considerations.

References

1. C. Yeh, "Propagation along moving dielectric waveguides," J. Opt. Soc. Am. **58**, 767 (1968)
2. C. Yeh, "Wave propagation on a moving plasma column," J. Appl. Phys. **39**, 6112 (1968)
3. W. Pauli, "Theory of Relativity," Pergamon Press, New York (1958)
4. A. Sommerfeld, "Optik," Akademische Verlagsgesellschaft, Leipzig (1959)
5. J. A. Stratton, "Electromagnetic Theory," McGraw-Hill, New York (1941)
6. C. Yeh, "Reflection and transmission of electromagnetic waves by a moving plasma medium," J. Appl. Phys. **37**, 3079 (1966)
7. A. Sommerfeld, "Electrodynamics," Academic Press, New York (1952)
8. M. L. Kales, "Modes in waveguides containing ferrites," J. Appl. Phys. **24**, 604–608 (1953); H. Gamo, "The Faraday rotation of waves in a circular waveguide", J. Phys. Soc. Japan **8**, 176 (1953); A. A. Van Trier, "Guided electromagnetic waves in anisotropic media," Appl. Sci. Research **B3**, 305 (1953); H. Suhl and L. R. Walker, "Topics in guided wave propagation through gyromagnetic media," Bell Syst. Tech. J. **33**, 579, 939, 1133 (1954); W. E. Salmond and C. Yeh, "Ferrite filled elliptical waveguides I. Propagation Characteristics and II. Faraday Effects," J. Appl. Phys. **41**, 3210–3226 (1970)
9. V. G. Veselago, "The electrodynamics of substances with simultaneously negative values of permittivity and permeability," Soviet Phys. Uspekhi **10**, 509 (1968)
10. J. B. Pendry, A. J. Holden, D. J. Robbins and W. J. Stewart, "Magnetism from conductors and enhanced phenomena," IEEE Trans. Microw. Theory Tech. **47**, 2075 (1999)
11. R. A. Shelby, D. R. Smith, S. C. Nemat-Nasser, and S. Schultz, "Microwave transmission through a 2-D isotropic, left-handed metamaterial," Appl. Phys. Lett. **78**, 489 (2001)

12. P. W. Milloni, "Science in Optics and Optoelectronics," Inst. of Physics Publishing, Temple Back, UK (2005)

13. J. B. Pendry, "Negative refraction makes a perfect lens," Phys. Rev. Lett. **85**, 3966 (2000)

14. D. Schurig, J. J. Mock, B. J. Justice, S. A. Cummer, J. B. Pendry, A. F. Starr, and D. R. Smith, "Metamaterial electromagnetic cloak at microwave frequencies," Science **314**, 977 (2006)

15. D. Felbacq, "Envisioning invisibility: Recent advances in cloaking," Optics and Photonics News 32–37 (2007)

16. J. B. Pendry, D. Schurig, and D. R. Smith, "Controlling electromagnetic fields", Science **312**, 1780 (2006)

17. I. V. Lindell and S. Ilvonen, "Waves in a slab of uniaxial BW medium," J. Electromagn. Waves Appl. (JEMWA) **16**, 303 (2002)

18. I. V. Shadrivov, A. A. Sukhorukov, and Y. Kivshar, "Guided modes in negative-refractive-index waveguides," Phys. Rev. E **67**, 057602 (2003)

19. H. Cory and A. Barger, "Surface wave propagation along a metamaterial slab," Microw. Opt. Tech. Lett. **38**, 392 (2003)

20. A. Alu and N. Engheta, "Guided modes in a waveguide filled with a pair of single-negative (SNG), double-negative (DNG) and/or double-positive (DPS) layers," IEEE Trans. Microw. Theory Tech. **52**, 199 (2004)

21. S. F. Mahmoud and A. V. Viitanen, "Surface wave character on a slab of metamaterial with negative permittivity and permeability," Prog. Electromagn. Res. **51**, 127 (2005)

22. B. I. Wu, T. M. Grzegorczyk, Y. Zhang, and J. A. Kong, "Guided modes with imaginary transverse wave number in a slab waveguide with negative permittivity and permeability," J. Appl. Phys. **93**, 9386–9388 (2003)

23. W. Shu and J. M. Song, "Wave Propagation in Grounded Dielectric Slabs with Negative Metamaterials," Progress in Electromagnetics symposium, Cambridge, MA, 246–250 (22–29 March 2006)

15

SELECTED NUMERICAL APPROACHES

Although the principal emphasis of this book is on the analytical solutions to canonical dielectric waveguide problems, the capability of numerical approaches to yield useful data for problems that had no analytical solutions must be recognized. The steady increase of computing power since 1960 has provided the impetus in increasing the use of numerical means to solve electromagnetic problems. In 1966, Yee [1] demonstrated the use of the finite difference numerical method (FD) to solve boundary value problems involving the Maxwell equations in isotropic media. Goell [2] in 1969 presented his circular-harmonic computer analysis to treat dielectric shapes that are close to a circle. The radially inhomogeneous circular dielectric waveguide problem was solved in 1973 by Dil and Blok [3] using the numerical integration technique, and in 1977 by Yeh and Lindgren [4] using the matrix multiplication technique. The finite element method (FEM) was first used by Yeh et al. [5] in 1975 to solve an arbitrarily shaped inhomogeneous optical fiber or integrated optical waveguide. An improved version was given by Yeh et al. [6] in 1979. Meanwhile, a forward marching, split-step fast Fourier transform technique, also called the beam propagation method (BPM[1]), was used successfully to treat the problem of wave propagation in a fiber with radially inhomogeneous index variation by Yeh et al. [7] in 1977. A few months later, Feit and Fleck [8] popularized this BPM approach. Subsequently, Yeh et al. [9–11], also showed how this BPM may be used to solve the single mode or multi-mode inhomogeneous fiber coupler problems, the fiber branches, tapers, or horns problem and fibers with longitudinal index variations. The BPM provides good results provided that any longitudinal reflections due to longitudinal variations may be ignored.

[1]The popularity of the BPM name appears to obscure the true origin of this technique, namely, BPM is identical to the well-known split-step Fourier transform method. BPM is simply a new name, not a new technique.

Starting circa 1980, the availability of inexpensive computing power encouraged rapid adaptation of numerical techniques to solve electromagnetic problems, resulting in the commercialization of computer programs based on the above mentioned techniques, such as FDTD, FEM, and BPM. Indeed, they are all commercially available now [12].

Other numerical techniques, such as the method of moments (MOM) [13], method of lines (MOL) [14], transmission-line method (TLM) [15], and extended boundary condition method (EBCM) [16, 17], have also been successfully developed to solve the Maxwell equations.

Here, instead of discussing in detail all available numerical techniques that have been used to solve the guided wave problems,[2] we shall provide illustrations to a few numerical techniques that can be used efficiently to yield the needed results for a family of guided wave problems of interest. Emphasis will be placed on the results from FEM, BPM, and FDTD techniques.

15.1 Outer Radiation Boundary Condition (ORBC) for Computational Space

All computational space is necessarily bounded. For calculations involving enclosed structures, such as, waveguides or cavity resonators enclosed by metallic walls, this condition would not present a problem. On the other hand, when open structures, such as antennas, scatterers, or surface dielectric waveguides, are involved, all external fields must satisfy the free-space radiation condition. This condition demands that no backscattered or reflected field could be generated from the outer computational boundary. To simulate this condition, the computational space should be made so large that it could emulate free-space or the outer boundary of the computational space is made into a perfectly absorbing boundary. One of the important goals of any computational technique is to minimize any backscattering or reflection from the outer computational boundary.

Many schemes have been proposed and tried with varying degrees of success. One of the most effective ones is that by Mur, called the Mur absorbing boundary [18]. His scheme is applicable to all the numerical techniques that we discuss in this chapter.

15.2 Finite Element Method (FEM) [5, 6]

The finite element method has been used successfully to treat many complicated structural and continuum mechanics problems. In 1975, Yeh et al. showed that this technique may be used to find the numerical solution to the problem of wave propagation along arbitrarily shaped inhomogeneous dielectric waveguides [5]. Later, an

[2]Detailed discussion will be beyond the scope of this text. Excellent references for these techniques are available.

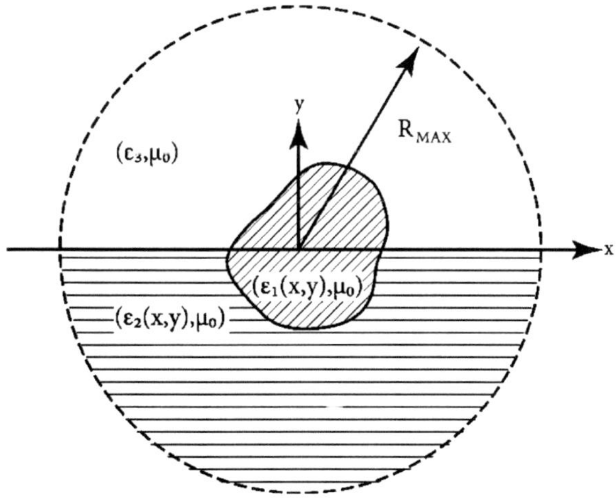

Figure 15.1. Generalized geometry of the problem

improved version was given by Yeh's group [6]. Since this approach is based on the numerical solution of the complete Maxwell equations with appropriate boundary conditions, the results are exact (to within the allowed numerical accuracy). A generalized geometry of the problem is shown in Fig. 15.1. There, an arbitrarily shaped inhomogeneous dielectric waveguide is immersed in several dielectrics. The guide is assumed to be uniform along its longitudinal z-axis. The dielectric inhomogeneity of the guide is a function of the transverse coordinates, that is, $\epsilon_1 = \epsilon_1(x, y)$, where ϵ_1 is the permittivity of the core region. The magnetic permeabilities of all regions are taken to be identical and equal to the free space value, μ_0. It is assumed that all fields outside the region bounded by the dashed boundary have decayed sufficiently that for all practical purposes no guided energy exists outside this region (see Fig. 15.1).

The governing equations for the longitudinal fields ($E_z^{(p)}$, $H_z^{(p)}$) of a guided wave propagating along the z-direction are

$$\left(\nabla_t + k_p^2\right) \left\{ \begin{array}{c} E_z^{(p)} \\ H_z^{(p)} \end{array} \right\} = 0, \tag{15.1}$$

where ∇_t is the transverse del operator,

$$k_p^2 = \frac{\omega^2}{c^2} \frac{\epsilon_p}{\epsilon_0} - \beta^2 \tag{15.2}$$

and the superscript (p) as well as the subscript p represent the pth region of the guide. The cross section of the guide is divided into P regions. In each of these regions, the medium is assumed to be uniform having a constant dielectric permittivity ϵ_p. The factor $e^{j\omega t - j\beta z}$ is assumed for all field components and suppressed throughout; ω is the frequency, β is the propagation constant, and c is the speed of light in vacuum. At the interfaces of the regions, appropriate boundary conditions must be satisfied by the fields, that is, the tangential electric and magnetic fields, which are $E_z^{(p)}$, $H_z^{(p)}$ $\tau_p \left[\gamma \left(\epsilon_0/\mu_0\right)^{1/2} \partial E_z^{(p)}/\partial s - \partial H_z^{(p)}/\partial n\right]$, and $\tau_p \left[\left(\epsilon_p/\epsilon_0\right)\left(\epsilon_0/\mu_0\right)^{1/2}(1/\gamma)\partial E_z^{(p)}/\partial n + \partial H_z^{(p)}/\partial s\right]$, with

$$\tau_p = \frac{\gamma^2\text{-}1}{\gamma^2 - \dfrac{\epsilon_p}{\epsilon_0}}, \tag{15.3}$$

$$\gamma = \frac{\beta c}{\omega}, \tag{15.4}$$

must be continuous from one region to another. The symbols $\mathbf{s}$ and $\mathbf{n}$ are, respectively, the tangential and normal directions along an interface and $\mathbf{n} \times \mathbf{s} = \mathbf{e}_z$, where $\mathbf{e}_z$ is the unit vector along the z-axis. All transverse field components are derivable from the longitudinal electric and magnetic fields as follows:

$$E_x^{(p)} = -\frac{j\omega\mu_0}{k_p^2}\left[\frac{\partial H_z^{(p)}}{\partial y} + \gamma\left(\frac{\epsilon_0}{\mu_0}\right)^{1/2}\frac{\partial E_z^{(p)}}{\partial x}\right],$$

$$E_y^{(p)} = -\frac{j\omega\mu_0}{k_p^2}\left[-\frac{\partial H_z^{(p)}}{\partial x} + \gamma\left(\frac{\epsilon_0}{\mu_0}\right)^{1/2}\frac{\partial E_z^{(p)}}{\partial y}\right],$$

$$H_x^{(p)} = -\frac{j\omega\epsilon_0}{k_p^2}\left[-\frac{\epsilon_p}{\epsilon_0}\frac{\partial E_z^{(p)}}{\partial y} + \gamma\left(\frac{\mu_0}{\epsilon_0}\right)^{1/2}\frac{\partial H_z^{(p)}}{\partial x}\right],$$

(1)

$$H_y^{(p)} = -\frac{j\omega\epsilon_0}{k_p^2}\left[\frac{\epsilon_p}{\epsilon_0}\frac{\partial E_z^{(p)}}{\partial x} + \gamma\left(\frac{\mu_0}{\epsilon_0}\right)^{1/2}\frac{\partial H_z^{(p)}}{\partial y}\right]. \tag{15.5}$$

According to the well-known Euler theorem, if the integral

$$I\left(E_z^{(p)}, H_z^{(p)}\right) = \sum_{p=1}^{P} \int\!\!\int_{S_p} f_p \left(x, y, E_z^{(p)}, H_z^{(p)}, \frac{\partial E_z^{(p)}}{\partial x}, \frac{\partial H_z^{(p)}}{\partial x}, \frac{\partial E_z^{(p)}}{\partial y}, \frac{\partial H_z^{(p)}}{\partial y} \right) dx dy \tag{15.6}$$

is to be minimized, then the necessary and sufficient condition for this minimum to be reached is that the unknown functions $E_z^{(p)}(x, y)$ and $H_z^{(p)}(x, y)$ should satisfy the following differential equations:

$$\frac{\partial}{\partial x}\left[\frac{\partial f_p}{\partial\left(\partial E_z^{(p)}/\partial x\right)}\right] + \frac{\partial}{\partial y}\left[\frac{\partial f_p}{\partial\left(\partial E_z^{(p)}/\partial y\right)}\right] - \frac{\partial f_p}{\partial\left(E_z^{(p)}\right)} = 0,$$

$$\frac{\partial}{\partial x}\left[\frac{\partial f_p}{\partial\left(\partial H_z^{(p)}/\partial x\right)}\right] + \frac{\partial}{\partial y}\left[\frac{\partial f_p}{\partial\left(\partial H_z^{(p)}/\partial y\right)}\right] - \frac{\partial f_p}{\partial\left(H_z^{(p)}\right)} = 0 \tag{15.7}$$

within the same region, provided $\left(E_z^{(p)}, H_z^{(p)}\right)$ satisfy the same boundary conditions in both the cases. One can show simply that the equivalent formulation to that of (15.1) and the required boundary conditions is the requirement that the surface integral given below and taken over the whole region should be minimized [19].

$$\begin{aligned} I = \sum_{p=1}^{P} \int\!\!\int_{S_p} \Bigg\{ & \tau_p \left|\nabla H_z^{(p)}\right|^2 + \gamma^2 \tau_p \frac{\epsilon_p}{\epsilon_0} \left| \frac{1}{\gamma} \left(\frac{\epsilon_0}{\mu_0}\right)^{1/2} \nabla E_z^{(p)} \right|^2 \\ & + 2\tau_p \mathbf{e}_z \cdot \left[\frac{1}{\gamma} \left(\frac{\epsilon_0}{\mu_0}\right)^{1/2} \nabla E_z^{(p)} \times \nabla H_z^{(p)} \right] \\ & - \left(\frac{\omega}{c}\right)^2 \times \left[H_z^{(p)2} + \gamma^2 \frac{\epsilon_p}{\epsilon_0} \left(\frac{1}{\gamma} \left(\frac{\epsilon_0}{\mu_0}\right)^{1/2} E_z^{(p)} \right)^2 \right] \Bigg\} dx dy, \end{aligned} \tag{15.8}$$

with $\left(E_z^{(p)}, H_z^{(p)}\right)$ obeying the same boundary conditions. Hence the Euler equations of the variational statement are

$$\delta I = 0. \tag{15.9}$$

Equation (15.9) provides the basis for generating the elemental waveguide characteristics. Let N_p denote the number of triangular subregions in a uniform dielectric region S_p and let $I_{p(n)}$ denote the integral in (15.8) for the nth triangle in S_p. Then (15.9) takes the slightly modified form

$$\delta I = \delta \sum_{p=1}^{P} \left(\sum_{n}^{N_p} I_{p(n)} \right) = 0, \tag{15.10}$$

where

$$\begin{aligned} I_{p(n)} = \int\!\!\int_{S_{p(n)}} \Bigg\{ & \tau_p \left|\nabla G^{(p)}\right|^2 + \gamma^2 \tau_p \frac{\epsilon_p}{\epsilon_0} \left|\nabla F^{(p)}\right|^2 \\ & + 2\gamma^2 \tau_p \mathbf{e}_z \cdot \left(\nabla F^{(p)} \times \nabla G^{(p)}\right) \\ & - \left(\frac{\omega}{c}\right)^2 \times \left[G^{(p)2} + \gamma^2 \frac{\epsilon_p}{\epsilon_0} F^{(p)2} \right] \Bigg\} \, dS_{p(n)}, \end{aligned}$$

with

$$\begin{aligned} G^{(p)} &= H_z^{(p)}, \\ F^{(p)} &= \frac{1}{\gamma} \left(\frac{\epsilon_0}{\mu_0}\right)^{1/2} E_z^{(p)}, \end{aligned}$$

where $S_{p(n)}$ is the triangular subregion of the nth element in regions S_p. In the present development, approximate longitudinal electric and magnetic fields, which are linear over the triangular region, will be expressed in terms of their values at the node. These approximate fields will be sufficient to assume interelement continuity. It is noted that although higher-order approximations for these fields can be taken, requiring a consideration of more generalized coordinates, it was not pursued here because the linear approximation, when used in conjunction with the efficient algebraic eigensolution technique, provides more than sufficient accuracy in the analysis.

To facilitate the development, let triangular coordinates $\xi_j (j = 1, 2, 3)$ be introduced as shown in Fig. 15.2. The coordinates ξ_j of a point $P = P(\xi_1, \xi_2, \xi_3)$ are defined as follows:

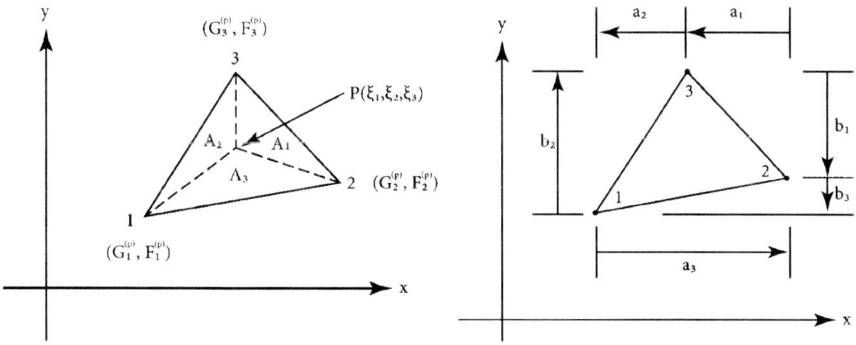

Figure 15.2. The triangular coordinates

$$\xi_j = \frac{A_j}{A}, \tag{15.11}$$

where A_j is the area of the subtriangle subtending the jth vertex and A is the total triangular area. Since $A = A_1 + A_2 + A_3$, ξ_j is constrained by the relation

$$\xi_1 + \xi_2 + \xi_3 = 1. \tag{15.12}$$

Introducing the notations

$$\{\xi^*\} = \begin{bmatrix} \xi_1 \\ \xi_2 \\ \xi_3 \end{bmatrix} \quad \text{and} \quad \{\xi^*\}^T = \langle \xi_1, \xi_2, \xi_3 \rangle \tag{15.13}$$

and assuming (x_j, y_j) as the Cartesian coordinates of the jth vertex of the triangle, one can show that $\{\xi^*\}$ is related to the Cartesian coordinates (x, y) through the relation

$$\{\xi^*\} = \frac{1}{2A} \begin{bmatrix} 2A_{23} & b_1 & a_1 \\ 2A_{31} & b_2 & a_2 \\ 2A_{12} & b_3 & a_3 \end{bmatrix} \begin{bmatrix} 1 \\ x \\ y \end{bmatrix}, \tag{15.14}$$

where

$$\begin{aligned} &A_{mn} = x_m y_n - x_n y_m \quad (m, n = 1, 2, 3), \\ &b_1 = y_2 - y_3 \qquad b_2 = y_3 - y_1 \qquad b_3 = y_1 - y_2, \\ &a_1 = x_3 - x_2 \qquad a_2 = x_1 - x_3 \qquad a_3 = x_2 - x_1. \end{aligned} \tag{15.15}$$

Denoting $G_m^{(p)}$ and $F_m^{(p)}$ $(m = 1, 2, 3)$ as the vertex values of the normalized longitudinal magnetic and electric fields, the approximate linear fields within a triangle can be stated in terms of the following interpolation formula:

$$\begin{aligned} G^{(p)}(x, y) &= \left\{G_m^{(p)}\right\}^T \{\xi^*\}, \\ F^{(p)}(x, y) &= \left\{F_m^{(p)}\right\}^T \{\xi^*\}, \end{aligned} \tag{15.16}$$

where

$$\begin{aligned} \left\{G_m^{(p)}\right\}^T &= \left\langle G_1^{(p)}, G_2^{(p)}, G_3^{(p)}\right\rangle, \\ \left\{F_m^{(p)}\right\}^T &= \left\langle F_1^{(p)}, F_2^{(p)}, F_3^{(p)}\right\rangle. \end{aligned} \tag{15.17}$$

The following derivatives of the field variables will be needed in subsequent derivations:

$$\begin{aligned} \frac{\partial G^{(p)}}{\partial x} &= \left\{G_m^{(p)}\right\}^T \left\{\frac{\partial \xi^*}{\partial x}\right\}, \\ \frac{\partial G^{(p)}}{\partial y} &= \left\{G_m^{(p)}\right\}^T \left\{\frac{\partial \xi^*}{\partial y}\right\}, \\ \frac{\partial F^{(p)}}{\partial x} &= \left\{F_m^{(p)}\right\}^T \left\{\frac{\partial \xi^*}{\partial x}\right\}, \\ \frac{\partial F^{(p)}}{\partial y} &= \left\{F_m^{(p)}\right\}^T \left\{\frac{\partial \xi^*}{\partial y}\right\}, \end{aligned} \tag{15.18}$$

where $\partial\xi^*/\partial x$ and $\partial\xi^*/\partial y$ can be obtained directly from (15.14); they are

$$\left\{\frac{\partial \xi^*}{\partial x}\right\} = \frac{1}{2A}\begin{bmatrix} b_1 \\ b_2 \\ b_3 \end{bmatrix}, \quad \left\{\frac{\partial \xi^*}{\partial y}\right\} = \frac{1}{2A}\begin{bmatrix} a_1 \\ a_2 \\ a_3 \end{bmatrix}. \tag{15.19}$$

The various terms in (15.10) can now be expanded in terms of the nodal values. For example,

$$\int\!\!\int_{S_{p(n)}} \tau_p \left|\nabla G^{(p)}\right|^2 dS_{p(n)} = \tau_p \left\{G_m^{(p)}\right\}^T [I_1] \left\{G_m^{(p)}\right\},$$

$$\int\int_{S_{p(n)}} \gamma^2 \tau_p \frac{\epsilon_p}{\epsilon_0} \left|\nabla F^{(p)}\right|^2 dS_{p(n)} = \gamma^2 \tau_p \frac{\epsilon_p}{\epsilon_0} \left\{F_m^{(p)}\right\}^T [I_1] \left\{F_m^{(p)}\right\},$$

$$\int\int_{S_{p(n)}} 2\gamma^2 \tau_p \mathbf{e}_z \cdot \left(\nabla F^{(p)} \times \nabla G^{(p)}\right) dS_{p(n)}$$
$$= 2\gamma^2 \tau_p \left(\left\{F_m^{(p)}\right\}^T [I_2] \left\{G_m^{(p)}\right\} - \left\{G_m^{(p)}\right\}^T [I_2] \left\{F_m^{(p)}\right\}\right),$$

$$-\left(\frac{\omega}{c}\right)^2 \int\int_{S_{p(n)}} \left|G^{(p)}\right|^2 dS_{p(n)} = -\left(\frac{\omega}{c}\right)^2 \left\{G_m^{(p)}\right\}^T [I_3] \left\{G_m^{(p)}\right\},$$

$$-\left(\frac{\omega}{c}\right)^2 \int\int_{S_{p(n)}} \gamma^2 \frac{\epsilon_p}{\epsilon_0} \left|F^{(p)}\right|^2 dS_{p(n)} = -\left(\frac{\omega}{c}\right)^2 \gamma^2 \frac{\epsilon_p}{\epsilon_0} \left\{F_m^{(p)}\right\}^T [I_3] \left\{F_m^{(p)}\right\}, \tag{15.20}$$

with

$$[I_1] = \int\int_{S_{p(n)}} \left[\left\{\frac{\partial \xi^*}{\partial x}\right\}\left\{\frac{\partial \xi^*}{\partial x}\right\}^T + \left\{\frac{\partial \xi^*}{\partial y}\right\}\left\{\frac{\partial \xi^*}{\partial y}\right\}^T\right] dS_{p(n)},$$

$$[I_2] = \int\int_{S_{p(n)}} \left[\left\{\frac{\partial \xi^*}{\partial x}\right\}\left\{\frac{\partial \xi^*}{\partial y}\right\}^T - \left\{\frac{\partial \xi^*}{\partial y}\right\}\left\{\frac{\partial \xi^*}{\partial x}\right\}^T\right] dS_{p(n)},$$

$$[I_3] = \int\int_{S_{p(n)}} \left[\{\xi^*\}\{\xi^*\}^T\right] dS_{p(n)}. \tag{15.21}$$

These integrals can be readily evaluated. Evaluating the expressions $I_{p(n)}$ in (15.10), using the expressions in (15.20), gives

$$I_{p(n)} = \{\theta_n\}^T [A_n] \{\theta_n\} - \frac{\omega^2}{c^2} \{\theta_n\}^T [B_n] \{\theta_n\}, \tag{15.22}$$

where $\{\theta_n\}$ is an ordered array of nodal values of $G^{(p)}$ and $F^{(p)}$ and

$$\{\theta_n\}^T = \left\{G_1^{(p)}, F_1^{(p)}, G_2^{(p)}, F_2^{(p)}, G_3^{(p)}, F_3^{(p)}\right\}$$

and $[A_n]$ and $[B_n]$ are symmetric matrices comprising the integrals in (15.20) and (15.21). Summing the contributions of all triangles over all the regions of the guide yields

$$I = \{\Theta\}^T [\mathbf{A}] \{\Theta\} - \frac{\omega^2}{c^2} \{\Theta\}^T [\mathbf{B}] \{\Theta\}, \tag{15.23}$$

where $\{\Theta\}$ is an ordered array of the longitudinal electromagnetic nodal variables and $[\mathbf{A}]$ and $[\mathbf{B}]$ are waveguide matrices. Hence, the variation of I in (15.23) gives the following algebraic eigenvalue problem:

$$[\mathbf{A}]\{\Theta\} = k_0^2 [\mathbf{B}]\{\Theta\}, \tag{15.24}$$

with

$$k_0 = \frac{\omega}{c}.$$

Solution of this eigenvalue problem will provide the required results on the propagation constants of various modes on a particular guide.

This finite elements method of solution can now be applied to several practical problems in integrated and fiber optics. It is important to recognize that the success of this technique depends intimately on how quickly and efficiently we may obtain the solution of the eigenvalue problem, that is, (15.24). We make use of a very efficient eigen solution technique [20] developed originally for mechanical systems. It permits a group of eigenvalues and eigenvectors to be extracted from the algebraic eigensystem in the frequency range of interest. The essence of the method is the reduction of the rank of the original algebraic eigensystem by means of a suitable subset of generalized coordinates. The eigenvalue analysis is conducted in the reduced space. An iteration scheme is augmented in order to achieve convergence. Convergence is assured because the entire process is merely an extension of the Stodola–Vianello method applied simultaneously to a group of eigenvectors rather than to only one at a time. To translate the eigenvalue problem posed by (15.24) to the frequency range of interest, it is necessary to regard only the eigenvalue parameter k_0^2 as being composed of two parts:

$$k_0^2 = k_c^2 + (\Delta k)^2, \tag{15.25}$$

where k_c^2 is the point of interest in the frequency range and $(\Delta k)^2$ is the incremental part which together with k_c^2 constitutes the time eigenvalue. Substituting (15.25) in (15.24) and rearranging terms yields

$$[\mathbf{A}^*]\{\Theta\} = (\Delta k)^2 [\mathbf{B}]\{\Theta\}, \tag{15.26}$$

where

$$[\mathbf{A}^*] = [\mathbf{A}] - k_c^2 [\mathbf{B}]. \tag{15.27}$$

The algebraic subsystem (15.26) can be solved by the same technique. In this case, the method will yield a subset of eigenvalues $(\Delta k)^2$ whose moduli are the smallest in the translated problem. Once they have been found, the true eigenvalues are recovered by means of (15.25).

In the computation, ten reduced coordinates were used and convergence of the two eigenvalues with the smallest moduli was sought before the iteration process was considered completed. By varying the dimensionless propagation constant γ over the range of interest, it was possible to trace the frequency spectra for the various modes in a very efficient computational manner.

To illustrate the versatility of the FEM, we introduce the following illustrations.

15.2.1 Circular Fiber

Exact modal solution exists for the circular step-index fiber [21, 22]. Comparison of the FEM results with the exact solution can be readily made. In Fig. 15.3, three dispersion curves for the dominant HE_{11} mode are displayed: the solid curve is the exact modal result; the dotted curve is obtained according to the FEM using 157 quadrilateral elements within one quadrant of the dashed boundary circle with $R_{max}/a = 14.0$, where R_{max} is the radius of the dashed circle and a is the core radius; and the dashed curve is also obtained according to the FEM using 588 quadrilateral elements within a quadrant of the dashed boundary circle with $R_{max}/a = 42.0$. Because of the presence of the two degrees of symmetry, one along the x-axis and the other along the y-axis, only one quadrant of the circular cross section needs be used. Imposing the boundary condition of either $H_z = 0$ or $E_z = 0$ on the x-axis or y-axis yields different modes of interest.

As expected, the accuracy of the FEM improves with a larger number of elements, and better agreement with the exact result is found at higher frequencies. One also realizes that at low frequencies, the field extends to a large distance from the core region, and so a large value of R_{max}/a as well as a larger number of

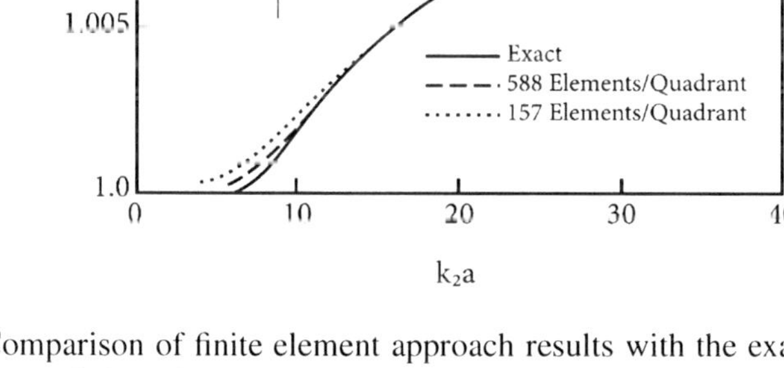

Figure 15.3. Comparison of finite element approach results with the exact results for the dispersion characteristics of the HE_{11} mode in a step index fiber [6]

elements are needed to achieve accurate results. Reasons for the discrepancies between the exact curve and the finite-element curves are summarized as follows:

- Insufficient number of elements were used, especially at lower frequencies.
- The element's boundary does not fit the curved boundary at the core-cladding interface.
- The trial functions, which are linear within each element, do not fit the exact solution.
- Round-off errors are present.
- Residual errors caused by the premature termination of the iterations used to evaluate the eigenvalues.

Another comparison can be made with the graded index fiber. Using 588 quadrilateral elements for one quadrant of the circular inhomogeneous fiber with core index

$$n(r) = n_i(1 - ar^2), \quad n_i = 1.53, \quad a = (n_i - n_O)/n_i, \quad n_O = 1.50,$$

and cladding index $n_O = 1.50$, the finite element approach was used to obtain the dispersion curve. Results are shown in Fig. 15.4. In the same figure the results obtained according to Yeh and Lindgren's staircase (10-layer) approach were also plotted [4]. The two curves are indistinguishable from each other.

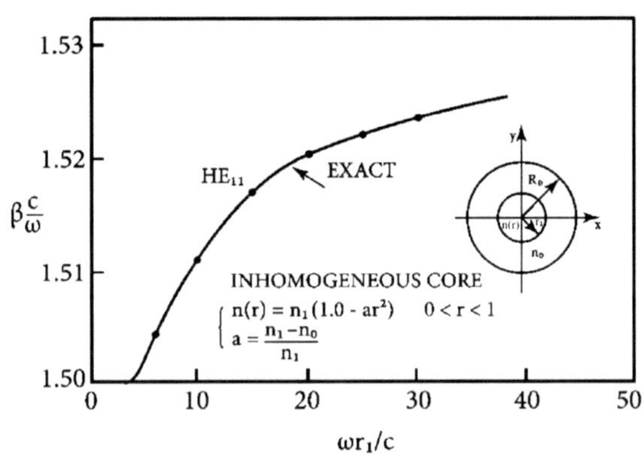

Figure 15.4. Comparison of the finite element approach results with the exact Yeh and Lindgren's results for the dispersion characteristics of the HE_{11} mode in a graded index fiber. Data are finite element results [6]

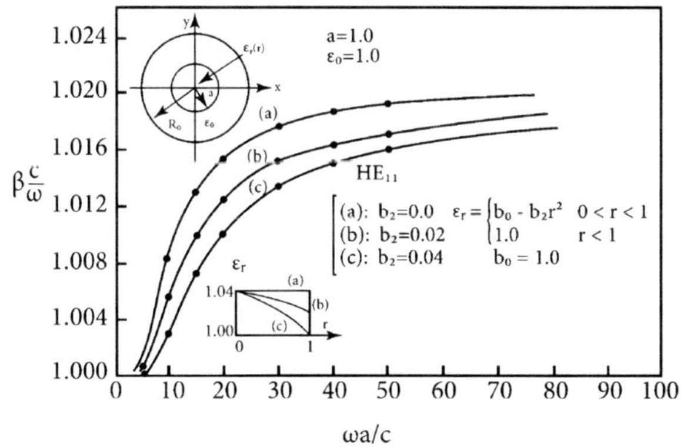

Figure 15.5. Comparison of the finite element approach results with the exact Dil and Blok's results for the dispersion characteristics of the HE_{11} mode in a graded index fiber. Data are finite element results [6]

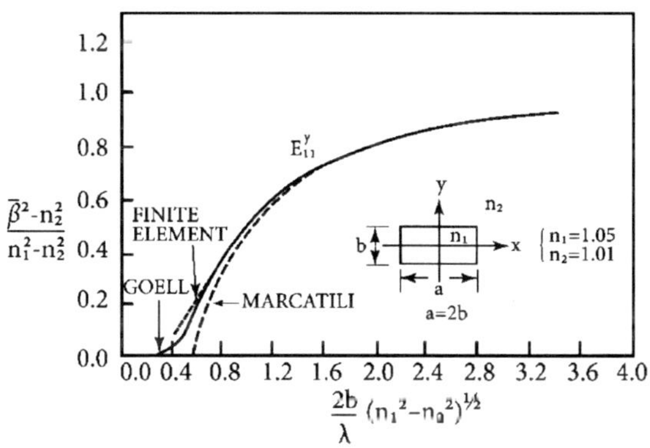

Figure 15.6. Comparison of the finite element approach results for the HE_{11} with Goell's results and with Marcatili's approximate results for the rectangular fiber guide [6]

Further comparison was carried out with Dil and Blok's differential equation approach [3]. The results are shown in Fig. 15.5. Again no noticeable differences can be seen.

15.2.2 Rectangular Structures

Goell [2], using the circular harmonic expansion technique, and Marcatili [23], using an approximate technique, obtained the dispersion characteristics of rectangular dielectric waveguides. Comparison is now made with the FEM results shown in Fig. 15.6. Again excellent agreement was found. The FEM result matches better with Goell's result. In the following figures, $\overline{\beta} = \beta / k_0$, where k_0 is the free space wave number.

Only Marcatili's approximate results are available for rectangular channel waveguide. Comparison is now made with the FEM results shown in Fig. 15.7. The agreement is very good at high frequencies. The two curves deviate from each other at lower frequencies. Since the FEM provides better coverage of the field outside the core region, one may conclude that the FEM results are more accurate.

The optical stripline is a planar waveguide with a strip of slightly lower index on a higher index thin film [24]. The cross-sectional geometry of the optical stripline is given in Fig. 15.8. It also shows the refractive-index distribution where n_1, n_2, n_3, and n_4 are the refractive index of thin film, substrate, strip, and air, respectively. These refractive indices satisfy one of the following relations for the stripline:

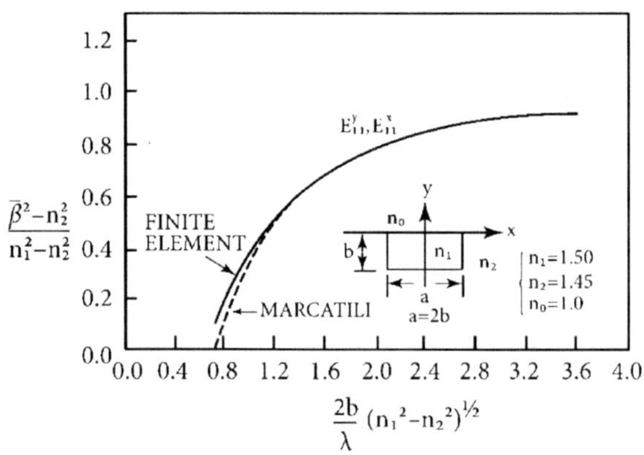

Figure 15.7. Comparison of the finite element results for the lowest order mode with Marcatili's approximate results for the channel waveguide [6]

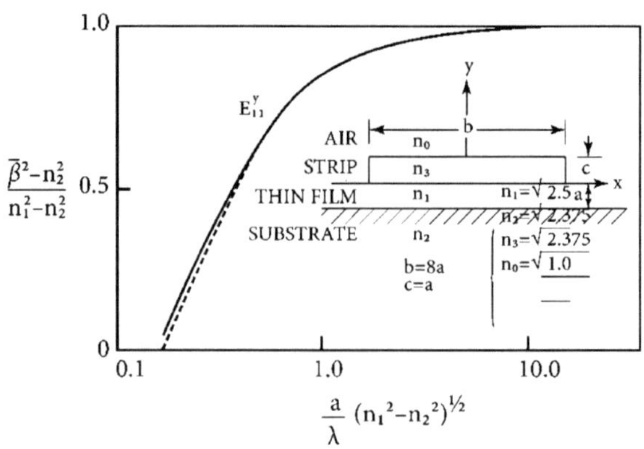

Figure 15.8. Dispersion curves for the dominant E^y_{11} mode in an optical stripline waveguide. The solid line represents results found according to the finite-element method, while the dashed line represents results found according to the vector variational method [6]

$$n_1 > n_2 \geq n_3 > n_4 \qquad n_1 > n_3 \geq n_2 > n_4.$$

In this study, the numerical data of the refractive indices are chosen to be $n_1 = (2.5)^{1/2}$, $n_2 = n_3 = (2.375)^{1/2}$, and $n_4 = 1.0$. This refractive-index distribution corresponds to the case where the thin film and the substrate are glass and the ambient is air.

The fundamental modes of this stripline guide are the E_{11}^y and the E_{11}^x mode. The principal transverse field components of the E_{11}^y mode are E_y and H_x, and those of the E_{11}^x mode are E_x and H_y. The present analysis involves solving only the E_{11}^x mode.

The finite element model employs 900 elements and 928 nodes in one-half of the cross section because of the symmetry on the y-axis. The dispersion curve for the E_{11}^xmode is computed with the boundary condition, and E_z = 0 on the y-axis. This curve is plotted in Fig. 15.8. The solid curve is the result of the finite element method, and the dotted one is that of the vector variational method. $\left[\left(\overline{\beta}^2 - n_2^2\right) / \left(n_1^2 - n_2^2\right)\right]$ vs. $(a/\lambda)\left(n_1^2 - n_2^2\right)^{1/2}$ is plotted, where λ is the free space wavelength and a is the thickness of thin film. These two curves are very close to each other. Therefore, the finite element method confirms the results of the vector variational method [24].

Figure 15.9 shows the power intensity distribution of the gray-scale taken at the normalized propagation constant $= 0.5$ and 0.8. This figure is almost identical

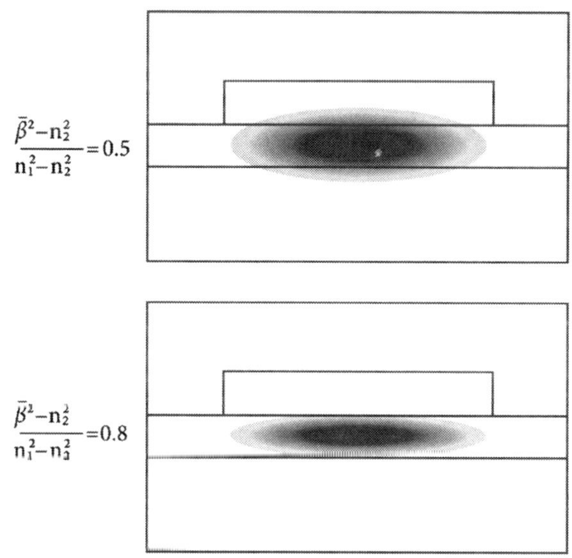

Figure 15.9. Gray scale plots for the power intensity distributions of the dominant E_{11}^y mode in an optical stripline waveguide at two different frequencies [6]

with those obtained by the variational method [24]. In Fig. 15.9, the intensity distribution is well-confined in the x-direction by the dielectric strip because the width b of the strip is larger compared with the thin film thickness a and the wavelength λ. It is also confined in the y-direction by the thin film. Therefore, the optical stripline minimizes scattering loss and undesired mode conversion caused by imperfections of the etched side walls of the optical waveguide.

As a rule of thumb, approximately 200 elements per quadrant are needed to achieve the desired three-figure accuracy for fields that are 95% confined within the core region and for the first three lower order modes.

15.2.3 Triangular Dielectric Guides

The propagation characteristics of the dominant mode in a triangular dielectric guide are obtained. It is assumed that this guide has an equilateral triangle core region with a refractive index $n_1 = 1.5085$ and a uniform surrounding region with a refractive index $n_2 = 1.5$.

The finite element model employs 689 elements and 659 nodes and requires 1,500K storage on the computer if all nodes are used. It was necessary to devise a special subroutine to input the rectangular coordinate positions of all the nodes, especially for the nodes at the boundary.

Finite elements of the rectangular type, which is the special kind of quadrilateral element, are used. Some triangular elements are also used to match the boundary of the triangular core region. The finite element in the surrounding medium increases in size with increasing distance from the origin. The subdivisions in the core region are uniform.

Since this guide has a symmetry line along the y-axis, the total field distribution needs to be calculated for only one-half of the guide. The boundary condition $H_z = 0$ is used on this line to obtain the dispersion curve for the E^y_{11} mode.

The normalized propagation constant is plotted in Fig. 15.10 against the normalized frequency $[(2b)/\lambda]\,(n_1^2 - n_2^2)^{1/2}$,where b is the height of this guide. The curve is plotted from $\left(\overline{\beta}^2 - n_2^2\right) / \left(n_1^2 - n_2^2\right) = 0.1$ since the active region below this point is too large to be contained by the present finite element model.

Figure 15.11 shows the intensity distribution for three different values of the normalized propagation constant. These figures closely match the anticipated intensity distribution.

The core region in Fig. 15.11 resembles an isosceles triangle rather than the original equilateral triangle because the computer printer increment is larger in the vertical direction than in the horizontal direction. This intensity test shows that the dispersion curve in Fig. 15.10 represents that of the E^y_{11} mode.

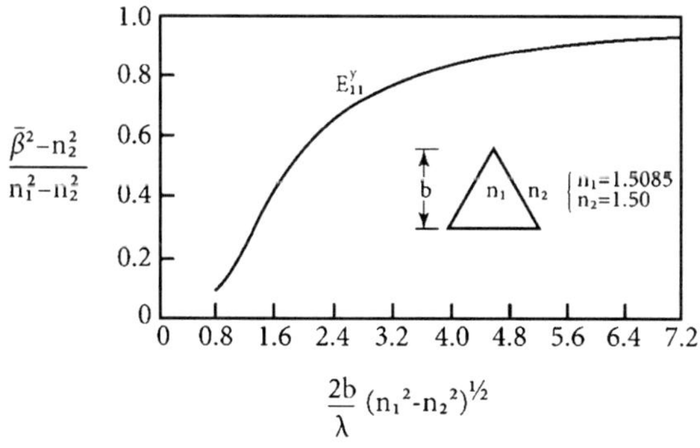

Figure 15.10. Dispersion curve for the dominant mode in a triangular fiber [6]

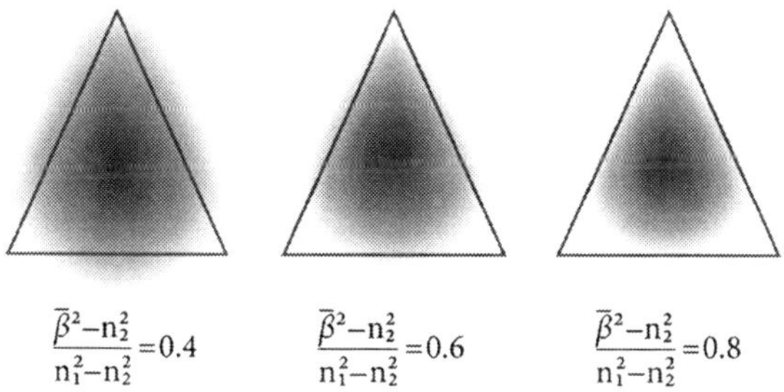

Figure 15.11. Gray-scale plots for the power intensity distribution of the dominant mode in triangular fiber at three different frequencies [6]

15.2.4 Elliptical Dielectric Guide

This guide consists of a uniform elliptical core region with a refractive index $n_1 = 1.5085$ and a uniform cladding region with a refractive index $n_2 = 1.5$. The aspect ratio of this guide b/a is 2, where a and b are the lengths of the major and minor axes, respectively.

The finite element model employs 615 elements and 658 nodal points. Only one-quarter of the entire cross section is covered by these elements, since this guide has two symmetry lines, one along the x-axis and the other along the y-axis. Most elements are of the rectangular type. Some triangular elements are also used to fit the elliptical core boundary. With the same boundary conditions on the symmetry lines, the dispersion curve of the E_{11}^y mode is obtained. Figure 15.12 shows the dispersion where the propagation constant $\left(\overline{\beta}^2 - n_2^2\right) / \left(n_1^2 - n_2^2\right)$ is plotted as a function of $\left[(2b)/\lambda\right]\left(n_1^2 - n_2^2\right)^{1/2}$.

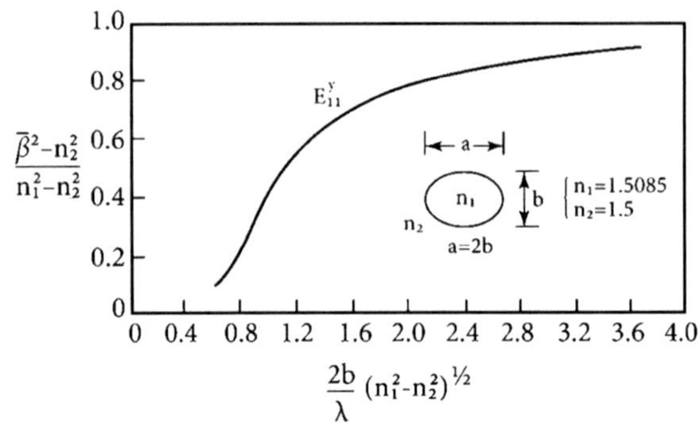

Figure 15.12. Dispersion curves for the dominant $_eHE_{11}$ mode in an elliptical fiber. FEM results are indistinguishable from the exact results obtained according to Yeh's exact formulation [6]

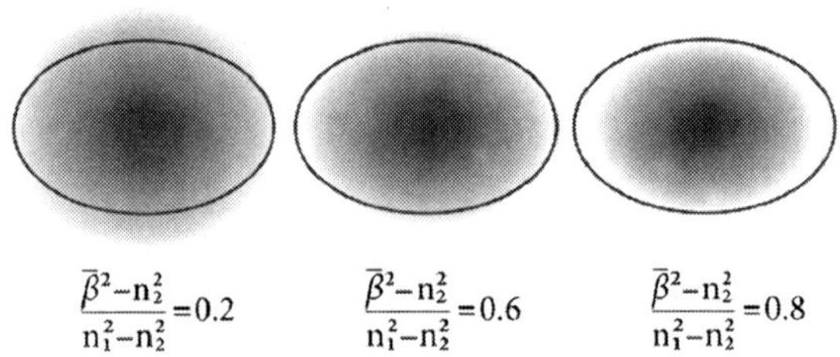

Figure 15.13. Gray-scale plots for the power intensity distributions of the dominant $_eHE_{11}$ mode in an elliptical fiber at three different frequencies [6]

The power intensity is also plotted on the gray scale. Figure 15.13 shows the intensity distribution for the normalized propagation constants 0.2, 0.6, and 0.8. This figure fits the anticipated behavior of the Poynting vector exactly, that is, the gray intensity is darkest at the center of the core region, and the field extends farther into the cladding medium near the cutoff frequency.

The results for elliptical fiber have practical importance since circular fibers commonly used in optical communication systems are easily deformed to various elliptical shapes. The present FEM can then be readily applied to these elliptical fiber problems having different aspect ratios.

15.2.5 Single Material Fiber Guide

Typically, optical fibers are constructed with a central glass core surrounded by a glass cladding with a slightly lower refractive index. The single material fiber is created by a structural form that uses only a single low-loss material.

Figure 15.14 shows cross-sectional views of two possible forms for the single material (SM) fiber. The guided energy is concentrated primarily in the central

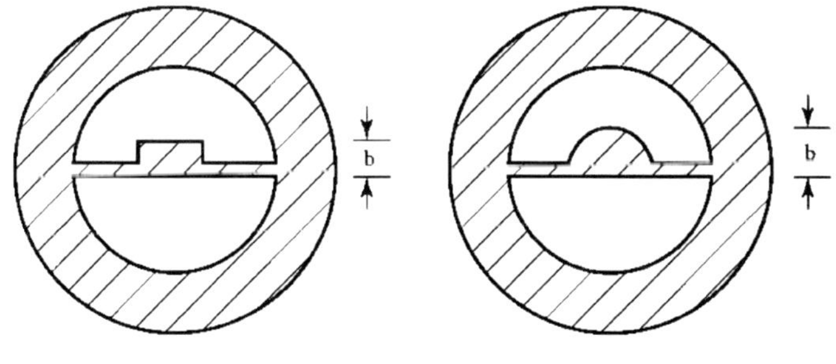

Figure 15.14. Cross-sectional view of single material fibers

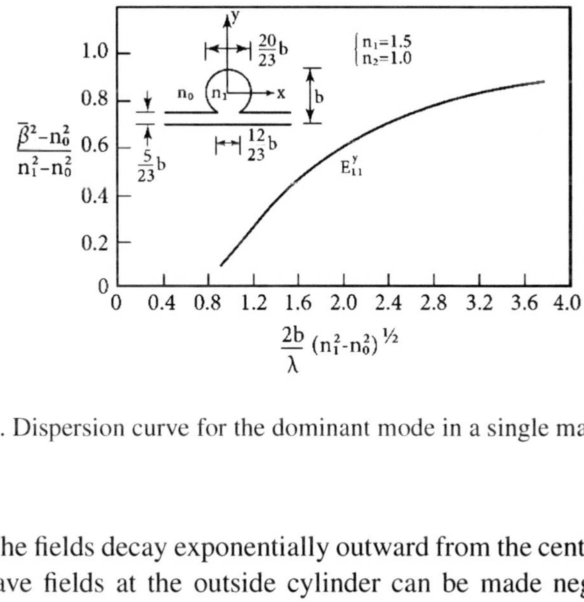

Figure 15.15. Dispersion curve for the dominant mode in a single material fiber [6]

enlargement. The fields decay exponentially outward from the central enlargement. The guided wave fields at the outside cylinder can be made negligibly small if spacing between the central enlargement and outer cylinder is sufficiently large.

Only the SM fiber with cylindrical core is investigated. The complete dimensions of the SM fiber used in the present study are given in Fig. 15.15. The slab has a thickness of $(5/23)b$, where b is the total height of the central enlargement. It is also assumed that this guide is made of glass material with a refractive index $n_1 = 1.5$. The refractive index outside the central enlargement is equal to 1.0. The finite element model of this guide has 913 elements and 969 nodes. These elements are employed in one-half of the cross section due to the symmetry line along the y-axis. The dispersion curve for the E^y_{11} mode is plotted in Fig. 15.15. In Fig. 15.16, the intensity distributions are plotted on the gray scale. Some errors appear in these figures, especially near the sharp edge. To correct these errors, the present FEM must be modified to allocate more fine subdivisions near this area. Since these errors occur only in a few elements out of several hundred elements, the dispersion

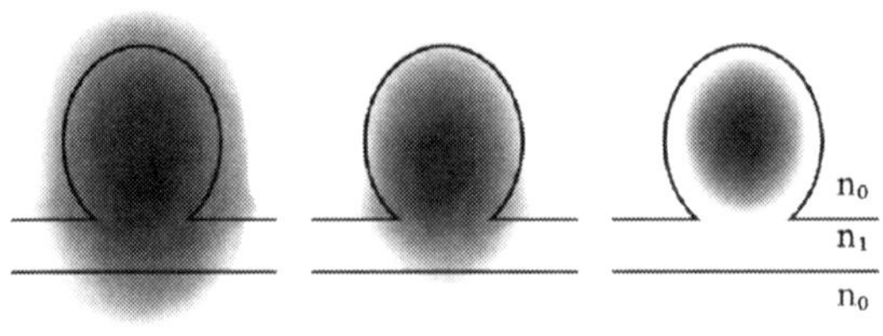

Figure 15.16. Gray-scale plots for the power intensity distributions of the dominant in a single material fiber at three different frequencies [6]

curve can be still considered accurate. These pictures are taken at three different values of the normalized propagation constant. The smaller this constant, the farther the fields extend out from the circular region. The gray-scale plot also shows that the fields do not extend far toward the outer cylinder, even near the cutoff; the main portion of the power intensity is well-confined within the central enlargement. Therefore, the assumption that the outer cylinder can be neglected is reasonable.

15.2.6 Concluding Remarks

We have demonstrated the versatility of the FEM in the treatment of wave propagation problems dealing with various dielectric waveguides and integrated optics waveguides. Not only are the propagation constants obtainable by this technique, but the field distributions are also readily available. Only when the operating frequencies are close to the cutoff frequency of the mode under consideration is the limitation of the FEM approached. For these cases, the mode field extends greatly beyond the waveguide core, and many more elements are needed to represent accurately the field region.

Although the illustrations given above are limited to the 2D case for which the dielectric constant variations as well as the guided structure shape variations are confined in the transverse (x, y) directions, the FEM can be generalized to treat problems with variations in 3D such as waveguide branches, couplers, horns, curves, longitudinal discontinuities, etc. It is noted that FEM is an exact numerical method satisfying the Maxwell equations and the boundary conditions.

15.3 Beam Propagation Method (BPM) or Forward Marching Split-Step Fast Fourier Transform Method [7]

The split-step fast Fourier transform method was a well-known numerical technique in applied mathematics and physics [25]. It had been used extensively in studying the beam propagation behavior in a linear atmosphere with statistically varying medium as well as in a nonlinear atmosphere caused by thermal blooming. The first adaptation of this technique to the inhomogeneous optical fiber problem was done by Yeh et al. in 1978 [7]. The name beam propagation method (BPM), which is identical to the split-step fast Fourier transform method, was coined and popularized in the literature. In a series of papers, Yeh et al. solved a number of problems dealing with beam propagation in homogeneous or inhomogeneous fiber guides and in dielectric waveguide branches, horns, tapers, or couplers [9–11, 26].

The BPM provides the evolution of a beam as it propagates down an optical fiber structure. Since this method is based on a forward marching algorithm, any reflections or backscattering must be ignored. Therefore, any variations along the propagating medium must be gentle. Even under these constraints, a large variety of optical fiber or integrated optical waveguide problems can still be treated by the BPM. Here, we shall present a few examples.

15.3.1 Formulation of the Problem and the Numerical Approach

Let us consider the case where the optical field can be derived from a scalar function $u(x, y, z)$ that satisfies the reduced wave equation (see Sect. 2.8)

$$\left[\nabla^2 + k^2 n^2(x, y, z)\right] u(x, y, z) = 0, \tag{15.28}$$

where $k = 2\pi/\lambda$ is the wave number and λ is the free space wavelength. $n(x, y, z)$ is the spatially inhomogeneous refractive index of the medium. If we write u as the product of a factor $e^{-jkn_0 z}$ that accounts for the rapid change in the phase of u along the direction of propagation and a complex amplitude $A(x, y, z)$, a further simplification of the calculational problem results. One then has

$$\left\{-j2kn_0 \frac{\partial}{\partial z} + \nabla_{\mathrm{t}}^2 + k^2 \left[n^2(x, y, z) - n_0^2\right]\right\} A(x, y, z) = -\frac{\partial^2 A(x, y, z)}{\partial z^2}, \tag{15.29}$$

where ∇_{t}^2 is the transverse Laplacian $\left(\partial^2/\partial x^2 + \partial^2/\partial y^2\right)$ and n_0 is a given constant, which represents the refractive index of some uniform medium. At laser wavelengths the complex amplitude, $A(x, y, z)$, varies much more rapidly transverse to the direction of propagation than it does along the direction of propagation. This enables us to make the paraxial approximation, wherein the term on the right hand side of (15.29) is neglected (in the Russian literature this is called the parabolic approximation). So the complex amplitude now satisfies

$$\left\{-j2kn_0 \frac{\partial}{\partial z} + \frac{\partial^2}{\partial x^2} + \frac{\partial^2}{\partial y^2} + k^2 \left[n^2(x, y, z) - n_0^2\right]\right\} A(x, y, z) = 0. \tag{15.30}$$

In addition to (15.30), the complex amplitude satisfies the initial condition on the fiber end, at $z = 0$,

$$A(x, y, 0) = u(x, y, 0) \tag{15.31}$$

and the boundary condition

$$A(\pm\infty, z) = 0. \tag{15.32}$$

As an example, if a truncated Gaussian beam is focused on one end of the optical guide, then

$$\begin{aligned} u(x, y, 0) &= u_0 \exp(-r^2/w^2) \quad \text{for } 0 \le r \le b, \\ &= 0 \qquad\qquad\qquad\quad\; \text{for } r > b, \end{aligned} \tag{15.33}$$

where $r^2 = x^2 + y^2$, w is the spot size of the beam, and b is the radius of the truncated beam at $z = 0$. Equation (15.30) is solved numerically by the BPM in the following manner. Let us write $A(x, y, z)$ in the form

$$A(x, y, z) = \exp\left[\Gamma(x, y, z)\right] v(x, y, z), \tag{15.34}$$

where $\Gamma(x, y, z)$ is a phase function associated with the medium inhomogeneities

$$\Gamma(x, y, z) = -\frac{jk}{2n_0} \int_{z_0}^{z} \left[n^2(x, y, z') - n_0^2\right] dz'. \tag{15.35}$$

The modified complex amplitude $v(x, y, z)$ then satisfies the equation

$$-j2kn_0 \frac{\partial}{\partial z} v(x, y, z) + e^{-\Gamma} \nabla_T^2 \left[e^{\Gamma} v(x, y, z)\right] = 0. \tag{15.36}$$

Although (15.36) does not look any easier to solve than (15.30), it is easier to solve numerically because, for sufficiently small increments in the z-direction and an appropriately chosen lower limit in the integral in (15.35), the value of $v(x, y, z + \Delta z)$ can be obtained to a good approximation by solving the simpler equation

$$\left(-j2kn_0 \frac{\partial}{\partial z} + \nabla_T^2\right) v(x, y, z) = 0, \tag{15.37}$$

with the initial condition

$$v(x, y, 0) = u(x, y, 0). \tag{15.38}$$

Physically, these equations approximate the propagation in the inhomogeneous medium by a two-step process at each z increment. First, we propagate the field

$u(x, y, z)$ at z to $z + \Delta z$ assuming that the intervening space is homogeneous. The effect of the inhomogeneities between z and $z + \Delta z$ is then accounted for by multiplying this solution by the phase factor $\exp(\Gamma)$. Equation (15.37) is solved by the fast Fourier transform technique. Replacing the Laplacian by its finite-difference equivalent but still retaining the z-derivative, the solution of (15.37) can be expressed in the form

$$v(m, n, z) = \sum_{m'=0}^{N-1} \sum_{n'=0}^{N-1} V(m', n', z) \exp\left[-\frac{j2\pi}{N}(mm' + nn')\right], \tag{15.39}$$

where $x = m\Delta x$, $y = n\Delta y$, and $V(m', n', z)$ satisfies

$$\left[-j2kn_0 \frac{\partial}{\partial z} + \frac{f(m', n')}{(\Delta x)^2}\right] V(m', n', z) = 0, \tag{15.40}$$

with the initial condition

$$V(m', n', z_i) = \frac{1}{N^2} \sum_{m'=0}^{N-1} \sum_{n'=0}^{N-1} v(m, n, z_i) \exp\left[\Gamma(m, n, z_i)\right] \exp\left[\frac{j2\pi}{N}(mm' + nn')\right]. \tag{15.41}$$

The function $f(m', n')$ is determined by the difference approximation used to represent the Laplacian in (15.37). For example, if $\nabla_T^2 v$ is approximated by the simple central difference expression

$$\begin{aligned}\nabla_T^2 v = {} & \frac{1}{(\Delta x)^2}\left[v(m+1, n, z) - 2v(m, n, z) + v(m-1, n, z)\right. \\ & \left. + v(m, n+1, z) - 2v(m, n, z) + v(m, n-1, z)\right],\end{aligned} \tag{15.42}$$

then $f(m', n')$ is

$$f(m', n') = -4\left[\sin^2\left(\frac{\pi m'}{N}\right) + \sin^2\left(\frac{\pi n'}{N}\right)\right]. \tag{15.43}$$

Note that the series in (15.39) is simply the discrete Fourier transform of the function $V(m', n', z)$ and thus can be evaluated numerically for a given $V(m', n', z)$ by a fast Fourier transform algorithm. Furthermore, the function $V(m', n', z)$ is readily determined from (15.40) as

$$V(m', n', z) = V(m', n', z_i) \exp\left[\frac{jf(m', n')}{2k(\Delta x)^2 n_0}(z - z_i)\right], \tag{15.44}$$

where $V(m', n', z_i)$ is given by the series in (15.41), which can also be evaluated by a fast Fourier transform algorithm. To summarize, we step from z to $z + \Delta z$ as follows:

- Take the inverse discrete Fourier transform of $u(m, n, z) = \exp[\Gamma(m, n, z)] \times v(m, n, z)$ by means of an inverse fast Fourier transform algorithm
- Multiply the result by $\exp\left[jf(m', n')\Delta z / 2k(\Delta x)^2 n_0\right]$ and take the discrete Fourier transform with a fast Fourier transform algorithm
- Multiply the result by $\exp[\Gamma(m, n, z + \Delta z)]$ to yield $u(m, n, z + \Delta z)$

This process is repeated until we have reached the desired z plane.

15.3.2 Gaussian Beam Propagation in a Radially Inhomogeneous Fiber

It has been well recognized that analytic solutions for the infinite Gaussian beam propagation problems exist only for certain specific radial index profiles. Even for these cases, the solutions are often involved and cumbersome to use. By simply specifying the transverse variation $n(x, y)$ one can readily calculate the propagation characteristics of an incident beam in this medium according to the BPM scheme. Some of the relevant questions that can be answered by these kinds of calculations are the following: (1) How well will this medium confine and focus the beam? (2) What is the focal distance? (3) How does the beam profile of an initial Gaussian beam change as it propagates in a medium with an index variation in $n(x, y)$?

Let us consider the specific case of an infinite Gaussian beam propagating in a radially inhomogeneous fiber that possesses a parabolic radial index profile as follows:

$$n(x, y) = n_a\left(1 - \delta\frac{r^2}{a^2}\right), \tag{15.45}$$

where $r^2 = x^2 + y^2$, r is the radial coordinate, n_a is the index at $r = 0$, and δ/a^2 is a given constant. Assuming an initial Gaussian beam of the form

$$u(x, y, 0) = \exp(-r^2/w_1^2), \tag{15.46}$$

one may obtain the required propagation characteristics of the beam. The results are summarized in Fig. 15.17, in which the normalized beam waist w_2/w_1, where w_2 is the beam waist along the fiber axis and w_1 is the beam waist of the input

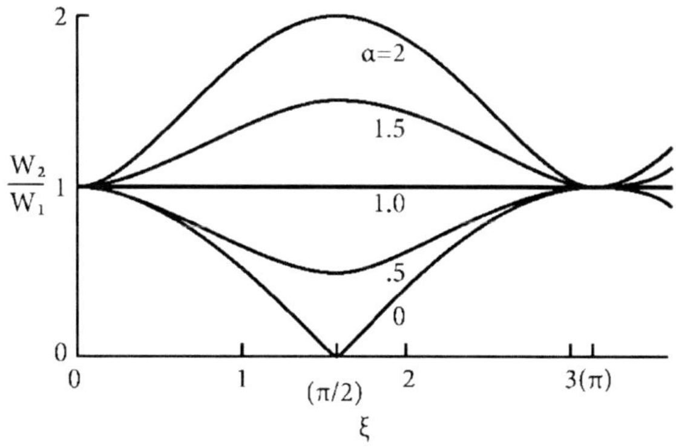

Figure 15.17. Normalized beam waist w_2/w_1 as a function of normalized axial distance ξ along the fiber axis for various α. w_1 and w_2 are, respectively, the beam waist at the input and that along the axis fiber [7]

beam, is plotted against ξ, the normalized axial distance along the fiber axis for various values of α. The coefficients ξ and α are defined as follows:

$$\xi = \frac{z\,(2\delta)^{1/2}}{a}, \qquad \alpha = \frac{2\lambda a}{\pi w_1^2\,(2\delta)^{1/2}}, \tag{15.47}$$

where λ is the free-space wavelength of the beam. The results shown in Fig. 15.17 are indistinguishable from the exact analytic solutions. To illustrate the intensity distributions of the beam as it propagates down the fiber guide, Fig. 15.18 is introduced for the case in which

$$\begin{aligned} n_a &= 2.0, \\ \frac{2\delta}{a^2} &= 8.0 \times 10^{-9} \quad (\mu\text{m})^{-2}, \\ \lambda &= 0.8 \quad \mu\text{m}, \\ \Delta z &= 878.1 \quad \mu\text{m}, \\ z_{\min} &= 17{,}562 \quad \mu\text{m}, \end{aligned} \tag{15.48}$$

where Δz is the step size and $z_{\min}$ is the distance from the beam entrance to the plane at which the spot-size is at its minimum value.

Consider the case of a truncated Gaussian beam propagating in a radially inhomogeneous fiber whose index profile is given by (15.45). The beam at the entrance of the fiber takes on the form given by (15.33) with $u_0 = 1$. Results of

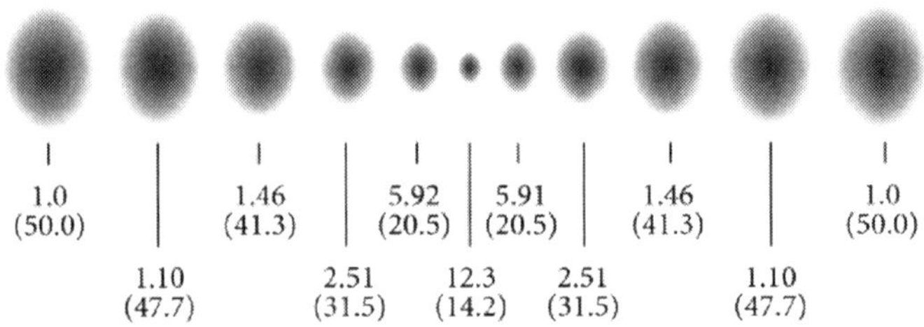

Figure 15.18. Selected gray-scale intensity patterns along the fiber axis for an infinite Gaussian beam propagating in a radially inhomogeneous fiber with parabolic index profile. The value in the brackets represent the beam waists in μm, while the other unbracketed values represent the highest intensity values for these patterns. Note that the Gaussian profile of the beam is retained throughout the propagation path. For the chosen parameters given in the text, it takes a 35,124-μm-long axial distance to complete one cycle as shown [7]

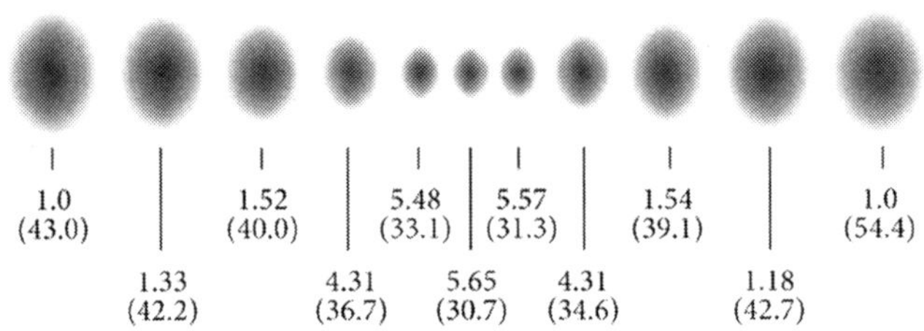

Figure 15.19. Selected gray-scale intensity patterns along the fiber axis for a truncated Gaussian beam propagating in a radially inhomogeneous fiber with parabolic index profile. The value in the brackets represent the beam waists in μm while the other unbracketed values represent the highest intensity values for these patterns. Note that the truncated Gaussian profile of the beam is not preserved along the propagation path. At some points along the path, the intensity at the center of the beam takes a dip. Owing to the presence of the diffraction effects, the beam is not completely symmetrical about a certain "focal point" where the beam achieves the smallest beam waist. The chosen parameters for this figure are the same as those for the infinite Gaussian beam case except the truncation radius is 75 μm [7]

the computation with the constants given in (15.48) and $b = 75$ μm are shown in Fig. 15.19. It can be seen that rather unusual intensity distributions are observed as the truncated Gaussian beam propagates down the fiber, and smoothing of the beam shape also takes place. Beam waist is defined as the second moment of the radial coordinates, that is, beam waist $=\left(\int_0^a u^2 r^3\, dr \Big/ \int_0^a u^2 r\, dr\right)^{1/2}$, where a is the outermost radius of the beam and u is the scalar field. Beam waist for a Gaussian beam is $w/\sqrt{2}$, where w is given in (15.33). One notes that analytic solutions are not available for the case of an initially truncated Gaussian propagating

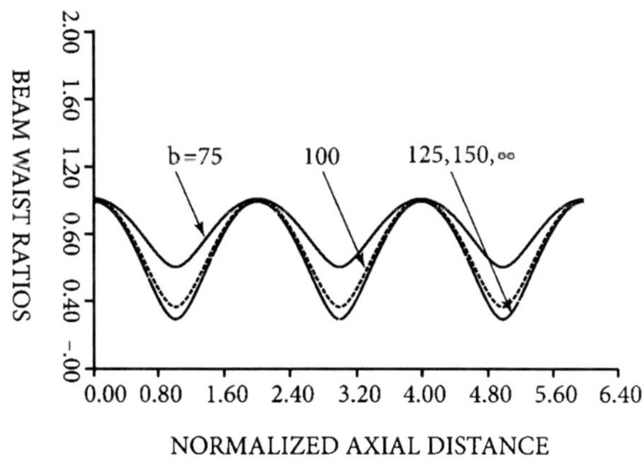

Figure 15.20. Normalized beam waist ratio (w_2/w_1) as a function of the normalized axial distance ξ for various values of truncation radius, b, in μm. The coefficient for the parabolic index profile is chosen to be $2\delta/a^2 = 8 \times 10^{-9}\,(\mu\mathrm{m})^{-2}$ [7]

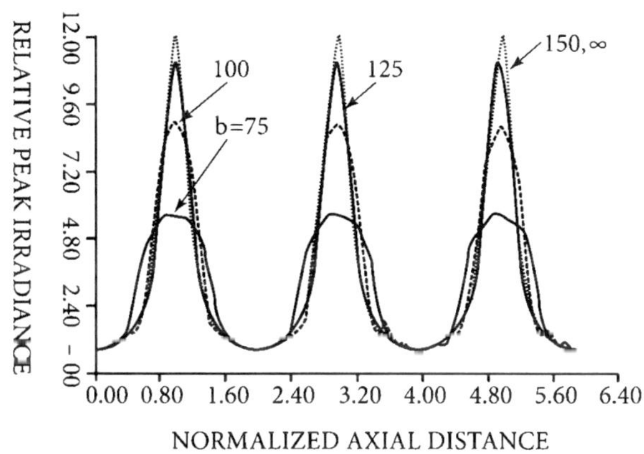

Figure 15.21. Relative peak irradiance as a function of the normalized axial distance $2\xi/\pi$ for various values of truncation radius, b, in μm. The coefficient for the parabolic index profile is chosen to be $2\delta/a^2 = 8 \times 10^{-9}\,(\mu\mathrm{m})^{-2}$ [7]

in a graded index fiber. Shown in Fig. 15.20 are plots for normalized beam waist ratios as a function of the normalized axial distance, ξ, for various values of beam truncation radius, b. It can be observed from these curves that truncation of an incident Gaussian beam alters significantly the way the beam waist changes as the beam propagates down the fiber. To see how the peak intensity of the beam varies as it propagates down the fiber corresponding to the case given in Fig. 15.20, we introduce Fig. 15.21. Truncation of an incident Gaussian beam clearly affects greatly the behavior of peak beam intensity along the fiber. Results of these plots reinforce

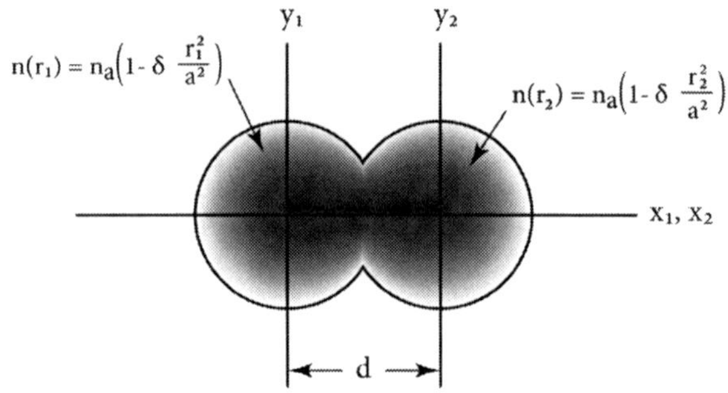

Figure 15.22. The fiber coupler

the importance of BPM since the results show that one cannot extrapolate readily from the infinite Gaussian beam case where analytic results are available to the truncated Gaussian beam case.

15.3.3 Fiber Couplers [9]

The geometry of an inhomogeneous fiber coupler is shown in Fig. 15.22. Two graded-index fibers with index variation given by

$$n(r_{1,2}) = n_a \left(1 - \frac{\delta r_{1,2}^2}{a^2}\right), \tag{15.49}$$

where $r_{1,2}$ are the radial coordinates of 1 and 2 fibers, respectively, and n_a, δ, and a are all known constants, are fused together as shown. The separation distance between the centers of the fibers is d. A Gaussian beam representable by

$$u(x, y) = u_0 \exp\left\{\frac{\left[-\left(x + \frac{d}{2}\right)^2 - y^2\right]}{w_1^2}\right\}, \tag{15.50}$$

where $u(x, y)$ is the scalar wave function of the beam, and u_0 and w_1 are given constants, is incident on one of the fibers. One wishes to learn how this beam evolves as it propagates down the coupled structure. In other words, the coupling distance and the beam shape will be obtained.

Before we proceed with the numerical computations, it should be recalled that we are dealing completely with total field quantities and not with the modes. In other words, we are interested in how the total field evolves as it propagates down the guiding structure; we are not interested in how each mode propagates.

Nevertheless, it is recognized that the total field may be decomposed into a set of orthonormal guided modes. For example, an incident Gaussian beam with a given beam width w may excite many modes in a parabolic-index-profile fiber when $\alpha > 1$ or when $\alpha < 1$, with

$$\alpha = \frac{2\lambda a}{\pi n_a w_1^2 (2\delta)^{1/2}}, \tag{15.51}$$

or may excite only one mode when $\alpha = 1$. Only the $\alpha = 1$ single mode will propagate down the parabolic-index-profile without experiencing the focusing and defocusing effects. For the $\alpha \neq 1$ case, the input Gaussian beam will experience focusing and defocusing effects. Another way to interpret the above phenomenon is that multimodes with different propagation constants are excited by the input beam when $\alpha \neq 1$, while only a single mode is excited when $\alpha = 1$. The stronger are the focusing–defocusing effects, the higher the content of different modes. Using this reasoning, one may expect that if two identical parabolic-index fibers were placed side by side with each other and if the beamwidth were so chosen that $\alpha = 1$ is obtained, one would expect the guided power to interchange between the two fibers in a periodic manner with a distinct coupling length. On the other hand, if α departs from unity, many modes are generated and many beat coupling lengths occur, so the guided power will no longer interchange among the guides in a simple manner. For the $\alpha \ll 1$ or $\alpha \gg 1$ case, even larger number of modes are excited. Depending on how the incident energy is distributed among the excited modes the back and forth (periodic) power exchange phenomenon could still prevail if the major amount of energy is distributed in several low-order modes. Only when the incident energy is distributed evenly among all modes and only in the limit of infinite number of modes will the input power be split in a 50–50 manner between the two fibers.

Case A: $\alpha = 1$

This is the single mode case. Choosing the parameters

$$n_a = 2.0, \qquad \lambda = 0.8\ \mu\text{m}, \quad \delta = 0.81 \times 10^{-6},$$

$$w_1 = 100\ \mu\text{m}, \quad a = 50\ \mu\text{m},$$

one has $\alpha = 1$. Shown in Fig. 15.23 is a typical picture of how a Gaussian beam evolves as it propagates down a coupling structure made with two graded-index fibers. One clearly sees the back-and-forth transfer of guided power from one guide to the other in a rather complicated manner. When all the power is transferred from one fiber to the other, the original Gaussian beam profile is recovered, provided that the separation of the fibers is relatively large. Small separation of the fibers tends to distort the parabolic-index profile so that the input Gaussian beam can no longer

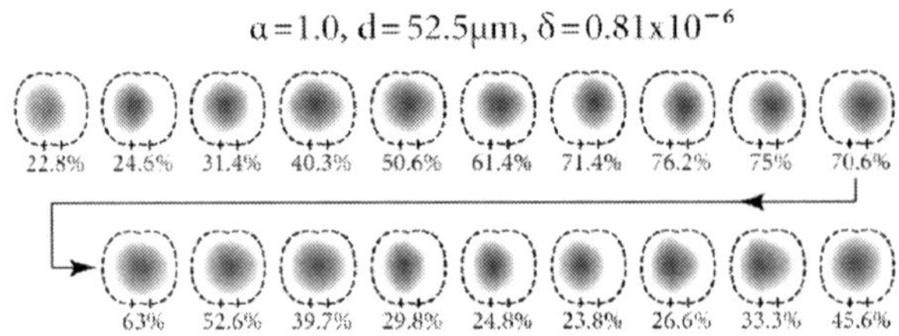

Figure 15.23. A cross-sectional view of the power density distribution for the guided wave. Note the power exchange phenomenon as the wave propagates along the coupled structure. The percent value indicates the percent of total power contained in the right-hand half of the structure (i.e., fiber 2). The distance between each frame is 24,682 μm [9]

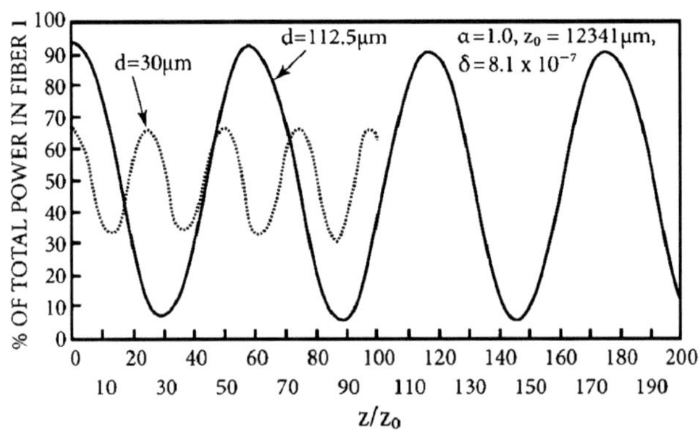

Figure 15.24. Percent of total power in the left-hand half of the structure (i.e., fiber 1) as a function of the axial distance along the coupling structure. This is the single-mode coupling case [9]

be considered as the single orthonormal mode of the guide composed of half of the coupling structure. Even though periodic power exchange can still be observed, the beam pattern tends to be somewhat distorted.

By integrating over the cross section of one of the fibers, one may obtain the amount of guided power contained in that fiber. This is shown in Fig. 15.24., in which the total guided power in fiber 1 is plotted against the normalized longitudinal distance z/z_0 along the fiber for various separation distances d. z_0 is a normalizing constant, which is equal to 12,341 μm for the example under consideration. It should be noted that the spot size of the incident Gaussian beam has been kept constant for the various fiber separations. For example, in the separation distance $d = 112.5$ μm case, about 7.14% of the incident beam power exists in the fiber 2 portion of the coupling structure, while 92.86% of the incident power is in fiber 1. When the separation distance $d = 30$ μm, about 33.3% of the incident beam exists in the fiber 2 portion of the coupling structure, while 66.6% of the incident power is in fiber 1. For this near single mode case, one may observe from Fig. 15.24 that the total guided power in fibers 1 and 2 exchanges in a periodic sinusoidal manner and that the coupling length, which is defined as the distance for which maximum guided power is transferred from one guide to the other, becomes shorter as the separation distance between the axes of the guides is shortened. At a separation distance of 112.5 μm, it take about 37 cm for the guided power to exchange among the fibers, while at a separation distance of 30 μm, it takes only 14.8 cm.

Case B: $\alpha = 0.4$

Turning our attention to the slightly multimode case with parameters

$$n_a = 2.0, \qquad \lambda = 0.8\,\mu\text{m}, \quad \delta = 5 \times 10^{-6},$$

$$w_i = 100\,\mu\text{m}, \quad a = 50\,\mu\text{m},$$

one obtains $\alpha = 0.4$. This means that even though the dominant power is carried by the dominant mode, other higher order modes are also excited. These higher order modes contribute to the focusing–defocusing character of the beam. It is anticipated that the coupling effects will no longer be uniform along the longitudinal distance, and stronger coupling exists when the beam is defocused. Coupling length can no longer be obtained unambiguously due to the presence of higher order mode coupling and thus the presence of beat coupling lengths. As an illustration, the power exchange phenomenon for this slightly multimode case is shown in Fig. 15.25. Because of the complex wave interaction that is taking place in this coupling structure, the beam can no longer retain the Gaussian shape as it propagates down the structure. It is nevertheless of interest to note that when the maximum amount of power is transferred to the other fiber, a Gaussian beam shape (although somewhat deformed) is recovered.

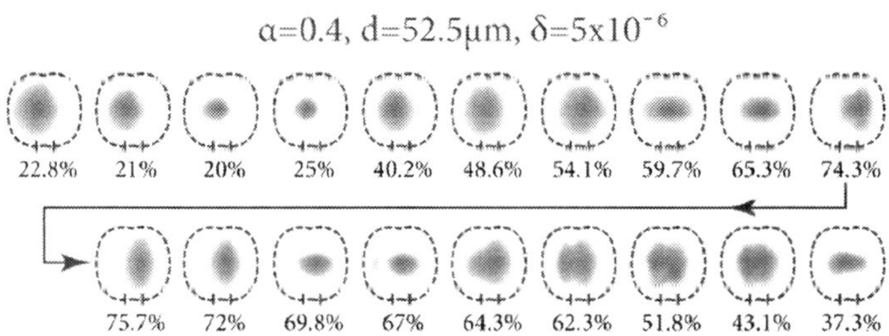

Figure 15.25. A cross-sectional view of the power density distribution for the guided wave. Note the power exchange phenomenon as the wave propagates along the coupled structure. The percent value indicates the percent of the total power contained in the right-hand half of the structure (i.e., fiber 2). The distance between each frame is 9,934 μm [9]

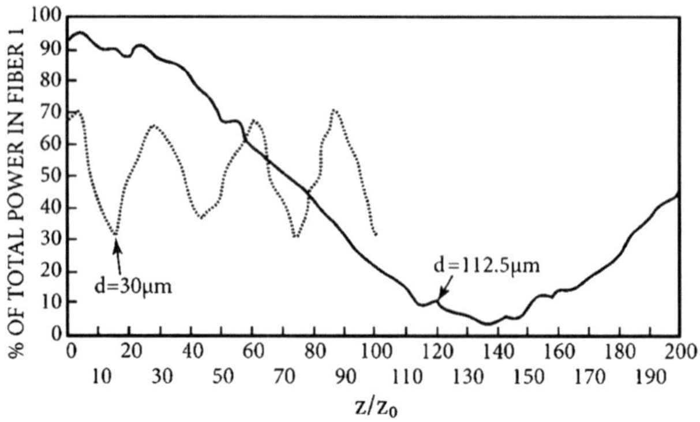

Figure 15.26. Percent of total power in the left-hand half of the structure (i.e., fiber 1) as a function of the axial distance along the coupling structure. This is the multimode coupling case [9]

In Fig. 15.26 we have plotted the total amount of power carried in fiber 1 as a function of the normalized longitudinal distance z/z_0, with $z_0 = 4,967\ \mu\text{m}$ for various separation distances. It is seen that unlike the single mode $\alpha = 1$ case, the power exchange among the fibers no longer varies sinusoidally as a function of the longitudinal distance due to the presence of multimode coupling. However, it appears that the periodic power exchange phenomenon is still retained. It is also noted that initial power transfer originates from fiber 2 to fiber 1 rather than the situation demonstrated for the $\alpha = 1$ case. It is still true that the coupling distance becomes shorter as the fiber separation becomes closer. One may also deduce from Figs. 15.25 and 15.26 that the distortion of the beam is much more pronounced when the fiber separation distance is small and when multimodes exist. From Fig. 15.26 it appears that the coupling length for the separation distance $d = 112.5\ \mu\text{m}$ case is about $69.5\ \mu\text{m}$, while it is $28.8\ \mu\text{m}$ for the separation distance $d = 30\ \mu\text{m}$ case. The fact that the coupling length is longer for the multimode $\alpha = 0.4$ case than the comparable situation for the single-mode $\alpha = 1$ case is somewhat deceiving. This is because the index profiles are different for these two cases. If the same index profile were used, the coupling length for the multimode case would be shorter than that for the single-mode case.

Case C: $\alpha < 0.4$

Using the parameters

$$n_a = 2.0, \qquad \lambda = 0.8\ \mu\text{m}, \quad \delta = 10^{-5},$$

$$w_1 = 100\ \mu\text{m}, \quad a = 50\ \mu\text{m},$$

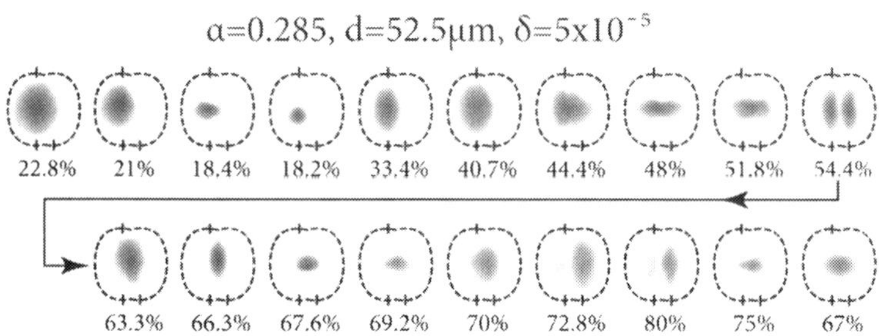

Figure 15.27. A cross-sectional view of the power density distribution for the guided wave. Note the power exchange phenomenon as the wave propagates along the coupled structure. The percent value indicates the percent of the total power contained in the right-hand half of the structure (i.e., fiber 2). The distance between each frame is 7,024 μm [9]

one obtains $\alpha = 0.285$. Based on the previous discussion, one concludes that when an incident beam is focused very sharply within the fiber, many higher order modes are generated. It is recognized that maximum power transfer (coupling) occurs when the beam waist is the largest. Since the incident beam is so sharply focused, within a given longitudinal distance the beam goes through many cycles of focusing and defocusing, which implies that there exist many occasions when maximum power transfer takes place. Shown in Fig. 15.27 is the several cycles of maximum power transfer.

Synchronous focusing and defocusing of the beams in both fibers can be observed. In other words, maximum power transfer tends to occur among similar beams in neighboring fibers even for the multimode case. For the chosen parameters, the incident Gaussian beam excited many modes in an unequal (amplitude-wise) fashion. It is therefore expected that the guided power would not be divided equally among the two fibers [27]. The back-and-forth power exchange phenomenon should still be present as one may see from Fig. 15.27. This type of wave coupling behavior highlights the importance of the beam characteristics of the incident beam. Not all modes are excited equally by an incident Gaussian beam. Hence, one cannot usually expect the split of the guided beam energy in a 50 50 manner even for highly multimoded fibers [28]. In fact, it seems most likely that the back-and-forth power exchange behavior will prevail even for highly multimoded fiber couplers. This fact adds to the complexity in the proper design of fiber couplers. It also means that the simple intuitive design idea, which assumes that the incident guided beam energy in a multimode guide is split in a 50–50 manner among two coupled multimode fibers, is not easily justified.

In Fig. 15.28 we have plotted the total amount of power carried in fiber 1 as a function of the normalized distance z/z_0, with $z_0 = 3,512\,\mu\text{m}$ for various

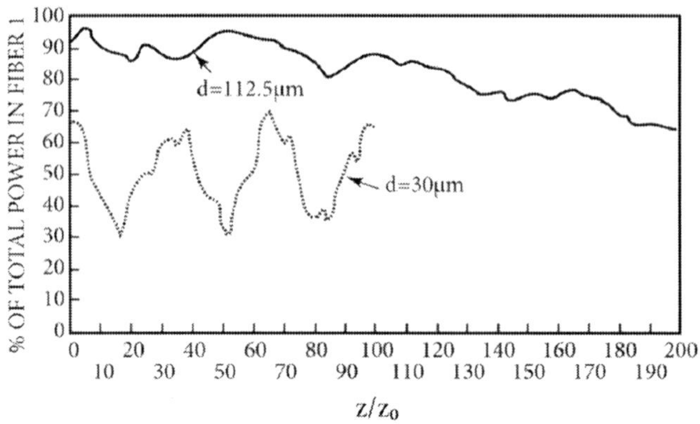

Figure 15.28. Percent of total power in the left-hand half of the structure (i.e., fiber 1) as a function of the axial distance along the coupling structure. This is the multimode coupling case [9]

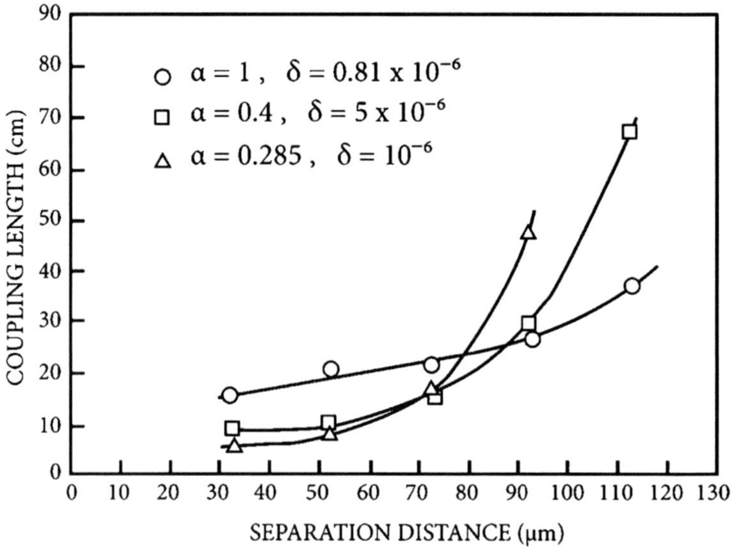

Figure 15.29. Coupling length as a function of separation distance for various single-mode and multimode cases for the fiber coupler [9]

separation distances. It is seen that the curves are even more ragged than those in Fig. 15.26 for the $\alpha = 0.4$ case. This behavior is again attributable to the presence of many higher order modes. The outstanding feature of these curves is that the periodic power exchange still prevails between the fibers.

Based on the above calculations one may summarize the results on the coupling length as a function of the separation distance for various values of α in Fig. 15.29.

It is seen that at a small separation distance for which the parabolic index profile of each fiber is substantially distorted, the coupling length for the highly multimoded guide is shorter than that for the single-mode guide. At a large separation distance for which the parabolic index profile for each fiber is preserved, the coupling length for the highly multimode guide appears longer than that for the single-mode guide. However, this observation is somewhat deceiving because the index profile for the multimode case is not the same as that for the single-mode case. When the same index profiles are used, the beam for the single-mode case occupies only a very small region in the center of the guide, while the beam for the multimode case expands (defocuses) and contracts (focuses) periodically so that stronger coupling occurs when the beam expands. Hence, the coupling length for the multimode case is shorter.

Several general observations may be made on the parabolic index fiber coupler that are consistent with our calculated results.

- For an on-axis Gaussian beam, periodic power exchange between two graded index fibers prevails always
- The coupling lengths are usually shorter when operating under multimode condition than under single-mode condition
- Coupling characteristics are critically dependent on the initial beam characteristics. Hence, the beam characteristics at the entrance of the fiber coupler are very important in order to predict accurately the coupling behavior of the coupler

15.3.4 Fiber Tapers and Horns [10]

The versatility of the BPM can best be appreciated by illustrations in which there are 3D variations in the refractive index, that is,

$$\begin{aligned} n(x,y,z) &= n_a\left[1-\frac{\delta r^2}{a^2(z)}\right] \qquad 0 \le r \le a, \\ &= n_a(1-\delta) \qquad\qquad r \ge a, \end{aligned} \tag{15.52}$$

where n_a and δ are given constants. Here, $a(z)$ is the radius of the circular taper or circular horn structure as a function of the longitudinal distance z (see Figs. 15.30 and 15.31). A Gaussian beam $u(x,y)$ is incident at $z=0$:

$$u(x,y) = u_0 \exp(-r^2/w_1^2), \tag{15.53}$$

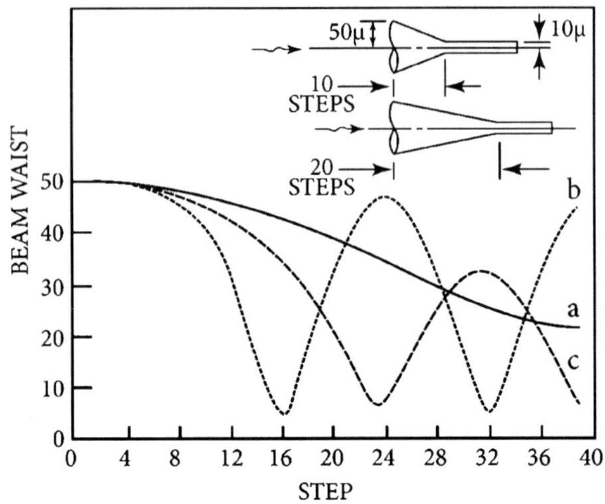

Figure 15.30. Beam waist as a function of axial distance for the following three tapered fibers: (**a**) no taper; (**b**) tapers in 10 steps from a fiber with a radius of 50 μm to a fiber with a radius of 10 μm; and (**c**) tapers in 20 steps from a 50 μm fiber to a 10 μm fiber. Each step size is 487 μm. The beam size is $\alpha = 0.42$ and $w = 50$ μm [10]

where w_1 is a given constant. It is of interest to learn how the beam evolves as it propagates down the structure. Figures 15.30 and 15.31 show the results obtained according to the BPM.

15.3.5 ω-β Diagram from BPM [11]

In this section we shall demonstrate how to generate the ω-β diagram for propagating modes using the BPM results. For structures with index profile given by

$$n(x, y) = n_a\sqrt{1 - 2\Delta f(x, y)}, \tag{15.54}$$

where n_a is a given constant, Δ is a profile height parameter, and $f(x, y)$ is the functional variation of the profile, the weakly guiding approximation applies when $\Delta \ll 1$. Under this approximation, wave solutions for these weakly guiding structures may be found from solutions of the scalar-wave equation (see Sect. 2.8). Physically it means the following:

- Modes are nearly TEM waves
- Spatial dependence of the transverse fields is governed by the scalar-wave equation
- Polarization properties of the guiding structure, as manifested by $\nabla_T \ln n^2$ in the vector wave equation where, ∇_T is the transverse del operator, are ignored

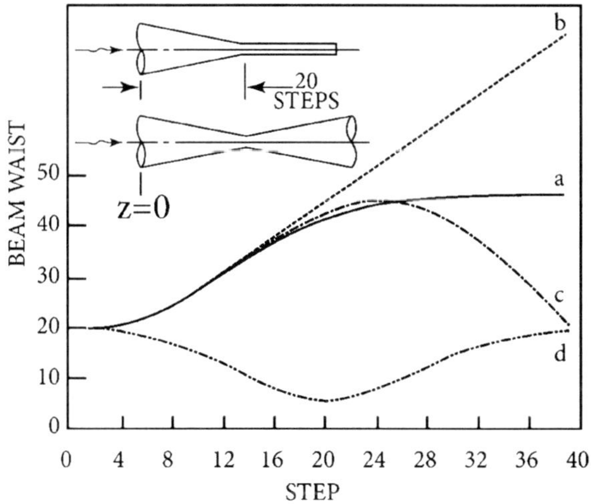

Figure 15.31. Beam waist as a function of axial distance for various taper transitions: (**a**) fiber with double tapers from a fiber with a radius 50 μm to a fiber with radius 10 μm to a fiber with radius 50 μm in 40 steps (see insert); (**b**) fiber of radius 50 μm with no taper; (**c**) fiber with radius 50 μm tapers to a fiber with radius 10 μm in 20 steps; and (**d**) fiber with radius 10 μm with no taper. Each step size is 487 μm. The beam size is $\alpha = 2.636$ and $w = 20$ μm [10]

Snyder has derived polarization corrections to the scalar propagation constant using the perturbation method [29]

$$\delta\beta_m \simeq \frac{a(2\Delta)^{3/2}}{2V} \frac{\int_A (\nabla_T \cdot \mathbf{E}_{tm})\, \mathbf{E}_{tm} \cdot \nabla_T f(x,y)\, dA}{\int_A |\mathbf{E}_{tm}|^2\, dA}, \tag{15.55}$$

with

$$\mathbf{E}_{tm} = E_{xm}\mathbf{e}_x + E_{ym}\mathbf{e}_y, \tag{15.56}$$

$$E_{xm} \sim \phi_m(x,y) \exp\left[-j\left(\beta_m + \delta\beta_{xm}\right) z\right], \tag{15.57}$$

$$E_{ym} \sim \phi_m(x,y) \exp\left[-j\left(\beta_m + \delta\beta_{ym}\right) z\right], \tag{15.58}$$

$$\left\{\nabla_T^2 + \left[k^2 n^2(x,y) - \beta_m^2\right]\right\} \phi_m(x,y) = 0\,, \tag{15.59}$$

and

$k \quad = 2\pi/\lambda,$

$\lambda \quad$ = free-space wavelength,

$$\begin{aligned}
\beta_m &= \text{scalar-wave propagation constant for } m\text{th mode,} \\
\mathbf{e}_x, \mathbf{e}_y &= \text{unit vectors parallel to } x, y \text{ Cartesian axes,} \\
V &= kan_0(2\Delta)^{1/2}, \\
\phi_m &= \text{transverse scalar wave-function for } m\text{th mode,} \\
a &= \text{core radius or half-width,} \\
m &= m\text{th mode number,} \\
\text{A} &= \text{infinite cross section.}
\end{aligned} \tag{15.60}$$

The BPM approach starts with the 3D scalar wave equation [7]

$$\left[\nabla^2 + k^2 n^2(x, y, z)\right] u(x, y, z) = 0, \tag{15.61}$$

where ∇ is the 3D del operator and u is the total scalar field, which is related to the vector electric field $\mathbf{E}(x, y, z)$ as follows:

$$\mathbf{E}(x, y, z) = \mathbf{e}_{\mathrm{p}} u(x, y, z). \tag{15.62}$$

Here, $\mathbf{e}_{\mathrm{p}}$ is a unit vector in the direction of the initial polarization of the wave. The scalar-wave equation (15.61), with the boundary condition on the initial surface, and the radiation condition at infinity completely specifies $u(x, y, z)$, from which one may obtain the electromagnetic field vectors $\mathbf{E}$ and $\mathbf{H}$ [7]. The evolution of the scalar-wave function $u(x, y, z)$ as the wave propagates along z, the direction of propagation, can be found according to the forward-marching fast Fourier transform technique (BPM), which has been described earlier. Several important characteristics of this technique are summarized in the following:

- To obtain the scalar-wave equation, see (15.61), the term $\nabla(\epsilon^{-1}\nabla\epsilon \cdot \mathbf{E})$, where $\epsilon = n^2(x, y, z)$ in the vector-wave equation, is discarded. This means that the polarization properties of the guiding structure are ignored. Hence, the waveguide is properly defined as the weakly guiding structure.
- Paraxial approximation (or the parabolic approximation) was used.
- In contrast with the usual modal approach, which deals with one mode at a time, this technique provides the evolution of the total field as it propagates down the guiding structure. It is recognized, however, that the total field may be decomposed into a set of orthonormal guided modes.

- The forward-marching characteristic of this technique implies that backward propagating waves are not allowed to exist (or are ignored), that is, no reflection along the direction of propagation is taken into consideration.

This BPM technique will now be applied to the problem of wave propagation along a number of circular or noncircular inhomogeneous and weakly guiding dielectric structures to yield information on the total field. Knowing the complete description of the total field one may derive the propagation characteristics for each normal mode on this given structure [11, 30].

All the information necessary for a complete description of the field can be extracted from the total scalar-wave function obtained according to the technique described previously. We note that the complex field amplitude $u(x, y, z)$ can be expressed as a superposition of orthonormal mode eigenfunctions as follows:

$$u(x, y, z) = \sum_m A_m \phi_m(x, y) \exp(-j\beta_m z), \tag{15.63}$$

where m is the mode number, β_m are the mode eigenvalues (or propagation constants), and A_m are the mode amplitudes, which are determined by a given incident field $u_0(x, y, z)$. The mode propagation constants β_m can be determined from a computation of the correlation function

$$P(z) = \int\int_A u^*(x, y, 0) u(x, y, z)\, dxdy,$$

$$A = \text{cross-sectional area}\ . \tag{15.64}$$

Using (15.63) and the orthogonality of the mode eigenfunctions $u_m(x, y)$, (15.64) can be reduced to

$$P(z) = \sum_m |A_m|^2 \exp(-j\beta_m z). \tag{15.65}$$

The Fourier transform of (15.65) with respect to z is

$$P(\beta) = \sum_m |A_m|^2\, \delta(\beta - \beta_m), \tag{15.66}$$

which suggests that the calculated spectrum of $P(z)$ will display a series of resonances with maximum at $\beta = \beta_m$ and peak values proportional to the mode weight coefficients

$$w_m = |A_m|^2\ . \tag{15.67}$$

In practice, only a finite record of $P(z)$ is available and that record must be enhanced with the multiplication by a window function $w(z)$ before the Fourier transform is computed. The resulting resonances in the spectrum $P(\beta)$, after multiplication by a window function, will thus exhibit a finite width and shape of the record length Z and the window function $w(z)$. A sample curve of $P(\beta)$ for a waveguide with a ten propagation mode is shown in Fig. 15.32. Since, in general, the resonant peaks do not coincide with the sampled values of β, errors will result in the values of w_m and β_m obtained from the maxima in the sampled data set for $P(\beta)$. For example, the maximum uncertainty in β_m so determined will be

$$\Delta\beta_m = \frac{1}{2}\Delta\beta = \frac{\pi}{Z}, \tag{15.68}$$

where $\Delta\beta$ is the sampling interval in $P(\beta)$. It is possible to reduce the uncertainties in β_m and w_m by lengthening Z, but it is far more efficient to first fit the correct lineshape function intrinsic to $w(z)$ to the sampled value of $P(\beta)$ that are closest to the resonances under consideration and to determine these β_m and w_m from the resulting lineshape fit. For the window function

$$\begin{aligned} w(z) &= 1 - \cos\frac{2\pi z}{Z} \qquad z < Z \\ &= 0 \qquad\qquad\qquad\; z > Z \end{aligned} \tag{15.69}$$

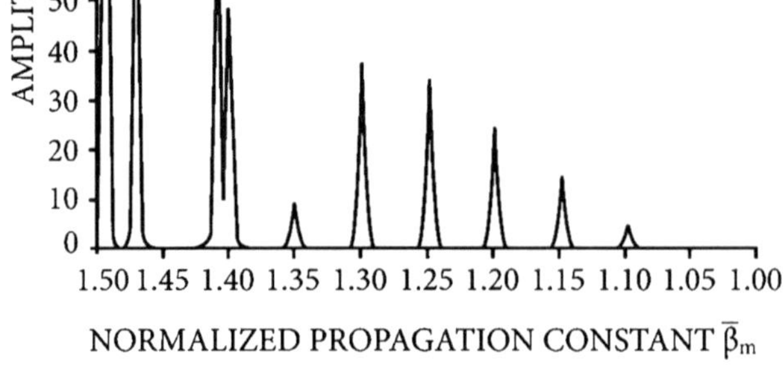

Figure 15.32. Modal power spectrum for dielectric waveguide with ten propagation modes [11]

used here, the normalized line shape corresponding to the harmonic z-dependence $\exp(-j\beta_m z)$ is

$$
\begin{aligned}
(\beta - \beta_m) &= \frac{1}{Z}\int_0^Z \exp\left[j(\beta - \beta_m)z\right] w(z)\, dz \\
&= \frac{\exp\left\{j\left[(\beta - \beta_m)Z + 2\pi\right]\right\} - 1}{j(\beta - \beta_m)Z} \\
&\quad -\mathcal{L}\frac{1}{2}\left[\frac{\exp\left\{j\left[(\beta - \beta_m)Z + 2\pi\right]\right\} - 1}{j\left[(\beta - \beta_m)Z + 2\pi\right]}\right. \\
&\quad \left. + \frac{\exp\left\{j\left[(\beta - \beta_m)Z - 2\pi\right]\right\} - 1}{j\left[(\beta - \beta_m)Z - 2\pi\right]}\right]. \qquad (15.70)
\end{aligned}
$$

Therefore, it is possible to represent $P(\beta)$ over the range of β that corresponds to guided modes as

$$
P(\beta) = \sum_m w_m \mathcal{L}(\beta - \beta_m). \qquad (15.71)
$$

By fitting $(\beta - \beta_m)$ to sampled values of $P(\beta)$ it is possible to compute extremely accurate solutions for w_m and β_m.

To study the accuracy of the BPM described above, we shall compare our results with those obtained according to other exact or approximate methods. Specifically, we shall compare the results for the following waveguides: (1) the step-index circular fiber; (2) the graded-index circular fiber; (3) the rectangular fiber; (4) the elliptical fiber; (5) the triangular fiber; and (6) the diffused channel rectangular waveguide.

15.3.5.1 The Step-Index Circular Fiber

The step-index circular fiber consists of a dielectric cylinder with homogeneous refractive index equal to n_1 surrounded by a cladding layer with a smaller index n_2. Because of pulse widening effects when two or more modes are propagating, this waveguide is usually used for single-mode operations. The cladding layer not only is used for mechanical support but also for selecting the n_1/n_2 ratio to be close to unity so as to allow for large core size even under single-mode operating conditions.

The exact solution for the propagating characteristics of the step-index fiber is developed by Dil and Blok [3]. For their computation, the refractive index was assumed to be

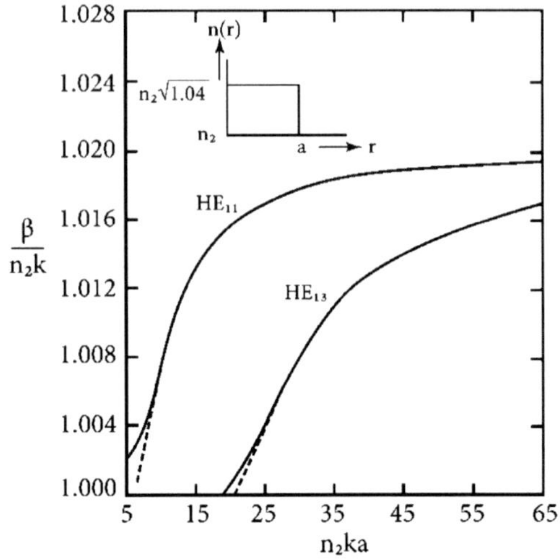

Figure 15.33. Dispersion curves for a step-index fiber. *Solid line*: Dil and Blok's results. *Dashed line*: BPM approach [11]

$$n(r) = \left\{ \begin{array}{ll} n_2\sqrt{1.04} & \text{for } 0 \le r \le a \\ n_2 & \text{for } r > a \end{array} \right\}. \tag{15.72}$$

The dispersion characteristics of the fiber is also computed by the BPM approach described in the previous section. In Fig. 15.33, the dispersion curves for the two modes are displayed where the solid line is the result reported by Dil and Blok and the dashed line is the result obtained by the BPM approach. As can be seen, the two methods produce almost identical results everywhere except near the cutoff region of the modes. This is due to the fact that near cutoff region the field is more loosely bound to the core region.

15.3.5.2 Graded-Index Circular Fiber

The graded index circular fiber has been analyzed by Dil and Blok [3] and the exact solution for the modal characteristics are obtained. Additionally, the finite element solution has been obtained by Yeh et al. [5, 6], which yields the exact result. In Fig. 15.34, the dispersion curves for the first five lower order modes of a graded-index circular fiber are shown when the refractive index is assumed to be

$$n(r) = \left\{ \begin{array}{ll} n_2\sqrt{1.04 - 0.04\frac{r^2}{a^2}} & \text{for } 0 \le r \le a \\ n_2 & \text{for } r > a \end{array} \right\}. \tag{15.73}$$

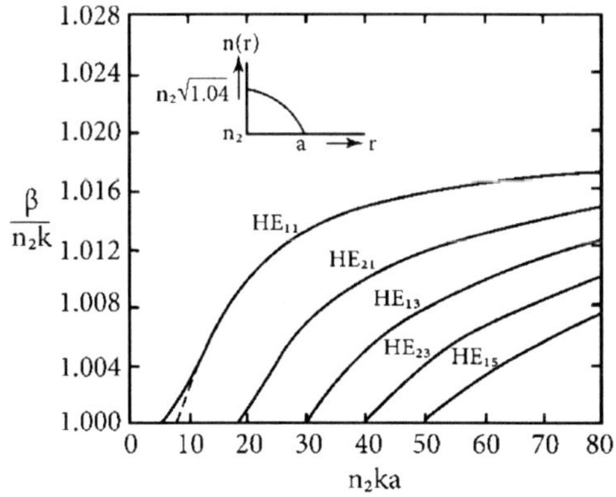

Figure 15.34. Dispersion curves for a graded-index fiber. *Solid line*: Dil and Blok's and finite element method's results. *Dashed line*: BPM approach [11]

In this figure, the solid line is the result reported by Dil and Blok and Yeh's finite element method and the dashed line is the one generated by the BPM approach. Again, it can be observed that the BPM approach produces accurate results everywhere except at the lower frequency range near the cutoff region of the dominant mode.

15.3.5.3 Rectangular Fiber

A rectangular optical fiber is a dielectric rod with rectangular cross section, surrounded by a medium with a refractive index less than the index of the rod. Several solutions for this optical fiber are available, such as Goell's circular harmonic approximate solution [2], Marcatili's approximate solution [23] and the finite element method solution [5,6].

To compare the results obtained by the BPM approach to the ones presented by Goell and Marcatili, a rectangular dielectric waveguide with core refractive index equal to 1.05 surrounded by a medium of index 1.01 is considered. The aspect ration of the guide a/b is assumed to be 2, where a and b represent the width and the height of the rectangular core, respectively. The dispersion curve for this problem is shown in Fig. 15.35, for the dominant modes E_{11}^y and E_{11}^x, which are almost degenerate for small core and cladding index differences. In Fig. 15.35, the normalized propagation constant $(\overline{\beta}^2 - n_2^2)/\left(n_1^2 - n_2^2\right)$ is plotted against the normalized frequency $2b(n_1^2 - n_2^2)^{1/2}/\lambda$, where $\overline{\beta} = \beta/k$, β is the propagation constant, and k and λ are the free space wave number and wavelength, respectively. The solid curve is obtained using Goell's computer solution of the boundary value problem. The dashed line is the Marcatili's approximate solution. The dotted line is the BPM approach solution. Goell's results are more reliable than those given

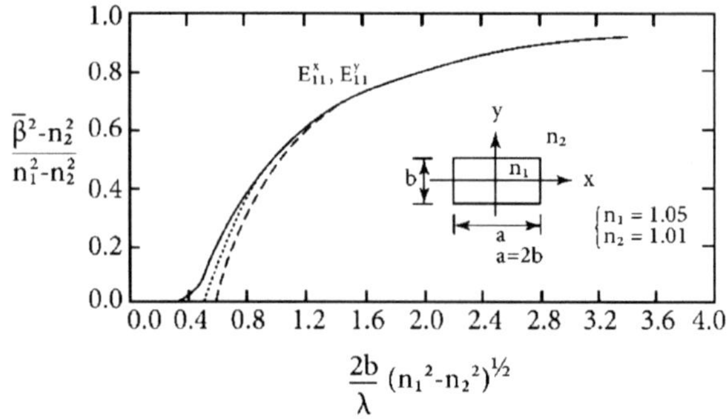

Figure 15.35. Dispersion curve for a rectangular fiber. *Solid line*: Goell's result. *Dashed line*: Marcatili's approximate result. *Dotted line*: BPM approach result

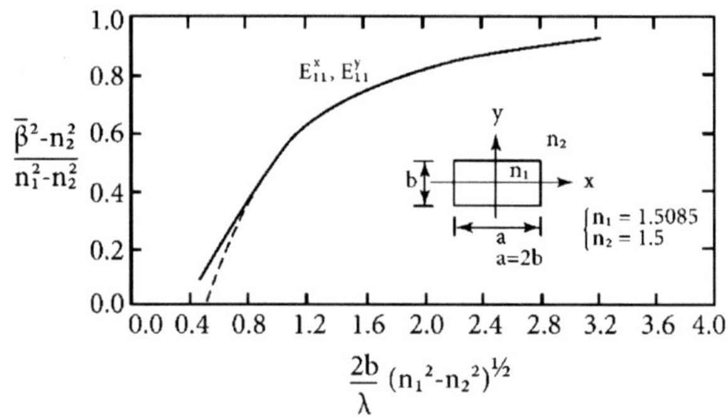

Figure 15.36. Dispersion curve for a rectangular fiber. *Solid line*: finite element method result. *Dashed line*: BPM approach result

by Marcatili's method, since the latter solution was obtained by assuming simple field distribution in the core region to match the fields along the four sides of the rectangular core. Therefore, comparing the curves of Fig. 15.35, we conclude that the scalar-wave approach is superior to Marcatili's approximate method.

To compare the results obtained by BPM approximation to those from the finite element method, a rectangular waveguide with core refractive index of 1.5085 and cladding index of 1.50 is considered. The dispersion curves for this problem are given in Fig. 15.36. The solid line shows the results from the finite element method and the dashed line shows the results obtained by the BPM approach. As can be observed, the two methods agree very closely everywhere except near the cutoff, where we expect the BPM approach to be less accurate.

15.3.5.4 Elliptical Fiber

An elliptical optical fiber is a dielectric rod with elliptical cross section and refractive index n_1, surrounded by a medium of smaller refractive index n_2. This optical fiber has practical importance since circular fibers commonly used in optical communication systems are easily deformed to elliptical shapes. This type of optical fiber was analyzed by Yeh [27, 31], who showed that for this class of dielectric waveguides all modes are hybrid modes. Later, Yeh et al. [6] computed the dispersion characteristics of elliptical fibers by the finite element method. The BPM approach can also be used to analyze this type of optical fibers.

Consider an elliptical optical fiber with core index $n_1 = 1.5085$ and cladding index $n_2 = 1.5$. The aspect ratio of the guide b/a is assumed to be 2, where a and b are the length of the major and minor axis of the core, respectively. The dispersion curve of the dominant modes for this problem is given in Fig. 15.37. The solid line shows the result obtained from the finite element method and the dashed line represents the result computed by the BPM approach. As can be seen, the two methods agree very closely almost everywhere except near the cutoff region.

15.3.5.5 Triangular Fiber

Consider a triangular optical fiber with the core refractive index $n_1 = 1.5085$ and the cladding refractive index $n_2 = 1.50$. Assume the fiber to have an equilateral triangular core region with a height equal to b. The dispersion curve of the dominant modes is given in Fig. 15.38. The solid line shows the results from the finite element method [6], while the dashed line shows the results obtained from the BPM approach. Similar to previous problems, the two methods are in close agreement everywhere except near the cutoff region, where we expect the finite element technique to yield more accurate results.

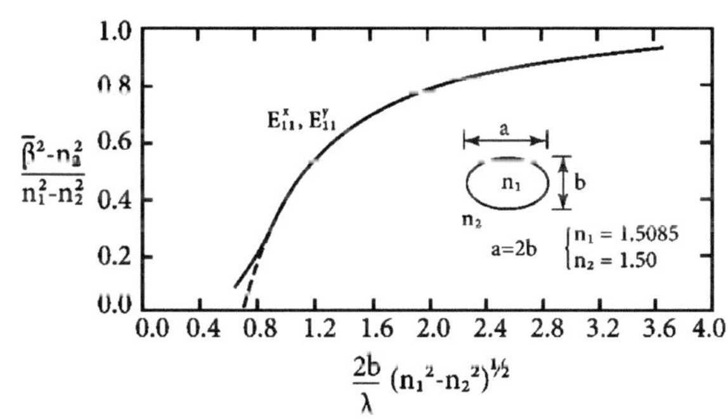

Figure 15.37. Dispersion curve for an elliptical fiber. *Solid line*: finite element method result. *Dashed line*: BPM result

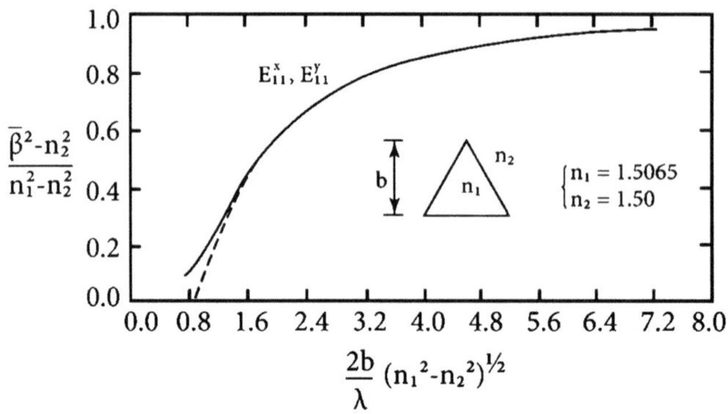

Figure 15.38. Dispersion curve for a triangular fiber. *Solid line*: finite element method result. *Dashed line*: BPM approach result

15.3.5.6 Diffused-Channel Rectangular Waveguide

A diffused-channel rectangular waveguide consists of a dielectric rod with rectangular cross section and refractive index n_1, embedded in another dielectric of slightly smaller index $n_2 = n_1(1 - \Delta)$, all surrounded by a medium which could be air or a dielectric medium with index n_0. For this class of optical waveguides, the Marcatili's approximate solution [21] as well as the finite element method solution [6] is available.

To compare the solution of the BPM approach with the solution obtained from the methods mentioned above, consider a diffused-channel waveguide with $n_1 = 1.5$, $n_2 = 1.45$, and $n_0 = 1$. The aspect ratio of the guide a/b is assumed to be 2, where a is the width and b is the height of the channel. The dispersion curves of the dominant modes for this problem are given in Fig. 15.39. In this figure, the solid line is the finite element result, the dashed line is the Marcatili's approximate result, and the dotted-dashed line shows the BPM approach solution. As can be seen, for this problem the BPM approach does not yield accurate results even for the range away from the cutoff region. This is because of the breakdown of the BPM approximation due to large index differences for this type of problem. In general, the BPM approach does not yield accurate results for problems where the index difference is larger than 10%. Fortunately, for most practical cases the index difference is only a few percent, which makes them suitable for treatment by the method described here.

15.3.5.7 Non-Axisymmetric Graded-Index Fiber

Consider a graded-index circular fiber, or the type discussed above, but with an index profile with its axis of symmetry shifted to one side by d. The geometry of this problem as well as its dispersion curves for the three lowest order modes

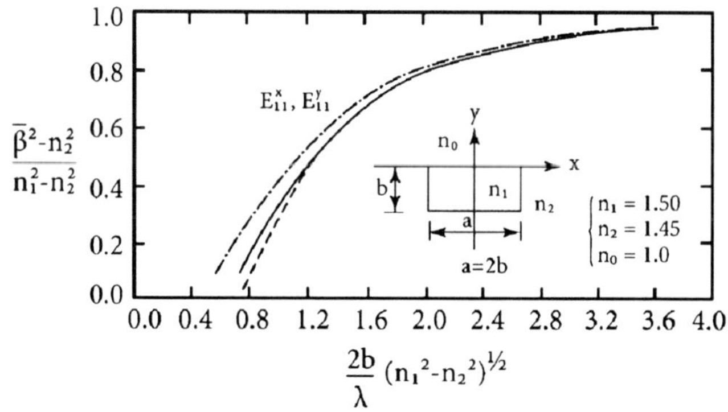

Figure 15.39. Dispersion curve for a uniform channel waveguide. *Solid line*: finite element result. *Dashed line*: Marcatili's approximate method result. *Dotted-dashed line*: BPM approach result

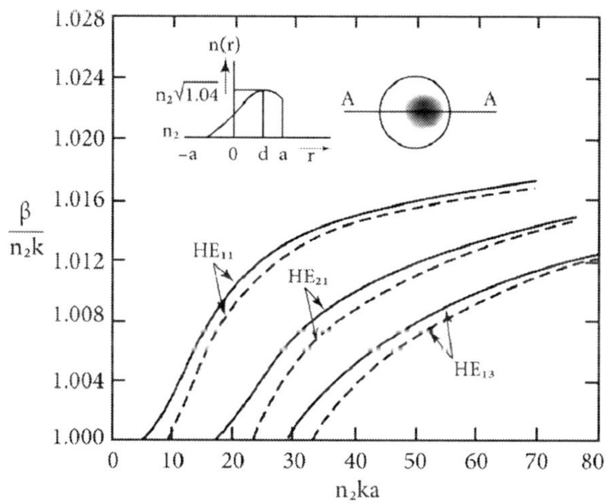

Figure 15.40. Dispersion curve for a graded-index circular fiber with shifted index waveguide. *Solid line*: results for $d = 0$. *Dashed line*: results for $d = a/2$

are given in Fig. 15.40. In this figure, the normalized propagation constant β/n_2k is shown vs. the normalized frequency n_2ka. The solid line is for the case where $d = 0$ and the dashed line is for $d = a/2$. It must be noted that the variation of the dispersion characteristics of this guide does not depend on d in a linear fashion, but that it changes much faster as d is increased from zero to a. The example discussed here envisions one of the possible ways that a graded optical fiber can be distorted due to fabrication errors.

15.4 Finite Difference Time Domain Method (FDTD) [1, 32]

Unlike the BPM, this finite difference time domain method (FDTD), like the FEM, is an exact treatment of the electromagnetics problem [1, 32]. The complete 3D Maxwell equations with the appropriate boundary conditions are solved in the time domain. These equations are expressed in a linearized form by means of central finite differencing. Only nearest-neighbor interactions need be considered as the fields are advanced temporally in discrete time steps over spatial cells of rectangular shape. Although FDTD is most suited to computing transient responses, it is still applicable to a single frequency or continuous wave (CW) response. For linear wave interaction problems, the CW wave response at all field points can be obtained over a wide band of frequencies by discrete Fourier transformation of the computed field-vs.-time waveforms at these points.

The Yee algorithm [1] is used to solve for both electric and magnetic fields in time and space using the coupled Maxwell curl equations. The **E** and **H** components are centered in 3D space so that every **E** component is surrounded by four circulating **H** components, and every **H** component is surrounded by four circulating **E** components. This way, the finite-difference expressions for the space derivatives used in the curl operations are central-difference in nature and second-order accurate, continuity of tangential **E** and **H** is naturally maintained across an interface of dissimilar materials if the interface is parallel to one of the lattice coordinate axes, and the location of the **E** and **H** components in the Yee space lattice and the central-difference operations on these components implicitly enforce the two Gauss' Law relations. Furthermore, the Yee algorithm also centers its **E** and **H** components in time in a leapfrog manner. In other words, all the **E** components in the modeled space are completed and stored in memory for a particular time point using previously stored **H** data. Then all the **H** computations in the space are completed and stored in memory using the **E** data just computed. The cycle begins again with the recomputation of the **E** components based on the newly obtained **H**. This process continues until time-stepping is concluded. This means that the leapfrog time-stepping is fully explicit, thereby avoiding problems involved with simultaneous equations and matrix inversion, the finite-difference expressions for the time-derivatives are central-difference in nature and second-order accurate, and the time-stepping algorithm is non-dissipative. The FDTD computer program is written according to Yee's algorithm. In the following illustrative examples, the FDTD method [32] will be used to compute the results [12].

15.4.1 Excitation of a Ribbon Dielectric Waveguide

As indicated in Chap. 11, a thin ribbon-shaped dielectric waveguide can offer much reduced attenuation for the guidance of a dominant ${}_{e}HE_{11}$ mode. It would be of

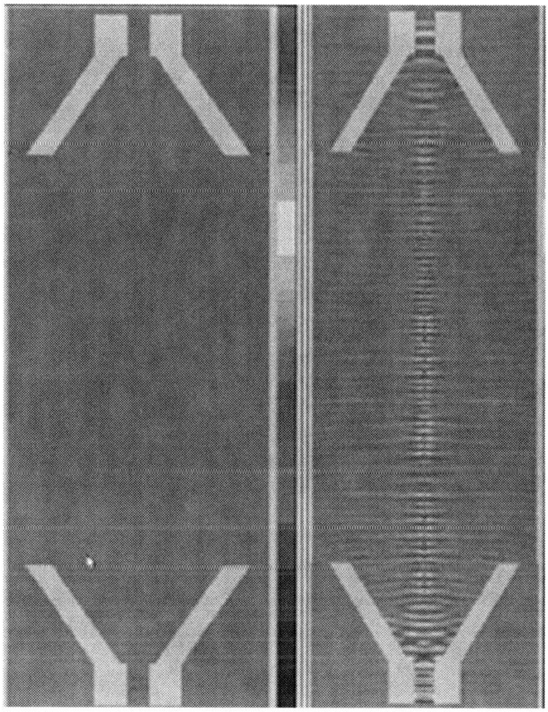

Figure 15.41. Excitation of a thin silicon ribbon using a horn. (*Left*) no ribbon and (*right*) silicon ribbon waveguide between the launching horns. The insertion loss is about 0.5 dB per transition

interest to learn how such a mode may be launched on this thin ribbon dielectric waveguide from a horn structure. Since it was impossible to analyze this problem analytically, FDTD numerical simulation was performed. This realistic simulation in the 2.25–2.75 THz range was done for a silicon ribbon with dielectric constant of 11.8 and loss tangent of 0.01. The simulation results are displayed in Fig. 15.41. It could be clearly seen that the ${}_{e}EH_{11}$ mode was launched successfully from the horn structure. The fact that FDTD could be used easily to treat this rather complex setup should be noted. [33]

15.4.2 Ribbon Waveguide Assembled from Dielectric Rods

One way to make a ribbon waveguide is to glue an array of quartz fibers together with an adhesive, such as cyanoacrylid. This structure was analyzed with the FDTD code. The results showing the successful guidance of the ${}_{e}HE_{11}$ are displayed in Fig. 15.42.

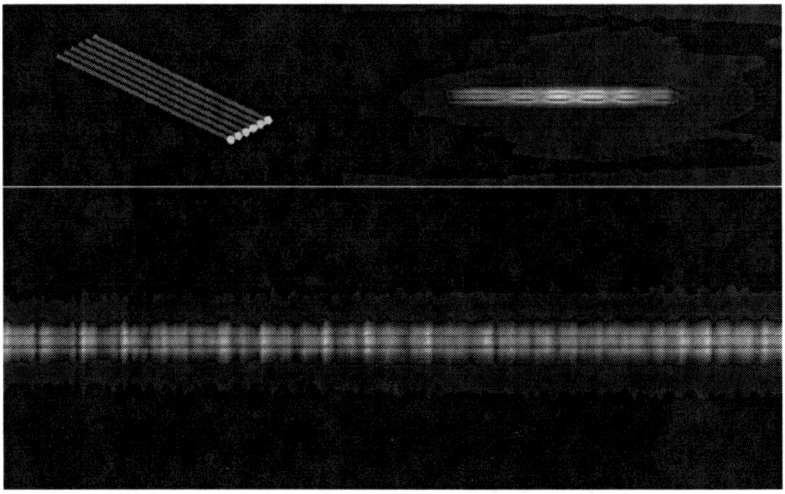

Figure 15.42. (*Top left*) Picture of a ribbon assembled from six quartz rods. Each rod is 80 μm in diameter. A 600 GHz signal is launched on the waveguide. (*Top right*) Cross-sectional view of the envelope of the transverse field; (*bottom*) side view of the envelope of the transverse field at the center of the structure as the signal propagates along the rod

15.4.3 Dielectric Waveguide Transitions [34, 35]

The FDTD method has also been used to analyze the following waveguide transitions:

(a) Tapered Transitions

To provide tighter guidance of the dominant wave on a thin ribbon structure in order to accommodate cornering and bending of the guide, a tapered transition region is envisioned consisting of a tapered low relative dielectric constant (polymer) on the thin alumina ribbon. The upper of the structure is shown in Fig. 15.43. To conserve computational resources, this structure, placed on an image plane, was analyzed. The full transition would have a symmetric structure on the bottom. FDTD simulation results showing the successful transition are shown in Fig. 15.44 with insertion loss of 0.3 dB.

(b) Inverse Tapered Transition

Another way to design a tapered transition section is to invert the taper as shown in Fig. 15.45. The outline of the top half of the structure is shown, starting from the bare alumina ribbon through the inverted transition to the polymer coated alumina, and using the same transition to the bare alumina ribbon. The dimensions of the

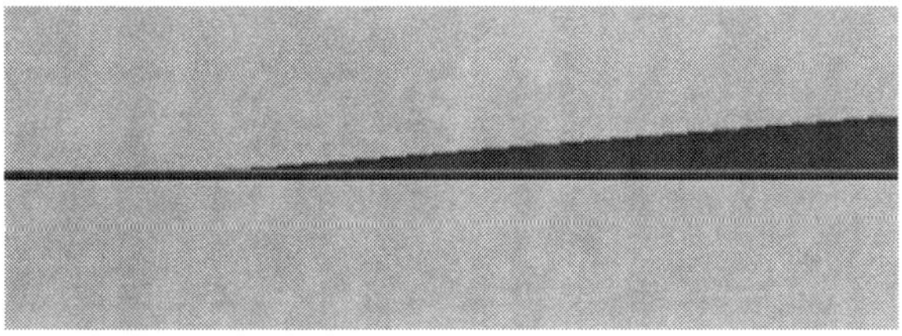

Figure 15.43. Image of the transition from a high relative dielectric constant (alumina) ribbon to a polymer (Teflon) coated alumina ribbon. The dimensions of the full alumina ribbon is $0.06\lambda_0$ (height) by $0.6\lambda_0$ (width). The polymer coating, of the same width as the alumina ribbon, transitions from a thickness of zero to a final height of $0.26\lambda_0$ at a distance of $10\lambda_0$. λ_0 is the free space wavelength

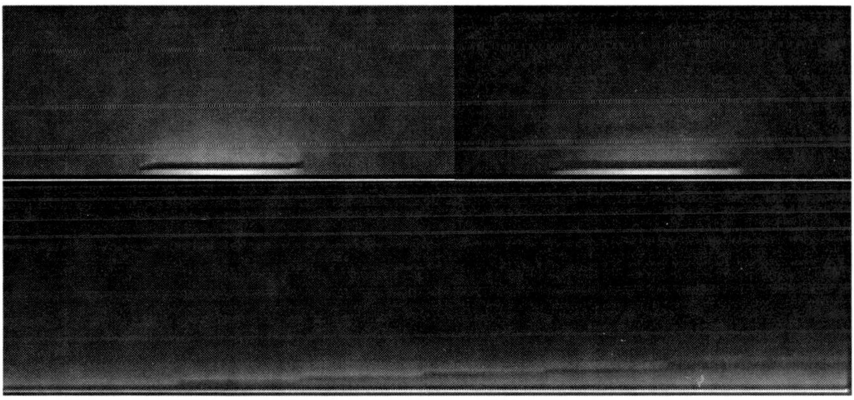

Figure 15.44. (*Top left*) Envelope of the transverse electric field on the bare alumina ribbon before the start of the transition; (*top right*) envelope of the transverse electric field at the end of the transition. At the end of the transition most of the guided power is confined within the polymer coating; (*bottom*) side view of the envelope of the transverse electric field as it propagates along the ribbon and through the tapered transition

full alumina ribbon is $0.06\lambda_0$ (height) by $0.6\lambda_0$ (width). The tapered sections have a full opening aperture of $2\lambda_0$. The polymer has the same width as the alumina and has a thickness $0.26\lambda_0$ at the center. Again FDTD method is used to simulate the propagation through this transition. The successful results are shown in Fig. 15.45 with insertion loss of 0.22 dB.

(c) Step-Index Transition

Strictly speaking, a step-index transition is a discontinuity, not a transition. It is nevertheless an important circuit element because it can be used to design filters for the ribbon-based circuits, provided that radiation loss is not excessive. A picture of a step-index transition is shown in the top of Fig. 15.46. The dimensions of the

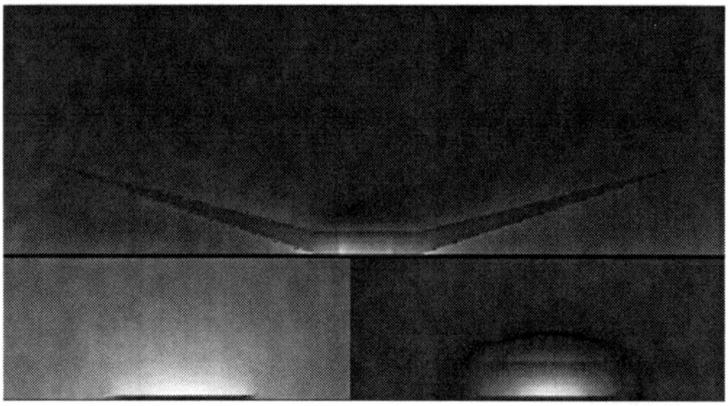

Figure 15.45. (*Top*) Envelope of the transverse electric field as the signal propagates from the alumina through the transition and polymer-coated section through the second transition and on to the bare alumina section; (*bottom left*) envelope of the transverse electric field before the signal enters the transition section; (*bottom right*) envelope of the transverse electric field at the center of the structure

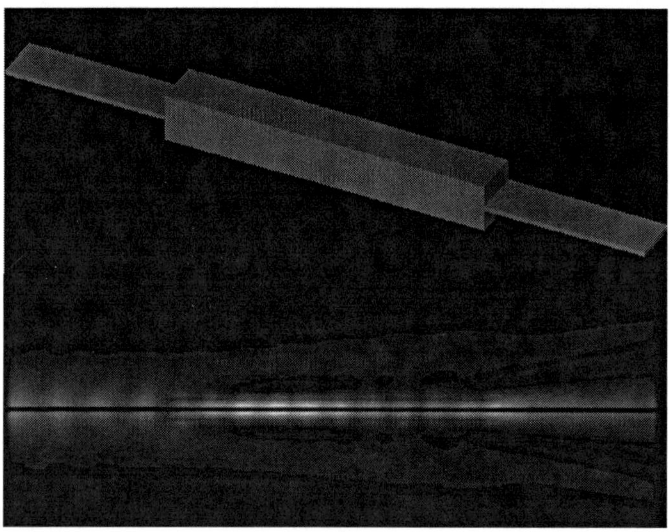

Figure 15.46. (*Top*) Picture of the step index transition and (*bottom*) envelope of the transverse electric field as it propagates through the structure

full alumina ribbon is $0.06\lambda_0$ (height) by $0.6\lambda_0$ (width) and the polymer thickness is $0.26\lambda_0$. Using FDTD, a simulation study is conducted. In Fig. 15.46, the bottom picture shows the envelope of the dominant mode $\mathbf{E}$-field perpendicular to the ribbon surface as the guided wave propagates through this transition from a bare high dielectric constant ribbon to a polymer-coated high dielectric constant ribbon and

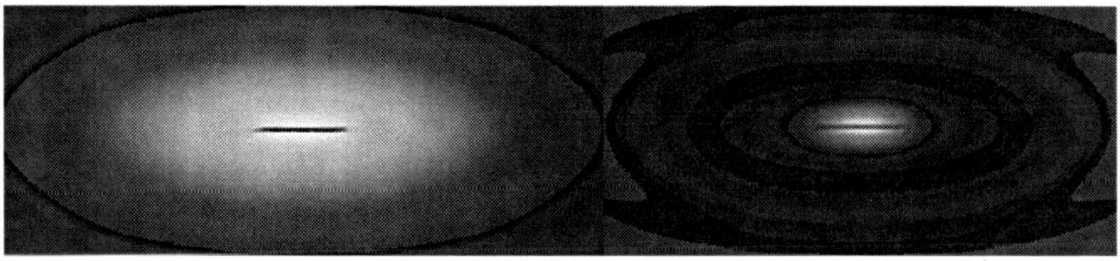

Figure 15.47. Envelope of the transverse electric field on the (*left*) alumina ribbon before the step transition and (right) in the polymer-coated section

back to a bare high dielectric constant ribbon. A standing wave is clearly seen at the input end of this transition showing a reflected wave. Although the significant reflective component is evident, it is encouraging to note from the S-parameter calculation that the radiation loss is not excessive. The total insertion radiation and reflection loss is approximately 0.5 dB. Therefore, the step index can be used as a circuit component, as in a filter design. Figure 15.47 shows the envelope of the transverse electric field on the alumina ribbon before the step and in the central polymer coated section.

(d) Microstrip Line and Ribbon Waveguide Transition

Microstrip lines of several forms (stripline, coplanar waveguide, microstrip, etc.) have been the backbone for wave propagation on planar microwave circuits for many decades. Despite their high-loss characteristics at submillimeter wavelengths, they are still being used due to a lack of alternative structures. To couple the available microstrip line geometry with the new low-loss ribbon structures, it is important to find an efficient and readily fabricated transition. For interchip or off-chip use, the simplest transition is to directly "butt-joint" the microstrip line with the ribbon waveguide. Because of inherent mismatches, this gives an unacceptable mismatch. However, if a butt joint is made between the microstrip and a polymer-coated ribbon, the situation is not nearly as bad. A sketch of such a transition is shown in top picture of Fig. 15.48.

Most of the transverse electric field for the microstrip is confined between its top conductor and its bottom conducting ground plane, and for the polymer-coated alumina ribbon over a ground plane the transverse electric field of the dominant mode is mostly confined within the polymer layer. The similarity of these transverse electric field patterns enables one to butt joint the polymer microstrip line to the polymer-coated high dielectric constant ribbon on a ground plane and to expect excellent field matching for a low-loss transition. In this simulation, a polyethylene substrate is used in the microstrip line and a polyethylene coating is used on the high dielectric constant ribbon. These dielectrics were chosen to get a closer impedance and phase velocity match. Computer simulation of the envelope of the transverse E field as it propagates from the microstrip line into the

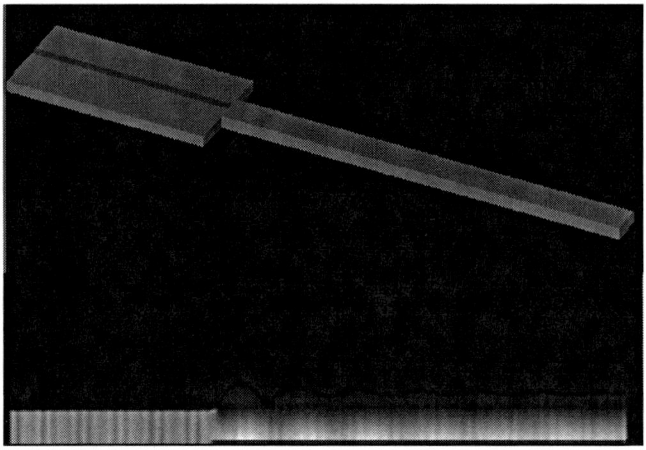

Figure 15.48. (*Top*) Picture of the microstrip to polymer-coated alumina ribbon transition. The substrate in the microstrip and the polymer coat on alumina is polyethylene. The thickness of the microstrip substrate is such that the height coincides with that of the polymer on the alumina. The thickness of the alumina is $0.03\lambda_0$ and the thickness of the polymer coat on the alumina is $0.26\lambda_0$. Both structures rest on a ground plane; (*bottom*) envelope of the transverse electric field as it propagates through the microstrip and polymer-coated alumina ribbon

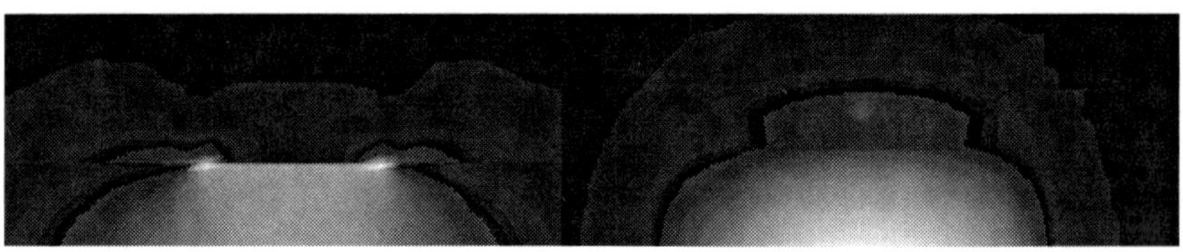

Figure 15.49. Envelope of the transverse electric field in the (*left*) microstrip guide and (*right*) polymer-coated alumina ribbon

polymer-coated high dielectric constant ribbon is shown in the bottom figure of Fig. 15.48. A standing-wave pattern is seen in the simulation. Radiated wave at the discontinuity can also be seen. This can be further reduced by optimizing the structural dimensions. A total loss (reflection and radiation) of 0.24 dB was found for this transition. The envelope of the transverse electric field in the microstrip and the polymer-coated alumina ribbon is shown in Fig. 15.49.

15.5 Concluding Remarks

The power of using a pure numerical approach in solving many problems that cannot be solved by analytical means is apparent. Nevertheless, it must be noted that the results obtained by numerical techniques are only as good as the understanding of the problem at hand and is usually guided by the analytic solution of a canonical problem. For the sake of discussion, let us use the example of a ribbon-like dielectric waveguide. The canonical problem of a ribbon-like structure is that of wave propagation on an elliptical dielectric waveguide. A subset of this problem is the problem of wave propagation on a circular dielectric waveguide. Some of the major findings that are applicable to the ribbon waveguide are the following:

- Only hybrid modes containing all six field components can exist on elliptical dielectric waveguide or on a ribbon waveguide.
- The dominant modes are also hybrid modes: One with its dominant electric field parallel to the major axis of the ribbon and one with its dominant electric field parallel to the minor axis of the ribbon.
- No evanescent mode can exist on the ribbon waveguide. Below the cutoff of a higher order mode, that mode simply no longer exists.
- Unless the excitation field exactly matches the total modal field of modes that can exist on the dielectric structure, unguided radiation fields will be launched and lost.
- Any disturbance or deviation (such as curves, bending, twisting, narrowing or broadening, blemishes, etc.) that occurs to a perfectly straight dielectric structure will generate radiated wave and/or other higher order guided wave if they can exist on this structure. No evanescent modes are generated, because they do not exist on a dielectric waveguide.
- Characteristic impedance for a given propagating mode used in transmission lines, microstrip lines, or metallic waveguides is usually defined as the ratio of transverse electric field to transverse magnetic field of that mode. When this definition is used for the hybrid mode on an elliptical or ribbon dielectric waveguide, the characteristic impedance becomes dependent on the transverse spatial coordinates of the ribbon. In the traditional microwave circuit design, the characteristic impedance of a transmission circuit must have a value that is independent of any spatial coordinates. This ambiguity must be resolved before the conventional circuit techniques can be used to analyze the ribbon waveguides that are based on the conventional transmission line theory.

Illustrations given in this chapter show the versatility and the power of the numerical approaches. Using them with fundamental understanding of the underlying principles can provide great benefits and insight in the solutions to many important practical problems.

References

1. K. S. Yee, "Numerical solution of initial boundary value problems involving Maxwell's equations in isotropic media," IEEE Trans. Antenn. Propag. **14**, 302 (1966)
2. J. E. Goell, "A circular-harmonic computer analysis of rectangular dielectric waveguides," Bell Syst. Tech. J. **48**, 2133 (1969)
3. J. G. Dil and H. Blok, "Propagation of electromagnetic surface waves in a radially inhomogeneous optical waveguides," Opto-Electronics **5**, 415 (1973)
4. C. Yeh and G. Lindgren, "Computing the propagation characteristics of radially stratified fibers – An efficient method," Appl. Opt. **16**, 483 (1977)
5. C. Yeh, S. B. Dong, and W. Oliver, "Arbitrarily shaped inhomogeneous optical fiber or integrated optical waveguides," J. Appl. Phys. **46**, 2125 (1975)
6. C. Yeh, K. Ha, S. B. Dong, and W. P. Brown, "Single-mode optical waveguides," Appl. Opt. **18**, 1490 (1979)
7. C. Yeh, L. Casperson, and B. Szejn, "Propagation of truncated gaussian beams in multimode or single-mode fiber guides," J. Opt. Soc. Am. **68**, 989 (1978)
8. M. D. Feit and J. A. Fleck Jr., "Light propagation in graded-index optical fibers," Appl. Opt. **17**, 3990 (1978)
9. C. Yeh, W. P. Brown, and B. Szejn, "Multi-mode or single-mode fiber couplers," Appl. Opt. **18**, 489 (1979)
10. C. Yeh, "Physics of Fiber Optics: Advances in Ceramics," Vol.**2**, B. Bendow and S. S. Mitra, eds., The American Ceramic Society, Ohio (1981)
11. C. Yeh and F. Manshadi, "On weakly guiding single-mode optical waveguides," J. Lightwave Tech. **3**, 199 (1985)
12. Quick-Wave-3D FDTD Software, QWED Sp. z o.o., Warszawa, Poland; Commercial programs for FDTD, FEM, BPM
13. R. F. Harrington, "Field Computation by Moment Methods," Series on Electromagnetic Wave Theory, IEEE Press, New York (1968)
14. Wolfram Mathematica Documentation Center, "The Numerical Method of Lines," Academic Press (2007); W. E. Schiesser, "The Numerical Method of Lines," Academic Press, New York (1992)
15. G. E. Mariki and C. Yeh, "Dynamic 3-D TLM analysis of microstrip-lines on anisotropic substrates," IEEE Trans. Microw. Theory Tech. **MTT-33**, 789 (1985)
16. P. Barber and C. Yeh, "Scattering of electromagnetic waves by arbitrarily shaped dielectric bodies," Appl. Opt. **14**, 2864 (1975)
17. L. P. Eyges, P. Gianino, and P. Wintersteiner, "Modes of dielectric waveguides of arbitrary cross sectional shapes," J. Opt. Soc. Am. **69**, 1226 (1979)
18. G. Mur, "Absorbing boundary conditions for finite-difference approximation of the time-domain electromagnetic-field equations," IEEE Trans. Electromagn. Compat. **23**, 1073 (1981)

19. S. Ahmed and P. Daly, "Finite-element methods for inhomogeneous waveguides" Proc. IEEE **116**, 1661 (1969)
20. S. B. Dong, J. A. Wolf Jr., and F. E. Peterson, "On a direct-iterative eigensoluti technique," Int. J. Num. Meth. Eng. **4**, 155 (1972)
21. J. R. Carson, S. P. Mead, and S. A. Schelkunoff, "Hyperfrequency waveguides – M- ematical theory," Bell Syst. Tech. J. **15**, 310 (1936)
22. E. Snitzer, "Cylindrical dielectric waveguide modes," J. Opt. Soc. Am. **51**, 491 ( )
23. E. A. J. Marcatili, "Dielectric rectangular waveguide and directional coupler fo- grated optics," Bell Syst. Tech. J. **48**, 2071 (1969)
24. F. Blum, D. Shaw, and W. C. Holton, "Optical stripline for integrated optical ts in epitaxial GaAs," Appl. Phys. Lett. **25**, 116 (1974); M. Ohtaka, M. Matsuh nd N, Kumagai, "Analysis of the guided mode in slab-coupled waveguides usin a- tional method," IEEE J. Quantum Electron. **QE-12**, 378 (1976)
25. R. H. Hardin and F. D. Tappert, "Applications of split-step Fourier met the numerical solution of nonlinear and variable coefficient wave equation," ev. Chronicle **15**, 423 (1973)
26. C. Yeh, L. Casperson, and W. P. Brown, "Scalar-wave approach for single ho- mogeneous fiber problems," Appl. Phys. Lett. **34**, 460 (1979)
27. J. A. Arnard, "Transverse coupling in fiber optics, Part IV. crosstalk," B ech. J. **54**, 1431 (1975); J. S.Cook, "Tapered velocity couplers," Bell. Syst. T 807 (1955)
28. K. Ogawa, "Simplified theory of the multimode fiber coupler," Bell. S **56**, 729 (1977)
29. A. W. Snyder and J. D. Love, "Optical Waveguide Theory," Ch Hall, London (1983)
30. M. D. Feit and J. D. Fleck, "Computations of mode properti fiber waveguides by a propagating beam method," Appl. Opt. **19**, 1154
31. C. Yeh, "Modes in weakly guiding elliptical optical fibers," Opt ctron. **8**, 43 (1976)
32. A. Taflove and M. E. Brodwin, "Numerical solution of steady gnetic scattering problems using the time dependent Maxwell's equat ns. Mi- crow. Theory Tech. **23**, 623 (1975); K. S. Kunz and R. J. Lueb ference Time Domain Method for Electromagnetics," CRC Press, Bo 992)
33. P. H. Siegel, private communication (2004)
34. C. Yeh, F. Shimabukuro, and P. H. Siegel, "Low-loss terah guides," Appl. Opt. **28**, 5937 (2006)
35. P. H. Siegel, S. E. Fraser, W. Grundfest, C. Yeh, and F. Sh le guide for in-vivo and hand-held THz imaging," Quarterly Prog PAR-03- 075, Caltech (2006)

SUBJECT INDEX

AUTHOR INDEX

Printed in the United Kingdom by
Lightning Source UK Ltd., Milton Keynes
140116UK00006B/18/P

9 780387 309293